d. *Absol.* or *ellipt.* for a church, sanctuary or institution dedicated to, or otherwise associated with, the saint.

Sanct Johnnis vnder the crag; 1388-9 MacRae *Early Sc. Texts* No. 2. Schir William Knollis, Lord Sanct Johnis of Jerusalem; 1491 *Treas. Acc.* I 166. At he lent the king to offer at Sanct Antanis; *Ib.* 198. The craft of hammyrmen gadderit at Sanct Leonardis; 1496 *Edinb. Hammermen* 14. In Sanct Mawarrokis; 1497 *Treas. Acc.* I 329. Spoilȝe of the trees of Sant Mergret Inche; 1534 *Chart. Coupar A.* II 138. He raid to Sanct Douthois his allane; *Diurn. Occurr.* 4.

e. In place-names.

Sanct Johnestoun, St. Johnston, Perth.

Ellipt. in *Sanct Androwis, Sanct Johnis, San(c)t Monanis.*

(1) In gud aray, Befor Sanct Jhonystoun com thai; BARB. II 250. A man .. brocht tythingis fra Sanct Johniston to the king; 1489 *Treas. Acc.* I 122. Four elnis of Sanct Johnestoun lyning; 1568 *Edinb. Test.* I 196b. Tua schipis command wnder saill by Sancttabbis heid; PITSC. I 228/18. In Sant Johnstoun; *Ib.* II 159/28. That they should be strictly confined by themselves in Sanct Colmis Inche; 1580 *Reg. Privy C.* III 313. Of all .. vessellis comeing within Dunnottar and Sanct Tobis head; 1647 *Acts* VI I 655/1. At Sant Johnes Clachen 11 March; 1672 *Kirkcudbr. Sheriff Ct. Deeds* I 319.

(2) To Sanct Androws he come in hy; BARB. I 353. In Pertht, to the king him self .. ; Item, the xv da of September, quhen the king past furth of Sanct Johnis; 1488 *Treas. Acc.* I 95. Except beneficis as ar at the gift of Sanctandris; 1510 *Reg. Privy S.* I 326/1. Evin at Sanctandrus [*Ch.* to Sanct-Androes] for to seik law; LYND. *Sat.* 1964 (B). The leif of the kingis graces jornais ar in Sanctandrois; 1542 *Inv. Wardrobe* 99. The Bischoperick of Sanctt Androis; PITSC. I 286/16. At the mercat croces of Dundie .. , Sanctandrois, Craill .., Sant Monanis, Weymis [etc.]; 1592 *Acts* III 615/1. Cuming fra Sanctandrois, Seres [etc.]; 1605 *Stirling B. Rec.* I 113.

(*b*) The byschop of Saynt Androus; 1389 Slater *Early Sc. Texts* No. 16. Gyf ony man .. be gane in pylgrymage .. in he haly land or to Saynt Jamys or til ony vthir haly stede [etc.]; *Burgh Laws* c. 29 (B).

B. *noun.* **1.** One recognized by the Church as having attained sanctity; a canonized person. Also *attrib.*, *possess.*, and *transf.*

Also predicatively in *sing.* without article, as, or as if, *p.p.* (like L. *sanctus*).

Also applied to the Sybil.

(1) The erle Thomas That syne wes sanct as men sayis; BARB. XVII 286.

transf. Thow [*sc.* Honour] stant ordant, as sanct, of grant maist wise; DOUG. *Pal. Hon.* 2139.

(*b*) Syne efter this Thomas That on this wis maid marter was Was saynct and myrakillis did; BARB. XVII 875.

(2) I wryt the lyf of sanctis sere; *Leg. S.* Prol. 28. Quhen my saule sall wend away .. Thow steid it quhair Thi sanctis is; *Fifteen Ois* 326. *Howlat* 988 (A). The sanctis in hevyn schynis as the sonne, HAY I 282/30. Haly sanctis and doctouris of the Cristyne faith; *Vert. Mess* (STS) 3. And of the glorius sanct Sanct Jhone the Baptist; 1475 *Edinb. B. Rec.* I 30. The sanct that thow art named efter; IRLAND *Asl. MS* 47/10. He .. deit at Carlele a sanct; *Asl. MS* I 262/13. Legendis of Scottis sanctis; 1507 Innes *Sketches* 273 n. Off sic hie feistis of saintis [*M.* sanctis] in glorie; DUNB. (OUP) 122/11. In hevin ȝe salbe sanctis full cleir, Thocht ȝe be knavis in this cuntre; *Ib.* 159/38. Off full few freiris that hes bene sanctis I reid; *Ib.* 165/28. BELL. *Boece* I iv. *Ib.* II 72. The haly marthiris war nowmerit in the catholog of sanctis; *Boece* 296b. The actis of sanctis, namyt the Marthologe; *Ib.* 356. Sweran dailie in common langage be God or Our Lady or othir sanctis; ABELL 103a. And now, in heavin, he is ane michtfull sanct; LYND. *Sat.* 2967 (Ch.). And quhat compositour .. ganestandis to do the samyn .. he sall [pay] ane stane of vax to the sanct unforgiffin; 1546 *Perth Hammermen* 56. LYND. *Mon.* 2630. Quhilkis did mok and scorne all invocation maid tyll sanctis; Q. KENNEDY *Tractive* 148. I wat Sanct Peter nor Sanct Paule Nor ȝit na sanct can saif ȝour saule; *G. Ball.* 193. And thair amongs these glorious sancts and pilgrims she sal be; FOWLER I 130/119. The kirk .. may erre in the canonizatione of ane sancte; KING *Cat.* in *Cath. Tr.* 207/4. 1588 *Haddington Presb.* in Mill *Mediæv. Plays* 253. Kirkes dedicat to sanctes; DALR. I 123/20. *Ib.* 121/7.

(*b*) As alsa the superstitious obseruaris of the festuall dayis of the santes sumtimes namit thair patronis; 1581 *Acts* III 212/2. Summondit .. for going to Doles or Carmichell kirk .. to seik help at santis; 1625 *Elgin Rec.* II 191.

(*c*) Be the sayntis of God, I wene I have rychtwis querele aganis thé; HAY I 266/6. Saintes; SKENE *Acts* (1597) Table s.v. *Halie-daies.* It is but a mint at the best that the saints mak that the lamb's praise; 1659 A. HAY *Diary* 96. In Saint Peiteris nam, Saint Paullis nam and all the saintis of hevin; 1662 *Crim. Trials* III 609.

(*d*) All haly virginis .. and .. all the sainctis and haly cumpany of hevin; 1525 *St. A. Formulare* I 269. We .. refuse .. his canonization of men, calling upon angelis or sainctis deperted; 1580-1 *King's Confession* in Dickinson *Source-bk.* III 34. James, the just, my peacefull king, hath fled To .. the sweet harmonies Of saincts and martyrs; 1633 LITHGOW *Poet. Remains* 79.

(3) To the laud and gloir of God, .. and thair patroun, and all sanctis; 1525 *Aberd. B. Rec.* I xl. Fra the feist off All Sanctis to .. Witsonday; BELL. *Boece* II 374.

(4) There he played .. at the sanct is found; URQUHART *Rabelais* I xxii.

(5) Sone hym [*sc.* Æneas] warnys Scibilla the sant; DOUG. VI ix 8.

(6) *transf.* To the blak deuill thairfoir I thé betetche With him remaine in hous ane hellis sanct; ROLLAND *Seven S.* 4002.

(7) *attrib.* and *possess.* That his conscience be examinat ther [anent] euerilk sanct day; 1525 *Aberd. B. Rec.* I xli. — All the historis of Scotland and Scottis sanctis lywis .. he [*sc.* Edward] gart birn and gart bring bukis of the ws of Ingland; ABELL 96a. That thair be nocht ane penny of the sanctis geir drunkyn in tyme cumming bot twa marks allenerlie; 1544 *Perth Hammermen* 49. Exponand thame auld sanctis lyvis; *Pedder C.* 19. And the auld sanctis dayis callit non sederunt to be dischargit frathynefurth; 1567 *Acts* III 41/2. The said J[ohn] M[enteyth] sall abstein .. fra all playing on the Sabbath, Yull day, Pasch day and the dayis callit of awld the sanctis dayis; 1596 Banks *Sc. Cal. Customs* III 239.

b. A representation of a saint.

For ane clath aboue the sanctis hedis; 1505 *Edinb. Hammermen* 53. 1516 *Ib.* 92.

c. *transf.* One held in the esteem or reverence usually offered to saints in another context.

A stranger is the sainct whome I adore; 1618 LITHGOW *Poet. Remains* 14.

2. A person of exemplary holiness of life. Also in proverbs.

Also ironical.

(1) Thought I dour wes .. I wes dissymblit suttelly in a sanctis liknes; DUNB. *Tua Mar. W.* 254. *Ib.* 444. Sum schippart slais the lardis scheip and says he is a sanct [: skant, cant, ant]; DOUG. VIII Prol. 44. I rak nocht quhidder fulys hald me devill or sanct [: grant]; Id. *Direct.* 71. No more he wate nor dois ane sancte Quhat thyng it bene to haue or wante; LYND. *Complaynt* 207. Gif him the habeit of ane freir The wyvis will trow .. He be ane verry sanct [: plant; *Ch.* saint]; Id. *Sat.* 4259 (B). Ȝe brether of religioun .. That day all creature sall ken Geue ȝe war sanctis or warldly men; Id. *Mon.* 5857. That may serue weill sic ane unsonsie sanct [: grant, obstant, want]; ROLLAND *Seven S.* 2694. Sum .. Thocht thay seme holy in deid that ar na sanct [: arrogant, want, dant]; 1570 *Sat. P.* xvii 84. For the late king, he lived a sanct, and died a martyre; 1650 *Misc. Maitl. C.* II 488. I found them by their profession to be sancts almost all, for each ane gaed uthers a good testimony which made me suspect them the more; 1659 Lorimer *St. Cuthbert's* 13.

(*b*) So moylie .. He luikit lyk ane sant [: skant]; MONTG. *Ch. & Slae* 112 (L). Those who mortall men wer cald And to the world did liwe lyke sants; FOWLER I 67/4. This earthlie sant this spak agane [etc.]; *Ib.* 80/87. In conscience it would vex ane sant, As holy as Mr. Andrew Cant; *c*1675 *Bk. Pasquils* 227.

(2) Suche are the now sants that oppresses Gods people; 1649 *Rec. Kirk Scotl.* 588.

(3) *proverb.* Of ȝung sanctis growis auld feyndis but faill; DUNB. (OUP) 61/35. So ȝe be nocht ane ouir ȝung sanct, And syne ane awld divill; LYND. *Sat.* 233 (B). A young sanct ane auld devill; CARMICHAEL *Prov.* No. 261. — King James .. Said that he [*sc.* King David I] was ane sair sanct to the croun; LYND. *Sat.* 2977 (Ch.).

b. *transf.* One's beloved. Only in CRAIG.

Blow winds, .. And secretly commend me to my sanct [: hant]; CRAIG II 46. Betwix my soveraigne and my sant [: want] I waist my wits; *Ib.* III 7.

3. In Biblical use, one of God's chosen people; by adherents of the Reformed Church, one of the elect, a member of the Reformed Church.

Graues war opnyt and mony bodijs of sanctis [*P.* seintes] that had slepit raise up; NISBET *Matth.* xxvii 52. I trou .. that thair is ane halie Chrissine kirk quhilk is the congregatione of sanctis that is of al Chrissine men; GAU 79/19. His mother is the .. most liberall to the sanctis afflicted that we have; 1559-60 KNOX VI 110. [The Kirk] is callit the communioun (not of prophaine persounis) bot of sanctis; 1560 *Conf. Faith* in *Acts* II 530/2. God, for thy grace, .. Ceis not to send thy sanctis sune support; *G. Ball.* 103. The .. mercy of God towartis his sanctis and elect; *Lyndesay Pref.* (STS) 401. To releif the necessitie of the pure sanctis that dwellit far from thame; FERG. *Serm. iii Malachi* in *Tracts* 70. All thingis that have occurred within this realme, during the tyme of this terrible conflict that hes bene betuix the sanctes of God and these bloody wolves; KNOX I 4. Ʒe chosen sancts that ar afflictit heir; E. MELVILLE *Godlie Dreame* 426. Scho suld be debarrit the communion of the sanctis at the tabill of the Lord till the presbyterie war advyisit with; 1613 *Fraserburgh Kirk S.* 21b (19 Aug.). Mr. Georg Cumming is appointed to haue a disput on the perseverance of the sancts and to hav his thesee in readines; 1658 *Dingwall Presb.* 298.

(*b*) Let me be with His justice cleid And conteit thai redemitt santt; *Montg. Suppl.* xxxv 88. The schedding of the bluid of the santes; 1651 NICOLL *Diary* 72.

(*c*) I beleue in .. the haly kirk vniuersall, the communioun of saintes; *G. Ball.* 4. The saints have a liferent-tack of the cross of Christ; RUTHERFORD *Christ's Napkin* 6. 1637 Id. *Lett.* (1675) 85. He hath made many black limbs of hell fair saints in heaven; Id. *Tryal Faith* (1845) 45. I thought you had ever a love to the saints and even to the poorest and sillyest who carried Christs image; 1649 *Last and Heavenly Speeches of Viscount Kenmuir* 4. I .. doe leave my soule to God and [or]daines my bodie to be buried amongst the saints; 1668 *Kirkcudbr. Test.* (Reg. H.) 12 Dec.

(*d*) [They are] permitted .. to prevayle .. over the veray saynctes of God; 1554 KNOX III 299. In daily sheading the blood of thy saincts; 1565 *Ib.* VI 311. When we ascend to that place where his angels and saincts glorified are; HENDERSON *Serm.* 332. MURE *Psalmes* XXX 4.

Sanct, Sant, *v.* [ME and e.m.E. *isonted* p.p. (Ancr. R.), *saynt* (*c*1460), *saint* (1553); SANCT *n.*] **a.** *tr.* To enroll among the saints, make a saint of, canonize. **b.** *intr.* To act as a saint, play the saint, pretend to be a saint. — **a.** The Erll Thomas, That syne wes sanctit [*E.* sanct] as men sais; BARB. XVII 286 (C). Men said .. this Thomas, That on this vis maid martir was, Was sanctit and myraclis did; *Ib.* 875. She [*sc.* the Church] sancted him (a sore one) for the croune; GARDEN *Kings* 58. — **b.** My former faith I can not weill recant, Nane I accuse, I come not heir to sant; 1571 SEMPILL *Sat. P.* xxviii 204.

Sancta Marie. (L. *Sancta Maria* = St. Mary, as an expletive.) — 'Sancta Marie,' said he, 'Bessie quhy makis thow sa grit dule [etc.]?' 1576 *Crim. Trials* I II 52.

Sancta sanctorum, var. of SANCTUM SANCTORUM. **Sancte,** var. of SANCT *adj.* and *n.* **Sanctefy,** var. of SANCTIFY *v.*

Sanctification, *n.* Also: **-atione, -acioun(e.** [e.m.E. *sanctificacyon* (1526), *-ation* (1582), OF *santificatiun* (*c*1120 in Larousse), eccl. L. *sanctificātiōn-*.] **a.** In Christian theology: The action, by the Holy Spirit, of making holy or sinless, the state of holiness or freedom from sin. **b.** Consecration, setting apart for a sacred purpose, keeping holy (a thing, the Sabbath).

a. And of this sanctificacioune [*sc.* by baptism] and halines writtis the apostile in [etc.]; IRLAND *Mir.* I 35/32. *Ib.* III 3/11 (see SANCTIFY *v.* a). This fulnes of grace of this glorious lady, seruit not onlie for hir auin sanctification, but also to obtein fulnes of grace [etc.] .. to vthers; HAMILTON *Facile Tr.* in *Cath. Tr.* (STS) 231/9. It were not amiss to exhort these persones to sanctification, and to keep the oath and covenant of God; 1609 *Sel. Biog.* I 90. Q. With quhat graces offers Chryst the sacrament to ȝou. A. With wisdome, sanctificationne, richteousnes and glorificationne; *a*1630 *S. Leith Rec.* 2 Ser. 283/1. Will you get among you all but one pint of faith, a gill of grace, and a mutchkine of sanctification; *Presb. Eloq.* (1694) 70.

b. Sanctification is the setting apart of a thinge for a holy or religious use; GILLESPIE *Eng. Pop. Ceremonies* III i 6. Acts [were] obtained for the carefull sanctification of the Lords day; *Review Bramble's Faire Warning* in *St. A. Kirk S.* 480 n.

Sanctifie, var. of SANCTIFY *v.*

Sanctified, -fiet, -feed, *ppl. adj.* [e.m.E. *sanctifiede* (*c*1485); SANCTIFY *v.*] Invested with sanctity; made holy. — Learne to be sad, silent, sober and sanctified: .. and pansing on heauenlie things; HUME 86/205. I wes thair mouth indeid, choissin be thame, sanctifiet be prayer; 1600 *Misc. Bann. C.* I 171. A man of all most spiritfull, Most sanctifeed and sage; GARDEN *Elphinstoun* 12. I mean of these who are elected, called, justified and sanctified; 1684 *Cloud of Witnesses* 216.

Sanctifiing, var. of SANCTIFYING *vbl. n.*

Sanctify, -fie, *v.* Also: **sanctifye, sanctyfie, sanctefy.** [ME and e.m.E. *seintefie* (Gower), *saynctyfye* (1413), *sayntefye* (Caxton), *sanctifie* (1526), OF *saintefier* (*c*1120 in Larousse), eccl. L. *sanctificāre.*] *tr.*

1. To invest (a person or thing) with sanctity, make holy.

Noye, his thre sonnis, and thare wyuis, .. Quhome God did blys and sanctyfie; Sayand, incres and multyplie; LYND. *Mon.* 1505. And be this also writtin, that the Lord hes sanctifiit His tabernacle [*sc.* the Virgin Mary]; WINȜET I 73/17. A faithfull man, cleane and holy, sanctifieth by the worde the creature of God; but .. a creature may not sanctifie any impure and unfaithfull man. But to the faythfull man all thingis ar sanctifeid; KNOX I 162. Did not the virgine myilde .. Bring furth to ws ane chylde Quha did ws sanctifie; ARBUTHNOT *Maitl. Q.* 94/148. That hellie spereit quha sanctifies all the .. faithfull; 1622 *Edinb. Test.* LI 198b. O Lord, sanctifye him; O Lord, putt over his youth without a mark or meinȝie; 1654 JOHNSTON *Diary* II 271.

2. a. *passive.* Of God, or an attribute of God: To be celebrated or manifest as holy (*in* or *to* a person or thing). **b.** *active.* To celebrate, honour (the rituals attached to the worship of a pagan deity).

a. The secund way the name of God is sanctified in ws throu remissioune of oure synnis for be the wertu of his haly name oure synnis are remytt; IRLAND *Mir.* I 35/36. In all His [*sc.* God's] warkis is sanctifyit His mycht; *G. Ball.* 116. God grant .. that in him [*sc.* James VI] [God] be ay glorifeit, Luifit, honourd, feird, and euer sanctifeit; HUME *Promine* 168.

b. Nor neuer clan or other nation by Lyke thame sall hallow, ne ȝit sanctify [L. *celebrabit; Ruddim.* sanctefy] Thy wirschip; DOUG. XII xiii 116.

3. To consecrate, dedicate, set apart for a holy or religious function.

He has ordanit thame [*sc.* the elements of the Eucharist] to be hallowit and sanctifiet be his hevinly wourdis that he has gevin to man for thar sanctificacioun; IRLAND *Mir.* III 3/10. Sex dayis to laubour, and the seuint day to sanctifie to the Lord; WINȜET I

28/13. Quhat scripture hes he or vther authoritie .. to sanctifie the Sonday to be the seuint day? *Ib.* 28/18. *Ib.* 115/4. That quhilk was sanctifyit and dedicatit to God; FERG. *Serm. iii Malachi* in *Tracts* 66.

4. To ward off the power of witches (from oneself, a thing), by invoking a blessing or other ritual act.

An quhan any sies thes strawes in a whirlewind, and doe not sanctifie them selues, we may shoot them dead at owr pleasour; 1662 *Crim. Trials* III 604. And to keip the eall from ws, that we haue no power of it, is to sanctifie it veill; *Ib.* 605.

(Sanctifying,) Sanctifiing, *vbl. n.* [e.m.E. *sanctifyinge* (1526); SANCTIFY *v.*] Honouring or setting apart as holy, consecrating. — The celebratioun and sanctifiing of the said sacrament of our Lordis body; WINȜET I 88/19.

Sanctio(u)n, *n.* [e.m.E. *sanction* (1563-87), OF *sanction* (14th c. in Larousse), L. *sanctiōn-*.] **a.** A law or compact supported by a specific penalty for its breach; or, a specific penalty appointed for the breach of a law or compact. **b.** Authoritative permission, authority, licence. — **a.** Gif ony Picht eftirwart made mencioun of pece .. he suld want the hede. Sum parte of the wise barouns imprevit baith this sanctioun & aith; *Boece* 362b. The sanction and pain of this divine law being by sin incurred; MCWARD *True Nonconf.* 316. — **b.** [The council] admitts .. Alexander Fleming to instruct the bayrnes of the hie schole .. in place of Mr. William Murdo, quha had the said sanctioun of befoir; 1594 *Edinb. B. Rec.* V 126.

Sanctitud(e, *n.* Also: **sanctytude.** [e.m.E. *sanctitude* (1616), L. *sanctitūdo*.] Sanctity, saintliness, holiness.

[The Pope] Sad in his sanctitud sekerly & sure; *Howlat* 85 (A). I may nocht suffys to se ȝour sanctitud sad; *Ib.* 96. Apostolis, martyres [etc.] .. The sound of thair excellent sanctitude Was hard ouer all the warld; LYND. *Test. Pap.* 776. He .. contendit with monkis preistis and bischoppis in halynes and sanctitude of life; *Boece* 312b. Amphibalus .. exceidit into sanctitude; STEWART 17977. *Ib.* 20822, 20828, etc. Specull of pulcritude, Thou seid of sanctitude; *Ave Gloriosa* 98. ROLLAND *Ct. Venus* III 797. The priestis, .. be thare fayned sanctitude, Abusit mony one man of gude; LYND. *Mon.* 2209. This moste tryumphant court of Rome .. That vmquhyle was .. the moste sapient sors of sanctytude, Bot now, allace! bair of beatytude; *Ib.* 4937. *Ib.* 2663, etc. And greit God in sanctitude Sall send .. Siclyke on ȝow; *G. Ball.* 181. Mort mvmlingis mixt with monye leis Sic sanctitude was Sathanis sorcereis; SCOTT i 93. Thair Catholik pietie confirmet the sanctitude of thair Catholiknes; DALR. II 391/3.

b. *pl.* As an attribute of more than one person.

That day ȝour faynit sanctytudis Sall nocht be knawin be ȝour hudis; LYND. *Mon.* 5862.

c. As an attribute of the Pope, hence, by metonymy, with the possessive pron. coming to stand as a designation of the Pope.

And all apperit to the Pape .. Salust his sanctitud with spirituale speike; *Howlat* 242 (A). Thus, all the warld .. His sanctitude thay do obey; LYND. *Mon.* 4312. Pope Celistene, Quhen that his sanctytude serene Did crown Henry the empriour [etc.]; *Ib.* 4596.

Sanct-like, *adj.* (e.m.E. *saintelike* (*c*1580) saintlike, resembling a saint.) — I doe mislike of those As sancts that bears a sanct-like shoe; GARDEN *Garden* 39.

Sanct Maruifis fair, var. of SUMERIVES FAIR(E.

Sanctorum. (Gen. pl. of L. *sanctus* (SANCT *adj.* and *n.*), appar. = 17th c. Eng. *sanctoral* (1641), mod. Eng. *sanctorale* (1872), 'that part of the breviary and missal which contains the offices proper for saints' days' (OED).) — Foure new half bukis, twa for symmer, and twa for wynter, contenand the temporale and sanctorum; *c*1555 *Crail Ch. Reg.* 65.

(Sanctrel,) Santrel, *n.* [Late ME *seyntrelle* (Prompt. Parv.), OF *sainterel* (1138 in Greimas), It. *santarello* (Florio in OED); SANCT *n.*] A little saint, a 'saintling'. — With a thousand other jolly little sancts and santrels; URQUHART *Rabelais* I xxvii 129.

Sanct Ruffus fair, var. of SUMERIVES FAIR(E.

Sanctuacie, *n.* (Appar. some irreg. or erron. deriv. of SANCT *n.* ? Saintliness, saintly forbearance.) — Which in dede is a thing that na sanctuacie can lesumlie digest with silence; 1590 *Cal. Sc. P.* X 433.

Sanctuar(e, *n.* Also: **sanctwar, sanctuair(e.** [North. and east-midl. ME *santuare, seyntwar(e* (Cursor M.), *saynt-ware* (Alexander), *saintuaire* (Gower), OF *saintuaire* (*c*1155 in Larousse), MF *sanctuaire* (*c*1380 in Larousse). Cf. SANCTUARY *n.*]

1. A holy or consecrated place; esp. a place reserved to priests and other religious initiates. Also *fig.*

Ane hallowit schaw on breid, as sanctuar, Plantit thai haue, and stabillit prestis thar; DOUG. V xii 181. Outwith al the sanctuar [L. *luco*] hy ȝow, hay; *Ib.* VI iv 48. Thar growys a gret schaw .. Quhilk is in wirschip haldyn .. As sanctuar [*Ruddim.* sanctuare]; *Ib.* VIII x 7. All the kirkis and sanctuaris [*M.* sanctuarijs; L. *loca sacra*] of Kent war suspendit with Cristin blude; BELL. *Boece* II 41. Bayth king and lord gaif thame [*sc.* the Druids] authoritie In sanctuar for to make ceremonie; STEWART 3404. Alaricus .. command hes gevin .. Fra all slauchter .. for till abstene In sanctuar [*heading* in the kirk]; *Ib.* 20748. Thair band with oblissing and aith On sacrament in sanctuar wes sworne; *Ib.* 27623.

fig. So Christ our Lord was not bound priest .. But Melchisedech .. high priest & King togidder .. And entred ones, in to the sanctuair By his owne blood, his priesthoode to declair, Where he offred his blessed fleshe also; NORVELL *Meroure* 15a.

b. *specif.* An inner, especially holy part of a temple or church; the area round the altar in a Christian church.

The *Compl.* quot. may belong in 1 above.

That he of the apostolis all In-to the sanctwar [can] ga, Quhen-euir he wald his prayeris ma; *Leg. S.* vii 65. King Fergus .. committit the cure hereof [*sc.* engraved tablets] to kirkmen to sett the samyn in sanctuare [L. *sacrario*] and thare diligentlie to be kepit; *Boece* 51b. The tempil .. the goldin alter, the chandleris of lycht ande al the goldin veschel .. the coupis tassis [etc.] .. ande al the goldin ornamentis of the sanctuar; *Compl.* 76/4.

2. A sacred place providing immunity from arrest for fugitives. = SANCTUARY *n.* 2.

Mony personis .. in trast that thai salbe defendit throw the immunite of haly kirk and girth and passis & remanis in sanctuaris; 1469 *Acts* II 96/1. Quhay com to ane sanctuaire or gyrth callit Navidaill; 1554 *Corr. M. Lorraine* 387.

Sanctuary, Sanctuarie, *n.* Also: **sanctwarie, sanctuari, sanctuarii, sanquary.** [ME and e.m.E. *sanctuary* (Rolle), *sentuarye* (Chaucer), *sainctuarye* (1481), *seyntuarye* (Wyclif), *seyntewarie* (Piers Plowman), *santuary* (*a*1586), L. *sanctuārium*, OF *saintuarie* (*c*1120 in Larousse). Cf. SANCTUAR(E *n.*]

1. A holy place; a church or temple; a shrine. Also *fig.* **b.** *Within sanctuarii*, in a holy place, on sacred ground.

Panthus .. Into his armys lappit to his breist The haly rellykkis of the sanctuary [*sc.* the temple of Apollo; *rh.* by and by]; DOUG. II vi 45. In quhat estait is sanctuary [*Ruddim.* -arie] and haly geir? *Ib.* 51. Behald Thou fra Thi sanctuarj [*P.* seyntuarie; *W.* sanctuary], fra the hie duelling place of heuen; NISBET *Ep. Ald Test.* xxxiv 15. Schaw Thi face on Thi sanctuarie quhilk is forsakin; *Ib.* xxxix 17. Sanctwaries; 1546 Pitsc. II 67/14. The preistis of it hes contemnit my lawe, hes defylit my sanctuarie [L. *sanctuaria*]; WINȜET I 14/11. Go furth of the sanctuary, for thou hes transgressit; 1564 KNOX II 450. The Lord behaldis ȝour knauerie .. As of his sanctuarie ȝe brak the vall; 1581 *Sat. P.* xliv 27. Within this Ile of Colmkill thair was ane sanctuarie or kirkȝaird; MONRO *W. Isles* (1961) 62.

fig. For thou art the chosin sanctuary of our Saluiour in quhom he hes lichtit [etc.]; *Lang Rosair* 159.

b. For violent handis putting on Schir Alexander Mernis his awn curat within sanctuarii of the kirk of Wyk; 1549 *Soc. Ant.* XI 89.

2. Passing into: A sacred place in which fugitives were by law or custom immune from arrest; also, once, applied to a statue. = GIRTH *n.*[1] 3.

The haly schaw Quhilk strang Romulus dyd reduce and draw In maner of franches or of sanctuary [L. *asylum*]; DOUG. VIII vi 69. He gaif to that ymage Off sanctuarie the priuilage, For quhat sum euer transgressour .. Seand that ymage .. Off thare gylt gat the kyngis grace; LYND. *Mon.* 2118. And gif it be found suddaintie, he salbe restored to the libertie of the girth and sanctuarie; SKENE *Verb. S.* s.v. *Schireff.* Suld I .. be aschamit to intre vithin the azil of thy blessed sanctuary; COLVILLE *Paraenese* 157. James Malcolmesone .. took occasion .. traterouslie .. to kill this Mackintoshie .. after which he flieth to the yle of the loch of Rothiemurcose, as a sanctuarie or defence for him; GORDON *Geneal. Hist.* 99. Aplecross .. was spared be all his name as a sanctwarie, and he, hearing that other inhabitants came to it as a refuge, he came; 1667 *Highland P.* II 42.

3. As the name of a particular piece of sacred ground or refuge of this sort.

That na persoun cut or peill ony of the Sanquary; 1549 *Elgin Rec.* I 98. The girth or sanctuarie of Halyrudhouse; 1636 CALDERWOOD VIII 144.

Sanctum sanctorum. Also pl. **sancta sanctorum.** [Late ME and e.m.E. *sancta sanctorum* (*c*1400), *sanctum sanctorum* (1558), late L. (Vulgate) *sanctum sanctorum* holy of holies, f. Biblical Gk. and (ultim.) Hebrew.] The Holy of holies of the Jewish temple and tabernacle. Also *transf.*, appar. the sanctuary in a Christian church. — Bosiliall nor Oliab .. Quhilk sancta sanctorum maid maist riche & deir; DOUG. *Pal. Hon.* 1454. — *transf.* Pay to Thomas Miller £74 for his work in tirring and theiking the queir and sanctum sanctorum; 1687 *Linlithgow B. Rec.* in *Ecclesia Antiqua* 92 n. 3.

Sanctus. (ME and e.m.E. *sanctus* (Wyclif), L. *sanctus* holy, 'the .. hymn beginning .. '*Sanctus, sanctus, sanctus*' .. which forms the conclusion of the Eucharistic preface' (OED).) — Furtht of ane mes bwik of Cadder .. one the baksyd w[r]ettin that the crucifix is one follovand sanctus; 1509 *Glasg. Univ. Mun.* I 46. Quhair day and nycht we sall not ceas Ay singand Sanctus sweit; *G. Ball.* 129.

Sanctwar, var. of SANCTUAR(E *n.* **Sanctwarie,** var. of SANCTUARY *n.* **Sanctyfie,** var. of SANCTIFY *v.* **Sanctytude,** var. of SANCTITUD(E *n.*

Sand, *n.* Also: **sande, sainde, saynd, sawn.** [ME and e.m.E. *sand(e,* OE *sand.*]

1. Sand as a material. Also *pl.* in same sense; freq. as used in building.

(*a*) He [*sc.* the conjurer] couth .. Mak .. Nobillis of nutschellis & siluere of sand; *Howlat* 788 (A). Wyth ledyng of sand; 1475 *Reg. Cupar A.* I 202. 1512 *Treas. Acc.* IV 279. For iij dosane of lyme and vj dosane of sand to poynt the palais; 1513 *Ib.* 528, etc. To xii verkmen this olk at the werk the quarellis and sand casting; 1529-30 *M. Works Acc.* (ed.) I 5, etc. For the casting and wynnyng of the sand; *Ib.* 36. Quhen this tempest was mytigat, the landis of Godwy .. with sand war all coverit and bedrevin; *Boece* 479b. Grete burdingis of erde and sand; BELL. *Livy* I 136/14. [Sowing corn there] quhar the sand blew; 1535-6 *Ayr B. Acc.* 76. Gabyounes [etc.] .. to beir sande; 1549-50 *Treas. Acc.* IX 364. Me thocht the king of farye had me tane And band me .. in ane lang raip of sand; LICHTOUN *Dreme* 8 (B). Necessaris to the gunhous .. viz. colis, wyre, hair, clay, sand and talloun for castin of certane grete hakbuttis of found; 1552 *Treas. Acc.* X 101. 1603 *Deer Presb.* in *Buchan Cl.* IV 182 (see SANDIT *ppl. adj.* a). To 8 wemen that bore sand and sklaitt for ten locht; 1617 *M. Works Acc.* (ed.) II 44. When .. the last pickle of sand shall be at the nick of falling down in your watch-glass; 1637 RUTHERFORD *Lett.* (1894) 387. As you break off their roots for use, lay their tops or sets in ground covered a little till the Spring for planting; .. these you spend not e're the frosts come, hard house among very dry sand, that you may have them when you will, rather as be barred from them by the frosts; 1603 REID *Sc. Gard'ner* 106.

(*b*) And thocht ouir synis monyfawld In nombir pas the saynd [: onderstand, staynd (= stand)]; CULLEN *Chron. Aberd.* 48.

(*c*) To poynd the deficients .. for not drawin staine & sawn to the calseys; 1656 *Dumfries Council Min.* 7 April.

pl. The saidis gummis and sandis proceding of the gold and silver mynis being affynit; 1575-6 *Reg. Privy C.* II 512.

b. *Deid sand,* ? sand of a firm and unyielding consistency (as used in the laying of foundations).

For the inbringing of deid sand to the laying of the said lystis; 1530 *M. Works Acc.* (ed.) I 36. Holand and wynnand clay and deid sand for bigging of the samen [oven]; 1531-2 *Ib.* 71. For deid sand to lay the fluris within the barris xxiiij s.; 1563-4 *Edinb. Old. Acc.* I 444. For fyve scoir nyntein laides of deid sand for laying the pavement in the counsallhous; 1627 *Ib.* II 206.

2. The sand of a shore, or of an expanse of sandy ground; a stretch of (chiefly shore) sand, the shore itself, a beach, also ? a sandbank. Also *pl.* in same sense. Also *se-sand,* SE *n.*

Small stanis of the sand [L. *littore maris*] He gadderit vpe in-to his hand; *Leg. S.* V 155. And as the king apon the sand Wes gangand wp and doun [etc.]; BARB. IV 632. Thai .. fand a kyrk in-to the sand; *Leg. S.* XXI 887. Before the listis .. thay come prekand throw the sand; *Alex.* II 1321. *Howlat* 209. HENR. *Fab.* 740. Quhill on ane sand the schip did brist and claif; DOUG. *Pal. Hon.* 1368. Our schippis .. we gart adres And lay almaist apon the dry sand; Id. *Æn.* III ii 135. The see was furth; the sand wes smoith & dryye; LYND. *Dreme* 115. Fywe thousand men war suckin be the sand; STEWART 33327. Throw aduenture [and] tempest of the se, .. ane schip .. wes brokin on ane sand; *Ib.* 36299. This ferriar .. did .. row .. quhill he come to that sand; *Ib.* 53941. Vpon the sand or craiges foreanent his landes; SKENE *Verb. S.* s.v. *Ware.*

pl. That he .. ly onerdyt in myddis of the sandis; DOUG. IV xi 75. Sa huge wilsum rolkis and schawd sandis; *Ib.* V xi 56. The flowand sey .. with his iawpys coverys in and owt The far sandis our the bay abowt; *Ib.* XI xii 68.

b. With the definite article as the name of a particular stretch of sand. Chiefly *pl.*

Freq., 'the Sands' adjacent to a place or belonging to a particular community.

pl. The forsaid masonys .. sall delyvir frely thaim at ony key of Abirden or ellis at the sandis at Laurence of Lethis hows; 1399 *Aberd. B. Rec.* I 377. 1446 *Reg. Dunferm.* 304. That xii personis .. sal set & dele the hale sandis tyl al .. induellaris; 1459-60 *Ayr B. Ct.* 14 March. 1506 *Treas. Acc.* III 203. Sum .. Wald ryid to Leith .. And wychtlie wallope ouer the sandis; LYND. *Complaynt* 179. The teyndis of the sandis of Mussilburghe; 1561 *Reg. Dunferm.* App. II 460. The tymmer in Leyth quhilk wes be the Lambeth fluid chiftit athort the sandis; 1598 *Edinb. D. Guild Acc.* 675. Accuisit of .. playing on the sandis in tym of preaching; 1615 *Fraserburgh Kirk S.* 43b (10 Jan.). Ane barne .. was exponitt & found in the sandis within the fluid mark; 1618 *Ib.* 83 (9 Dec.). 1625 *Justiciary Cases* I 28. In respect the Inglish schipes haid chaist in ane barque .. on the sandis; 1651 *Aberd. B. Rec.* IV 120. A vessell .. driuen in vpon the plaine sands a litell be-east the Black Craigs; 1655 LAMONT *Diary* 84. A worm called lug digged out of the sands at low water; 1683 *Acc. of the North Side of the Coast of Buchan* in *Coll. Aberd. & B.* 100.

sing. John Wat ʒid to the sand .. to gather bait; 1621 *Fraserburgh Kirk S.* 124 (11 July).

c. *To leave or put* (a person) *to the lang sands,* appar., to leave someone in a difficult situation.

Udney transacts for the haill, pays himself, and leaves Pitreichy to the lang sands; 1671 M. P. Brown *Suppl. Decis.* II 539. The accomplishment seemed hard at hand; yet how quickly were they [*sc.* the people of Egypt] put again to the long sands (as we say,) when they .. were put to wander in the wilderness fourty years in end; J. BROWN *Life Faith* I (1726) 42.

d. In other place-names.

Recium meorum trahencium super Joymersandes; *c*1240 *Lindores Chart.* 84. Burch-in-the-sand Men callis this toun; BARB. IV 203. Oure Solouay sand Ane Inglis ost com in Scotland; *Leg. S.*

xl 1087. De terris de Sandhalch; 1435 *Antiq. Aberd. & B.* III 582. Sandfurde; 1449 *Reg. Dunferm.* 310. Cum piscariis de Newscaris et Sandis de Edin; 1517 *Treas. Acc.* V 102. [In the] sandegat [of Ayr]; 1518 *Prot. Bk. Gavin Ros* 42. For carage hyer .. furtht of the castell of Edinburght to the lang sandis of Leith; 1544-5 *Treas. Acc.* VIII 360. Of James Williamsons rig in the sandhalff; *c*1616 *Montrose Treas. Acc.* (Mary Hospital Rental) MS. Betwixt the sand port and the port of the Cittiedaill; 1682 *Edinb. B. Rec.* XI 35.

3. *Attrib.* and *comb.* **a.** With *cart, chist, hous,* **b.** With *beirar, ledar,* **c.** With *flure,* appar. = 2 above, **d.** In *sand-colour, -colored,* sand-coloured. Also SAND-BED, -GLAS, -MAN, etc.

a. Ane stirrop to ane sand cart; 1530 *M. Works Acc.* (ed.) I 40. *Ib.* 41, 48. For twa laid of brume to mend the sande housis; 1555-6 *Edinb. Old Acc.* I 185. Samuell Hoyle coppersmith .. tua sand chists; 1668 *Edinb. Test.* LXXIII 224b.

b. To the brether Ruidmen sandledderis for i[c] vi dosane vi ladis sand; 1536 *M. Works Acc.* (ed.) I 192. 1538-9 *Ib.* 244. Sand ledaris; 1540-1 *Ib.* 270. *Ib.* 288. [For sand] to John Horne, sand ledar; 1592-3 *Ayr B. Acc.* 177. Tuelf sand and lyme beiraris; 1629 *M. Works Acc.* (ed.) II 297.

c. In parochia de Ruthven, Ruthventoun [appar. = Ruthwell in Dumfriesshire.] nuncupat cum lie salt cott, sand flure et marresia earundem; 1542 *Reg. Great S.* 606/2. The towne and lands of Revell [= Ruthwell] with the saltcott sandfluires and hous stead thairof; 1633 *Acts* V 146/2.

d. Ane govne of sainde coeiller schairge; 1640 *Tailor's Acc. Bk.* B 33b. Ane gray sand colored naig; 1670 *Kelso Baillie Ct.* 70b.

Sand, *v.* [ME and e.m.E. *sond* (Chaucer), *sand* (1453); SAND *n.*] *tr.* **a.** To run (a ship aground). Also *fig.* **b.** To inundate (land, crops) with sand. Cf. SANDING *vbl. n.* **c.** Of a dam: To be silted up or damaged by the action of water and water-borne sand and detritus. **d.** *intr.* Of a sea-wall or embankment: To become silted up or suffer damage as in prec.

a. The tvme schippis .. brokin and sonkin and stikking to the ground war sandit; *Boece* 457b.

fig. I am sanded, and my love is sanded, [*1894* stranded] and I find not how to bring it on float again; 1637 RUTHERFORD *Lett.* (1664) 357.

b. Mutch of ther corne sanded by breaking foorth of the waters; 1659 LAMONT *Diary* 118. The teinds of Furvie (now almost sandit); 1675 *Rec. Univ. Aberd.* 350.

c. Damm which the Town had upon the same water, which still was sanded and broke with speats and torrents of the water, [and further that they] .. have not any going milne at present; 1661 *Dumfr. & Galloway Soc.* IV 62.

d. The aucht dayis expensis in bakking of bulwark in Leyth .. The aucht dayis expensis to dyk up the bulwart quhen it sandit .. for .. schip-timmer to be baking to the bulwart; 1555-6 *Edinb. B. Rec.* II 320.

Sand, p.t. of SEND *v.*

Sandal, -el, *n.*[1] [e.m.E. *sandell(e* (*c*1485), *sandal(l)e* (1526), ME and e.m.E. *sandalie* (Wyclif), *sandalyne* [sic] (1493), OF *scandale* (*c*1170 in Larousse), med. L. *sandalum* (12th c. in Latham), L. *sandalium.*] A sandal. **a.** As a simple form of foot-wear. **b.** A decorated half-shoe forming part of the dress of a sovereign or a prelate. — **a.** Nocht a bag, nore breid, nouthir money in the belt, Bot schod with sandalis [*P.* sandalies]; NISBET *Mark* vi 9. Preistis, tak ȝour staffe, And preiche the Euangell on ȝour feit, And set on sandellis full meit, Bot cast ȝour pantonis of; *G. Ball.* 195. — **b.** Septyr, ryng and sandalys Browdyn welle on kyngys wys; WYNT. VII 1999. [Baculus deauratus cum duabus mitris cum sandalibus et cirothecis; 1501-2 *Reg. Episc. Aberd.* II 171.]

Sandal, -ile, *n.*[2] [Late ME and e.m.E. *sandell* (*c*1400), *sandal* (*c*1450), *sandole* (1588), OF *sandal* (1256 in Larousse), med. L. *sandalum* (*a*1250 in Latham): cf. SANDALIE *n.* and SANDRE(E *n.*[1]] Sandalwood. — Perfumand also al claithis in priuat lugeingis with the reik of sandal [etc.]; SKEYNE *Descr. Pest* 19. The principal preseruatiue cure of the pest is .. in tyme of symmer .. sandile of all sortis, orange appillis; *Ib.* 20.

Sandale, Sandal(l, *n.* Also: **-ell, -ele, -eill, -ill.** [ME and e.m.E. *cendal* (*a*1225), *cendel* (Cursor M.), *sendal* (Piers Plowman), OF *cendal* (1160 in Greimas), med. L. *cendallum, cindalium, sendallum, sandalum,* etc. (1216, *c*1200, 1225 and 1303 all in Latham).] Sendal, a fine silken material. **b.** A garment of sendal.

Thare he of chance a ymag fand Of Jhesu Cryst .. But mannis handis mad .. In sandale; *Leg. S.* xxiv 159. Salaryne Armit in armour gude and fyne All couerit in fyne sandale [: stale]; *Alex.* I 1707. Syne thair hors sa richelie trappit war With .. cleth of gold or siluer or sandale [: haill]; HAY *Alex.* 1345. Thair mycht nothir silk nor sandall hald agane; *Ib.* 1369. Ane sark scho had of silkyn sandele quhite; *Ib.* 7208. xlij Flemis ell of sandeill; 1546 *Aberd. B. Rec.* I 234. The sylk and sandell Thé to eis, Ar hay, and sempill sweilling clais; *G. Ball.* 50. Thair was ane body of ane ȝoung chyld found in ane kaise of leid wumpillit wpoun sandell [*1728* -al, *1814* -ill]; PITSC. I 61/7.

b. Into the renk prekit Lyoun, All helit in ane sandale broun; *Alex.* I 1190.

Sandalie, *n.* [Cf. SANDAL *n.*[2] and SANDRE(E *n.*[1]] ? Sandalwood. — Inuarte customes .. Sandalie the pund; 1597 *Bk. Rates* 10.

Sandall, var. of SANDALE *n.*

Sandapilariane, *n.* (f. L. *sandapilāri-us* a bier-carrier f. *sandapila* a common kind of bier.) — BIRNIE *Kirk-b.* vii (see POLLINCTOR *n.*).

Sand-bed, *n.* Also: **-bedde, sambed; sinbed.** [Late ME and e.m.E. *sandbedde* (*c*1475), *sand-bed.* Written as one word or two.] A stretch or low mound of sand, chiefly or only one in or by the sea or other expanse of water; a sand-bank. Freq. as a place-name; also *attrib.*

(1) *sing.* Thai come on ane sand bed, traisting the samyn firme land, becaus it wes vncoueritt with watter; BELL. *Boece* (M) II 318. The soldeouris .. at Kingorne war resauit be the marinar quhilk on the nycht at eb see landit thame one ane sand bed; *Boece* 592b. Hold the south syd of the entres to avoid a sand bed; LYND. *Rutter* fol. 5 (B). We were once teddered on a sand-bed, had the wind been as it was the day before .. our veshell had been dung in shards; 1641 BAILLIE I 355. The tounes commoun ground callit of old the mildam, now the green, from the kill doune to the sand bed; 1653 *Peebles B. Rec.* II 9. 1662 LAMONT *Diary* 155. [For] cutting ane sandbed .. to cawse the current off the loch called Achenreoch ryn from the staiding dam to a low moss; 1670 *Kirkcudbr. Sheriff Ct. Processes* No. 75 (26 May). 1682 *Peebles B. Rec.* II 102.

(*b*) To the officeris for bringing over the beastes to the sinbed on ane mercat day; 1645-6 *Dumfries Treas. Acc.* MS p. 20.

attrib. For a key to the sandbed mylne doore; 1640 *Dumfries Treas. Acc.* MS 9b. For the sandbedmill gyle 2 dalls att 3 s.; 1649 *Ib.* p. 31. For thake to the sambed mille; *Ib.* p. 41.

(2) *pl.* xl s. for skaithe in his foirstair and sandbeddis; 1596 *Kirkcudbr. B. Rec.* I 329. Ad aquam de Nith prope lie sandbeddes; 1641 *Reg. Great S.* 361/2. James Allaster is dilate of cursing & sweiring befoire his trade on the sandbeds vpon the day of the electioun; 1654 *Dumfries Kirk S.* 19 Oct.

Sand-blind(e, *adj.* (*n.*). [Late ME (15th c.) and e.m.E. *sande-blynde.*] Half-blind, dim-sighted; short-sighted. Also as *noun.*

adj. And sand blind ene ar schamefull commounly; HAY *Alex.* 10194. Drumlanrick being sum thing sand blind and saw nocht weill; PITSC. I 347/24. *Myops,* sand blind; CARMICHAEL *Etym.* 21. He loukethe aglied and is verie sandblind; 1603 *Reg. Privy C.* VI 580. The blinde man, who seeth nothing, and .. the sand-blinde man, who seeth little; D. LINDESEY *Heavenly Chariot* 34. Those who are sand-blind cannot see far aff; HENDERSON *Serm.* 204.

noun The stany wring the stane and sand blind; ROWLL *Cursing* 61 (M).

Sanddy-bed, var. of SANDY-BED *n.*

(Sande, *n.*) see SAYND(E *n.* **Sande,** var. of SAND *n.* **Sandé,** var. of SANDY *adj.* **Sande-hill,** var. of SAND-HILL *n.* **Sande-hour,** var. of SAND-HOUR *n.* **Sandeill,** var. of SANDALE *n.* **Sandel,** var. of SANDAL *n.*[1]

Sand-ele, Sandell, *n.* Also: **sand-eil(l, -eile, -eel, -iell; sandill.** [ME and e.m.E. *sandel* (1307), *sandhell* (1338), *sandeel(e* (1425-6), *sawndel* (Prompt. Parv.), *sandil* (1558), *sand-eele* (1671); SAND *n.* and ELE *n.* Also in the mod. dial. with reduced second element as *san(d)al, -il, -le, sannal.*] A sand-eel, the small eel-like fish found on sandy shores. Also *attrib.* with *bait, polk* (POKE *n.* 3 b).

(*a*) To gather wilkes, cokles, lempets, mussels, sand-eiles, small fish or baite; SKENE *Verb. S.* s.v. *Ware.* Ane cleir glancing plaitt .. full .. with quhyttingis or sandeillis schyning and glittering thairintill; 1609 *Aberd. Eccl. Rec.* 70. That quhaso evir vpon the Sabboth day salbe fund raiking sandelis or gathering enie other sort of baite .. they sall pay [etc.]; 1616 *Fraserburgh Kirk S.* fol. 79a (27 Nov.). GORDON *Geneal. Hist.* 5. The sand-eel is of the shape of a fresh water eel, about a foot or some more in length; 1661 Symson *Descr. Galloway* App. 149.

(*b*) Dischargis the haill toun fra passing to the sandells during this seisoun; 1584 *Kirkcaldy B. Rec.* 99. In taking of sandellis on the sabboithe day; 1623 *Fraserburgh Kirk S.* 23 July. The rest that was withe tham gaddering sandellis; *Ib.* There is a place in Southerland .. wher the inhabitants doe tak a kynd of fish called sandells; GORDON *Geneal. Hist.* 5. The fisch they make greatest cont of are that they call the sardine, which seimes to be our sandell; 1665-7 LAUDER *Jrnl.* 62. To my wife to help to pay cockles and sandills; 1689 *Foulis Acc. Bk.* 113.

uninfl. pl. The Devill cutt them as small as sandell that ever shall seek mor sandiell baitt; 1649 *Cullen Kirk S.* 17 June (= Robertson *Cullen Ch. Ann.* 31).

attrib. [Ane artikill] anent the sandeil poilkis in the west; 1595 *Conv. Burghs* I 469. Aganis the brughis of Air, Irewin .. for suffering the fischeris quha fischis in thair watteris to fische with sandeill polkis to the grit distructioun of the hering fry; 1602 *Ib.* II 151. — Sandiell baitt; 1649 (see *uninfl. pl.* above).

Sandele, -ell, var. of SANDALE *n.* **Sandeman,** var. of SANDMAN *n.*

Sander, *n.* [e.m.E. *sander* (1627); SAND *n.*] A workman working with or delivering sand; appar. = *sand-beirar,* SAND *n.* 3 b. — Lyme and sanderis xviii wirking that weik .. Lyme and sand beiraris xviii working that weik; 1629 *M. Works Acc.* (ed.) II 298.

Sandery, var. of SANDRE(E *n.*[1] **Sandey,** var. of SANDY *adj.*

Sand-glas, *n.* Also: **-glass(e.** [e.m.E. *sand(e) glass(e* (1556). More often written as two words.] A sand- or hour-glass. = GLAS *n.* 2 (4), SAND-HOUR *n.*

As *pl.* also without inflexion.

sing. Ane sand glas to mesour the houris witht, iij s.; 1540-1 *Treas. Acc.* VII 491. Ane sand glas to sett besyde the freir in the polpet cost iiij s.; 1554-5 *Edinb. B. Rec.* II 355. His book in the a hand and sand-glass in the uther; MELVILL 273. For ringing the bell and keiping the sandglas; 1622 *M. Works Acc.* (ed.) II 144. 1640 *Dundonald Par. Rec.* 465. A standing iron to the sand-glasse; 1680 Craven *Ch. in Orkney* 91. For making ane caise to the church sandglass; 1694 *Boharm Kirk S.* 25 March. For an iron standard to a sand glasse in the church; 1697 Cramond *Ch. Cullen* 137.

infl. pl. For tua sand glasses to the king, iiij s. viij d.; 1505-6 *Treas. Acc.* III 180. 1506 *Ib.* 337, etc. For tua sand glassis ane haill hour glas and ane half hour glas; 1615 *M. Works Acc.* (ed.) I 356. Tua creills containing sand glasses viz. ane eight hours: ane six hours: & ten four hours: six half hours & eightein half minutts glasses; 1681 *Blackness Customs* 31b.

uninfl. pl. For iiij sandglas for hir [*sc.* the ship]; 1512 *Treas. Acc.* IV 454.

Sand-hill, *n.* Also: **sande-.** [Late ME *sond-hylle* (Prompt. Parv.), e.m.E. *sand-hill (a*1603), OE *sondhyll.*] A sand-hill, sand-dune. — The landis callit Sandehillis lyand in .. Kirkurd within .. Peblis; 1495 *Acta Conc.* I 415/2. It sall nocht be lesum .. to scheir ony bent vpon the commoun sandhillis; 1597 *Prestwick B. Rec.* 83. The pulling of bent, broom or juniper off sand hills .. the same being the natural fences of the adjacent countries to the saids hills; 1695 *Acts* IX 452/2.

Sand-hour, *n.* Also: **sande-.** [SAND *n.* and HOUR *n.*] **a.** An hour-glass. = SAND-GLAS *n.* **b.** The sand necessary to measure an hour. — **a.** For twa sand houris to the schippis; 1513 *Treas. Acc.* IV 531. — **b.** Rycht as the sande-hour in the glasse, Elykewyse dois thare tyme heir passe; LAUDER *Off. Kings* 33.

Sandi-, -ie, varr. of SANDY *adj.* **Sandiell,** var. of SAND-ELE *n.* **Sandile,** var. of SANDAL *n.*[2] **Sandill,** var. of SANDALE *n.* and SAND-ELE *n.*

Sanding, *vbl. n.* [Cf. SAND *v.* b] The inundating of (land) with sand (by sea or wind). — Quhairof sixteen chalders .. is defalcat for inundatioun and sanding of the landis be watteris and wound; 1561 *Reg. Episc. Morav.* Pref. xix.

Sandit, *ppl. adj.* [e.m.E. *sanded* (Shakespeare); SAND *n.*] **a.** Inundated with sand, covered with sand by the action of the wind or sea. Cf. 1603 quot. in SAND *n.* 1 (*a*). **b.** ? Sand-coloured. **c.** Of horses: Suffering from some disease or condition of the feet, ? sandcracks or splitting of the hoof. — **a.** Both kirk and kirkyard is ourcassin with the sand, and therefour Mr. David Robertson ordaint to voyce the building of the new kirk .. [That stern company sitting in the] sandit [kirk are determined (etc.)]; 1603 *Deer Presb.* in *Buchan Cl.* IV 182. — **b.** Ane reid sandit sow; 1525 *Wigtown B. Ct.* 164b. — **c.** For hailling of a hors that wes sandit in the feitt; 1618 *M. Works Acc.* (ed.) II 105.

Sandman, *n.* Also: **sande-, sanman.** Pl. **-men.** [SAND *n.* 1 and MAN *n.* 1 e; cf. 19th c. Eng. *sand-man* (1821) id. Written as one word or two.] One who supplies sand for building, etc.; a digger or transporter of sand. Cf. SAND *n.* 3 b. — To the sandemen for thriescoir laid of sand; 1614 *M. Works Acc.* (ed.) I 341. To the sand men for four lead of deid sand; *Ib.* 343. 1615 *Ib.* 361. 1616 *Ib.* II 3. The sandmen wer dischairged to break the touns grund without leive of the Counsell; 1658 *Edinb. B. Rec.* IX 120. David [Sharp], sanman; 1673 *Greyfriars Interments* 582.

Sand-poke, *n.* Also: **-poik, -pock.** [Late north. ME *sand-poke* (1415-16), *sandepoke* (1421-2).] A sand-bag. Also *attrib.* — To advys vpon the ballesting of fische bottis quhither with sand pokis or with staines; 1611 *Conv. Burghs* II 319. For furnesching of sex laid saidellis .. ten pair of sand pockis; 1618 *M. Works Acc.* (ed.) II 88. Sand pockis; 1652 *Peebles B. Rec.* II 194. [For Edinburgh Castle] of canvas for sand pocks; 1696 *Sc. Hist. Rev.* XIV 96. To David to buy seckine for sand pocks; 1703 *Foulis Acc. Bk.* 313. — *attrib.* To the couper for ten girthis to the sand poik mouthes; 1618 *M. Works Acc.* (ed.) II 99.

Sandre(e, -drie, Sand(e)ry, Saundre, *n.*[1] [ME *zawndre* (north., 1388), *sandrey* (15th c.), e.m.E. *sa(u)nder* attrib. (1481-9), ME and e.m.E. pl. *saundres* (north., 1329-30), *zaunders* (north., 1390), *saundres* (Gower), *sawnderys* (*c*1430), *sanders* (1598), OF *sandre, sandle* (Hatz.-Darm. s.v. *Santal*), med. L. *sandrium, saundrium, zandria* (1377, 1372 and 1453 all in Latham); cf. SANDAL *n.*[2]] Sandalwood. — Onerat se de ij libris sandre; 1331 *Exch. R.* I 392. Et in empcione duarum librarum de sandery [8 s.]; *Ib.* 402. In .. saundrio [*ed.* sanudrio]; 1371-2 *Ib.* II 370. Per empcionem de .. j libra de savndre; 1375 *Ib.* 507. 3 li. trousell cost 12 g. li.; 3 li. sandry cost 14 g. li.; 1495 HALYB. 41. Empt. $\frac{1}{2}$ libra sandre vij s.; 1528 *Household Bks. Jas. V* 173. Empt. de Vxore Lethame erga Natale j libra sandrie pricium xiiij s.; *Ib.* 174. Sandree; 1529 *Ib.* 180.

Sandre(e or **Saudre(e,** *n.*[2] [?] The name of some cloth, evidently not SANDALE *n.*, despite the appar. formal similarity (-*n* may however be confused with -*u*). Appar. only in *Treas. Acc.*, 1511-12: only *gray sandre.* — To be the king ane schamvart and ane gret parplait, vij elnis and half ane quartar gray sandre; 1511 *Treas. Acc.* IV 192. For his hois of gray sandre xx s.; *Ib.* 255. To be the king ane coit iiij elnis gray sandre, price elne xxiiij s.; 1511-12 *Ib.* 199. To be the king ane nycht gowne vij$\frac{1}{2}$ elnis gray sandre, price of the elne xviij s.; *Ib.* 200. Sandree; 1512 *Ib.* 244.

Sandrie, var. of SANDRE(E *n.*[1]

Sandy, -ie, *adj.* Also: **sandey, sandi-, ? sandé.** [ME and e.m.E. *sandy* (Wyclif), *sondy* (*c*1440), OE *sandig* (*c*1000).] Consisting of, or covered in sand; sandy.

See also SANDY-BED *n.*

A hammere & small sandy stanys; *Seven S.* 2019. The place of the feild was sandy ground; *Boece* 447. The ground .. is sandy and stanerie; 1579 *Reg. Privy C.* III 129. This kynde of fishe .. in cleir water delytes, and in sandie furdes; DALR. I 50/6. *Ib.* 100/14. I began to lope upe upon a litle sandie bray; MELVILL 21. Ane little laich mane [sandey] ile; MONRO *W. Isles* (1961) 63. The house newlie builded by him upon the sandy mollen; 1683 *Dumbarton B. Rec.* 93.

b. As the first element of place-names, as *Sandy knowis, -landis.*

Also in the surname *Sandilands,* q.v. in Black *Surnames.*

Willelmus de Douglas .. confirmasse Jacobo de Sandilandis .. omnes terras meas de Sandilands; 1348 *Misc. Spald. C.* V 245. He schupe him in the nycht to far To Sandylandis; BARB. VI 397. Jacobo de Sandilandis; 1384 *Reg. Episc. Aberd.* I 169. Terris vocatis ly sandelandis; 1531 *Ib.* 399. He sew na seid in to hir Sandelandis; *a*1568 SEMPILL *Sat. P.* xlvii 40. — Nixt the sandy knoys; *c*1505 *Reg. Cupar A.* I 261. Sande Knowis; 1550 *Ib.* II 97. — The Sandieburne; 1632 *Lanark B. Rec.* 324.

c. *Comb.* in *sandie-snowted,* appar., with a snout covered in sand, or of a sandy colour.

A sandie-snowted hog; *Prognostication.*

(Sandy-bed,) Sanddy-bed, *n.* = SAND-BED *n.* — Lyke .. to ane schip .. enforsed .. betuix maist daingerus sanddy beddis; WINȜET I 3/8. Sum .. fleing fra the sanddy beddis, speidis .. to brek .. the schip on the .. rokis; *Ib.* 3/17.

Sane, var. of SAIN(E *v.*, SAY *v.*[1], SEN *adv.*, *prep.* and *conj.* and SON(E *adv.* **Sanesyne,** var. of SENSYN(E *adv.* **Sanetie,** var. of SANITE(E *n.*

Sang, Song, *n.* Also: **sange; songe, soung.** [Chiefly north. and north. midl. ME *sang* (Cursor M.), midl. and south. *song* (Layamon), OE *sang, song.*]

See also EVINSANG *n.*, PLAIN-SANG *n.* and *priksang* PRIK *adj.*[2] a.

1. The action or activity of singing; singing as an art; vocal music. **b.** The singing of a bird or birds.

(1) In all my tyme Ik hard neuer in sang na ryme Tell off a man that [etc.]; BARB. III 178. Of ane hour hale the space Vith gret delyt of angel sange Fed scho was; *Leg. S.* xvi 803. Warldly sang and vanyte; *Ib.* xxvi 656. Thay maid .. lauchter and sang; *Alex.* II 5071. Dame Natur has .. Ascendit .. with solace & sang; *Howlat* 943 (A). Wes nowthir solace nor sang thair sorow to soft; *Gol. & Gaw.* 1055. DUNB. (OUP) 147/16. With hir ythand sweit sang and caralyng; DOUG. VI xvi 25. With hevinly and delitious sang; BELL. *Boece* I 233. LYND. *Depl. Magd.* 89.

(*b*) To angell song and hewinlie armony; *K. Hart* 312. Wyth sowne of clarioun, organe, song and sence; DUNB. (OUP) 109/22.

(?) Ther maister of grammer and sang; 1561 *Rentale Dunkeld.* (SHS) 342. Prowyding thar be na uther scoill teachit in this toun, bot sang onlie; 1582 *Kirkcaldy B. Rec.* 72.

b. To the swete sang of foules seire; *Troy-bk.* II 1654. And for till heir .. Thair [*sc.* the birds'] cairfull sang and lamentatioun; HENR. *Fab.* 1877. The pyot .. sesit hir sang & maid na glewe; *Seven S.* 1962. The fader that kest his son in the se for the birdis sang; *Ib.* 2562 heading. For all thar trast was on the cokis sang; *Fyve Bestes* 216. I hard a merle with mirry notis sing A sang of lufe; DUNB. (OUP) 60/4. Id. *Twa Mar. W.* 7. KENNEDY *Pass. Christ* 422. Tender twystis trymlyt on the treis For byrdis sang; DOUG. XII Prol. 244. The lark discendis .. Syngand hir complyng sang; *Ib.* XIII Prol. 35. LYND. *Meldrum* 1013. The dou croutit hyr sad sang; *Compl.* 39/17. ROLLAND *Seven. S.* 8421.

(*b*) Throw birdis songe with opine wox one hy; *Lanc.* 13. Quhill al the wood resonite of thar [*sc.* birds'] songe; *Ib.* 66. *Bann. MS* 229b/42. O blisit burd quhois soring song [etc.]; *a*1624 *Edinb. Univ. MS* La. II 319.

2. Something sung; a song, as composed, sung, played, etc.; something expressed in song or ? rhyme.

(*a*) Sangis ȝa of lychery Vile & als dewylry; *Leg. S.* xviii 933. & yhone Sibile sang [*ed.* saug] can say [L. *Sibilla quoque sic ait:*] .. That God that hangit one the tree .. happy is he; *Ib.* l 439. He cleikit vp ane hie ruf sang; *Peblis to Play* 57. The sang is sueit quhen that the sound is suyth; *Regim. Princ.* 4 (Fairf.). I will put on my haly-dayis clais .. Syne chant this sang, 'Wes neuer wedow sa gay!'; HENR. *Fab.* 515. Mercurius .. Setting sangis and singand merilie; Id. *Test. Cress.* 243. I mete A mirry man .. Singand this sang that [etc.]; Id. *Age & Yowth* 7. Quhen .. all .. maneris pertenand to him ar concordand .. and consonaunt as the voces and the notis in the sang; IRLAND *Mir.* II 110/29. Sangis to mak undir the levis grene; DUNB. (OUP) 141/28. Scho playit sangis so duilfull to heir; *Ib.* 147/23. Thay that prouokis ony ewil desir .. with sangis or wordis or foul takine; GAU 16/10. Sonyt war trumpettis, .. sangis and rymys war rehersit and song with merynes [etc.]; *Boece* 161. Prentaris .. that .. prentis .. ballattis sangis blasphematiounis rymes; 1551 *Acts* II 488/2. [He could] Sangis set with diuers tunis expres; 1567 *Sat. P.* iii 43. 1572 SEMPILL *Sat. P.* xxxviii 33. For passaig of the sangis to and fro; *Art of Music* 6. How manfully he behauit him self, certaine sangs asweill in France it self, as in Scotland beiris witnessing; CHARTERIS *Wall. Pref.* 173. HUME 69/32. To answer for .. singing of superstitious and prophane sangis; 1609 *Stirling Presb.* in Hume xxxiv.

(*b*) For singing of bawdry songis; 1591 *St. A. Kirk S.* 705.

(*c*) James Arthour, kirk officer, being delait for singing off nevyeir soungis on nevyeiris ewin; 1649 *Echt Kirk S.* in *Strathbogie Presb.* xiii.

b. Freq., a religious or spiritual song.

(*a*) This haly sang and orisone [*sc.* the Paternoster] .. that Jhesus withe His haly mouthe [sueitlie sang]; IRLAND *Mir.* I 32/24. The angellis sall sing the haly chansoune and sang of pes; *Ib.* 119/25. With sangis and ympnis, and othir divine cerimonyis; BELL. *Boece* I 232. The thankfull sangis of Marie and Zachary; NISBET I 9. Q. KENNEDY *Compendious Ressonyng* (ed.) 177/21. Be singing of the psalmes and spiritual sangis; *G. Ball.* 1. Followis ane sang of our coruiptit nature; *Ib.* 24. Sinnaris, vnto my sang aduert, Quhilk Christ [etc.]; *Ib.* 34. *Ib.* 21, etc. HUME 11/15.

(*b*) Followis the song of the Virgin Mary callit magnificat anima mea dominum; *Bann. MS* I p. 22 colophon. The bischopis, abbottis [etc.] .. maid great solemnitie .. witht mese songis and playing on the organis; PITSC. I 379/26.

c. A poetic composition or narrative, more generally.

(*a*) This falyhyd fra he deyd suddanly, This sang wes made off hym for-thi; WYNT. VII 3618. Thy scharp sugurate sang Virgiliane; DOUG. II Prol. 29. Ȝhe Musys now, .. Entone my sang; *Ib.* VII xi 5. To thee, my Lord Regent, I turne my sang; 1567 *Sat. P.* vi 129.

(*b*) Your iust disdaynes of reasoun now enclynes To cast my songs asyde and thame to ryve; FOWLER I 138/8.

d. *transf.* A piece of music played on or characteristic of a particular instrument.

O dulfull harpe with mony dolly stryng Turne all thi mirth and musik in murnyng And ces of all thi subtell sangis sweit; HENR. *Orph.* 136.

3. In various *fig., transf.,* and *allusive* uses.

To mak na sang of, to say nothing about, keep silent about. (Cf. *to mak a sang* (= fuss, to-do) *about, of* or *ower* something, in the mod. dial.).

To sing the samin or *ane sang,* to express the same opinion, to declare agreement.

(1) Off all hys thowcht he mad na sang Bot prewaly [etc.]; WYNT. VI 1983.

(2) And that ȝe think vnressoun or wrang, Wee al and sundrie sings the samin sang; *Prestis of Peblis* 142. Gyf ye sing all ane sang concerning the iudiciall lawes of Moyses; 1580 HAY in *Cath. Tr.* (STS) 63/15.

(3) How sayis the sang? Thair salbe mirth at our meting ȝit; *Peblis to Play* 247. Had thair bein mair made of this sang Mair suld I to ȝow say; *Ib.* 255. Wa was the hart of the emprys Scho held chalmer with lytill sang, Gretand and murnand; *Seven S.* 797. Till our rymis be rung And our mistonit sangis [*B.* songis] be sung; LYND. *Sat.* 75 (Ch.). That mony heris and knauis His Worde .. and dois not thairefter, bot turnis it in ane sang of thair mouth, thair hartis being geuin tyl auarice; WINȜET I 45/6. Thay .. persuade him [etc.] .. The king .. heiris thair sueit sang plesantlie; DALR. II 64/9. Thou has gotten the woman's song laid, as thou promised .. Many pretty men has thou putten down both in ships and boats; thou has gotten the woman's song laid now .. I have keeped thee from doing many ill turnes. Thou has now laid the woman's song; 1633 *Kirkcaldy Kirk S.* in *Coll. Witchcraft* 114.

(*b*) The .. gospel .. is a set of pleasant melodious songs that may .. refresh us till we come unto the city where we shall all sing the song of the Lamb. What a song is liberty to captives .. Ye would all think salvation .. a sweet song .. Many listen to the song of justification, but they will not abide to hear out all the song; BINNING *Wks.* 597.

(4) The songe Of sousinge seas; FOWLER I 380/53.

Sang, p.t. of SING *v.*[1] **Sangaster,** var. of SANGSTAR(E *n.*

Sang-buird, *n.* [SANG *n.* and BUIRD *n.*] Appar., a part of an organ; ? = *sound-board.* — Ane pair of fyne organes, weill furnishit with their sang buird and all their tungis; 1591 *Misc. Spald. C.* III 160.

Sang-buk(e, *n.* Also: **-bwke.** [e.m.E. *song-boke* (1506), *-booke* (*a*1586), OE *sang-bóc.*] A book of songs, a song-book. — To Wilȝeam Sangstare .. for a sang bwke he brocht to the king; 1489 *Treas. Acc.* I 114. Off sang bukis with v messis v bukis; *c*1500 *Coll. St. Salvator* 159. Pro vno libro missarum dicto precett sangbuk xxxii s.; 1515 *Reg. Soltre* 177. [2 quires of paper] for to mak sang bukis; 1540-1 *Ayr B. Acc.* 85.

Sange, var. of SANG *n.* **Sangin(g,** p.p. of SING *v.*[1] **Sangis-scole,** var. of SANG-SCULE *n.* **Sangistar, -er,** varr. of SANGSTAR(E *n.*

Sangler(e, -eir, -ier, *n.* Also: **sanglar; sang(w)eler.** [Late north. ME *synglere* (Morte Arth.), e.m.E. *sanglier* (1575), OF *sengler, sanglier* (*c*1200 and *c*1300 in Larousse), L. *singulāris* solitary, late L. as *noun* a boar separated from the herd.] A wild boar.

(*a*) Lyke to the strenthy sangler [*Ruddim.* sanglere] or the bor; DOUG. X xii 47. Pro pastu certorum suum vocatorum sangleris in parco de Falkland; 1542 *Exch. R.* XVII 513. To schut out or cut of alsua the wyld sangleris—that is the proude schismatikis and obstinat heretikis; WINȜET I 45/14. He .. to the death in rageing furie smyt The sangler strong, the tygar or the boir; J. STEWART 96/594.

(*b*) Ane tapestrie of the historie .. of the huntis of the sangleir; 1561 *Inv. Q. Mary* 144. Wyld beistis, sik as hart, .. beiris, bairis, sangleiris, wyld swyne; BISSET I 28/34.

(*c*) The sanglier .. is callit in sum placis porc and othir sum sanglier. First he is richt ferce wnclene & ennuyse best; *Loutfut MS* 16b. With lion, sanglier and with beere; JAMES VI *Poems* I 134/379. The uraith .. Quho sollaime lyke a sanglier makkis her aufull teeth to crakk; *Ib.* 185/665.

(*d*) To Lauchlane McLauchlane and the cumpanie that wes with him in the Ile of Sanda at severall tymes looking out the sanglares; 1643 *Argyll Reg. Acc.*

(*e*) As to the sangeleris and cokis of Ynd, I sall provid for the samin; 1541 *Sc. Hist. Rev.* VI 157. For expensis maid .. upoun the sangweleris quhilk come furtht of France; 1541 *Treas. Acc.* VII 461.

Sang-scule, Song-scuill, *n.* Also: **sange-, sangis-** and **-skule, -schull, -scuil(l, -scuyll, -skuill, -schuill, -schuiil, -schuyll; -scole, -skole, -schole, -scholme, -scoill, -schoill; -scoule, -scowl, -schoull; -scool(l, -scoole, -school(l)(e.** [SANG *n.* and SCULE *n.*[1] Cf. e.m.E. *song-schoole* (1593), Du. *zangschool,* Sw. *sångskola.* Cf. also SING-SCOLE *n.*] A singing-school.

Normally one attached to a church or other religious foundation, and specialising in the teaching of ecclesiastical singing and music, as well as providing some general elementary education.

(*a*) Maister John [Kendale] has giffin our the sang skule to [etc.]; 1451 *Ayr B. Ct.* 22 Oct. Richard Boyl salhaf .. the sang scule of Abirdene as principale maister of the samyn; 1483 *Aberd. B. Rec.* in *Cart. S. Nich. Aberd.* II 335. And that the childe Cristy Narne .. was passit fra the sang scule to the gramer scule and has maide nay seruice in the queir as yit; 1503 *Ib.* 342. The rector chori .. [to] ken the barnis of the sang scule, and keip all divine service .. als wele as ony uthir rector chori .. in Scotland; 1527 Hay Fleming *St. Andrews* (1953) 19. Sang schull; 1540 *Ecclesia Antiqua* 286. [To] put this man to sang scule or playing, .. dalians or dansing, [It is als] incontrair his complexioun [etc.]; ROLLAND *Ct. Venus* Prol. 203. Sculemaister of the sangscule of the kirk of Glasgw; 1567 (1590) *Glasgow Chart.* II 540. Jhone Suentoun Mr of the sangschull .. exponitt .. thatt he had greatt travaillis in teaching of the ȝoutheid of this toun in musik and in taking wp of the psalmes; 1578-9 *Perth Kirk S.* MS 12 Jan. Set in uerray dulce musike be Andro Kempe sumtyme maister of the sang skule of Sanctandrous; *Wode's Psalter* (Tenor) 127.

(*b*) One feriall dayis, quhane he is teiching the sang scuil; 1525 *Aberd. B. Rec.* I xli. Sang schuyll; 1537 *Reg. Episc. Aberd.* I 412. Sang scuyll; 1539 *Glasgow Prot.* IV 118. The hall consell .. assignis to Robe Portair and Robe Nicholsoun .. fourtty s. .. sua thai continew the sang scuilis and serf in thair queir; 1541 *Aberd. B. Rec.* I 177. Sang skuill; 1546 *Ib.* 239. Sang scuill; 1558 *Ib.* 313. For instructioun of the ȝouth in the art of musik and singing quhilk is almaist decayit .. oure souerane lord .. requeistis .. to erect and sett vp ane sang scuill [etc.]; 1579 *Acts* III 174/2. The cuir .. of ane maister of sang schuill [is] to instruct .. the youth of the .. brucht in the art and sciens of musick; 1595 *Elgin Rec.* II 397. Sang schuiil; 1612 *Aberd. B. Rec.* II 309 (see SCULE *n.*[1] 1 c (2)).

(*c*) Sir Jhone Fethy .. to haif thair organis and sang schole for instructioun of the men of gudis barnis; 1544 *Aberd. B. Rec.* I 207. Sang skole; 1552-3 *Edinb. Old Acc.* I 71. Sang scole; 1560 *St. A. Kirk S.* 40. Sang schole; 1574 *Aberd. Eccl. Rec.* 16. Understanding quhow necessar it is to haif ane sang scole within this burgh .. hes feit .. Johne Buchan to serve in the office of ane maister of ane sangschole as als to uptak the psalmes in the kirk; 1583 *Haddington B. Rec.* (Robb) 17 June. Jhone Ros, maister of the sang scole .. askit God mercy .. for his using and playing of ane part of the comode and play in St. Leonardis College; 1595 *St. A. Kirk S.* 813. Jhon Roull, as maister of the sang schole, to caus the best of his scholaris sitt besyid him self, about the pulpeit, to help to sing the Psalmes on the Saboth dayis; 1599 *Ib.* 908. To the redare and master of the sangscholme [*sic*] xl li.; 1612-13 Edgar *Hist. Dumfries* 257.

(*d*) To teche ane sang scoill conforme to his conditioun; 1552 *Elgin Rec.* I 118. Sang schoill; 1574-5 *Misc. Maitl. C.* II 39. The charge .. of maister of sang schoill to instruct .. the ȝowthe .. of Elgin .. teching [them] .. to play vpoun all musicall instrumentis [etc.]; 1597 *Elgin Rec.* II 398.

(*e*) Sange scoule; 1496 *Aberd. B. Rec.* in *Cart. S. Nich. Aberd.* II 339. Sang schoull; 1518-19 *Ib.* 353. Sang scowl; 1605-6 *Montrose Treas. Acc.* 2b.

(*f*) John Swinton, master of the sang school, is found not qualified for that office; 1582 *Perth Kirk S.* in J. P. Lawson *The Book of Perth* 145. Maister Willeam Skene, maister of the sang school, sall serve bayth the kirkis in uptacking of the psalme theirin, everie Sonday, the ane kirk be himselff, and the uther be the doctour, or sum of his discippullis; 1597 *Aberd. B. Rec.* II 157. Sang schooll; 1598 *Ib.* 174. Sang scooll; 1603 *Elgin Rec.* II 398. David Cowy [is] to attend and teitche the young bairnes in the sang schooll to reid and wreit; 1620 *Ib.* 164. Sang school; 1621 *Sc. Ch. Hist. Soc. Rec.* I 45. And that thair is necessitie of ane sang schoole for musick being taucht heir of auld; 1627 *Orkney Rentals* III 35. To the readar and maister of the sang schoole, iiijxx lib.; 1633-4 *Misc. Maitl. C.* II 43. The master or doctour of the musick schoole .., the said master and doctour of the sang scoole; 1643 *Aberd. B. Rec.* IV 6. He discharge it reiding of scriptures or singing of psalmes at lykewalkis .. and preiudgit the maister of sangscool of his commoditie; SPALDING II 66.

(*g*) Grammer scoolis, song scoolis, and vther scoolis givin wp, and the barnes had hame to thair parentis; SPALDING I 225.

b. *Specif.*, a building or room set aside or specially built for use as a singing-school. Also *attrib.*

(*a*) Payit for vj geistis [etc.] .. quhilk wer delyverit to the biging of the sang skule; 1553-4 *Edinb. B. Rec.* II 285. For male of ane chalmer to be ane sang scole; 1577 *Glasgow B. Rec.* I 462. Six sparis to be kippellis to the sang scoole; 1595 *Edinb. D. Guild Acc.* 579. Spargeing the sangscoole in the new kirkȝaird; *Ib.* 587. Sang scoole; 1599-1600 *Misc. Spald. C.* V 72. Sklaites .. to theik ane sang scooll newlie maid wp be him of ane hous within .. Innes cloiss; SPALDING II 188. [To] buy timber to be seats and wreitting tables for the sang or musik scool; 1657 *Elgin Rec.* II 289.

(*b*) To Duncan Finlaye, for the maill of ane chalmer to be sangis-scole [*Glasgow B. Rec. I 463* to the sang scole], xl s.; 1579 *Glasgow B. Rec.* (MC) 116.

(*c*) If he will set us George Donaldsones barne for ane song scuill; 1644 *Elgin Rec.* I 282.

attrib. Ane pair of bands .. to the sang scole dur; 1554-5 *Edinb. Old Acc.* II 37. [To] have .. tuenty merkis yeirlie for payment of the sang scole maill; 1579 *Edinb. B. Rec.* IV 128.

Sangstar(e, Songster, *n.* Also: **sangster, -air; sangistar, -er, sangaster.** [North-east midl. ME *sangester* (Manning), ME and e.m.E. *songster* (Wyclif), OE *sangestre, -ystre* a female singer.] A person who sings, a singer, freq. a professional singer. See also PLAIN SANGSTAR *n.* Also as a surname.

Jacobo Sankstar [charter witness]; 1452-3 *Reg. Episc. Aberd.* I 273. Frater Andreas Sancster; 1486 *Ib.* II 300.

(*a*) With al delytis and plesans of women sangstaris and othir delytis; *Wisd. Sol.* (STS) 78. The aldirman, ballieis, consale and communitie .. promittit to gif Robert Huchonsone, sangstar and master of the organis, alsmekile fee as thai gaf him this ȝer bigan; 1489 *Aberd. B. Rec.* in *Misc. Spald. C.* V 30. To Wilȝeam sangstare of Lythqow for a sang bwke he brocht to the king; 1489 *Treas. Acc.* I 114. Our continuale familiar seruitor Robert Huchosone sangstar .. to singe .. mes .. and ale other hours and divine seruice; 1496 *Aberd. B. Rec.* in *Cart. S. Nich. Aberd.* II 339. Sangstair; 1501 *Treas. Acc.* II 127. Thare perpetuale chaplanreis .. salbe gevin .. tile sangstaris that can syng plane sang ande prik sang; 1508 *Aberd. B. Rec.* in *Cart. S. Nich. Aberd.* II 347. In the meyn tyme to sing & saye in the qweyr amang the laiff of the sangstars; 1519 *Dumfries B. Ct.* fol. 8 (5 Oct.). Fidlaris, sangstaris and siclik men of vile reputation; BELL. *Boece* II 20. For tua new keyis to the sangstaris chalmer durris; 1532 *M. Works Acc.* (ed.) I 103. George Cultis, sangster; 1532 *Treas. Acc.* VI 92. Missall, ane of the quens gracis sangstaris; 1565 *Canongate Kirk S.* (ed.) 35. To the sangstar that takis up the psalmes befoir and eftir the precheing; 1573 *Ayr Chart.* 109. All menstrallis sangstaris and taill tellaris not avowit in speciall seruice be sum of the lordis of parliament .. salbe .. pvneist as strang beggaris and vagaboundis; 1574 *Acts* III 87/2. To the violeris and sangsteris at the kingis entrie .. the sovme of thre pundis for thair dayes wages; 1579 *Edinb. B. Rec.* IV 124. Ane great number of idle persons without lawfull calling, as pypers, fidlers, sangstars, sorners, pleasants; 1596 *Bk. Univ. Kirk* III 874. Gif ewer he be fund heirefter ryoting and intertening ane number of pypers and uthir sangsters, to pay [etc.]; 1608 *Kinghorn Kirk S.* 10. [Not to] giwe ony meatt or drink to these sangsteris [*sc.* singers of New Year songs] or lat thame within thair houss; 1612 *Aberd. Eccl. Rec.* 78. That the minister give up the names of idle sangsters [*Bk. Univ. Kirk III 1121* songsters] and minstrellers within the paroche, to the end they may be called and punished as idle vagabounds; 1616 Calderwood VII 225. To the sangster Gowdie .. xij s.; 1620 *Grant Chart.* 326.

(*b*) The said William sal present ane sufficient cheplain that be ane sangaster and chorist; 1490 *Aberd. B. Rec.* in *Cart. S. Nich. Aberd.* II 338. The .. Counsale .. hyrit and feit Wat Strathin sangaster to syng and help to hold dyvin seruice within the said kirk; 1505 *Ib.* 343. Sangistar; 1516 *Treas. Acc.* V 85. 1537 *Ayr B. Acc.* 27. Sangister; 1561 *Treas. Acc.* XI 85. Blak stamyng to be hois to nyne pagis and ane sangistar; 1562 *Ib.* 164. To pay tham xvij s. for the sangistaris that luigit in thair hous; 1568 *Crail B. Ct.* MS 1 June.

b. Also applied to a bird.

Oscen, qui ore canit, a sangster; a singing foule shewing things to come; DUNCAN *App. Etym.*

c. *transf.* A poet.

That made that Homer, who a songster bene, Albeit a beggar; JAMES VI *Poems* I 23/93. But thou my deir one .. The great gods holy songster had receiued; *Ib.* 35/307.

Sangsterie, *n.* [SANGSTAR(E *n.*, appar. on the analogy of *chanterie* CHANTORIE *n.* beside CHANTOUR *n.*] A chantry. = CHANTORIE *n.* 1. — [To remit all presbyteries,] sangsteries, chaplaries [.. to the ministers serving .. the cure of the said kirk]; 1597 *Cal. Sc. P.* XIII 142.

Sanguine, Sanguene, Sanguane, *adj.* Also: **sanguyne, sangwyne, sanguein, sangweyn, sangwane, sangune.** [ME and e.m.E. *sanguyn* (Wyclif), *sangwyn* (Chaucer), *sangweyne* (Trevisa), *sangewyn* (1494), *sanguyne* (1526), OF *sanguin* (*c*1138 in Larousse), L. *sanguineus* of or pertaining to blood, containing blood.]

1. Blood-red; also appar. used of other more or less red colours.

Also *broune sanguane,* rendering, in DOUG., L. *lutea* 'golden yellow', *sanguineum ostrum* blood-red purple, and, as *noun, ferrugo* the colour of iron-rust.

(1) His colour was sangweyn; *Wall.* IX 1932. The v stone is callit a loys a sanguein stone or synamer; *Loutfut MS* 54a. Of huge weght and bludy sangwyne [*v.r.* sanguyne] hew [L. *loricam .. Sanguineam*]; DOUG. VIII X 61. Turnus .. in hys sanguane bludy scheild .. Kest schynand fyry bemys lyke fyre flaucht; *Ib.* IX xii 21. As fresch Aurora .. Ischit of hir safron bed .. With sangwyne cape; *Ib.* XII Prol. 16. Flowris .. sa mony diuers hew, .. Sum greyce, sum gowlys, sum purpour, sum sangwane, Blanchit or brovne, fawch ȝallow mony ane; *Ib.* 107. Pretius stonis .. of euerie sort and hew, Quhyt, purpur, sanguine, cramsie, broune and blew; J. STEWART 51/133. Hir comelie cheeks .. Of rid and vhyt ymixt, Ar lyk the sanguene jonet flour Into the lillie fixt; MONTG. *Misc. P.* XXXV 39. He wos .. sanguine in colloure bothe of hyd and haire, colerique of nature; 1611 *Reg. Panmure* I xxxi.

(2) Within hyr rosy cartis cleirly schane Aurora vestit into brovn sanguane [L. *Aurora .. fulgebat lutea*]; DOUG. VII i 4. And he hym self in brovne sangwane weill dycht [L. *ferrugine clarus*], .. Schaftis he schot; *Ib.* XI XV 11. As quha byspark wald the quhite evor Indane With scarlet droppis or with brovn sangwane [L. *sanguineo .. ostro*]; *Ib.* XII ii 38.

2. Of the physiological 'complexion': Sanguine. Also, in astrology: Tending to produce sanguineness of 'complexion' or temperament.

The naturis of the sternis and planetis ar contrary; for we se that ane is calde, ane othir is hate [etc.], .. the tane is sangwyne, the tothir is malancolius; HAY I 76/26. Gemini .. is of sangune complextione; *Medical Recipes* 87b.

Sanguineane, *adj.* (*n.*). Also: **-ian.** [ME *sanguinien* (Ayenb.), *-yene* (*c*1400), e.m.E. *sanginean* (1655), L. *sanguine-us* (cf. SANGUINE *adj.*).] **a.** Blood-red. **b.** = SANGUINE *adj.* 2. **c.** A person of a sanguine temperament. — **a.** Fair Angeliqve .. Imploying all hir diligence .. his lyf for to reskew, Reiosing so in his sanguinian hew; J. STEWART 73/9. — **b.** The planetis aspectis and complexiounis .. ar thir four: in compt first phlegmatike: The secund is sanguineane siclike; ROLLAND *Ct. Venus* Prol. 12. — **c.** No corpulent sanguinians make me feare; W. ALEXANDER *Julius Caesar* 2081. For I am naturally a sanguineane and used to be merrily disposed befor his majesties sufferings; 1660 *Laing MSS* 314. They were affected thereanent, According to their temperament. Sanguinians did only laff, Cholerick melancholians chaff; COLVIL *Whig's Suppl.* (1681) II 20.

Sanguinolent, Sanguynolent, *adj.* [Late ME (*c*1450) and e.m.E. *sanguinolent*, L. *sanguinolentus* full of blood, bloody.] Blood-red. — Lyke as the comete stern sanguynolent [*Ruddim.* sanguinolent; L. *cometae Sanguinei*], With hys red cullour tryst and violent, Schynys; DOUG. X V 141.

Sanguis draconis, *n.* [e.m.E. (1599), med. L. (1220 in Latham).] Dragon's blood, the botanical substance used by apothecaries. — Inuarte Custumes .. Sanguis draconis the pund; 1597 *Bk. Rates* 10.

Sanguisug, *n.* [e.m.E. *sanguisuge* (Skelton), L. *sanguisuga.*] A leech. — As sanguisugs .. Cleaues to the skin quhile they be full of blud; HUME 73/175.

Sangweler, var. of SANGLER(E *n.*

(Sanicle,) Sanycle, *n.* (Late ME and e.m.E. *sanicle, sanycle* (15th c.), OF *sanicle* (12th c. in Hatz.-Darm.), med. L. *sanicula* (13th c. in Latham) the umbelliferous plant or the medicament derived from it.) — Geif him .. sanycle to purge the wound. And if the wound be one heid .. than geif to drynke no sanycle .. or geif him pympernell [etc.]; 15.. Comrie *Hist. Sc. Medic.* I 189.

Saning, var. of SAINING *vbl. n.* **Sanit,** p.t. of SAIN(E *v.*

Sanite(e, *n.* Also: **sanitie, sanetie.** [ME *sanite* (1432-50), e.m.E. *sanitie* (1586), OF *sanité* (1190), L. *sanitās.*] Health. — The saidis personis that takis the said cure of sanitie vpoun thame [*supra* the quhilkis takis vpoun thame to hale the said contagious infirmitie, *sc.* grandgor]; 1497 (*c*1580) *Edinb. B. Rec.* I 72. Mary .. Thou suete wyne tre & well of sanitee [*Bann. MS* I p. 6/13, sanite, *Ib.* 40a/13, sanetie]; *Ch. & M. Prints* vi b 13.

Sank, var. of S(C)HANK *n.* **Sankstar,** var. of SANGSTAR(E *n.* **Sanman,** var. of SANDMAN *n.* **Sannachy,** var. of SCHENACHY *n.* **Sanquary,** var. of SANCTUARY *n.*

Sans, *prep.* Also: **sance.** [ME and e.m.E. *sans* (14th c.), *saunz* (*c*1320), *saunce* (*c*1400), OF *sans, senz* (*c*1167 and *c*1050 in Larousse).] Without. Only in verse; chiefly in DOUG. in collocation with words of French origin. Also SANS-FAIL(L. — Geffray Chauceir as A per se sans peir [*L.* sance pere] In his vulgare; DOUG. *Pal. Hon.* 919. And bot my myght rasistit thame, sans dout Thai hadbene brynt; Id. *Æn.* II x 72. My marred mynde sans hop of grace; FOWLER I 280/162.

Sans, var. of SENSE *v.* **Sansable,** var. of SENSIBIL(L *adj.* (*adv.*). **Sansar,** var. of SENSUR *v.*

Sans fail(l. Also: **-fale.** [ME and e.m.E. *sauntz fayle* (*c*1325), *sanzfaile* (*c*1380), *sauns faille* (*c*1384), etc., *saunz faile* (*c*1530), OF *sans faille* (11th c. in Tobler-Lommatzsch).] Without fail, without doubt. Only as a rhyming tag in DOUG. — Pomponius quhais fame of lait sans faill [: Laurence of the vale; *L.* fale] Is blawin wide; DOUG. *Pal. Hon.* 911. Of thé writis Macrobius sans faill [: Saturnaill; *v.rr.* fail]; Id. *Æn.* I Prol. 67. *Ib.* 128, etc.

Sant(e, Santt, varr. of SANCT *adj.* and *n.* and *v.* **Santrel,** var. of SANCTREL *n.* **Sanudrio,** see SANDRE(E *n.*[1] **Sanycle,** var. of SANICLE *n.* **Sanyne,** var. of SAINING *vbl. n.* **Sanyt,** var. of SAINIT *ppl. adj.* **Sanȝe,** var. of SENȜE *n.*[2]

Sap, Sop, *n.*[1] Also: **sape, saup.** [ME and e.m.E. *sap* (? *c*1377), *sappe* (1523), OE *sæp.*]

1. Sap; the vital juice of plants. **b.** *transf.* Moisture more generally; *specif.* the juice in meat.

(1) This tre may happyn for to get The kynd rwte .. And sap [*C.* And sape, *W.* And the sap] to recovyr syne Bath [the] leyff and flewowre fyne; WYNT. VII 551. Spryngand herbys .. War .. cuttit soyn, To get thar mylky sap [*Sm.* sop] and vennom blak; DOUG. IV ix 85. The auld [plant] .. That pith inlaikis, sap, sapour and substance; ROLLAND *Seven S.* 1488. There little sap [*MS No. 1311* sop] in dry peis hooles; FERG. *Prov.* No. 842. What odds find the saints in hard trials when they feel sap at their roots; 1640 RUTHERFORD *Lett.* (1894) 593.

(*b*) Ane medecyne .. ȝe cowth mak it With .. the sop of the sege; HENR. *Practysis* 29. Quhan herbs and flours .. dois vpcast Sueit sauorus smell in sop; J. STEWART 54/52. As dewie drops distels .. On vallowit veids .. That syn vith sauorus sop dois sprout on spray; *Ib.* 171 § 1.

b. Thay eit, for common, flesche half raw; for the saup is maist nurisand in that maner; BELL. *Boece* I lvi. *Vligo,* the naturall sop; DUNCAN *App. Etym.*

2. *fig.* Vital energy; goodness (in soil).

Sap and sorrow, may be derived from the earlier *sop of sorrow* (see SOP *n.*[1] 2 b).

(*a*) The Scottismen .. drinkis out the sap of ȝour cuntrie, and soukis out your merche; 1616 *Crim. Trials* III 587. He wes lyk ane dry witherit kill stik and na sap in him; 1619 *Perth Kirk S.* MS 31 Aug. Like lying lea of lean and weak land .. whill it gather sap for a better crop; 1637 RUTHERFORD *Lett.* (1894) 466. Take him to you .. for much sap and sorrow we have had since he came amongst us; J. KIRKWOOD *Plea Before the Kirk* (1698) 15.

(*b*) Being planted in him, than cummis that fresche sop out of him into thee, and transformes thee; ROLLOCK *Wks.* I 358.

Sap, Sapin, *n.*[2] [ME and e.m.E. *sapin* (1323-4), *-yn* (Caxton), OF *sap, sapin* (11th c. and 1100 in Larousse), L. **sappus, sap(p)inus.*] A fir- or pine-tree. — [The barnacle goose] is baith fische & foull and bredis in .. rottin treis and auld mastis of schippis quhilkis ar maid of treis callit sap [F. *sap*] and for that caus mony folkis callis thaim sapins [F. *sapins*] .. and thai ar federit and fleis as foullis .. and signifies at he that first bure him in armes wes of the natur of watteris, of fische & of foullis and be lordschip he mycht say he wes king of the sey becaus he flew abone al warldlie power and be his natur had dominatioun abone the hiest lordis as the sap [F. *sap*; *Lindsay MS* sapin] dois abone the heiast treis; *Loutfut MS* 34b.

Sape, Sope, *n.* Also: **saip(e, saipp, sayp; saep; seap(e, seapp; sep(e; seip(e; seep; sop(p; soip; soap(e.** [Early and north. ME *sape* (*a*1200), midl. and south. ME and e.m.E. *sope* (Ancr. R.), *soap* (1635), OE *sápe.*]

1. Soap.

For further examples see BARRELL *n.* 1, FIR(I)KIN *n.* and KINKIN *n.*

(*a*) Of a kyst .. wyth sape in it; *Acts* I 306/2. Tappand wyne walx terr and sape; *Ib.* 336/2. 1454-5 *Edinb. Chart.* 80. Hawf ane pund of butter and tua schillings worth of sape to creische the towes; 1655-6 *Peebles B. Rec.* II 198.

(*b*) Colis or assis oley or steil saip or ony uthir merchandice; *Ship Laws* c. 2 (H). Saipp; 1468 (*c*1580) *Edinb. B. Rec.* I 23. Bocht in Medylburche .. 6 barell of sayp, ilk barell cost fre abord 15 s. 4 g.; 1497 HALYB. 59. Saip; 1530 *Edinb. B. Rec.* II 21. To sell richt deir, & by gud chaip, And mix ry meill amang the saip; LYND. *Sat.* 4056. 1547 *Stirling B. Rec.* I 48. The foure lasts saip laitlie cumin in ane Hameburgh schip; 1549 (*c*1580) *Edinb. B. Rec.* II 145. Suppoise thow wesche thi self with saip and [etc.]; HAMILTON *Cat.* 23. The .. counsale grantis licence to Johnne Gaittis, Inglisman, to brew saip within .. this burgh; 1554 *Edinb. B. Rec.* II 205. 1557 *Peebles B. Rec.* I 237. 1573 *Treas. Acc.* XII 350. 1577 *Edinb. Test.* V 306b, 1580 *Ib.* VIII 193, etc. (see KINKIN *n.*). For ule and saip to cresche the Madin with; 1583 *Soc. Ant.* VII 552. 1584 *Edinb. B. Rec.* IV 360. Half barrell of saipe, .. ane cutting of saip price iij li. xvj s.; 1591 *Edinb. Test.* XXIII 43b. For saip to wesche with, vij sindrie tymes getting at ilk tyme a pund and quarter of a pund at xxvij d. the pund; 1592 *12th Rep. Hist. MSS* App. VIII 117. 1596 *Edinb. B. Rec.* V 168. Saipe; 1604 *Shetland Sheriff Ct.* (ed.) 126. Saip of the Low Cuntreyis; *Bk. Rates* (Halyb.) 326. 1616 *M. Works Acc.* (ed.) II 5. For saip to the portcoolies, a pund monthlie; 1650 *Stirling B. Rec.* II 305.

proverb. Saip and clein claith begiles manie man; CARMICHAEL *Prov.* No. 1332.

(*c*) And quhou schou schynes lyk ony saep, My ladye with the mekle lippis; DUNB. (OUP) 106/9.

(*d*) Seape; 1570 *6th Rep. Hist. MSS* App. 646/2. ij kempkynnis of seap; 1580 *Dundee Shipping L.* 199. Thair is monopollis purchest be certane privat persones .. lykas .. making of seapp be John Nasmeth chirurgean; 1610 *Conv. Burghs* II 300.

(*e*) v lib. wyecht of sep; *c*1590 *Thanes of Cawdor* 196. Ane bark .. contenand .. fyve barrellis sepe; 1614 *Dundee Shipping L.* 242.

(*f*) That he sall nocht sell na kynd of stapill guid sic as salt, hemp, lint, woll, seip nor na vther sort of merchandreis; 1581 *Elgin*

Rec. I 159. That na gangrell man or woman sell ether butter seip or scheis .. bot with the commowne vechtis; 1583 *Crail B. Ct.* MS 21 May. Seipe; 1593 *Brechin Test.* I 114.

(*g*) For seep to wesche the baskits; 1631 *Buccleuch Household Bk.* 29 Oct.

(*h*) For ane pompe sope [6 s. 8 d.]; 1595 *Skipper's Acc. (Morton)* 9a. For butter and sope to the cairtis; 1616 *M. Works Acc.* (ed.) II 52. For 4 ounces sope to raze me with, 3 s.; 1678 CUNNINGHAM *Diary* 106. 17 s. & 6 d. as the price of ane feirkin of sope; 1696 *Dumfr. & Galloway Soc.* 3 Ser. XXXIII 136. A receat for the galshot or gulsit, Take a net great and a litle more of black sope with a yoak of a black hens egge [etc.]; 16.. *Nat. Lib. MS* 22.2.11, penult. p.

(*i*) Two kinkings sopp; 1659 Craven *Ch. in Orkney* II 197. For 3 pwnd and half of sop for stifin; 1687 *Weaver's Acc. Bk.* 24 June. Sopp; c1698 Innes *Sketches* 548. For fish 11 s., spice 1 s., sop 3 s. 8 d. [etc.]; 1702 *Household Bk. Gr. Baillie* 66.

(*j*) ij last soip; 1588-9 *Wedderb. Compt Bk.* 220. Soip; 1627 *Glasgow Chart.* II 582.

(*k*) Lint, irne, soap [etc.]; *Bk. Rates* (Halyb.) 281. For xii pond soap to rub the great cabell; 1617 *M. Works Acc.* (ed.) II 41. To summond our nichtbors for hame-bringing of foran soap; 1623 *Aberd. Council Lett.* I 206. *Ib.* 205. For soape to wesche the siluer veschell; 1631 *Buccleuch Household Bk.* 3 Oct. For soape to the kairt wheils; 1646 *4th Rep. Hist. MSS* App. 510/1. 18 daill of soap [£5 15 s.]; 1690 *Foulis Acc. Bk.* 125. For 4 unce castle soap; 1691 *Ib.* 138.

b. Also with qualifying terms: *Gray, quhite, sueit sape.*

Cf. e.m.E. *swete sope* (1611-12), late ME and e.m.E. *white sope* (c1425).

Thair tedderis wer maid weill grit to graip With silkin schakillis and sowlis of quhyte saip; LICHTOUN *Dreme* 64 (B). Tuelf litill ferikynnis of gray saip; 1574 *Edinb. Test.* III 137b. Tak the frothe of quheit saip and straik on the sair; 1588 *Sc. Ant.* IV 37. 2 unce seap, .. 2 unce cleren seap; 1597 *Wedderb. Compt Bk.* 1. Tua kinkingis of sueit saip; 1609 *Edinb. Test.* XLV 233.

2. a. *Attrib.* with *ballan(d)is, dische, silver*: Of or for soap. **b.** *Comb.* with *brewar(e, man.*

Later Sc. has *saipman* (1813) a soap-maker.

Also SAPE-BALL, -HOUS, etc.

a. Ane pair of saip ballandis with ane half pund wecht [etc.]; 1560 *Stirling B. Rec.* I 74. Ane pair of saip ballanis set to x s. viij d.; 1573 *Edinb. Test.* II 335b. — Ane lawer ane handfatt .. ane saip dische; 1644 *Edinb. Test.* LXI 60. — Margaret Dowchaell .. lavendare to the kingis maiestie .. [got] .. oulklie alsua x s. of saip siluer; 1575-6 *State P.* (Reg. H.) No. 96.

b. The toun of Campheir hes imposit certane imposts .. vpoun the saip payet be the saip brewares; 1612 *Conv. Burghs* II 363. 1620 *Edinb. Test.* L 293b. — Cristopher Barnes saipman; 1616 *Edinb. Test.* XLIX 164b. Christopher Barnis saipman in Campheir; 1624 *Brechin Test.* IV 175.

Sape, var. of SAP *n.*[1]

(Sape-ball,) Saip(e)-, Soap-ball, *n.* [e.m.E. *sope-ball* (1601); SAPE *n.* and BALL *n.*[1]] A ball or cake of soap. — Thrie dussone of sweitt saip ballis pryce of all thrie pundis .. fyve dussone of comoune saip balles; 1632 *Edinb. Test.* LVI 23b. Saipe ballis; 1643 *Ib.* LX 239b. To him to buy pouder, a jesmie glass, a soap ball [etc.]; 1690 *Foulis Acc. Bk.* 128.

(Sape-boiler,) Sope-boyller, Soap-boyler, -boyeller, *n.* [e.m.E. *sopeboyler* (1594).] A soap-boiler. (Cf. *saip-brewar(e,* SAPE *n.* 2 b.) — Francis Muire, sope boyller, .. his calling in makeing and boylling of sope; 1668 *Glasgow B. Rec.* III 112. Robert Douglas soap boyeller in Leith; 1676 *Edinb. B. Rec.* X 275. Soapboyler; 1695 *Acts* IX 491/2.

(Sape-hous,) Saip-, Sopp-, Soap-hous(e, *n.* [e.m.E. *sope-house* (1648); SAPE *n.* and HOUS *n.*[1] 2.] A building in which soap is made. Cf. SAPERY *n.* and SAPEWORK *n.* — For careing and wirking ilk last saip out of the saiphous of the toun to the schip; 1612 Rooseboom *Sc. Staple* App. 112, p. cxxxi. His loft in the Dubraw above the soap house; 1655 *S. Leith Rec.* 102/2. For cariage of daillis from the sopp house .. to [etc.]; 1691 *Glasgow B. Rec.* III 513.

(Sape-makar,) Saip-maker, *n.* [North. e.m.E. *sape maker* (Cath. Angl.), e.m.E. *sope-maker* (1558).] A soap-maker. — Johne Gleig saipmaker in Dundee; 1586 *Digest Justiciary Proc.* L 21.

(Sape-making,) Sopmaking, *vbl. n.* [e.m.E. *sope-making* (1603).] Soap-making. — [An] act of parliament in favour of those that used the manufactorie of sopmaking in Scotland; 1670 *Edinb. B. Rec.* X 79.

(Sapery,) Soparie, *n.* Also: **sopperie, soprie, soaperie, -arie.** [SAPE *n.* Also in the later dial.] A soapworks. Also *comb.* in *sopperie-works*, and *attrib.* with *yaird.* — In ansuer to the desyre of them that intendis to build a soparie, .. the said magistrates [etc.] .. grantis them that libertie; 1673 *Glasgow B. Rec.* III 173. Ane peice waist ground in Candleriggs, opposite to the soaparie; 1692 *Ib.* IV 45. Standing in his soprie or uorkhouse at Leith oyle and ashes; 1695 *Edinb. Test.* LXXX 122b. — *comb.* and *attrib.* For incouradgement of the sugarie and sopperie works; 1689 *Conv. Burghs* IV 95. Candlerig Street on the west, the soaperie yaird on the north; 1695 *Glasgow B. Rec.* IV 158.

Sapes, pl. of SOP *n.*[1]

(Sapework,) Sopework, *n.* Also: **soapwork.** Pl. **sopewarkis, -workis, -workes, sopworks.** [SAPE *n.* Also *soap-work* in 18th c. Sc. dial.] The process or business of manufacturing soap; manufactured soap collectively; a soap-factory. Also *pl.* in same senses. Cf. SAPERY *n.* — The preiudice whilk the decay of the sopeworkis hes occasioned to the kingdome; 1649 *Acts* VI II 300/2. Report .. concerneing the sopewarkis; 1650 *Ib.* 596/2. The sope-workes of Leith; BALFOUR *Ann.* IV 68. Acts of parliament to be revivit for sopworks and sterch and sort wyveing; 1661 *Aberd. Council Lett.* IV 154. She wes stented for her sopework at Leith; 1670 *Edinb. B. Rec.* X 79. The act of parliament anent the manufactorie of soapwork; 1676 *Ib.* 275. For establishing a manufactory in Leith of soap work, sugar, purslin, lame works, and stearch; 1695 *Acts* IX 365/1. *Ib.* 491/2.

Sapheir(e, varr. of SAFER *n.* and SAPHIR(E *n.* and *adj.*

Saphena, Sophene, *n.* [ME and e.m.E. *saphena* (Trevisa), *sophena* (c1400), *sopheyne* (1541), OF *saphène* (1256 in Larousse), med. L. *saphena* (c1210), *sophena* (a1250), of Arabic origin.] The saphena or saphenous vein. — The vayn that is ewin vnder the ankleth, that is callyd the sophene; c1420 *Liber Calchou* 450. In man opinnand the vayne callit mediana, in vemen saphena; SKEYNE *Descr. Pest* 28. Gif benethe the secreittis sic diseis appeir, the interioure vaine of the fute callit saphena moist be opinnit; *Ib.* 33.

Saphik, var. of SAP(P)HIK *adj.*

Saphir(e, *n.* and *adj.* Also: **saphyr(e, -ier(e, -er(e, -eir(e; sapphir; saffyr, -er(e; safyr, -er(e, -eir; saiffer, -ar, saifer; saufir, -er; salphir.** [ME and e.m.E. *saphir* (a1272), *safir* (Ayenb.), *saphire* (Wyclif), *safyre* (14th c.), *sapphire* (1477), all nouns, also *saphir* (Lydgate), *saphire* (Milton), adjs., OF *saphire* (c1119 in Larousse), *safir* (12th c. in Littré), L. *sapphīr-*, f. the Gk.]

A. *noun.* **1. a.** A sapphire, the precious stone. **b.** In general or collective use: Sapphire. **c.** *comb.* and *attrib.* **d.** *transf.* and *fig.*

For further examples see the Indexes to the various vols. of *Treas. Acc.*

a. In suprema le vche discendendo per medium a gret saphiere tua mid rubijs a mid saphier et thre litill saphiris; 1464-5 *Reg. Episc. Aberd.* II 163. A croce of gold .. with five sapphiris .. a grete perle [etc.]; 1482-3 *Acta Aud.* 131*/1. A rycht riche & noble saphire; *Loutfut MS* 43b. Saphir; 1511 *Treas. Acc.* IV 315. 1511-12 *Ib.* 331. Anew Of rubies red and saphirs blew; BUREL *Queen's Entry* 264. Saphyre; 1640 *Black Bk. Taymouth* 346. Ye may suppose it to be a blew saphir; 1665-7 LAUDER *Jrnl.* 28.

(*b*) [Scho] With a saffyr sanyt his face and eyne; *Wall.* VII 94. Safyr; 1498-9 *Acta Conc.* II 297.

(*c*) The ches of sapheris war; *Alex.* II 3707. Spar halkis .. With eyne .. circulit as sapheris; *Howlat* 333 (A). HENR. *Thre Deid Pollis* 28. STEWART 11896. Sapher; 1538 *Treas. Acc.* VI 415. Saphere; 1540-1 *Ib.* VII 428. Thre ringis of gold ane with ane saphere; 1550-1 *Dundee B. Ct.* II fol. 43 (6 March). Sapheris; PITSC. I 360/6.

(*d*) Dicht as a dyademe .. Circlit on ilka syd with a sapheir; *Howlat* 343 (B). For all the ruif .. Hang full of plesand lowpit sapheiris [*L.* saphyrs] cleir; DOUG. *Pal. Hon.* 1913. With rubeis reid and mony sapheir blew; STEWART 11898. Set full of roobies and sapheiris blew; *Clar.* II 1437. 1558-9 *Edinb. B. Rec.* III 30. Ane vther gold ring with ane sapheir formit like ane harte; 1561 *Crim. Trials* I I 416. Ane ring with ane blew sapheir price thre score crownis of the sone; 1566 *Crawford Mun. Invent.* II 89 (22 May). 1579 *Inv. Wardrobe* 279. Ane tablett sett with ane sapheir; 1592 *Melville Chart.* 133.

(*e*) Ane roll with ringis, ane with a grete saffere; 1488 *Treas. Acc.* I 82. A grete safere set in gold; *Ib.* 84. Saffer; 1495 HALYB. 12. Safer; 1539 *Ex. Processes* (Reg. H.) No. x (Scott v. Scott & McDowalls).

b. Iape [etc.] .., Saphir of assay, & berial clere; *Leg. S.* vi 281. Of siluer and saphir schirly thai [*sc.* banners] schane; *Gol. & Gaw.* 22. In the apill ar mony fyne stanys, the smarag, jacynct, saphire [etc.]; *Boece* 430. Quhill all that vattir cleir as saphir scheine Vas drumlie trublit; J. STEWART 93/501.

(*b*) Dicht as a dyademe .. Circulit on ilk syde with the sapheire [: cleire]; *Howlat* 343 (A). The beryallis .., Rubeis and sapheir, Precious stanis that weir; *Gol. & Gaw.* 995. *Rauf C.* 464. Moir semly na is the sapheir one to seyne; DUNB. (OUP) 104/36. In armour bricht .. Cirklit with sapheir and mony targe and schield; STEWART 12513. *Ib.* 12705. ROLLAND *Ct. Venus* I 108. Off culloure lyke the sapheir blew; LYND. *Mon.* 307. NORVELL *Meroure* 16a. Safeir; 1618 *Test.* in *Sempill P.* p. xlvii.

(*c*) And the colour is liknyt til a richt noble precious stane the salphir the quhilk excedis of hir beaulte [etc.]; *Loutfut MS* 11a.

c. Ane ring of gold with fyne safer stane worth j^c li.; 1542 *Acts & Decr.* I 140b. Ane gold ring with ane blew sapheir stane thairin; 1594 *Edinb. Test.* XXVII 290b. Ane saiffer stane estimat to vij crounis; 1609 *14th Rep. Hist. MSS* App. III 91. Saiffer; 1611 *Logie Par. Hist.* 65. — Tua gret ringis .., ane saphir and ane amerant ringis; 1505-6 *Treas. Acc.* III 177. To my dochter .. the blew saifer ring; 1574 *Edinb. Test.* III 113.

d. Of the diamont & the saiffer or of othir thingis as the price of the iustis; *Loutfut MS* 44a. Fairweill my weilfair may .. Fairweill the saiffar of assay; *Bann. MS* 225b/11.

2. ?A colouring agent of the blue colour of sapphire.

Ane hundreth weight of saphire; 1647 J. HOPE *Diary* (1919) 140.

B. *adj.* Of the blue colour of a sapphire. (Only in verse, chiefly of the sky).

The cristall air, the sapher [*B.* sapheir, *M.* saufer] firmament; DUNB. *G. Targe* 37. Luk on the saufir firmament; *Bann. MS* 229b/29. My voice surmontis the sapheir cludis hie; SCOTT xv 26. Begaried is the saphire pend [*sc.* the sky], With sprainges of skarlet hew; HUME 27/65. Hir saphir veins lyk threids of silk; MONTG. *Misc. P.* xxxv 55. The sunne .. faintly iourneyes vp Heauens saphire path; DRUMMOND II 15/45.

Saphroin, -o(u)n, varr. of SAFFRO(U)N(E *n.*

Sapience, -ens, *n.* Also: **sapyens.** [ME and e.m.E. *sapience* (Piers Plowman), *sapyence* (1509), OF *sapience* (*c*1120 in Larousse), L. *sapientia* good sense, wisdom, knowledge.] Wisdom; intelligence; knowledge. Also as an attribute **b.** Of God, **c.** Of a pagan deity, and **d.** Personified.

God .. has gevin thame sapience to teche otheris; HAY II 78/24. To be of hye witt and of grete sapience; *Ib.* 89/9. *Ib.* 72/1. Bot quhen our mynd is myngit with sapience; HENR. *Orph.* 469. Sapience; *Bernardus* 6. Be this ȝe may weill ken That foly is no sapience; *Colk. Sow* I 425. Thare [*sc.* the common people's] hedis hes na sure sapience; *Contempl. Sinn.* 391 (Harl.). Richt sa in governyng he schew his sapiens; *Ib.* 1146 (Asl.). Sapiens; *Ib.* 838. Sapiens; IRLAND *Mir.* I 69/19. Sapyens; *Kal. Sheph.* (1503) a iiii b. To speik off science, craft, or sapience [etc.]; DUNB. (OUP) 203/1. With sapience of the prudent Salomone; LYND. *Test. Pap.* 480. [Falsat:] .. I am sa full of sapience That sumtyme I will tak a trance; Id. *Sat.* 864. Athenes, the vmquhile fontane of sapiens; *Compl.* 21/7. The prudent Salomon .. in the 7 cheptor of his beuk of sapiens; *Ib.* 154/24. ROLLAND *Ct. Venus* II 938. Trew sapience and science baith, his hart delits to leir; HUME 22/158.

b. The fyre of hell be godlie sapience .. Sall pvnis synnaris; *Contempl. Sinn.* 1270 (Asl.). He [*sc.* God] has gouernit his kyrk and pepil be wertu and sapiens; IRLAND *Mir.* I 6/2. Wisdome is gottin throu gud lif & clene consciens, for the devin sapiens sais [etc.]; *Ib.* 12/25. The infynit wisdome and sapience of God; *Ib.* II 139/24. *Ib.* I 97/18, 21, etc. O hie supernale father of sapience; DUNB. (STS) lxx 9. God .. gaue, be diuyne sapience, Tyll euery ster thare influence; LYND. *Mon.* 693. The faderis power the sonis sapience The vertew of the haly gaist devyne [etc.]; *Bann. MS* I p. 6/24.

c. Faire Phebus is the god of sapience [: eloquence]; HENR. *Orph.* 425. Sen ȝe ar all seuin deificait Participant of deuyne sapience [: recompence]; Id. *Test. Cress.* 289. Mercurious .. god of eloquence .. In arte exparte and full of sapience; LYND. *Dreme* 396. Appollo, that the poietis callis the god of sapiens; *Compl.* 43/22. Sa Minerue wald me sapience send; LYND. *Meldrum* 28.

d. Thai saw the souuerane lady, Dame Sapience, be werray ressone .. inclyne to mercy; IRLAND *Mir.* I 117/12.

Sapienciale, var. of SAPIENTIALE *adj.*

Sapient, *adj.* (*n.*). [Late ME and e.m.E. *sapient* (1471), OF *sapient* (1180 in Greimas), L. *sapient-* wise, knowing, sensible.]

1. Wise, sagacious; also, learned, adept in study. **b.** Characterised by or pertaining to wisdom. **c.** As *noun*: A wise person, a sage.

The sapient and wys Salamon sais [etc.]; IRLAND *Mir.* II 8/34. A quene suld be richt werraye sapient; *Bk. Chess* 479. Sanct Paule, that doctour sapient; LYND. *Dreme* 597. Gar lordis send thare sonnes .. To seik science, and famous sculis frequent; Syne, thame promoue that war moste sapient; Id. *Test. Pap.* 1031. He wes honest of maneris .. in sermon maist sapient [etc.]; ABELL 65b. That into Elgin .. Sall be ane sute of clark[i]s sapient, Togidder with ane prudent precident; LYND. *Sat.* 3828 (Ch.). *Compl.* 5/10. Sche .. patt hir husband in beleue .. That he suld be als sapient As the gret God omnipotent; LYND. *Mon.* 933. With .. the rycht sapient Quene Saba; *Ib.* 5675. *Clar.* I 349. The wyse sages seuin, Quha was that tyme maist sapient vnder heuin; ROLLAND *Seven S.* Prol. 117. The sapient Salomon, with wemen was confoundit; LAUDER *Minor P.* v 9.

b. Devyne doctowris .. Sayis thair ar hid .. Grave materis wyis and sapient Vnder the workis of poyetis gent; *Bann. MS* 298b/5.

c. Conformand til ane adagia of ane of the seuyn sapientis of Rome; *Compl.* 127/34.

2. Having a taste or savour, sapid.

After the orig. sense of L. *sapĕre.*

All odors .. Of euerie substance sapient, the sapor and the taist .. the mouth will try in haist; HUME 19/75.

(Sapientiale,) Sapienciale, *adj.* [e.m.E. *sapyencyall* (*c*1485), *sapientiall* (1568), OF *sapiencialz* n. pl. (*c*1400 in Larousse) the 'wisdom' books of the Bible, L. *sapientiālis* of or belonging to wisdom.] Containing wisdom, as applied to the 'wisdom' books of the Bible. Also *absol.* — In the ald testament ar foure maneris & wayis of bukis, legall, historiall, sapienciale & prophetall; *Asl. MS* I 327/7. — *absol.* Sapienciale [books are] the epistlis of Paule, the prophetall, the prophesis of the ald testament, Sanct Jhon in the apocalips; *Asl. MS* I 327/10.

Sapin, var. of SAP *n.*[2]

Sapins, *n. pl.* (OF *sapins*, appar. nonce usage in F. source of Loutfut, barnacle geese). — *Loutfut MS* 34b (see SAP *n.*[2]).

Saplene, var. of SAPLING *n.*

(Sapless,) Sopelesse, *adj.* [SAP *n.*] Sapless, lacking in sap. — My moisture as a sopelesse sheard Is quite consum'd and ceast; MURE *Psalmes* xxii 15.

Sapling, Sipling, *n.* Also: **saplyn(e, sapplyn(e, saplene; sipline, -ein, sipplyn, syplin, -yn(e, sippilling, syppilling, -ylling, sepplyn, seplene, suppline.** [North. ME and e.m.E. *sappelynge* (1415), *sapling* (1543), *saplin* (1607), *-lyn* (1681), perh. f. *sap* SAP *n.*[1]; also north. e.m.E. *sipling* (1610), with obscure change of vowel, but possibly by influence f. SIPE *v.* Cf. OF *souplin* a shoot of a tree (Godef.).] **a.** A sapling, freq. *specif.* of oak, also (once) hazel-wood. Also *attrib.* Also (once) saplings, collectively. **b.** A length of wood; a staff. Also *attrib.* and *comb.* **c.** *fig.*

a. (1) A litil mote .. Quhar hepthorn buskis .. grew .. And evin saplynnys [*E.* syplinnis, *Ruthv.* syplynnys, *1553* sepplynis; L. *hastilibus*] of myrthus, the tre funerale; DOUG. III i 47. Within the rynd And stalwart sapplyn or bark of cork tre [L. *libro et silvestri subere*]; *Ib.* XI xi 43. Williame Ros .. sellis and disponis .. the wood and growand treis of saplingis and birk; 1607 *Fam. Rose* 294. All .. persones .. dwelling in the said townes .. sall keip thair haill nolt, hors [etc.] .. from .. quhair thair is any apperance of birk or saplenes; 1665 *Forbes Baron Ct.* 260.

(*b*) The said James cuttit the number of 12 sipleinis of aik .. ilke pece therof furth of the overwode of Colstoun; 1631 *E. Loth. Antiq. Soc.* II 133. For the wrangous cutting and destroying .. of seven dussones of young hasel siplines; 1643 *Ib.* 147. John Rae .. being accusit of cutting greenwood and timber .. confessit cutting ane oak sipling; 1665 *Holmains Baron Ct.* 8 March.

(*c*) [They] ar ordained to sight the vodis of Bithnie, Putachie [etc.] .. and seplenes of Logie; 1661 *Forbes Baron Ct.* 228.

(2) *attrib.* Williame Ros .. bindis and obleissis him .. to caus carie and transport .. fourscoir bollis sapling bark; 1607 *Fam. Rose* 295.

(3) *coll.* Tuaye horse laidis of the greyne young saplyne, saucht, birk, and of the vther young growtht; 1586 *Inverness Rec.* I 307.

b. (1) Thare wiffis syne come efter thame rynnand Ilkane ane grete sapplyne into thare hand; HAY *Alex.* 12098.

(*b*) And ane hail sipplyn [*E.* suppline, *Ruddim.* syplyne; L. *myrtum*] of a gret myr tre .. In maner of a speir in hand scho bair; DOUG. VII xiii 85. Four akin syppillingis to be extreis to the stane cartis and sand cartis .. xiii s. iiii d.; 1529-30 *M. Works Acc.* (ed.) I 26. For viii akin syppyllings of xx futis for ane new stane cart; 1530 *Ib.* 48.

(2) *attrib.* For four othir sippilling sparris for lymmerris to the said cartis; 1529-30 *M. Works Acc.* (ed.) I 26. For a sipling gird to the mill 14 s.; 1686-7 *Montrose Treas. Acc.* MS.

(3) *comb.* With grite sipling treis gaif him mony bauch, bla and bludie strykis; 1611 *Reg. Privy C.* IX 146.

c. Shall these siplings and sprowtings [*sc.* of episcopalianism] be taken up and set with guiding again in the Lord's vineyard? SCOT *Course of Conformity* 43.

Sapour, *n.* Also: **-oure, -owr, -or.** [e.m.E. *sapor* (1477), *sapour* (1638), L. *sapor.*]

1. Flavour, taste, savour. Also *fig.*

Lyke as a strand of water or a spring Haldis the sapour [*Asl. MS II 155/23* -oure] of his fontall well; HENR. *Orph.* 23. Eftir the wourdis ar fourmyt thar is na wyn bot the sapour and colour of it; IRLAND *Mir.* III 22/28. Quhen it [*sc.* a herb] is transportit or plantit in ony othir part, it is of litill sapor or gust; BELL. *Boece* I xxxvii. Blak cokis and hennis, nocht unlike to ane fasiane, baith in quantite and sapoure of thair flesche; *Ib.* xlii. Thai .. war content of sic sapoure and gust as nature propirlie to the metis had gevin; *Boece* 477. Thay sall feill sic ane smell .. And in thare mouth ane taist .. Off sweit and supernaturall sapowris; LYND. *Mon.* 6158. The sapor and the taist .. the mouth will try in haist; HUME 19/75.

fig. Thow flagrant flour Off sueit sapoure, Haill, deir dochter of Ann; *Ave Gloriosa* 36. Tha ar cum drye with lust and carnall heit, Because tha want the sapour of the spreit Of Christ Jesus; LAUDER *Minor P.* i 110.

2. Sap of a tree; also, juice, more generally.

(1) The auld [tree] .. That pith inlaikis, sap, sapour and substance; ROLLAND *Seven S.* 1488. The widderit branche [has no] .. Jusse or sapour of the tre, Quhen it is cuttit; LAUDER *Minor P.* i 107.

(2) My propyne com from the pres fute hait .. In fresch sapour new from the berry run; DOUG. V Prol. 54. [The rudy cullour vmquhile] as sapour [*C.* purpour] schene [.. fadys quyte owt of hir face]; *Ib.* XI xv 118 (E).

Sapour, var. of SOPOUR *n.*

Sappar-stone, *n.* ? Irreg. var. of, or ? erron. for, *sparr-stone* SPAR *n.*[2] Cf. also SAXE(E)RE-STANE *n.* — There was found a peice of broun sparr .. somewhat like unto sugar candy, which after it was broke, had in it an ounce of gold .. which sparr stone as I suppose is called the sappar stone in foreign nations and provinces; 16.. *Macfarlane's Geog. Coll.* III 43.

Sap(p)hik, *adj.* [e.m.E. *saphic* (1585), OF *saffique* (*c*1340 in Larousse), L. *sapphicus*, f. the Gk.] Of or in the metre(s) or verse-form(s) associated with Sappho. — And sum of thame .. playit and sang Sa plesand verse .., Metir saphik and also elygie; DOUG. *Pal. Hon.* 801. Notabillie depaincteng it furth in sapphik verse; DALR. I 245/27.

Sapphir, var. of SAPHIR(E *n.* and *adj.* **Sappis,** pl. of SOP *n.*[1] **Sapplyn(e,** varr. of SAPLING *n.*

Sappy, -ie, Soppy, *adj.* [Late ME and e.m.E. *sapy* (Prompt. Parv.), *sappie* (*a*1555) (of a tree, etc.) full of sap, (1563) full of substance, (1562) juicy, succulent, *soppie* (1611) full of sops, late OE *sæpig* (1100) full of sap.] **a.** Full of moisture, moist, wet. **b.** Of meat or fish: Full of sap, juicy, succulent. Also *fig.* **c.** Of part of a tree: Full of sap.

a., b. To se the soill that wes richt sessonabill, Sappie and to resaue all seidis abill; HENR. *Fab.* 1719. Quharof the beys wrocht thar hunny sweit .. In sappy subtell exalations; DOUG. XII Prol. 140. The talloun of thair [*supra* ky and oxin] wambis is sa sappy, that it fresis nevir; BELL. *Boece* I xxix. Furth of fresche fludes, .. Rais sappie subtil exhalatiounis; HUME *Promine* 74. The cherrie .. Whose sappy sloking .. Might melt into his mouth; MONTG. *Ch. & Slae* 1377 (Wr.). Quhen thay [*sc.* salmon] ar sa sappie and sa fatt and gudlyke; DALR. I 50/30. Because than thay [*sc.* beef] war mair sappie and bettir than; *Ib.* 90/2.

fig. What a sappie life were it to the merchand or the tradesman, when walking before his shop-door [etc.] .. Then the meditation of Him shall be sweet; *c*1660 *Sel. Biog.* I 269. A substantial sappy thing [*sc.* a bishoprick]; *Jus Populi* 437.

(*b*) Thou letst all goe like the smoake of tobacco, or like a soppy billow, which flees from the shell of the walnut, and straight doth vanish in the aire; 1609 GRAHAME *Anat. Hum.* 6b.

c. Fat and green and fruitful may ye be in the true and sappy root; 1637 RUTHERFORD *Lett.* (1894) 525.

Sapros. Var. of CYPRES *n.* (cypress). (Cf. ME *sypres* (15th c.) and SYPAR *n.*) — 2 lang burdis, 1 rundaile of sapros, 1 chalmer burd; 1565-6 *Prot. Bk. Thomas Johnsoun* 97.

Saps, pl. of SOP *n.*[1] **Sapulture,** var. of SEPULTUR(E *n.* **Sapyens,** var. of SAPIENCE *n.*

Saquhis, pl. of SAUCH(E *n.*

Sar-, Ser-. In many instances the spelling given in dictionary entries represents an expansion of the MS abbreviation for *(s)ar-/(s)er-*.

Sar, var. of SAR(E *n.*[2], *adj., adv.* and *v.*, SAVO(U)R *v.* and SER *v.*

Saracen(e, -sen(e, -zen(e, *n.* and *adj.* Also: **Sarasein, Sarracen(e, -zen(e, Sarazan(e; Sar(r)asin(e, -syn(e, -cine, -zine, -zyn(n, -zyne; Sarrezen(e; Sarisen(e, -scene, Saryzine, -zyn(e; Sarisiane; Sariazine; Sarzen(e, -zan(e; (Sazarene).** [ME and e.m.E. *sar(a)zin* (both Cursor M.), *Saracen, sarzyn* (both Trevisa), *sarysyne* (Manning), *Sarrazin* (*c*1330), *sarsyn* (*a*1375), *sarasene* (Piers Plowman), *sarseyne* (Lydgate), *sarezen* (*c*1449), *sarasin* (Caxton), *sarson* (1519), OF *sarrasin* (1080 in Larousse), OE *Sarracene* pl., late L. *Saracēni* pl.]

A. *noun.* **1.** A Saracen; an Arab or other Moslem opponent of the Crusaders.

(*a*) That was than in the haly land On Saracenys warrayand; BARB. I 140. Sarazenis [*E.* Saryzynys]; *Ib.* XX 348 (C). Sarasenys; *Ib.* 422. Sarasenis; *Leg. S.* xxxiii 905. Sarazanis; *Ib.* xxxvi 788. Thar ar mony Saracenys [*W.* Sarazenis, *C.* Sarzanys], Ysmalytys, and Agarenys [etc.]; WYNT. I 927. *Howlat* 484. Suppos the Sarazenis and othir mistrowand folk hald nocht the poyntis of the Ewangele; HAY I 104/32. We are nocht his sonis be natur, bot be creacioune and ymage, as are the Saracenis ore the laf; IRLAND *Mir.* I 26/36. *Asl. MS* I 315/23. Sarazene [*B.* Sazarene, *M.* Saracine], symonyte provit pagane pronunciate; KENNEDY *Flyt.* 525 (Ch. & M.). Bot vicious men denude of charitie As feinȝeit fals flattrand Saracens; LYND. *Sat.* 2305 (Ch.). Id. *Mon.* 5208. Finally thou art made as ane pagan, Saracene or infidele and sall perische for evirmair; HAMILTON *Cat.* 50. Saraseinis; *Clar.* III 322. DALR. I 340/4. The Sarazens are descended of Esau; LITHGOW *Trav.* IV 161. His son kill'd a More, who came in with some Sarazens to infest Galloway; MACKENZIE *Science Herauldry* 90.

(*b*) Gret Machomete, That Sarracenys thaire prophete Held; WYNT. V 5668. Sarracenes [*W.* Sarazenis, *C.* Sarzenes]; *Ib.* VI 250. Bot as to mak generale were agayn the Sarrazenis; HAY I 105/11. Sa the Sarrazene sperit at the Jowe quhat lay he lyvit apon; *Ib.* II 159/17. Infidelis as paganis Jowis Machometistis Sarracenis; IRLAND *Mir.* II 82/32. We cursis .. all thai that helpis the Sarrazenis aganis Cristyn men; *Rathen Manual* 27/2. The Sarrazenis ar circumcidit; ABELL 52a. HAMILTON *Cat.* 255.

(*c*) Sarazynys; BARB. XX 478 (E). Sarazynns; *Leg. S.* xxxiii 948. Alexander .. Sittand at the ches playand With ane Sarasyne; *Alex.* II 2991. Sarazine; *Rauf C.* 856. The strang willis of Sarazynis; *a*1500 *MS Rawl. Q. b. 4* fol. 91b. Sarasinis; *Clar.* III 43. The Sarrasin, vith bittir brag and blast Hint out his sourd; J. STEWART 18/105.

(*d*) Sarrezenis; *Leg. S.* xxvi 1119.

(*e*) On Sarisenis [*E.* Saryzynys] thre derenȝeis did he; BARB. XIII 324 (C). The hey king of Balmeryne With mony a mody Saryzine [*C.* Sarasyne]; *Ib.* XX 404 (E). Sariscene; LYND. *Mon.* 5790.

(*f*) Sum sayis ȝe ar a verry loun .. Sum sayis ane fals Sarisiane [: certane]; LYND. *Sat.* 2195 (B).

(*g*) Fore Jow, Sariazine & Cristine man In thare degre lufit hyme thane; *Leg. S.* xxi 653.

2. A pagan, a non-Christian more generally.

Sen God of mycht inclynit wes Till a Sarasyne [*sc.* Alexander the Great] throuch grace; WYNT. I 612 (W). Julius Cesar .. Saxounis [etc.] .. All borne Sarazenis vtterlie; STEEL *Roy Robert* 72. I reffer the expositione of this text to the vniuersal Cristianite to iuge quhiddir that Inglismen be Sarrasyns or Cristin men; *Compl.* 164/29.

B. *adj.* Saracen; of or pertaining to the Saracens.

Ane othir king, prince or emperour Sarrazene; HAY I 96/36. *Ib.* 246/2, etc. Wrait nocht in Caldye language .. Nor ȝit in to the language Sarazene; LYND. *Mon.* 574. Than sum began for to speik Grew, Sum Duche, sum language Sarazyne; *Ib.* 1777.

Saraly, var. of SARRALY *adv.*[1]

Sarcinett, Sars(e)net, Sas(i)nett, Sessenot, Sesnet(t, Sessnatt, Sesnit, *n.* Also: CESNAT. [ME and e.m.E. *sarsinet* (1378), *sarsenet* (1394), *sarcenet* (1477), *sarsynet* (1483), *sesenet* (1521), *sarsnet* (1610), AF *sarzinett*, prob. dimin. of *sarzin* SARACEN(E *n.* and *adj.*: cf. MF *drap sarrazinoiz* (1567 in Tobler-Lommatzsch), med. L. *pannus saracenicus* 'Saracen cloth'.] Sarcenet, the silken material. **b.** *Sessenot-taffetie, sesnetis taffeties, taffata-sarsenet, taffetas sarcinett*, ? a taffeta of this variety, a specially fine silken variety of taffeta (also e.m.E. *taffeta sarsnet* (1610 B. Jonson *Alchemist* II ii)). — 3 ells reid sasinett for 2 branches at £4 the elne is £12 .. 6 quarters blew sesnett for ane branch £6 .. 6 quarters ȝellou sasnett for ane branch £6; 1635 Balfour *Her. Tracts* 117. — **b.** (1) Nyne ellis sessenot taffetie at xlvij s. vj d. the elne; 1624 *Edinb. Test.* LII 321. Four scoir .. elles panis and sesnetis taffeties; 1631 *Ib.* LV 136. Sessnatt; 1640 *Black Bk. Taymouth* 349. Sesnit; 1643 *Edinb. B. Rec.* VIII. 420. — (2) I pray you send me fortie ells of taffata sarsenet such as will not run & scatter in the threads for thinnes nor will not frett by stiffnes; 1642 *Caldwell P.* 94. His [*sc.* King James VI's] skin vas als softe as tafta sarsnet; BALFOUR *Ann.* II 108. And breye it in a brasen morter into ane fyne pouder be searceing it in a pottingars box throcht a taffetas sarcinett; *Alchem. MSS* V.

Sard, *v.* [ME and e.m.E. *serdyn* (*c*1425), *sard(e* (1530), to seduce (a woman), OE (Northumb., once) *serð* imper.: cf. ON *serða* to have sexual intercourse with, MLG *serden*, MHG, early mod. Germ. *serten*.] *intr.* Of a man: To play the seducer, to copulate with women. — I trow Sanctam Ecclesiam Bot nocht in thir bischops nor thir freirs Quhilk will for purging of thir neirs Sard vp the ta raw and doun the vther; LYND. *Sat.* 3028.

Sard, p.t. of SER *v.* **Sardanis, -us,** varr. of SARDONICE *n.* **Sarde,** p.t. of SAR(E *v.*

Sardiane, *n.* [ME and e.m.E. *sardyn* (Wyclif), *sardiners* (? read *sardines*) (13th c.), *sardyne, sardone* (both Mandeville), *sardyne* (Tyndale, 1526; cf. 1881 (revised) *sardius*), cf. 18th c. Eng. *sardion, -ian* (1728 and 1741, Chambers *Cycl.*), F. *sardonie, sardoine* (11th and 12th c. in Larousse) (cf. SARDONICE *n.*), f. or f. as L. *sardius* (Vulgate), f. the Gk. *Sardis*, see SARDIS.] A sard or cornelian. — With brycht & schenand preciuse stanys, As sardiane, topias fyne, jaspis; *Leg. S.* vi 279.

Sardis, Sardix, *n.* [ME and e.m.E. *sardis* (Gower), appar. f. the name of the town Sardis: see SARDIANE *n.*] A variety of precious stone; a sard or cornelian. — Sardis is .. of ane saffrone colour; *Lindsay MS* 46b. Give their be a rid sardix heir, it hath direictly of that same very bigness another sardix answering to it their; 1665-7 LAUDER *Jrnl.* 28.

Sardius, only in NISBET after Purvey. — NISBET *Apoc.* xxi 20 (see SARDONICE *n.*).

Sardonice, Sardonis, Sardanis, -us, *n.* [ME and e.m.E. *sardonyse* (14th c.), *sardonix* (Wyclif), etc., L. *sardonyx*.] Sardonyx, the variety of precious stone consisting of onyx alternating with strata of sard or cornelian. — Into the palyce .. adorned with riche stones, Jasp, beriall and sardonis; *Alex.* II 7638. War all the buirdis maid of maist riches, Of sardanis [*L.* sardanus], of jasp [etc.]; DOUG. *Pal. Hon.* 1916 (E). The fonndementis of the wall .. war ournit with al precious staan. The first .. iaspis .. ; the fijft sardonice [*P.* sardony, *Wy.* sardonix], the sext sardius [*P.*, *Wy.* sardius] [etc.]; NISBET *Apoc.* xxi 20.

Sardonien, *adj.* (e.m.E. *Sardonian* (1586), f. L. *sardonius* (of laughter, etc.) sardonic, also Sardinian, f. Gk.: a 'Sardinian plant' (L. *herba Sardonia*) 'was said to produce facial convulsions resembling horrible laughter, usually followed by death' (OED).) — Sardonien parcell sennoune drauer; JAMES VI *Poems* I 129/149.

Sare, *n.*[1] (Appar. the Slavonic word, Russ. *tsar*, Croatian, Boh., Pol. *car*, etc., king, sovereign, emperor, ultim. repr. L. *Cæsar*: cf. e.m.E. *Czar* (1555), *Tzar, Tsar* (*a*1670) the emperor of Russia.) — Beȝond the Gret Se In the prowince of the Sare Quhar-to the Cayne is tributare; *Leg. S.* xxxvi 757.

Sar(e, Sor(e, *n.*[2] Also: **sair(e, sayr; soir; soare.** [ME and e.m.E. *sar* (south-west., Layamon), *sor* (midl. and south., *c*1250), *sare* (north., Cursor M., Rolle), *sore* (Trevisa), *soor(e* (*c*1400), OE, ON *sár*.]

1. a. Physical pain or suffering; also, an instance of this.

(1) He [*sc.* the Emperor Nero] that mycht nocht thole sic sare; *Leg. S.* ii 760. Till al thai that come thare With ony seknes ore with sare; *Ib.* iii 852. *Ib.* xxvii 606. Feyll Sotheroun said at Wallace feld na sayr; *Wall.* XI 1402. His wound to wesche in sobering of his sair; *K. Hart* 238. Suppoise the devill maid that graith .. to asswage my sair, I wald have tane it; 1584 SEMPILL *Sat. P.* xlv 341.

(2) Throw sairis smart he [*sc.* Lazarus] had ane peirles pyne; *G. Ball.* 40.

b. A physical injury or wound; a sore place on the body, a sore caused by disease, an ulcer, etc.

(*a*) Mekly vald scho wesch thar fet & with softe sawis thare saris bet; *Leg. S.* xlvi 102. A full deip wond and a grewos sayr [: thar]; Doug. xi xv 115. Quhill thai war haillit of thair sair[is] sund, Of brokin bone and mony bludie wound; Stewart 6721. Remeid for the ringworme. Tak the frothe of quheit saip and straik on the sair; *a*1595 *Misc. Spald. C.* II xxx.

(*b*) Hert is on the hurd and hand is on the sore; Henr. *Orph.* 409 (Ch. & M.). He . . healed our languors by his sores; King *Cat.* App. 26. He had to renew the mynd of melancolik persons besyds the using of his skill in haling of sores and wounds; 1591 *Warrender P.* (SHS) II 159. They confessed they . . use . . holy words for healing of shotts and sores; 1632 *Perth Kirk S.* in *Chron. Perth* 97. This we say thryse over, straiking the sor, and it becomes heall; 1662 *Crim. Trials* III 609.

(*c*) Tak plaister maid of wormewod with fresche butter or hoggis creische . . and presse tham togidder and la the same to the soir; *a*1595 *Misc. Spald. C.* II xxx.

c. Affliction, disease or illness, or an instance of this.

If not merely further examples of above senses.

(1) *sing.* Thair is na salue may saif thé of thy sair; Henr. *Test. Cress.* 411 (Ch.). My self is sauf fra seiknes and fra sair [*B.* soir]; Id. *Age & Yowth* 51 (M). Had ȝe rather saue Alexanders life, And helit of sair nor haue ȝour sonnis leuand; Rolland *Seven S.* 10210. Medecein His sair to cuir and dollour to remeid; *Montg. Suppl.* vii 12.

(*b*) There are words instituted for transferring of the sore or sicknes on other persons; Kirk *Secr. Commonw.* (1964) 412.

(2) *pl.* Of the pure he wes refuge . . And curit thair sairis with greit comfort; Lynd. *Meldrum* 1542. And all ȝour sairris callit secreit swnȝeis; Maitland *Maitl. F.* 286/54. The said Bessie Quhyt tauld hir that hir mother . . could charme sik sairis; 1608 *Dundonald Par. Rec.* 170.

2. *transf.* and *fig.* In applications of sense 1. **a.** Mental suffering, emotional distress or affliction, or an instance of these. **b.** Applied to non-material things. **c.** In proverbs.

a. Bot all . . Micht nocht ameis my greuous panefull sair; Doug. *Pal. Hon.* 663. Albeit thow sich ane thousand syth It will nocht sauf thé of thy sair; *Maitl. F.* 252/12. For no sair Nor sorrow can I soup; *Maitl. Q.* 234/30. To seik for salue of her that gaue the sair; *Philotus* 446. Its sche must hail my sair; Mure *Early Misc. P.* v 26.

pl. Goddis gret pyte . . That off all sarys is medycyne; Wynt. viii 2434. To meys Eneas thochtis and hys sarys [L. *curas*] Thus he begouth to speke; Doug. viii i 36. ȝit tyme may cum may mend my sairis Thocht I be ald; Maitland *Maitl. F.* 329/9.

b. Still in the sessoun all things remanit thair Perpetuallie but outher noy or sair; Doug. *Pal. Hon.* 1424. Tuiching that maledie and sair, quhilk . . at this present maist cheifly vexis and anoyis it [*sc.* the Church]; Ferg. *Serm. iii Malachi* in *Tracts* 62. Sal I gar this wound sa gret bleid agane? Sal I rub vpe this aulde sair? Dalr. I 110/31.

c. Flattera[r]is flok fordwart! for I [haif] hard tell, ȝe had ane saw richt sicker for all sairis; Charteris *Lyndesay Pref.* Adhort. 59. Than was I legat licent be the Paip With dispensatiounis, sawis for euerie sair; 1571 Sempill *Sat. P.* xxviii 106. Better auld debts nor auld saires; Ferg. *Prov.* No. 171. A salve for al sairs; Carmichael *Prov.* No. 172. Na thing hes na sair; *Ib.* No. 1202.

3. Affliction, suffering, more vaguely or generally; mental suffering, ? sadness, despondency; also, an instance of these.

(1) Efter lang pane, . . Till all his sair he soucht na saw bot ane; Kennedy *Pass. Christ* 257. Chryist . . Salwe of all sair and sweit succour; *Bann. MS* 38b/3.

(*b*) And euery wicht his awin suete or sore [: more] Has maist in mynde; *Kingis Q.* § 182. *Ib.* § 187. To wyn out off bondage Scotland agayn fra payn and felloun sor [: tharfor]; *Wall.* xi 303. Memore of sore, stern in aurore; Dunb. (OUP) 6/53. I felt as lying in a sound no sore; Craig i 7.

(*c*) Loo, this my dame dois work my lesting soir [: thairfore]; *Montg. Suppl.* xxiii 13.

(*d*) Lyf giwes no more To cuir my inward soare; Mure *Early Misc. P.* xi 31.

(2) Thus al day gan be sor Of thocht my gost with torment occupy; *Lanc.* 74.

(3) The mair feilling of the soares of sinne, the graitter missoure of grace, for it is by grace that we can feill sinne sair; 1596 Melvill 366.

b. *To do* (another) hurt, in senses 1 and 2.

Fore-thi, my sone, do thaim [*sc.* kindness, honesty, etc.] na saire; *Ratis R.* 1333. Gyf a man had don hym sare, Lauchful ore evyne pwniscioune [etc.]; *Ib.* 1541.

c. Expression of distress or grief (cf. 2 a above), lamentation.

In *to mak sare* (Mak *v.*[1] 30 b).

Til hyre he sad thusgat, Gret sar makand, sare pyte, Sayand: 'wais me, douchtir, for thé [etc.]'; *Leg. S.* xxxiii 145.

Sar(e, Sor(e, *adj.* Also: **sair(e, sayr(e, saier; ser, seir, sear; soir(e; soor(e; soar.** [Early and north. ME *sar* (Layamon), *sare* (Cursor M.), midl. and south. ME and e.m.E. *sor(e* (1297), OE *sár*, ON *sárr*.] Sore.

1. Of (a part of) the body: Suffering or in pain from wounds, sores or other injury; injured or painful; hurt. Also *proverb.*

(*a*) He fand his harme sa sare & wa; *Leg. S.* xxii 677. Half ane elne of quhite carsay to be the king a mittane till his sare hand; 1495 *Treas. Acc.* I 257. 1528-9 *Reg. Morton* I 3. Sare be his shins, and's kail ay cauld; *a*1660 *Sanny Briggs* in *Watson's Coll.* I 38/62. With a sare skin, baith black and blae; *Mare of Colinton* 159.

(*b*) And all was sound before was sair; *Sir Eger* 370. For vii elne of quhyte to be logouris to the king, the tyme his leg wes sayre; 1489 *Treas. Acc.* I 149. Castand him doun quhill his body wes sair; Kennedy *Pass. Christ* 375. To the draucht trumpet that hed ane sair lip; 1507-8 *Treas. Acc.* IV 102. For in this deidlie liff al our bodis . . ma be maid sair and wondit and slane; Gau 66/16. Quhilk causit ane bluid fal in my leg that was sair befoir; *c*1550 *Balcarres P.* 316. Pitsc. I 222/10. Hir awin pape beand sair . . scho send for the said Issobell; 1608 *Dundonald Par. Rec.* 171. [He] had stowin sa meikill . . as maid the said Johnne ane sair craig to beir it hame in ane basket vpon his head; 1613 *Fraserburgh Kirk S.* fol. 20b (5 Aug.). Therfor it fell out weill for your hors' sair bak; 1631 *Sel. Biog.* I 360.

(*c*) To ane of the borrowmen for helping of ane sore leg gottin in the wark; 1615 *M. Works Acc.* (ed.) I 364. Ane sore bak and . . schudder blaid; 1628 *Maxwell Mem.* I 481. To J. Quantanes man that had a sor finger hurt in our work xx s.; 1632 *Glasg. Univ. Mun.* III 483. Charme . . for a sore leg; 1650 *Maxwell Mem.* I 356. The disciples . . had neither sore limb nor lith; *a*1665 W. Guthrie in *Sel. Biog.* II 75.

(*d*) Remember my soir bodie, Sa woundit and bludie; *G. Ball.* 157. His maiestie tuik sa soir ane fuit that he sowned with it; Bisset II 178/25.

(*e*) Never a soar finger during the haill labour; Melvill 6.

proverb. Touch me not on the sair heel; Ferg. *Prov.* No. 851.

b. *specif.* Of the throat or the teeth: Inflamed or the like and in pain.

Emerides or the sair als; Rowll *Cursing* 51 (M). For ane irn to byrn sair teith, ij s.; 1507 *Treas. Acc.* IV 77.

c. *specif.* Of the head: Aching.

(*a*) The decoctione of it is remeid for ane sair hede; *Compl.* 67/12.

(*b*) Bessie Spence hes ay ane sore head, it is ay pained; 1643 *Orkney Witch Trial* in *Misc. Abbotsf. C.* I 177. That he haid ane sor head with the megrum; 1661 *Elgin Rec.* II 298.

d. *A sore skin*, ? a beating.

In every .. country that he geed to, he was ay for to get a sore skin; M. BRUCE *Six Dreadfull Alarms* 18. Till he made him to ken it by a sore skin; Id. *Sermon in Edinb. Tolbooth* 10.

e. *To lauch* (one's) *hert sair* (also, *quhill* (one's) *hart grew soir*), to laugh one's heart sore, to laugh very heartily. (Cf. 7 b.)

God lukit & saw hir lattin in & lewch his hert sair; *Kynd Kittock* 20 (Ch. & M.). Quha wald not lauche quhill his hart grew soir; 1571 *Sat. P.* xxix 24.

2. Of a person or animal: Suffering from wounds or sores; injured, hurt; freq. in collocation with *seke*. **b.** *Absol.* as *noun*. **c.** *transf.* Of a period of time during which a person experiences pain, illness or suffering.

(1) The sair man [*supra* ane woundit man, ane man .. Sa sair woundit] him beheld; *Prestis of Peblis* 527. *Ib.* 572. Thocht he sair was, [he] fersly dyd persew; DOUG. XII xii 168. I am so crukit, auld and sair; LYND. *Compl. Bagsche* 12. Ane scabbit hors will feill quhair he is sair; 1570 *Sat. P.* xxii 65.

(*b*) And to the barras magrie him full sore Him buire; *Clar.* v 2350.

(2) Quhene he men mad sek & sare; *Leg. S.* ix 17. To sek and sar folk gyfand heile; *Ib.* xxvii 1212. *Ib.* x 113, xxxi 319.

(*b*) Al that euire ware sek & sayre; *Leg. S.* vi 367. Hele me, that is baith seke and sair; *Arundel MS* 253/446. All medicinaris .. quhilk takis on hand to help seik or sair men; HAMILTON *Cat.* 103. To day ane man is fresche and fair, To morne he lyis seik and sair; *G. Ball.* 30. To .. visite .. the puir quha sall happin to be seik or sair, laymet or hurt; 1585 *Edinb. B. Rec.* IV 420. He is lyeing seik and sair .. Let him lye intill his bed—let him lye intill it seik and sore; 1662 *Crim. Trials* III 609.

b. [Miracles] done at his grawe to sare & seke; *Leg. S.* xxxvi 565. *Ib.* xl 765. The sair can say, 'Be him that can vs saue'; *Prestis of Peblis* 555. To wesye oft seik folk & the saire; *Bk. Chess* 1691. *Asl. MS* I 304/15. For he left nowthir seik nor sair Unslane; DUNB. (OUP) 161/19. Oure nacioun .. producit the seik or sare to the strete; *Boece* 73b. He haillit the seik, sair, lamit, and blinde; *G. Ball.* 67.

c. Isabel asked how the said James did and the said Janet answered that he had ane sore summer; 1662 *Soc. Ant.* XXII 222.

3. Of various forms of physical assault, instruments of physical attack or implements of restraint: Inflicting hurt or pain, esp. of a severe or harsh sort, passing into an intensive use, severe, harsh. Also *fig.*

(*a*) In Philippis, with wandis sar, He wes firste doungyne; *Leg. S.* ii 9. The angel .. Band vs with chenȝeis fel & sar [: rare]; *Ib.* iv 113. Gyfand hyme mony dintis sare; *Ib.* xxvi 692. *Ib.* xlvi 201. Quha-sa sare strakis can strike; *Alex.* II 1855. Off wenyme and stangys sare; WYNT. I 658. *Ib.* 1616. Thai band thaim fast with wedeis sad and sar; *Wall.* III 215. *Ib.* v 383, 845. *Remembr. Passion* 311. Thay hurlit him heir and thair And gaif him mony buffat sair; DUNB. (OUP) 7/21. Quhilk of the goddis sa cruel flammys sayr Held from Troianys? DOUG. IX iii 2. [He] straik .. hir with his nevis and feit .. mony sair bauch straikis; 1576 *Crim. Trials* I II 61. The Devil .. gave you ane sair stroke on the right shoulder; 1662 *Soc. Ant.* XXII 226.

(*b*) He said it was the sorest nip he had ever felt; 1649 *Sc. N. & Q.* IV 28. Isabel Murray .. got a sore knock upon the back with a stone; SINCLAIR *Satan's Inv. World* 152.

fig. A dum pastour or a wicious, is a sair scurge on the peple; WINȜET I 101 marg.

4. Of an action, manner of acting or attitude in regard to another, of a person's treatment of or by another: Severe, harsh; extreme, passing into a more general expression of emphasis, complete, entire.

(*a*) For trubling of thaim in the wrangus and sair taking of thair custum; 1546-7 *Stirling B. Rec.* I 46. All godlie men feare .. Godis sair displeasure; 1561 *Bk. Univ. Kirk* I 12. To absteyn fra all fylthines .. under pan of disciplyn in saier maner; 1564 *St. A. Kirk S.* 224. To him .. vill scho navayis fald, Bot rather die, hir haittrend is so sair; J. STEWART 28/484. Ther is laike to be ane sare preseqution agens all manaier of Catholiques; 1655 *Blairs P.* 92.

(*b*) Nor Gods seir wraith abone ȝour heidis to haif; 1570 *Sat. P.* xiii 205. He movit the provest, with sear rubbing of the ga of his conscience, to [etc.]; MELVILL 125.

(*c*) God schawis the richt With soir vengence; DUNB. (OUP) 107/30. Goddis soir displeasur to fall upoun you; KNOX II 317. Quhen as tha thole soir persecutioun; LAUDER *Minor P.* i 136. And the actioun to be interit in all convenient diligence that we may anes have a decision thairof in sore contradictione [to previous occasions when no decision had been taken]; 1624 *Aberd. Council Lett.* I 227. Men would think the greatest hill .. a light burden .. to get themselves hid from the peircing view of a slighted .. mediator. Oh, but that will be bitter and sore to bide; DURHAM *Comm. Rev.* 25.

b. *specif.* Of a (legal) penalty, punishment or sentence.

Gif he may nocht find that man inditit he aw to arreste all his gudis .. vnder soir borowis; *Acts* I 346/2. Lat ws to borch our men fra your fals law .. Deyll nocht thar land, the wnlaw is our sayr; *Wall.* VII 436. Thai sall for this suffer sair pvnicioun; *Contempl. Sinn.* 381 (Asl.). *Ib.* 620. Than was all maist writtin the sentence sair; DOUG. *Pal. Hon.* 769. Sayr; Id. *Æn.* II x 23. Than to the woman, for hir offence, God did pronunce this sore sentence; LYND. *Mon.* 1052. His death hathe .. the soore sentence of the law .. broken; NORVELL *Meroure* 10b. And quha that this seir sentence small esteims [etc.]; 1570 *Sat. P.* xiii 149. Quhilk double sentence wes baith suith and sair; MONTG. *Sonn.* xxxiii 8. Judgement is at the door and indeed your sentence will be sore to bide; 1681 *Cloud of Witnesses* 96.

c. Of verbal argumentation, accusation, etc.: Stern, harsh; strongly urged, forceful; intense.

Arreste throw sare praying Went to warne Alexander; *Alex.* I 2076 heading. Him selff began a sair cusyng to mak; *Wall.* VI 400. For the letters (sayis he) ar sair and strong bot his bodilie presence is waik, and his speich is of na value; ROLLOCK *Wks.* I 405. Ȝour word, as suord, is sair; My breist is persit with uyolence; *Montg. Suppl.* x 29. Sche enterit sa upon me with sa soar threatnings; MELVILL 19. Soar; *Ib.* 33.

d. Of a person: Severe, harsh, extreme (in attitude, action, etc.); bitter, hostile; hard (to understand, argue against or deal with). Also *proverb.*

The Romane pepill quhilk daily invadis us with insufferabil injuris, ar sare ennimes; BELL. *Boece* I 88. Ye ar oure sair for me, (said the quene); KNOX II 285. To Sandelandis ȝe wer our sair to schame hir; *a*1568 SEMPILL *Sat. P.* xlvii 9. My creditoris are verray sair in craving their dettis; *c*1575 *Waus Corr.* 114. Schir Johne Hotham, younger (a sore enemy to the king); SPALDING II 211. He is a very sore party for the sinner, that stands in opposition to him; RENWICK *Serm.* 170.

proverb. Ane ill payer maks a sore craver; Ferg. *Prov.* MS No. 142. A sair flyter was never a gud fechter; CARMICHAEL *Prov.* No. 200.

5. Of **a.** Wounds or hurts, **b.** Pains or diseases, **c.** Of the pain itself, also *specif.*, the *painis of dede*, **d.** Of other bodily afflictions: Severe, grievous, painful, distressing, hard to bear.

BELL. quot. in a may belong in 1 above.

a. Bot thai that gan thaim abandoun .. to woundis sar; BARB. XVII 643. Clengeand thi hortis that are sare; *Leg. S.* xxii 357. Sair; *Gol. & Gaw.* 727. The Bruce him hurt at the returnyng thair, Wndyr the hals a deip wound and a sayr; *Wall.* x 338. Thir deip wondis & sair stundis I thoill for thé; *Makc. MS* XV 17. He rasyt hym vp apon hys wondit thee .. thocht the violens of hys sair smart Maid hym onfery; DOUG. X xiv 69. As precius balme revertis thingis sare And makis thaim of rot impacient; BELL. *Boece* I xii. Into his syid ane deip wound and ane soir; STEWART 29501. Quhan ve com to owr awin shapes again ve will find the scrattis and rywes on our skins werie sor; 1662 *Crim. Trials* III 611.

b. Thay war weseit with suddane soir seiknes; *Colk. Sow* II 34. God tuke wengeans .. With sic ane schot in till his wame That he had seiknes saire [etc.]; *Seven S.* 748. Expell dolour, expell diseisis sair; DOUG. *Pal. Hon.* 1018. This emprice tuke ane meruellous malady That scho behuid for to tak bed .. With sair seiknes sa troublit was hir corce; ROLLAND *Seven S.* 76. *G. Ball.* 33. The langer spairit, the plaigue is ay the sairest; 1572 SEMPILL *Sat. P.* XXX 135. That sair seiknes, named the sueit of Britannie, cam nevir till ws; DALR. I 5/18. Perceiueng his helth with sair and hauie seiknes to be persewit; *Ib.* 234/16. Margaret McKonnell in Park is .. with sore and heavie sickenes; 1672 *Kirkcudbr. Sheriff Ct. Processes* No. 157 (5 Nov.).

c. And suddand deith with ithand panis sore; HENR. *Fab.* 1137. He tholit the sairast painis of dede; HAMILTON *Cat.* 151. A desease, comonlie called the Irish aygo, which was a terrible sore paine of the head; 1650 LAMONT *Diary* 20.

d. In tyll tyme off hungar sare; WYNT. II 355. Till hir that sufferyd thrystys sare And paynys hard quhen scho hym bare; *Ib.* V 581. *Ib.* 3123. Vse derth o Lord or seiknes and hungir soir; HENR. *Prayer* 25. Gif that ane man .. Wald think to sare and passing sufferance On his a syde to ly but libertee In ane soft bed [etc.]; *Contempl. Sinn.* 1251 (Harl.). [To] refreshe his people opprest with sair hungre; DALR. I 72/8. Manie men speiks of my meikle drink, bot few of my sair thirst; CARMICHAEL *Prov.* No. 1112.

6. a. Of battles, fighting, etc.: Fierce, hard-fought.

The folk of Grece .. Him reskewit with fechting sare; *Alex.* I 2656. A sar assay forsuth thar mycht be seyn; *Wall.* IV 628. Still saxte dayis at sar sailȝe thai baid; *Ib.* XI 18. *Ib.* III 158. Sare and bludy in the begynnyng was this were; *Boece* 37. The batell committit wes sair bot the wictorie maist inclinit to the Scottis; ABELL 73b. The sair battellis .. that vas cruelly exsecutit on Scottis men; *Compl.* 92/1. Thair vas ane soir carmishe betuix thame of Leith and the men of Edinburgh; PITSC. II 284/24. Nathir was this his lyfe without mekle blude and sair slauchtir; DALR. I 161/10. Manie sear battels and greivus tentatiounes did my God uphauld me in; MELVILL 80. Ye must run, strike, fight, and endure sore bickerings, before the crown be gotten; DICKSON *Wr.* 133. The Picts assisting the Scots, a sore overthrow was given to the Britains; SPOTSW. *Hist.* (1655) 6. M'Kenzie .. overtook them at Inverchoran quher he gave them a saire skirmish; 1667 *Highland P.* II 24.

compar. The salt was all the sarar [*M.* sairar] suth to sayn; DUNB. *G. Targe* 198.

b. Of weather and the seasons: Severe, bitter. Also *fig.*

Gret cald thare-for of ser snaw I haf tholyt; *Leg. S.* XVIII 1003. In to ane coif he crap Fra sair wedder and frostis him to hap; HENR. *Fab.* 1292 (H). Ane sarar wynd into Scotland .. Blew neuer na tyme; STEWART 46970. *Ib.* 56207. ROLLAND *Seven S.* 565 (see c (*a*) below). And soft this seasoun, quhilk hes bene schairp and sair; MONTG. *Misc. P.* li 4. For I had sore weather betwixt Aberdeen and this; 1665 *Sel. Biog.* I 460. That day .. they had the sorest storm that they had felt; W. ROW *Blair* 143.

fig. So shall ye stand solid, as on a rock, albeit the blast be sore; DICKSON *Wr.* 66.

c. Of other afflictions, impositions, etc.: Oppressively severe, heavy, burdensome; bitter, distressing, hard to bear. Also *transf.* to a person.

(*a*) Fra Pers I com, of presoun seir [: bachelare]; HAY *Alex.* 974. Ane sair stonay and stour at thair hartis standis; *Gol. & Gaw.* 1056. The sayr grewans ramaynyt in his entent; Los off his men and lusty payne off luff; *Wall.* V 714. Be thar chansler the tothir pees was bwn, And that full sair our forfadris has fwn; *Ib.* IX 664. And that me think ane coumpt our sair; DUNB. (OUP) 133/15. Geif sick thing war tresoune the kingis hienes mycht euir quhen he wald charge ony lord .. to do ane thing .. and quhen he left it vndone forfalt him .. quhilk war oure sare and dangerous exemple for all the kingis liegis; 1528 *Acts* II 323/1. That gratius God abone, Ane sarar plaig sould sone amang thame send; STEWART 36235. HAMILTON *Cat.* 98 (see d (1) below). Vpon the sey thay sufferit greit perrell .. be sair storme and trauell; ROLLAND *Seven S.* 565. Than ȝe knew thay wer ȝour fais, Bot now thay cum in freindis clais, Quhilk is ane sairer sey; 1572 *Sat. P.* XXXI 78. My [*sc.* Lady Scotland's] torment sair constrynis me this to speik; *Ib.* XXXIII 275. With walkryfe labour and expensis sair; 1573 SEMPILL *Sat. P.* XXXIX 342. Eftir sair trauell quhen mony dangeris he escaped had; DALR. I 2/23.

(*b*) Gif ȝe taistit had befoir Of Strivilling toun the panis soir; DUNB. (OUP) 75/86. The Scottis prescribit war .. out of Albione .. With soir travell; STEWART 20243. Ay scho thankit God .. Of all hir trubillis and hir chansis soire; *Clar.* III 908. He will be put to a sore pinch before this victory be gotten; DICKSON *Wr.* 134. The suddenness of a destruction makes it to be so mickle the sorer; HENDERSON *Serm.* 98. Wishing that God Almightie may neaver trist yow with such a sore cros; *c*1685 *Annandale Corr.* 315.

transf. King James the First .. said that he [*sc.* King David I] was ane sair [*1821* soir] sanct for the croune, as he wald signify that he dotat the Kirk our richelie and [etc.]; BELL. *Boece* (M) II 185.

d. Of labour, etc.: Involving physical effort, hardship, etc.; heavy, burdensome, severe; also, an instance of this.

(1) For I haue oft tymes swet in seruice full sair; *Rauf C.* 637. For the other tua hors, quhilkes makes xiij hors, be resoun sho was oure sair ane draucht to the xj hors; 1547 *Treas. Acc.* IX 90. Quhay takis ouer sair mail .. or puttis thair cottaris to ouir sair labouris; HAMILTON *Cat.* 98. Sore laubouryng, tyll thy browis sweit; LYND. *Mon.* 1089. Vnto his grace declair My walkryf nichtis and my labour sair; BELLENDEN *Bann. MS* 366b/5.

(2) They would work a sore work, if they would aye work as they do at that time when their master is beholding them; HENDERSON *Serm.* 339.

e. Of a person: In severe straits, hard-pressed. (Cf. esp. SAR(E *adv.* 7 c.)

Sir, I pray ȝow stay, for ȝour freindis is not heir, and ȝe will be soir to heiff ado with menne; *Hist. Kennedy* 46.

7. a. Pertaining to grief, sorrow or regret as felt or expressed; sorrowful, sad. Also *proverb.*

(*a*) Bot Cristine men .. Faste cuth gret with ser sichinge; *Leg. S.* I 666. Scho .. Mad sare chere; *Ib.* XXXIX 322. *Alex.* I 835, 2379. He .. mad hys confessiown Wyth gret and sare contrityown; WYNT. V 2914. Witht sair regrating for my syn; *Fifteen Ois* 318. He gave a sare sob, with a grete siche; HAY II 8/19. This scheip .. I .. hard .. mak sair lamentatioun; HENR. *Fab.* 1285. *Ib.* 1555. Thai grat with mony saire allace; *Seven S.* 918. To Scottis and Pichtis this day was sare and dedelie; *Boece* 263b. The mater is to me sa sad and sair, That I may not sa lang abstene thairfra; ROLLAND *Seven S.* 4472. *Ib.* 164. Sair it was to se ȝoure prince with murther prest; Sairar, I say, him, in his place possest, The deid that did; 1567 *Sat. P.* IV 106, 107. My sorrow and my sichingis sair; SCOTT X 53. MAITLAND *Maitl. F.* 429/1. Makand hevye sair dule with hir self; 1576 *Crim. Trials* I II 51. This mater maid Mr. Ninian verie inviet with the hæretickis, and verie saire; DALR. II 467/32. MONTG. *Misc. P.* lii 2.

(*b*) And soir Contritioun .. My visage all in watter drownit; DUNB. (OUP) 10/99. My soir torment Dois moir augment; *Maitl Q.* 106/77. *Ib.* 206/141. Thir [*sic*] is ane soir mynd to all our hartis; 1622 *Aberd. Council Lett.* I 196. The greatest persecutor or malignant will have sore missing of His favour in that day; 1680 *Cloud of Witnesses* 51.

proverb. Efter drinking commes winking, after winking sair forthinking; CARMICHAEL *Prov.* No. 492.

b. *fig.* Of the heart: 'Sore', distressed, unhappy. Also *comb.*

And he with sar hart tauld him sone; BARB. V 40. Of ned his hart suld have ben sar That fellone jugment to heyre; *Leg. S.* I 654. The monkis .. Of hyre dissese had hartis sare; *Ib.* XXX 620. The motheris .. Sall haue lang tyme thare hartis sare; *Alex.* I 3160. Scho had na lufe that for hir vailȝeit, All-thocht sare hartis hir assailȝeit; *Ib.* II 2344. HAY *Alex.* 1688. *Wall.* I 337. I sich without sair hert or seiknes in body; DUNB. *Tua Mar. W.* 446. DOUG. IV vi h. of ch. *Ib.* XIII iii 106. My hairt will ay be sair Gif [etc.]; *Freiris Berw.* 266 (B). I leif hir, than, with hart full sore; LYND. *Meldrum* 1479. Wyth sorowfull mynd and sayr hart, bursting owt wyth tearis throw [etc.]; 1565 *St. A. Kirk S.* 235. *Maitl. Q.* 50/8. Ȝour lust & languore I lament With hairt richt soir; *Montg. Suppl.* ix 14. Giff I haid a dammist sear hart, God knowes, luiking for na thing bot the burning of the haill place; MELVILL 85. The Lord knows I have a sore heart to mention these things; 1681 *Cloud of Witnesses* 92. There have been few sore hearts in Scotland for the want .. of the gospel; RENWICK *Serm.* 284. My daughter has a sore eye, and a sore heart too; *c*1691 *Melville Corr.* 239.

comb. Quhilk sycht to se .. maid me sair hartit; 1570 *Sat. P.* x 394.

c. Causing or involving distress or regret; unhappy, unfortunate. Also *proverb.*

Ane sare cunnand me think this is! *Alex.* II 2272. The subcharge off thy seruice is bot sair; HENR. *Fab.* 346. This sair exampill .. Sowld caus all men fra wicket vycis fle; Id. *Thre Deid Pollis* 15. The feird [Muse] endytis .. Sair tragedeis; DOUG. *Pal. Hon.* 861. Of Troy the sayr rewyne; Id. *Æn.* III vii 10. Ane myserabill and sare bargane; *Boece* 45. Bot in the end thai gat ane sair faw; ABELL 94a. The Prince James decessit afoir him twa dayes .. quhilk was ane sair deid to this realme; *Diurn. Occurr.* 24. Sair is the sang Poure folkis amang; MAITLAND *Maitl. Q.* 163/12. Whow grait and sear ar the consequents that insewes on your fact; MELVILL 208. Bvt to aware that sore sin of profanation; BIRNIE *Kirk-b.* xviii. A sore matter for a sinner to be corrected, and yet to go light-farrand under it; DICKSON *Wr.* 87. Our cloaths are thinn, our purses are right bare, To bide these two, great sir, it is right sare; 1687 *Fugitive Poetry* II xl 4/108. It will be a soor metter if still acounts and anwelrents shall swell to principell soums; 1695 *Douglas Corr.* 286.

proverb. It is a sair lyfe, to be lang a maidin and syne a preists wyfe; CARMICHAEL *Prov.* No. 971.

Sar(e, Sor(e, *adv.* Also: **sair(e, sayr(e; sear; ser; soir(e, soyr; soar.** [Early and north. ME *sare* (Layamon), *sare, sair* (Cursor M.), midl. and south. ME and e.m.E. *sore* (*c*1200), OE *sáre.*]

1. a. Of the infliction of physical injury, as by wounding, striking, gripping, etc.: So as to cause great injury and/or pain; severely; painfully; seriously, badly. Also *proverb.* and *fig.*

Very common with *wound* v.

(*a*) Feyle men dede and woundyt sar; BARB. II 354. *Ib.* XVII 733. And with scharpe schurgis befte rycht sare; *Leg. S.* iii 613. Thane the edryse bate tham sa sare, Til [etc.]; *Ib.* xi 311. & resone wald that his throt ware Vith a snar cord hangyt ful sare Fra quham be tresone com that voice That [etc.]; *Ib.* xii 288. As the bochoure wes mendand The scho, he hwrte rycht sare his hand; *Ib.* xiii 82. The tormentouris .. Tuk yrne kamys that bat sare; *Ib.* xxxvii 256. Sare woundyt; WYNT. V 4373. The fader and the moder wald have lerit him thair craft and dang him oft rycht sare thareto; HAY II 151/1. Throw his hawbrek he woundit him full sare; Id. *Alex.* 3191. Buffat and scurgit at ane pillar, With rude knoppis smytand sair, And vthir mony innomarabill pane; *Fifteen Ois* 43. *Boece* 66b, etc.

(*b*) Sayre; *Leg. S.* xxii 647. Fle this suddane schot Quhilk smytis sair [*Bann.* soir] withoutin resistence; HENR. *Fab.* 790 (H). Sayr; *Wall.* XI 175. Scho bait him sair so did he hir; *Seven S.* 438. Thai band him sair quhill his fingeris wer haw; KENNEDY *Pass. Christ* 1041. In this battel Chappel is strukne .. with a brasen nale, his heid il birst and sair hurt; DALR. II 328/12.

(*c*) Tholome also was woundit sore [: befoir]; HAY *Alex.* 3194. And to Sir Gawane thei haith o lytter maad, Ful sore ywound; *Lanc.* 2148. Na grettar payne mycht be Than thai with in insufferit sor to duell; *Wall.* VII 443. DOUG. X xiv 12. He .. pullit at the arrow .. The knight full sore [*ed.* sorelie] schrinkit at the stound; *Clar.* IV 2118.

(*d*) The breir byndis me soir; *Compl.* 64/26. The Inglis men being sa soir handilit thairat [*sc.* in the battle]; LESLIE 96.

compar. Before the houre fare sarare we Ar brynte thane [etc.]; *Leg. S.* xi 417. Evir the sarar this ern strenys hys gryp; DOUG. XI xiv 72. Wemen .. , the sarare thai war woundit, war mare egir in cruelte; *Boece* 166b. The far sarar it sat him; *Christis Kirk* 164 (M).

superl. The grewhound .. The sarest bytar of that menȝie; *Rob Stene* 11.

proverb. An old hound bytes sair; FERG. *Prov.* No. 33. He is sairest dung when his awn wand dings him; *Ib.* No. 388. Its a sair dung bairn that dare not greit; *Ib.* No. 550. Kem sindle, kem sair, they are evil eased kemmis ilk day; CARMICHAEL *Prov.* No. 1007. The emot bytes sairer nor the clok; *Ib.* No. 1433.

fig. Thi teris brynis me so sare That in hel leware me ware; *Leg. S.* xxviii 441. Sic a tempest sodanly Ourtuke ws and so sore gan smyte That [etc.]; *Troy-bk.* II 2525. The wyndis ar scharp and snell and sare bytand; HAY II 132/7. HENR. *Fab.* 535, etc. MYLL *Spect.* 282/21. Lord hald thy hand that strikken hes so soir; DUNB. (STS) lxx 33. Quhen ane man is in penuritie It [*sc.* avarice] sowis soir without saturitie; STEWART 13862. And luiffis dart thirlit his hart so soir; *Ib.* 26197. The stang of conscience broddit him so soir; *Ib.* 43394. Quhair pryde hes credence .. he [*sc.* God] scurges sair; ROLLAND *Seven S.* 6325. *G. Ball.* 127.

b. Of the effect of disease or other bodily affliction or injurious agent: Severely, grievously; painfully.

Thame hungerit alsua weill sair; BARB. XIV 432 (C). *Leg. S.* xxix 521. Thare-for my-self on wil fare & with feile feuir wond hire sare; *Ib.* xxxii 284. Na man wald be content that ony othir lettit him to visyte his childe and he war sare seke; HAY I 225/22. Now Dede .. him handillis full sair in euery part; KENNEDY *Pass. Christ* 776. Quhiddir is it mereit mair To gif him drink that thristis sair; DUNB. (OUP) 123/7. Grit pane and wo .. thay suffir sair for us; *Ib.* 196/16. *Ib.* 72/3. The gout .. fel so sare in his juncturis and lethis of his body; BELL. *Boece* II 123. Lucius Anthenous legate, sare with infirmite vexit; *Boece* 168. I wes sayr handillit witht ane suddane sairnes on my kne; 1534 *St. P. Henry VIII* V 14. This Lucius .. With eild and seiknes viseit wes so soir; STEWART 15106. This king Duffus with greit infirmitie Soir vexit wes; *Ib.* 35751. DALR. I 151/1. Was he [*sc.* Job] nocht sair with byles and bruiks and pouertie oppresst? HUME 44/108. Bot Doctour Ross micht not flie, becaus he wes lying sore seik; SPALDING I 151.

superl. Thairefter the humoris, puttand sairest at the naturall humiditie of the hart; SKEYNE *Descr. Pest* 4.

c. With verbs of punishing, tormenting, harassing: Severely.

(1) Thai wikit men avysit ware Hou thai wald punyse hym sarare; *Leg. S.* xxxviii 398. Sare; *Ib.* vi 87. Sar; *Ib.* xi 214. And [princis] salbe fer sairar punist of thair faultis; HAY II 146/35. IRLAND *Mir.* II 55/26. Quhilk fyre .. sall saire pvnis baith body & the spreit; *Contempl. Sinn.* 1227 (Asl.). ROLLAND *Seven S.* 1619.

(2) Thair shot was great and greauit thame sare; *Alex.* I 1625. How thai ware Anoyid in tyll Egypt sare; WYNT. II 398. The Troian pepill sair trublit hes thow; DOUG. XIII i 72. NISBET III 317/5. 1567 *Sat. P.* iv 66. *G. Ball.* 41. Soire; MONTG. *Ch. & Slae* 227 (W).

d. *To by* (an action) *sare*, to pay dearly for, in physical suffering.

War it nocht that he war Sic a catyve he suld by sar Hys wordys that war swa angry; BARB. XVIII 514. Cristine gyf I ne ware, Thu suld by that strak [ful] sare; *Leg. S.* xix 314.

e. (To bleed) to a grievous extent, grievously.

Mony off thaim war bledand wondyr sar; *Wall.* V 69.

2. a. Of armed conflict, a combatant, etc.: Fiercely, ferociously. Also *proverb.*

[He] bad thaim .. schut hardely Amang thar fayis and sow thaim sar [*H.* saile them saire]; BARB. XVI 395. A batell fell Wes dwne bathe sare, and scharpe, and snell; WYNT. VII 1980. Sadly and sair, Thoght thai war astonait, in that stour stithly thai stude; *Gol. & Gaw.* 574. Full feill thai slew that sarest on thaim socht; *Wall.* XI 98. *Ib.* 406. MAITLAND *Maitl. F.* 311/90. [Wpoun the] plaice sarest chairgit; PITSC. I 97/16 (I). Tha brashte it sa sair that the vtter barnkin wal tha wan; DALR. II 193/16.

proverb. As sair [*MS* sor] fights wranes as cranes; FERG. *Prov.* No. 77.

b. Of pursuit: Fiercely, closely, hard. Also *fig.*

Wpon Wallace thai folowed wondyr sayr; *Wall.* VII 698. The happy goishalk .. With swyft weyngis persewis wonder sair The silly dow; DOUG. XI xiii 167.

fig. Bot ȝit the sammyn myschance persewis thame sayr; DOUG. I v 25.

3. With expressions of fear, dread, anxiety, doubt, astonishment: Dreadfully, terribly, to an extreme degree; greatly.

For further examples see DREDE *v.* 2 and 3.

Sa dred thai sar the gayne-cummyng Off Schyr Robert; BARB. II 453. Bot thai war skownrand wonder sar Sa fer in-to Scotland for to far; *Ib.* V 201. Sayr; *Ib.* I 440. For thay war thair Stonyit, and that richt wonder sair; *Alex.* I 2062. Thai dowtyt sare that he walde wyn Thare landys; WYNT. II 1101. The burges, the quhilk is sare disesit of his barne, takis hede to na weris; HAY I 224/17. I may not eit, sa sair I am agast; HENR. *Fab.* 319. Sor; *Wall.* V 664. Thir feirfull wonders gart me dreid full soir; DOUG. *Pal. Hon.* 759 (E). Vlixes .. With new crymys begouth affray me sair; Id. *Æn.* II ii 68. Nor ȝit my deid sa sair I not suspekkit; ROLLAND *Seven S.* 3338. Quhairat the said regent and lordis wer sair amazed; *Diurn. Occurr.* 129. The Inglischemen was soir agaist at the comming of the Scottis men; PITSC. I 405/8. The guid nychtbouris .. come verry feirfull sumtymes, and fleit hir verry sair; 1588 *Crim. Trials* I II 163. Thay feired nocht litle, bot verie sair, that .. al sulde succeid verie ill; DALR. I 77/21. Whairat the toune of Elgin wes sore effrayit; SPALDING I 194.

4. With verbs of **a.** Causing, **b.** Feeling, and **c.** Expressing, grief, sorrow, regret, shame or repentance: Sorrowfully, bitterly, greatly, intensely. Also *proverb.*

a. *Alex.* II 4749 (see GREVE *v.* 3 (*a*)). Thus throw the body he lay stikit thaire At quhilk Nicolas his hart was grevit sair; HAY *Alex.* 1606.

b. I trow weill .. that him sair repent sall he, That [etc.]; BARB. IX 469 (C). My hart it wald haue mislykit sare; *Alex.* II 3130. That thai war samyn ourset thairfor I murn sore; *Howlat* 524 (A). All helpyt nocht thocht I repentyt full sor; *Wall.* IX 194. And of this talk the king schamit richt saire To pvnis hir [etc.]; *Bk. Chess* 1559. Grit God is grevit, that me rycht soir forthinkis; *Want of Wyse Men* 54 (B). DUNB. (OUP) 18/87. STEWART 17059. That ȝe haif gottin my grote full sair I rew; LYND. *Sat.* 2254. *Ib.* 2691. Id. *Mon.* 598. Baneist is faith now euerie quhair, And sair forthinkis me; *G. Ball.* 177. Quhilk [*sc.* Dumbarton Castle] was givin over be appoyntment; quhilk doing thaj rewit full sair for the governour never keipit ane word of his promessis; *Diurn. Occurr.* 42. I begoude to weirie soar of my lyff; MELVILL 24. But his hart bet him thairefter oft and sear, that he sould haiff .. usit a piece of dissimulation; *Ib.* 43. Bot trewlie quhilk [*sc.* events] my hart pitied fallon sear; *Ib.* 313. *Ib.* 366 (see SAR(E *n.*[2] 3 (3)).

c. He fel one hym, gretande ful sare; *Leg. S.* i 102. And with his ene sair coud he weip; *Alex.* II 124. WYNT. II 1446. He sobbit and he sighit sare; HAY *Alex.* 3595. Quhen scho had said, the king sichit full sore; HENR. *Orph.* 127. So soir the silly pig quhrynit; *Colk. Sow* II 117. The ȝyng childring .. Stude al on raw .. Abowt the tresour quhymperand wondir sayr; DOUG. II xii 15. Tha crabit on euerie syde so soir, Accusand vther; STEWART 39858. Ane bony boy was soir makand his mone; 1567 *Sat. P.* iii 3. That they may sare lament his dead; 1574 DAVIDSON *Three Reformers* 117. A woman .. complainting sair, that the erle .. had slayne her housband; DALR. I 323/29. I harde two kealwyffes sore complaine; 1639 *Bk. Pasquils* 80. These conceats .. made him soar sighe; HERRIES *Mem.* 74.

proverb. Gute, gravel [etc.] .. garris manie person grone ful sore; CARMICHAEL *Prov.* No. 578.

5. With verbs of **a.** Aggrieving, angering, vexing or annoying: Severely, greatly.

For it anoyis me wondyr sar [: ar] [etc.]; BARB. V 230. Tharfor anoyit sair [*E.* swa] he wes; *Ib.* VI 6 (C). And the knychtis war wraithit sare; *Alex.* II 4222. Gyf ony debate be amang the schypmen .. and thai be sare grevit the maystyr aw to .. ger it be amendyt; *Ship Laws* c. 27 (B). He belevit That he the goddis rycht sair haid agrevit And maid his testament and his devise; HAY *Alex.* 2206. *Ib.* 1959 (see 7 a (*a*)). Scho than als hait as fyre, Aggrevit sor, inflambit in felloun ire; DOUG. XI xiii 136. LYND. *Meldrum* 584. The King of Scotland being sare movit in his minde for that his nobilyte wald not consent to invade Ingland [etc.]; LESLIE 164. The Duke of Somersate sair offendet that his horsmen had gottin sik a brashe; DALR. II 313/9.

b. Accusation, condemnation, reproach or threat: Severely, harshly.

Bot schortly thei sall be sar accusit; *Lanc.* 1659. [They] Accusyt hir sar of resset in that cas; *Wall.* IV 715. A wyf repruffit him richt sor That he was borne bot of a bastardis blud; *Bk. Chess* 1269. This Wortigerne thairof soir blamit he; STEWART 24899. Auerrois .. Repreuvit sair Aristotell and sour; ROLLAND *Ct. Venus* Prol. 300. Wa worth thy wit, thy banis he may sair ban; Id. *Seven S.* 2062. *Maitl. F.* 240/3. Sair boistit thay my husband commoun-weill; 1572 *Sat. P.* xxxiii 45.

c. Disapproval, displeasure, envy, etc.: Strongly, greatly.

Thair huredome haitit He rycht sair; *G. Ball.* 181. For gif thow do thow sall be sair invyit; MAITLAND *Maitl. F.* 22/50. And pauchtie pryide richt sair I doe detest; ARBUTHNOT *Maitl. Q.* 122/107. My poor boy as sor enuied as euuer I was; *c*1580 Q. MARY in *12th Rep. Hist. MSS* App. VIII 9. Quhilke quhen the queine harde [she] was sair displeiset; DALR. I 258/17. The Inglismen, sair discontent, .. prouokis him to battell; *Ib.* 283/2. Quhome wicked personis inviet sa sair; *Ib.* 308/21.

6. With verbs of working or expending effort: Hard, diligently; with great effort or strength, strongly.

Sore laubouryng tyll thy browis sweit; LYND. *Mon.* 1089. He is ane pover young man laubourand and travelland sair for his leving; 1567 *Reg. Privy C.* I 517. He that wrings sair, thirsts out blude; CARMICHAEL *Prov.* No. 765. His wyff .. vrought it [*sc.* clay] werie sore, lyk rye-bowt; 1662 *Crim. Trials* III 605.

7. a. With other verbs, more or less as a simple intensifier: Severely, strongly, greatly.

(*a*) The baudrik brak .. Than was King Alexander rycht sair agrevit That he his armour haid sa sair apprevit; HAY *Alex.* 1960. I wonder sair and fast in mind did stair; DOUG. *Pal. Hon.* 472. Ful sayr constrenyt am I Art magyk to excers or sossary; Id. *Æn.* IV ix 41. I faillit ser to ȝow and ȝour wiff calling you ane cukcald; 1547-9 *Liber Offic. S. Andree* 139. *Dum Wyf* 27. Ane vther quair .. Anents wemen, not tuitching thame sa sair; ROLLAND *Seven S.* Prol. 112. It did abound sa sair in deidly sin; *Ib.* 34. Why art thou thus with thought ou'r-set sa saire? HUME 34/4. Of quhais burning thay ar nocht sair solist; DALR. I 98/10. In releiving of his mikle and sear occupied mynd; MELVILL 147.

(*b*) Sor mowit in his entent His fadris dede to wenge him giff he mocht; *Wall.* IV 536. Nocht to paye and to be craiffit sor Mor schame it is than [etc.]; *Bk. Chess* 1612. Off all sic wirdis he wes full dum, So soir he wes agast; DUNB. (OUP) 155/153. I knaw that I haif sinnit soir; *G. Ball.* 37. I grant that I haif faultit sore To stok and stane geuand his glore; *Ib.* 61. That sycht Quhair I haif set my hairt so foley soir; *Bann. MS* 221b/9. Bot euer moir Tormentit soir In hellis fyire; MAITLAND *Maitl. Q.* 164/34. [If I get no relief [*sc.* of debts] at this Whit Sunday I am] soir wrakkit; 1595 *Cal. Sc. P.* XI 596. The fyre with heitt .. Dois sor consume and burne; *Montg. Suppl.* xxxiv 98. We ar sore bwssiett with executiounis; 1616 *Haddington Corr.* 132.

proverb. A gentil horse sould not be over sair spurred; CARMICHAEL *Prov.* No. 119.

b. Common with LANG *v.*[1] 1, q.v. for further examples.

His dochter .. langing sare to se hir fader; *Boece* 124. To speik with hir he langit wonder soir; STEWART 60518.

c. (Placed) in, or (thrown) into, severe straits or difficulty; (to press) hard (upon).

To think On the king that wes sa sar stad That bot [twa hunder] with him had; BARB. III 363. Quhat kin thing wes lufe, Quhilk now settis thé so sair; SCOTT XX 20. How sair saevir the weiris had ouirthrawin thame; DALR. I 84/15.

8. Greatly, much (*aganis* one's *will*).

Suppois it was richt soir aganis thair will; STEWART 23245. 1567 *Sat. P.* iii 232. I am troubled, and that soir aganis the desyre of my inward man; 1572 KNOX VI 483. This deliuerance [*sc.* the surrender of Edinburgh castle] wes maid sair aganes the will of the inhabitantis; *Diurn. Occurr.* 107. *Ib.* 119. LITHGOW *Trav.* X 449. 1644 *Sc. Eccles. Soc.* 306.

9. a. (Burdened or encumbered) grievously, greatly. Also *fig.*

In hevy wayt frog stad and chargyt soyr [L. *gravatum*]; DOUG. VI v 132. Sum [*sc.* bees] sueitlie hes the hony socht Quhill thay war claggit soir; MONTG. *Ch. & Slae* 64 (L).

fig. With sinnis I am laidnit soir; *G. Ball.* 220.

b. Of mental oppression or the like: Grievously, heavily.

Sone, thir tythings sytts me sor; *Wall.* I 439. Tha saw appeir sa grit perrell thairin, Be ane exempill quhilk sat thame richt soir; STEWART 32051. Thair deid so soir it lay into his thocht; *Ib.* 42021. [He] never despyseth the sobbis of the soir afflictit; 1554 KNOX III 155. The trouble of the spirit wil oftimes .. trouble the whole estate of man: as though he were takin with sum agonie or sharp fevar... every man whom the Lord so visits, will not bee alike sore troubled .. but .. according to the modificatioun of the Lord; HUME 107/31.

Sar(e, Sor(e, *v.* Also: **sair.** P.t. and p.p. **sarit,** etc.; **sarde,** etc. [North. ME *sare* (Cursor M.), e.m.E. *sore* (Caxton), OE *sárian*, 'to be pained or grieved' (OED); the transitive sense may be due to the influence of the cognate ON *særa.*] *tr.* **a.** To wound or injure; to hurt. Also *fig.*

A possible further example may be DOUG. XI xiii 170, see SARRIS *v., pres.*

The *Alex.* quot. may, rather, belong in SAR(E *adv.* 1 a (*b*).

pres. Richt as it war ane baittit bair Quhan the houndis byte sould him sair Garres thame fle; *Alex.* I 2854. Than but delay apoun [him] all thai schot, Preiffand thair pith, quha fastest couth him sair; KENNEDY *Pass. Christ* 443. Thai .. Saris his senonis & stoundis all his wanis; *Ib.* 450.

(*b*) Quhat wemen micht nocht sor thair wame; *Bann. MS* 286b/19.

p.t. He socht in sa sadly, quhill sum of thame he saird; *Rauf C.* 656. Twa sarde [*1570* sairit] him maist that cruell war and keyne; *Wall.* II 58.

p.p. Quhan men war sarit, thair confort thair suld tak; HAY *Alex.* 1210. *Ib.* 3145, 6179, etc. Quhen scho [*sc.* the wolf] was sarit with the houndis, scho ruschit on the king; BELL. *Boece* II 109. Mos and merres, thik forestis and woddis, maist sikkir refuge quhen thai war sarit be inemyis; *Boece* 148b.

fig. Ilk day me saris and grewis [*R.* me sare aggrevis]; WYNT. IX Prol. 40 (C).

b. To sorrow or grieve for, to regret.

Deploring, and soring, Thair ignorant estaits; BUREL *Pilgr.* II 333.

Sare, var. of SAVO(U)R *v.*

Sarely, Sorelie, *adv.* Also: **sare-, sairlie, -ly(e; searlie; soirly; soarlie.** [Early ME *særliche* (Layamon), ME and e.m.E. *sarely, sorely* (both *c*1400), *soreli* (*c*1450), OE *sárlíce*; SAR(E *adj.*] = SAR(E *adv.* in various senses.

For a possible further example, see *Alex.* I 2221 in SARRALY *adv.*[1] (*b*).

1. Of the infliction of physical injury or the effect of disease or illness: Grievously; painfully.

(1) Gif a man be hurt sarely be ane othir; HAY I 94/15. The cheuailleire of sey .. is ane monstre in the sey .. in form of ane armyt knycht .. And has armes & hand stark quharwith he strikis sairly; *Loutfut MS* 32b. For baith hir tendir handes War strenȝeit sairly boundin hard with bandes; DOUG. II vii 78 (Ruddim.).

(2) Cuthbert Greig, .. sueitting and brocheing in ane vncouth seiknes, sairlie handillit, quhairby he was nocht able to gang; 1622 *Crim. Trials* III 526.

2. Of punishing or tormenting: Severely.

Sairlie for that gret trespas God tuke wengeans on Ypocras; *Seven S.* 745. When she telled any of them things she was sairly tormented by them; 1588 *Sc. N. & Q.* I 106.

3. Chiefly with verbs of sorrow, regret or reproach: Sorely, bitterly; sadly, sorrowfully.

(1) Cadall .. towart the hevin lifting his handis, blamyt sarelie fortune and the goddis; *Boece* 87. Seand thi panis, sorelie weip I walde; *Arundel MS* 256/52. I speik not but pruife, quhilk I may sairlie rew; 1567 *Sat. P.* iv 157. He lamented sairlie that the realme had bene .. owerthrawin witht civill weairis and discoird; PITSC. I 128/1.

(2) Rycht soirly [*G. Ball.* 61 sorelie] musing in my mynde; *Compl.* 64/35.

4. More or less as a simple intensifier: Grievously, greatly. Also *fig.*

Obignie .. wha haid sa soarlie oppressed the king's majestie [etc.]; 1583 Melvill 164. I have broken and transgressed, maist unthankfullie, sinning searlie at all tymes against my guid God; 1596 *Ib.* 364.

fig. Hes not mony throw inlak of techement in mad ignorance .. sairlye stummerit? WINȜET I 5/25.

Sarély, var. of SARRALY *adv.*[1]

(Sarenes,) Sarnes, Sorenes, *n.* Also: **sair(e)-, searnes.** [Early ME *særnesse* (Layamon), north. ME *sarnes* (Cursor M.), *sarenes* (14th c.), e.m.E. *sorenes* (1495), OE *sárnys.*] **a.** Pain, painfulness, as a condition. **b.** An instance of this, a pain, an ache. **c.** (A sort of) injury or hurt, more generally.

a. Weill neir I may na sairnes feill; *Alex.* II 10856. The strenth of his ȝouth had wrassilit with the sairnes of the pane; BUCH. *Detect.* (1727) 64.

(*b*) Be grait incressing searnes in my syde; 1601 MELVILL 496.

(*c*) Nor sorenes found I never a pyle, While I came here within a myle; *Sir Eger* 425.

b. Sairnes; 1534 *St. P. Henry VIII* V 14 (see SAR(E *adv.* 1 b). The said horse .. crepit on his wambe .. trocht the sairnes was in his guttis; 1571-2 *Inverness Rec.* I 211. 1584 *Waus Corr.* 283. The sorenes in the lims and nervis that so dois men molest; FOWLER I 88/64. Scho .. seithit kaill rwntis in watter for ane ill or sairnes scho had in hir head; 1602 *Dundonald Par. Rec.* 7. Ane grit dollour and sairnes in his richt leg; 1615 *Crim. Trials* III 318.

c. The fewir, agow, the sarnes of e, The fellone byll; *Ratis R.* 179. Is nowdir fevir nor fell .. Seiknes nor sairnes .. Bot I can lib thame and leiche thame; HENR. *Practysis* 19. To seike folkis or in to sairenes syne; *Prestis of Peblis* 155 (A).

Sarf(e, varr. of SERV(E *v.*[1]

Sarg(e, Serg(e, *n.* (*adj.*). Also: **sair(d)ge; s(c)harge, s(c)hairge; serdg(e; seirg(e; searg(e, shearge; serige; sereich; cierge; sirge;** (? SURAGE *adj.* or *n.*). [Late ME and e.m.E. *sarge* (Wyclif), *seerge* (Lydgate), *serge* (1599), OF *sarge*, MF *serge* (*c*1175 and 1360 in Larousse), pop. L. **sarica*, L. *sērica*, belonging to the Seres (i.e. Chinese).]

A. *noun.* Woollen fabric (also, occas., silk) of various sorts and manufactures; serge. Also *pl.*, chiefly, indicating different varieties of the material, also, occas., different quantities.

sing. Of quhit sarge ij s. standis; *c*1500 *Coll. St. Salvator* 153. Cappis nigris de panno laneo vel de ly sarge; 1505 *Comm. Univ.* IV

(Aberd.) App. 137. ij syed jaceits ane of tham of sarg and the twy of say; 1530 *Balmerino and L. Chart.* II 32. For twa pecis of fyne sarge for hosing; 1539 *Treas. Acc.* VII 57, etc. Ane werdur bed with twa courtingis of sarge; 1551-2 *Dundee B. Ct.* II fol. 134b (22 Feb.). 1580 *Coll. Aberd. & B.* 391. 1667 LAUDER *Jrnl.* 180. 4 ells of sarge at 14 s. the ell; 1684 *Edinb. City Archives* Letters Vol. II (25). A bed hung with gray sarge. 3 knups upon the top of it lyned with gray sarge; 1685 *Soc. Ant.* LVIII 361.

(*b*) Thre pendikillis to the hie alter, ane of greyne byrgh satyne, ane of blew sairge, and ane of reide sairge; *c*1555 *Crail Ch. Reg.* 64. Twa wallis of reid sairge for the ymages in lentrone; *Ib.* 65. To weive twentie elnes of sairge; 1659 *Melrose Reg. Rec.* I 228. Siven ells of litted and wacked sairdge; 1691 *Kirkcudbr. Test.* (Reg. H.) 22 May.

(*c*) Three quarters of scharge thre shillings nynpence; 1665 *Kirkcudbr. Sheriff Ct. Deeds* I 47. 4 ells 3 qtrs fine broad sharge for lyning the coat all throw; 1694 *Lett.* in *Annandale P.*

(*d*) Maid to ȝour wyfe ane govne peitticoit & schankis of blacke schairge; *c*1640 *Tailor's Acc. Bk.* B 42.

(*e*) To be him ane gowne and ane hugtoune, vj elnis serge; 1533-4 *Treas. Acc.* in *Crim. Trials* I 282. For the thrid pece of claith of gold fresit, sett apoun serge; 1538 *Treas. Acc.* VII 4. xxvij elnis of reid, ȝallow, and grene serge .. to be ane cannabie to Lady Jane; *Ib.* 89. To be stremaris to the kingis schippis, ix score elnis reid and ȝallow sey .. and till complete the samin, xv elnis serge; 1540 *Ib.* 310. viij elnis stamyng of serge, the elne iij li.; 1562 *Ib.* XI 180. Ane furnesit bed, viz., xl elnis of grene serge, xx li.; 1565-6 *Ib.* 461. j elne ij quarteris of reid armosene serge, price of the elne iij li. x s.; 1572 *Ib.* XII 308. Ane elne incarnat serge to be schankis, price thairof v li.; 1575 *Ib.* XIII 61. Reid crammosie serge tobe thame mandillis; 1580 *Ib.* MS 12. For mounting of each man in his blew cloath coat weel lyned with sufficient whyt stuff or serge; 1672 *Dumfriesshire Valuation Bk.* MS 15 April. To James Moody for weaving 26 ells of serge to my wife; 1676 CUNNINGHAM *Diary* 68.

(*f*) Blew serdg to lyne Geordies coat; 1680 *Foulis Acc. Bk.* 59. To Meg to pay for 10 ell serdge to be her Lissie and Grissie 3 under pitticoats black £15; 1690 *Ib.* 122. For the use of serdge to hing around at Raevelston £2 4 s.; *Ib.*

(*g*) xiiij ellis blak seirge at iiij l. xvij s.; 1588 *Cal. Sc. P.* IX 671.

(*h*) To lyn a clock to my lord iiii els unwalket searg at xxviii s. the ell; 1586-7 *Acc. Lady Bellenden* 1. [To] cause William Douglas, the taillor, .. line the coat well with good searge; 1676 *Douglas Corr.* 276. For whit searge; 1679 *Sc. Ant.* IX 108. 5 yeardis 3 qrteris of milde searge; 1687 *Berw. Doc.* fol. 1, No. 7. 5½ ells finn searge; 1688 *Douglas Bequest* 24 July.

(*i*) Ane govne of fyne blak chamlett of silk of cierge; 1579 *Edinb. B. Rec.* IV 115.

(*j*) Ane old wastcoat of sereich worth iiij lib.; 1664 *Edinb. Test.* LXXI 258b.

(*k*) Sirge for plying the 2 vestis; 1667-9 *Home Clothing Acc.* MS (Earl of Home) I.

pl. Sex vlnas de serges; 1342 *Exch. R.* I 504. The saidis strangearis ar cum within this realme to exercise thair craft .. in making of searges growgrams fusteanis [etc.]; 1587 *Acts* III 507/2. For the alledgit stowth of ane pakatt contening seargeis and uther coistlie wairis of the schipe; 1603 *Shetland Sheriff Ct.* (ed.) 97. Stuffis of thrie quarteris breid and xxij elnes of lenth .. Seargeis, double and single; 1620 *Reg. Privy C.* XII 339. Ane hundreth .. ellis of severall sortis of seargis at xlj s. x d. the ell; 1642 *Edinb. Test.* LX 41b. On ilk elne of narrow clothe, sarges, and vther worset or hair stuffis; 1644 Spalding II 312. His maiestie .. doth prohibite & discharge any weaver .. to make any searges vnder the breadth of an ell & a naill; 1661 *Acts* VII 256/2. At the exporting of a pack of cloathe woollen or sarges etc. to be payed by the bwyer sex schilling Scotts; 1670 Ure *Rutherglen* 43. For litting by weight such as small fingrin searges, which will taick four ell to ane pund; 1697 *Dundee B. Laws* 561.

b. With qualifying terms, specifying the place of origin, etc.

Sarge de aras, darras, darrest (= of Arras), *(de) bowas, bowes, deboyes* (= of Beauvais), *millanis* (= of Milan), *of Orliance* (= of Orleans), *paddoway* (= of Padua) *serge.*

Sarge dropp(i)e (= spotted: see DROPPIT *p.p.* and *ppl. adj.*).

Serge de soye (= silk serge: MF *sarge de soye* (16th c. in Huguet).

Also SARGE-DASKET *n.*, SARGE DEMEEN and SARGEDENIN.

(1) Ane steik sarge darrest contenand xxvj ellis; 1568 *Edinb. Test.* I 153. Sex elnis of serge darras; 1574 *Ib.* III 150. Sex elnis of sarge de aras price of the eln xx s.; 1582 *Ib.* X 331. — x elnis .. of blak sarge bowes, .. vj elnis .. sarge bowas; 1584 *Edinb. Test.* XIV 132b. Ane half ell of searg of bovas to be a nother pair of schankis; 1586-7 *Acc. Lady Bellenden* 4. Ane steik of serge de bowas; 1590 *Edinb. Test.* XXI 286b. Thrie peices of serge deboyes; 1633 *Ib.* LVI 358. — Serge of Florence to be ane cote and ane cloke; 1552-3 *Treas. Acc.* X 163. Of sarge of florense to be ane goune and ane cloik mulis and schiuine; 1566 *Facs. Nat. MSS* III li. My joup of sarge of florence pasmentit; 1583 *Edinb. Test.* XIII 43b. xx ellis seirge Florance; 1588-9 *Cal. Sc. P.* IX 673. The best schairge of Florence; 1610 *Edinb. Test.* XLVI 176. — iij elnis ij quarteris of serge of Millen to be ane cloke, the elne v li.; 1566 *Treas. Acc.* XII 11. Seirge of Millane; 1579 *Ib.* XIII 271. Ane elne .. of blak sarge millanis, .. thrie elnis rid sarge millanis at xxxv s. the elne; 1597 *Edinb. Test.* XXX 163. — Of Orliance (see (2) below). *Ib.* 224. — Ane cloik dowblet & breikis of rid paidaway shairge; 1640 *Tailor's Acc. Bk.* B 38. *Ib.* 43. 14 ell of black paddouay serg for the eayll; 1661 *Funeral Acc.* i 3. — Searges of Scots makeing ilk fiftie ells; 1661 *Acts* VII 253/2. — Als much stirline searge as lyne the said .. cloak; 1684 *New Mills Manuf.* 83.

(2) vj elnis .. of serge of Orliance alias serge droppe; 1561-2 *Treas. Acc.* XI 106. 1 blak cannope of sarge droppre [*sic in pr.*]; 1565-6 *Prot. Bk. Thomas Johnsoun* 97. I leue .. to my sister Isobell Clark my goun of sarge droppie worth iiij li.; 1585 *Edinb. Test.* XIV 233b.

(3) Remanes of serge desoye xiij elnes; 1611 *Edinb. Test.* XLVI 252.

(4) Ane pece of blak counterfutit serge; 1575 *Edinb. Test.* III 360b.

B. *adj.* Of material, yarn or garments, etc.: Made of serge.

(1) Thrie steikis of serge droggie; 1570 *Edinb. Test.* II 79b. Serg yairn; 1672 *Aberd. Trades* 296.

(2) [9½ oz.] of serge pasmentis for the quenis grace dowble gownis .. vij li. xij s.; 1567 *Treas. Acc.* XII 50. Hir serge gair cuill and the serge bodies effearand thairto; 1591 *Edinb. Test.* XXIII 39. Ane Flanders shairge suitt; 1614 *Rep. Hist. MSS* I 136. Ane gray serge house goune; 1649 *Edinb. Test.* LXIV 250b. Twa red serige bedis, .. ane grene serge bed; 1651 *Buccleuch Mun.* 290. [James Huntar to pay to Janet Young, 58 s.] for to buy ane shearge waistcoat; 1663 *Melrose Reg. Rec.* II 66. Ane searge pittecot; 1669 *Edinb. Test.* LXXIII 292b. Serg stokings and grytt stokings; *a*1670 *Old Ross-shire* I 143. To Ketherin McClellane .. one rid serg peticoat; 1671 *Kirkcudbr. Test.* (Kirkcudbr. Sheriff Ct.) 17 June. Twentie aught shilling aught pennies for twa sarge aprons; 1676 *Kirkcudbr. B. Rec.* MS 19 July. Searge bead with shelonigs on it £12; 1679 *Old Ross-shire* I 127. A stand of new seirg cloaths with a gunne and one new kep; 1680 *Kirkcudbr. Test.* (Reg. H.) 23 July. Searge; 1685 *New Mills Manuf.* 105.

Sarge, var. of SERCH *n.* and SERGE *n.* **Sargeantship,** var. of SERGEANDSCHIP *n.*

Sarge-dasket, Serge-dascat, *n.* Also: **serg-** and **-dascote, -dascot(t, -dascoit, -dascoth(t, -dascocht, -(de)-dasco, -daskat, -daskane, -dasky(ne.** [SARG(E *n.* A b and ? *Aerschot* (in Brabant).] ? Serge of Aerschot. Also *attrib.*

xx elnis of serge dasco to draw furtht thair hois; 1561 *Treas. Acc.* XI 84. Serge de dasco, xxv s.; *Ib.* 86. To gif hir grace tailȝeour ten elnis of serge dascocht; *Ib.* 93. Serge dascoit; *Ib.* 95. Serge dascotht; 1561-2 *Ib.* 153. Serge dascot; 1562-3 *Ib.* 238. Serge dascote; 1566

Ib. 503. Sevin elnis of serge daskyne; 1567 *Edinb. Test.* I 34. xj elnis of serge Dascoth walkit; 1569-70 *Treas. Acc.* XII 185. Serge dasket; 1570 *Canongate Ct. Bk.* 234. Serge daskat; 1570 *Edinb. Test.* II 79b. Serge dascat; 1575 *Ib.* III 361. Ane gowne of sarge dasky and my wirsett coitt; 1585 *Ib.* XIV 280. Serge dascott; 1589 *Treas. Acc.* MS 157b. iij ellis ane quarter unwalkit sarge dasket; 1592 *Thanes of Cawdor* 214. Fyve elnis of sergdaskat at xxxvj s. the elne; 1615 *Edinb. Test.* XLVIII 341b.

attrib. Ane pair of new serge daskane breikis with buffingis & schort hois; 1573 *Edinb. Test.* III 1.

Sarge demeen, Searge de Mein. (Appar. metathetic f. **sargedenim*, SARGEDENIN, serge of Nîmes; but perhaps the 1681 example should read *de Niem.* Cf. SARG(E *n.* A b.) — That kind of searge which is comonly called searge de Mein; 1681 *Acts* VIII 361/1. An ell ¾ of Irish freeze, 19 ell ¼ of sarge demeen; 1688 *Reg. Privy C.* 3 Ser. XIII 310.

(Sargedenin,) Shergedenin. (e.m.E. *serge denim* (1695), *serge de Nim* (1703), F. *serge de Nîmes* or *Nismes* (1742 in OED s.v. *Denim*), serge of Nîmes in S. France. Cf. SARG(E *n.* A b and SARGE DEMEEN.) — Your shergedenin suit; 1681 *Thanes of Cawdor* 354.

Sargen, -entt, varr. of SERG(E)AND *n.*

Sargone, *n.* (e.m.E. (1598), F. (1574-90 in Huguet), f. L. *sargus* (in e.m.E. 1591) a fish of the sea-bream variety, pop. believed since classical times to have been sexually attracted to goats.) — This Cupide hidd .. that ioines .. the goate uith sargone pairtrichis uith does into one pairte; JAMES VI *Poems* I 119/40.

Sariand, var. of SERG(E)AND *n.* **Sariazine,** var. of SARACEN(E *n.* and *adj.* **Sarie,** var. of SARY *adj.* **Sariely,** var. of SARRALY *adv.*[2] **Sarin,** var. of SERING *vbl. n.*

Sariot, *n.* (? OF *sarrot* (1276 in Larousse) (F. *sarrau*) a loose-fitting robe with sleeves, worn over the outer garments for protection.) — Apone a coursere .. he sat .. In till a sariot [*v.rr.* gowne] schapin wyde, .. With brokin letteris on that govne; WYNT. IV 219 (W).

Saris, var. of SARRIS *v., pres.* **Sariscene, -sen(e, -siane,** varr. of SARACEN(E *n.* and *adj.* **Sarit,** p.p. of SAR(E *v.* **Sarjand, -ent,** varr. of SERG(E)AND *n.*

Sark, Serk, *n.* Also: **sarke, shark; sairk; syrk.** [North. and midl. ME *serc* (Cursor M.), *serke* (Manning), *scherk* (14th c.), late ME *sarke* (*c*1440), north. e.m.E. *sark* (1515), OE *serc*, ON *serk-r*. Also in mod. Sc., as *sark, serk*, and in mod. north. Eng. dial., as *sark*.]

For further examples see the Indexes to the various volumes of *Treas. Acc.* s.v. *Shirt(s)*.

1. The male or female body-garment worn nearest the skin; a shift, chemise, shirt. Also *fig.*

(1) Al nakit, bot sark & breke; *Leg. S.* xl 1080. Buft brawlit hois, coit, dowblet, sark, and scho; 1572 *Sat. P.* xxxiii 369.

(*b*) In serk and mantill eftir hir I went In to this garth; DUNB. (OUP) 142/46. Thoght I haif cair undir cloke the cleir day quhill nyght, 3it haif I solace undir serk quhill the sone ryse; Id. *Tua Mar. W.* 471.

(2) *indef. art.* A wennomus sark; *Asl. MS* I 317/7. I se him want ane sark [: clerk, bark, merk]; DUNB. *Flyt.* 223. *Compl.* 80/27. Ane gud new sark; *Wyf Awcht.* 43. The forme of it .. wes lyke ane litill bodie .. cled in quhyt lening lyk a sark; 1601 *Aberd. Eccl. Rec.* 184. The said Margaret .. ressaueit for waigeis, ane sark and ane pair schone; 1623 *Perth Kirk S.* MS 14 May. Ane cassine sark worth xij s.; 1643 *Kirkcudbr. B. Rec.* II 703.

(*b*) A doublat and a serk; 1483 *Acta Aud.* *112/1.

(*c*) To .. pay James Cryus vij s. of siluer & ane syrk; 1536 *Linlithgow B. Ct.* 13 Nov. Ane bonet ane syrk; 1562 *Carte Northberwic* 86.

pl. Sewine sarkis schene He gert lay a-pone the grene; *Leg. S.* xx 240. Find me ane dog .. Hes maid sa mony bludie sarkis; LYND. *Compl. Bagsche* 88. Where are ye commanded to dissagyse your selues lyke players .. with shauen crownes, long wide sarkes aboue your clothing [etc.]; FERG. *Answer* in *Tracts* 20. In the ministers .. we thinke .. vnseimlie .. all kynd of .. variant hews in sarks .. as red, blew, 3eallow; 1575 *Bk. Univ. Kirk* I 335. [To] buy ane sutte of cloathes, .. also sailing cloathe as well four sarkes to him; 1650 J. M. Beale *Fife Schools* 143.

(*b*) For unbleichit claith to be serks to the lard; 1652 *Caldwell P.* 106.

(3) *def. art.* Withtout al clathis bot the sark, quhilk al was brynt the sleiffis except; 1577 *St. A. Kirk S.* 424. [He] was constranit .. to mak calffing to his boit of the sark was on his bak; 1580 *Inverness Rec.* I 281. Luiking at the breist of the sark he tauld that the seiknes was nocht cum as 3it to his heart; 1623 *Crim. Trials* III 556.

(*b*) Serk; HENR. *Bludy Serk* 81. Hercules .. was .. slane be his lady Dyonera throw the inuennomyt serk scho maid him to weir; MYLL *Spect.* 278/17.

pl. Othyre thir serkis tak 3e lo And [etc.]; *Leg. S.* xx 244.

(4) *possess.* Hir sark suld be hir body nixt, .. Hir kirtill suld be [etc.] .., Hir gown suld be [etc.]; HENR. *Garmont* 9. Vnlusum was his likame dicht, His sark was all bludy; Id. *Bludy Serk* 62. In haist his weid off castis he: .. Held on his sark, .. and syn lap in the flud; *Wall.* IX 1181. To Luke of the wardrop to feche Sanct Margaretis sark to the quene; 1511-12 *Treas. Acc.* IV 334. On fut I sprent into my bair sark; DOUG. XII Prol. 269. Your cord and lowsie coit and sark; *a*1540 GLENCAIRN in Knox I 74. And thairefter riffing of his flesch, clays and sark; 1542 *Elgin Rec.* I 70. Of 3allow taftais wes hir sark, Begaryit all with browderit wark; LYND. *Meldrum* 125. Furth scho did hir marke In treases with hir kirtill and hir sarke; *Clar.* III 484. Sawis seindill saffroun in sawt for thair sarkis saik; MONTG. *Bann. MS* 163a/8. Thay gloir in thair weill ruffit sark [: wark]; *Maitl. F.* 439/88. Ane man .. come to hir with ane bairnis sark to be charmt; 1617 *Reg. Privy C.* 2 Ser. VIII 346. Efter that tuik the watter with the bairnes sark .. and kaist baith in the burne; 1623 *Perth Kirk S.* MS 16 May. She got up in her sark and said, .. do not come near me for I'm naked; *Presb. Eloq.* (1693) 102.

(*b*) King Robert Brucis serk; 1488 *Treas. Acc.* I 83. Gudis of ayrschip .. his hat, his bonet, his best serk; 1534 *Selkirk B. Ct.* (ed.) 144. [He] cast ane nicht gowne ouer his serk, and as he was half nakit ran to her in haist; BUCH. *Detect.* (1727) 4.

(*c*) As his bludie sairk beiris witnes; 1584 *Reg. Privy C.* III 708.

pl. Ye gaif twa fair cleane sarkis of your husbandis .. to be inchantit be thame; 1591 *Crim. Trials* I II 250. To change thair wyfis .. as oft almaist as thay do thair sarkis; HAMILTON *Facile Tr.* 435.

(5) *proverb.* (see KIRTIL(L *n.* 2 (1) (*c*)).

(6) *fig.* Sone thai wald denude thame of sark of symony and restore the frut of sic benefice; ABELL 16b.

b. With defining terms specifying a function, type of material, place of origin, etc.

For some further examples, see HARDIN *n.* and *adj.* 2 (2), LININ *adj.* and *n.* B 3 (*b*), LOSIN(G *n.* 4 a, and LOSINIT *adj.*

(1) Ane gray bonat, ane nicht courchre and hus sark; 1521 *Stirling B. Rec.* I 13. For x elnis of claith for nycht sarkis to the kingis grace; 1522 *Treas. Acc.* V 193. Lynyng claith to be ryding sarkis; 1544 *Ib.* VIII 292. [To provide the apprentice with] ane new per of shone with ane straiking serk; 1643 *Melrose Reg. Rec.* I 100.

(2) My steming sark & rokket was laid doun; 1571 SEMPILL *Sat. P.* xxviii 69. With linnen sarke, and no more was she cledde; *Pope's New-Year Gift.* Ane yellow wollin shark; 1649 *Bk. Dunvegan* 122. When the hair sarkes war put wpon the wyfes [*sc.* witches] in the touboth; 1650 *Dumfries Treas. Acc.* Ane syd Holland cloth sark; SPALDING II 388.

(3) [A] noble seme was on his sark of Rence [*sc.* Rennes]; ROLLAND *Ct. Venus* I 127. Ane hieland syd serk of yallow lyning pasmentit with purpour silk and silver; 1578 *Inv. Wardrobe* 215.

c. (*In*) *sark alane* (*alone*), wearing only a 'sark'.

Freq. as a penitential form of dress.

He to cum on Sounday with ane walx candill in sark alane and ask the provest forgevenis; 1504 *Edinb. B. Rec.* I 100. The sayd Schir James sall cum sarkallane to the mercat croce .. and than sall offer ane nakyit schowrd .. to the sayd Erle of Lennox; 1530-1 *Maxwell Mem.* I 261. He .. sark allane, for mair mereit and meid, And barfeit als to Canterberrie 3eid; STEWART 43396. Thai ordane the said Besse .. to gang sark alane afore the processioun vith ane vax candill in hir hand of ane pound to be offerit to the haly blud lycht; 1538 *Aberd. B. Rec.* I 155. Sayand the said Margarat Baffour vas ane huyr and ane wyche and that sche ȝeid widersonnis about mennis hous sark alane and hir hayr abone hir eyne for wyche craft; 1545 *Elgin Rec.* I 84. 1548 *Aberd. B. Rec.* I 258. 1551 *Cupar B. Rec.* 1 June. [The bailies] decernit that .. he sall come to the mercat cross sark alane, his head discoverit, and, upon his knees, tak his whinger be the point and deliver the same to John; 1563-4 *Old Dundee* II 358. 1566 *Prestwick B. Rec.* 70. His barneman being threscheand and laubourand sark allane, he crewellie invadit him; 1566 *Reg. Privy C.* I 460. According to hir [*sc.* Queen Mary's] comandment [he] past to hir chalmer sark alane onlie coverit with his nycht gowne; Buch. *Indict.* 34. Cheissit him sark alane and naikit, fra the said hous; 1594 *Misc. Spald. C.* II 123. 1608 *Perth Hammermen* xxxvi. The minister, who wes lying sleiping in his bed, .. rais, sark allane, and preast to have relevit the said Jonnet; 1616 *Reg. Privy C.* X 488. That scho saw theme naikit sark alone in the said Elspeth bed; 1622 *Perth Kirk S.* MS 21 Oct. 1628 *Reg. Privy C.* 2 Ser. II 236. He went out and stayit the space of ane hour or mair sark allone; 1633 *Orkney Bp. Ct.* MS 87.

d. *Attrib.* (also *possess.*) with *collar, hand,* etc.; and *comb.* with *cote*, and, in *pl., washar.*

Also Sark-wylycote *n.*

attrib. (possess.) iij boltis of Parys rubanis .. for sark collaris and handis; 1538-9 *Treas. Acc.* VII 144. Small linning to band the .. sarkis and sarkhandis; 1599 *Ib.* MS. — Scho fischit all the Spanȝe seis With hir sark lape befor hir theis; *Crying of Play* 70. And that the crook hand took his sark lap and hinderd him that he took not that woman; 1673 *Kingarth Par. Rec.* 87. — xv sark nekkis price xxj s.; 1575 *Edinb. Test.* III 328b. I leue to Walter my bruther .. sex of my best sark neckis and alsmony sark handis; 1582 *Ib.* XI 33b. He then presented himself to the heading-stock, his eyes being covered with a napekine .. , his sark neck removed be him and doublet neck laid doun over from his shoulders; 1584 *Misc. Bann. C.* I 105. Certane sarkis neckis, and schone; 1594 *Misc. Spald. C.* II 127. His halie day claythtis as hat cloik breikis hors doublet sark schone sark nekis with al thingis belangand thairto; 1595 *Edinb. Test.* XXVIII 236. — Ane packet of sairkis slevis and hois for thame [*sc.* dolls]; 1578 *Inv. Wardrobe* 238. Tua auld sarksleiffis; 1615 *Reg. Privy S.* MS 21 Nov. Ane combing cloath and ane pair of lining serk sleivs; 1648 *Edinb. Test.* LXIV 52.

comb. Quhils the erle Bothuel was in Italy his sarks washar sayd .. that he to quhome belonged these sarks .. suld be beheded by his prence; 1591 *Warrender P.* II 161. Fyne small plaiding to be his sark coittis; 1600 *Treas. Acc.* MS 68b.

2. = Sarking *vbl. n.*^{2} 2.

For .. the fluringis of the gret turngres heid the heid and tymmer prik ruiff and sark of the samyn; 1539-41 *M. Works Acc.* (ed.) I 281.

Sark, *v.* [Back-formation f. Sarking *vbl. n.*^{2}: cf. also late north. ME (once) *serked* p.p. (gl. *camisiatus, jnterulatus*) (Cath. Angl.), and Sark *n.* 2.] *tr.* To furnish (a building or a roof) with sarking (Sarking *vbl. n.*^{2} 2).

The capitanis hall quhilk is presentlie theikit with leid and is sarkit abone with deillis; 1583 *M. Works Acc.* (ed.) I 312. And the haill tymer work of the heid of the said towr to be tirrit, mendit, new sarkit and thickit againe .. ; .. the haill liberarie to be tirrit [etc.] .. , new sarkit with daillis and of new theikit againe; 1623 *Rec. Univ. Aberd.* 283. To caus repair the samen [roof] .. and to sark the samen swa far as is repaired; 1646 *Edinb. B. Rec.* VIII 93. In [*pr.* is] respect that the kirk was only sarked as yet; 1657 Cramond *Ch. Urquhart* 38. 1664 *Glasgow B. Rec.* III 31. 1668 *Old Ross-shire* I 192. The south quarter A third part of it is rooffed and sarked redie for the lead and slaite; 1674 Mylne *Master Masons* 187. 1677 *Inverness Rec.* II 275. 1689 *Edinb. B. Rec.* XII 10.

Sarke(i)n, varr. of Sarking *vbl. n.*^{1}

Sarkful, *n.* (Sark *n.*; only here and in the later dial. (to promise, in later dial. also to give, someone) *a sarkful of sare banes* = a sore body resulting from a painful beating.) — Sche .. depones that the said Cristiane Watsoune promeist hir ane sarkful of sore bainis; 1622 *Falkirk Par. Rec.* I 34.

Sarkin, var. of Sarking *vbl. n.*^{2}

Sarking, *vbl. n.*^{1} Also: **sarken, -ein.** [Sark *n.*; cf. also late north. ME *serked* p.p. (see Sark *v.*).] **a.** *Sarking of lining* (= linen), also *lining claith, cloth*, etc.: A length of (chiefly or only linen) cloth sufficient to make one 'sark' (= shift or shirt).

Also with ordinal numeral (in this instance without inflexion) in multiples of this length.

(1) Ane sarking of lynning claith; 1579 *Edinb. Test.* VII 287. And xx s. for the price of ane sarking of small lynning; 1588 *Kirkcudbr. B. Rec.* I 237. Scho leuis ane sarking of vnbleitschit lynning to James Richemont hir broder; 1590 *Edinb. Test.* XXII 122b. Scho leuis .. to .. Johne Lintoun ane elne of small quhite and ane sarking of lynning; 1597 *Ib.* XXXI 117. [The said John paying yearly] ane sarking of small lyning yeirlie with ane dussan fowles halff hens halff cappones; 1666 *Kirkcudbr. Sheriff Ct. Deeds* I 279. The same to be xxvi merks with the said sarking of clothe or threttie fyve schilling; 1671 *Rothesay B. Rec.* 206. With a sarking off linen; 1675 *Kirkcudbr. Sheriff Ct. Deeds* II 45.

(2) To content & pay to James Kennedy .. the sowme of tuenty merkis xiij s. iiij d. for the rest of his stentis wiccarage teindis and fyve sarking of lining [etc.]; 1575 *Carrick Baillie Ct.* 2 June.

b. As a qualifying term: (Linen, also cloth) as used for making 'sarks'. Also *ellipt.*

(1) Foure dosane of sarking lynning price of the dosane xxxvj s.; 1575 *Edinb. Test.* III 400b. Ane .. pece of sarking lynneng of eln breid; 1576 *Ib.* V 21. Ane hundretht elnes of round sarking lynning price of the elne vj s. viij d.; 1594 *Ib.* XXVII 193. xx elnis sarking lynning at ix s. the elne; 1616 *Ib.* XLIX 168b. 28 s. Scots for $3\frac{1}{2}$ ells of sarking lyneing; 1638 *Hawick Ann.* 42. 1671 *Rothesay B. Rec.* 205. Sarken lining 2 lib. 6 s., to the wabster 7 s.; 1674 *Inv.* in *Sc. Ant.* I 178. Restand to James McKaig soomtym servant of fies £5 10 s. money forsaid and thrie ellns of sarkein lining; 1676 *Kirkcudbr. Test.* (Reg. H.) 12 Jan.

ellipt. Twa webbe of lining, ane of sarking, and ane uther smaller for scheitts; 1630 *Bamff Chart.* 223.

(2) Reddendo .. 20 ulnas panni linei lie sarking lyning clayth; 1615 *Reg. Great S.* 488/1. To pey the sowme of sex pundis money for ilk dozen off fourtein dozen elnis of sarking lining clayth; 1620 *Urie Baron Ct.* 31.

(3) xli elneis of sarking claytht at iiij s. the elne; 1596 *Edinb. Test.* XXIX 321. Quhairvpone order wes givin out to serche the countrie for gray hydis, and gray claith, and sarking cloth; Spalding I 377.

c. Appar. shortened from b: ? Sarking cloth, sarking linen.

For 4 yds. fine cloth .. 10 drop of silk .. 8 ells sarking for lining [the mortcloth]; 1695 *Coldingham Kirk S.* in *Hawick Arch. Soc.* (1904) 29.

Sarking, Serking, *vbl. n.*^{2} Also: **sarkin, serkin, serkne.** [Late north. ME and north. e.m.E. *sarking*, in sense 1 below (1568-9), *sarkyng-nale* (1464-5), *sarkyn-boord(e* (1597); appar. f. Sark *n.* (cf. id. 2) perhaps as being the innermost layer or underpart of the roof. (Cf. also Sark *v.*)]

1. The furnishing of (a building or the roof of a building) with sarking as in 2 below.

(*a*) i^{c} suadyn burd for sarkin and completing of the platform ruf; 1532 *M. Works Acc.* (ed.) I 95. For skaffalting fluring dovelling sarking of the leid ruiffis sentreis and coumys making; 1535-6 *Ib.* 180. Giffin to the scletter Pa. Maklarin for theking certen breachis in the kirk .. and sarkin of ane syd of the queir xx merks. .. Giffin to the vright for dressing the dealis to the sarking of the queir ii merks 4 d.; 1637 *Alyth Kirk S.* in *Sc. N. & Q.* 1 Ser. XII 44. 1646 *Rec. Old Aberd.* II 22. For the sarking of the mill five hunder dobell pllenshor [nails]; 1680 *Irvine Mun.* II 287. To 4000 double nails for sarking the kill at Oldmilns; 1694 Donaldson *Cramondiana* 42.

(*b*) ij^m dure nalis to the saidis roundis serkin and for syle and skaffolting; 1531-2 *M. Works Acc.* (ed.) I 80. To Wm. Sommervell, for serking the Lyme Craig, £3; 1675 *Trans. Glasg. Archæol. Soc.* I 436.

2. 'Rough boarding, plain edged or rebated, nailed on top of rafters and to which slates are nailed' (G.L. Pride *Glossary of Scottish Building* (1975) s.v.).

For further examples see LATH *n.*

(*a*) Bocht fra ane Duchemanne in Leith xxix^xx iiij dalis .. to be sarking, lofting and scaffaling; 1539 *Treas. Acc.* VII 217. 1554-5 *Edinb. Old Acc.* I 144. vj dalis, to be sarkin to the massoun luge; 1555-6 *Ib.* II 55. The said George Blak hes oblist himself to furnish .. unto the theiking with sklaits, ane ruif of timmer, with the close sarking of deal; 1560 *Old Dundee* II 186. For mending of the haill sarking of the queir bow stringe; 1562-3 *Edinb. Old Acc.* II 175. They found ruiff of the kirk .. sufficient, except it be in sylling and sarking; 1638 *Rec. Univ. Aberd.* 410. And ther yet to be made use of for sarking to the kirk; 1653 *Ellon Par.* 142. For every rood of sklater-work, [a slater] is to have ten pounds Scots, he furnishing pins to the work, and boring the lath to the balk, and the sarking to the top; 1656 *Edinb. Justices Peace* 409. 400 nailles to call on the sarking of the knock hous; 1665 *Stirling Common Good* 35. He .. did contryve and make his escape .. by breaking .. throw the sarking and rooffe of the said tolbuith; 1671 *Reg. Privy C.* 3 Ser. III 375. The whol tofalls will neid new sarking; 1710 *Stirling B. Rec.* II 124.

(*b*) For ane rude of burd to be serkin to the saidis roundis; 1532 *M. Works Acc.* (ed.) I 111. 1554-5 *Edinb. Old Acc.* II 41. For sawin of twa gestis to be clattis, and xxxij dalis to be serkne; 1555-6 *Ib.* 57. For naillis to him [*sc.* the slater] for fessying on the serking agane; 1562-3 *Ib.* 173. A parcell of old daills that were the old serkin of the kirk; 1712 *Lanark B. Rec.* 285.

3. *Attrib.* with *burd(e, naill.*

(1) For ane rud of serkin burd to the est syd of the grete hall; 1531-2 *M. Works Acc.* (ed.) I 107. For certane sarking burd of Scotis ayk; 1535-6 *Ib.* 189.

(2) For ii^m and vi^c sarking naill for the round at x s. ilk hundreth xiii lib.; 1628-9 *M. Works Acc.* II 286.

Sarkle. Appar. var. of CERCLE *n.* ?A (more or less) circular barrier or fence. — For making of ane burd and treistis, and the sarkle about the uter nether tolbuyth quhar the justice sittis; 1553-4 *Edinb. Old Acc.* I 109.

Sarkning, *vbl. n.* (Altered f. SARKING *vbl. n.*[1] b.) — To pey to him xxij elnis sarkning lyning he gaif hir to bleiche; 1609 *Kirkcudbr. B. Rec.* II 57.

(Sark-wylycote,) Sark-wyl(l)iecoat, *n.* Also: **-wilicoat, -welicott, -weilli(e)cott.** [SARK *n.* 1 d; WYLYCOTE *n.*] ?An undergarment consisting of a petticoat and bodice. — The pannel was seen .. conferring at the foot of her own yeard in her sark wyliecoat and bare legged with the Devil; 1634 Black *Sc. Witches* 15. Ane baskett for lynning cloithes .. ane sarkwelicott; 1639 *Edinb. Test.* LIX 96. Ane quhyt sark wylliecoat .. ane reid sark wylliecoat; 1645 *Ib.* LXI 199. 6 elne of pleding to be hir ane sark wilicoat and ane under wilicoat; 1645 *Sc. Hist. Rev.* XXX 148. Tua weillicotts .. tua sark weilliecotts; 1658 *Brechin Test.* VI 157.

(Sarment,) Serment, *n.* [ME and e.m.E. *sarmente* (Trevisa), OF *sarment* (*c*1119 in Larousse) a vine-shoot, L. *sarment-* twigs, etc.] A vine-shoot. — S. Aug[ustine] comparit hereseis vnto serments or sneddings of the vyne; COLVILLE *Paraenese* 45.

Sarmon(d, varr. of SERMO(U)N(E *n.* **Sarnes,** var. of SARENES *n.*

Sarp, var. of S(C)HARP *adj., adv.* and *n.*[2]

Sarp(e, Serp, *n.* [Late ME and e.m.E. *sarpe* (1416), *serpe* (1438), of unknown origin.] A type of ornamental wreath or chain worn round the neck or suspended from the shoulders. — At nane vthir weir broudry na perle na bulȝeonne bot aray thaim at thar awin list in all vthir honest aray as serpis beltis vches & chenȝeis; 1429 *Acts* II 18/1. Pro coleriis longis, videlicet serpis vulgariter nuncupatis; 1438 *Exch. R.* V 66. He .. tuke fra his [*sc.* the knight's] hals a grete wreth of golde callit a sarp be sum men and put it about his hals the quhilk is callit in Franche a redorte; HAY I 46/14. With sarpis of gold and chenȝeis thai him band; Id. *Alex.* 6771. In a blac coffre .. the grete sarpe of gold contenand xxv schaiffis with the fedder betuix; 1488 *Treas. Acc.* I 86.

Sarplar(e, Serpl(a)ith, *n.* Also: **serplar(e, -air, -er; sarpleth, serplath, -eth, serplarth, serpla(y)t; sirpleth, -ith(e; surplaith, -at; sorpclaith;** abbrev. **serp.** [ME *sarpler* (north., 1353-4), *sarpuler* (Chaucer), *sarp(e)lere* (Lydgate), e.m.E. *sarpler* (*a*1513), *sarplier* (1565), AF *sarpler* (1321-2 in OED), OF *sarpilliere, cerpiliere, serpilliere* (late 12th c., *c*1340 and 1403 in Larousse), med. L. *sarpellarium*, etc., *serpalarius* (1171 and 1208 in Latham); cf. MDu. *sarp-, serpelier(e* 'a quantity of wool of a certain weight' (Verwijs and Verdam).

-aith, etc., the later Sc. ending, may have developed through *-arth*, perhaps, as in the case of *sorpclaith*, also influenced by CLATH(E *n.*]

1. A bale of wool, skins or cloth, of a more or less standard amount, latterly authoritatively defined as amounting to eighty stones of weight.

[Sarplarium; 1417 *Exch. R.* IV 270. *Ib.* 275. Serplarium; 1483 Latham s.v. *Sarpellarium*.]

(*a*) For ilk sarplare of wole and skynes, of friemen .. iiij d., .. of ilk twn, of friemen iiij d. .., and of all vther thingis, pock, pack and barrell, proportionablie ansuerand to the sarplar and to the twn; 1428 *Edinb. Chart.* 63. Of ilke sarplar of skynnis x s. and of ilke sarplar of woll ix s.; 1447 *Aberd. B. Rec.* MS IV p. 495 (20 Oct.). It is sene speidfull .. that thar saill na persounis bot hable and of gud fame and at he haif at the lest thre serplaris; 1457 *Acts* II 49/1. Serpler; 1458 *Reg. Great S.* 142/2. Serplare; 1464-5 *Acts* XII 30/2. [They] consentit .. to compere .. one Friday next tocum for the pley of the serplare of woll; 1478 *Acta Conc.* I 11/1. Serplare; 1479 *Ib.* 28/1. The merchandis sall .. bring in of ilk serplar of woll skynn hidis and claith .. four vnce of brynt siluir; 1483 *Acts* II 166/1. Our souerane lordis lettrez .. twiching thre serplaris of skynnis; 1492 *Acta Conc.* I 252/1.

(*b*) Serplair; 1447 *Dundee Chart.* 24. Of strangeris and wnfremen foure pennys, and of al wthire thingis, pok, pak, and baraile proportionable ansuerand to the serplare and the town; 1458 *Reg. Great S.* 143/1.

(*c*) For ilk pak of woll lattin out of the loft, cairtyng and weying of the samyn, iiij d., and of ilk serplarth takyn furth of the loft, carting and weying vj d.; 1520 (*c*1580) *Edinb. B. Rec.* I 198.

(*d*) Ande of vthir gudis that aw na custum or that aw custum eftir the fraucht of the serplaith that is to say it at payis as a serplaith in fraucht sal bring thre vnce of bulȝeon hame; 1436 *Acts* II 23/2. Of ilk freman .. of ilk serplaith wol or skynnis iiii d. .. and of all vther sic like thingis, baith polk, pak, and barell, proporcionably ansuerand to the serplaith and to the ton; 1482 *Edinb. Chart.* 167. Ilk serpleth of woll, ilk last of salmond, ilk iiii^c claith, four uncis of brynt silvir; 1538 *Acta Conc. Publ. Aff.* 473. Ilk serpleth gudes to Danskyne and the eistar seais to be estimatt be the tun; 1582 *Conv. Burghs* I 140. The serplaith of lambskins con. 8000 iiij ounce; the serp. of cunning skins con. 16000 iiij ounce; ilk serplaith of futfelles con. 4000 iiij ounce; SKENE *Verb. S.* s.v. *Bvllion. Serplath* .. conteinis foure score stanes for the Lordes of Councel in anno 1527 decerned four serplaithes of packed wooll to conteine sexteene score stanes of wooll .. The merchandes vsis to pay fraucht for their gudes to Flanders be the sek, to France, Spayne and England be the tunne, and to Danskene and the Easter seas be the serplath; *Ib.* s.v. Ilk serplaith of cunning skynes contenand xvj^m; 1597 *Bk. Rates* in Halyb. cxvi. Serplith; PATERSON *Sc. Arithm.* 7.

(*e*) The said Thomas left with me in Brugis ij serplaytis of woll; 1499 HALYB. 225.

(*f*) Tane fra the sammyn [*sc.* warehouse], x sarpleth of pakkit woll; price of the sarpleth xl li.; 1495 *Treas. Acc.* I 220.

(*g*) For the fraucht of xix s. gr. and xviij s. gr. the sirpleth of woll and skin; 1518-19 (*c*1580) *Edinb. B. Rec.* I 186. Because the merchandice passis nocht conform to the saidis ackis be sirpleth and vtheruis; 1554 *Ib.* II 203. Les nor he haif free thre sirpleth of guidis; 1570 *Reg. Privy S.* VI 165/2. Sirplithe; HUNTAR *Weights & Measures* 3. The sirplith of goods which was the common fraughting of merchandise betwixt this countrey and the eastern countreys is esteem'd to weigh eighty stone-weight or 1280 pound weight; MACKENZIE *Observ.* 119.

(*h*) Thomas Mar sal deliver .. ane surplat [*pr.* susplat] of woll in the Feir contening alsmony wauis [*pr.* wanis] and nalys of wecht as the surplat [*pr.* susplat] of woll that was the said Alexander Redes wecht; 1489 *Aberd. B. Rec.* I 416. Ane surplaith of kirsp, ane of lane, ane of holane claith; 1542 *Acts & Decr.* MS I 141.

2. *transf.* A wrapping or sack for a bale of wool or skins.

(*a*) Malcum Alderman .. has .. prefit .. the deliuering as efferis of [4] canwes serplatis of the said maister Robertis; 1522 *Dundee B. Ct.* I fol. 100b (30 July). Fourty aucht elnis canves to be serpleth to cary the kingis woll [£3 4 s.]; 1537 *Treas. Acc.* VI 333.

(*b*) ij^{c}xxx elnis barres canwes to be sorpclaithis to cary the kingis gracis woll fra Selkirk, and to pak the samyn to be send to the sey [£17 5 s.]; 1535 *Treas. Acc.* VI 261.

Sarpleth, var. of SARPLAR(E *n.*

Sarrabossa. A 'word' in a sailor's shanty sung while weighing anchor. The meaning of the whole is uncertain, but the first element may be F. *serrer* (SARRIS *v., pres.*) to hold tightly, tighten, make fast. For further discussion see POURBOSSA, and the references given there. — Ane marynel cryit and al the laif followit in that sam tune, 'caupon caupona .. sarrabossa, sarrabossa', than thai maid fast the schank of the ankyr; *Compl.* 40/24.

Sarracen(e, varr. of SARACEN(E *n.* and *adj.*

Sarraly, -lie, Sarrély, *adv.*[1] Also: **saraly, sarély, sairly.** [ME *sarreliche* (*c*1330), *-ly* (Manning), *sarrilich* (14th c.), f. as SARRAY *adv.*] Close together, in close formation.

Alex. I 2221 may be a further example of SARELY *adv.*

(*a*) The king .. Saw fyrst cummand thar fyrst eschele Arrayit sarraly and weile; BARB. VIII 222. The formest off his mengne Enbrasyt with the scheldis braid And rycht sarraly [*C.* sarray] togydder raid; *Ib.* 296. In myddis thaim the king thai bar And ȝeid about him sarraly; *Ib.* IX 177, etc. Heir cumis Clarus full sturdely Armit, and all his company All sarraly ridand in battale; *Alex.* II 1199. Sarraly; *Ib.* 1611. The Scottismen in gud array Togidder knyt thaim sarrely [*C., R.* apertly]; WYNT. VIII 2565 (W). Sarraly; *Ib.* 2249 (C). Ryd sarralie, skaill nocht thi men oure wyde; HAY *Alex.* 452. And sarralie thai followit one the chase; *Ib.* 3227.

(*b*) To reill thame ane horne thai blew And syne into ane sop thai drew .. A point than peirtly can thai ma And sarely thare fais socht. For thai wald venge thame gif thai mocht; *Alex.* I 2221. Thar leyff thai laucht and past but delay Rycht saraly and [*MS* far alyand] in a gud aray; *Wall.* IX 1274 (McD). All .. with a pavys of tre .. abuf thar hedys .. Sa sairly [*Sm.* sairly, *1553, Ruddim.* surely] knyt that maner embuschment Semyt tobe a clos volt quhar thai went; DOUG. IX viii 113.

(Sarraly,) Sariely, *adv.*[2] [Appar. OF *serreement* 'vivement, rapidement' (12th c. in Greimas). Cf. 14th c. ME *sarri* in K. Sisam *Fourteenth Century Verse and Prose* (1970) 255, on which Sisam comments: 'prob. OF *serré, sarré*, in the developed meaning 'active', 'vigorous', seen in the adv. *sarréement*'. Cf. SARRAY *adv.*, SARRALY *adv.*[1]] ? Vigorously. — Byrdis smale As turturis and the nychtyngale Begouth rycht sariely [*C.* meraly] to syng; BARB. V 5.

Sarrasin, -syn, varr. of SARACEN(E *n.* and *adj.*

Sarray, *adj.* (Perh. a blend or confusion of late north. ME *Serres* (*a*1400-50), e.m.E. *Seres* (1580), L. *Sēres* 'a people of Eastern Asia (the modern Chinese), celebrated for their silken fabrics (Lewis & Short), and ME (Wyclif) and e.m.E. *Siry, -é, Syry* Syrian, L. *Syrius*, and cf. *Serrie* Syria (*Bann. MS* 240b/78).) — Ane actoun .. The quhilk was made of plewan and assyse Off Sarray silk plett full of vlypyis; HAY *Alex.* 14739.

Sarray, Sarré, *adv.* (F. *serré* (*c*1112 in Larousse), in close order, p.p. of *serrer* (SARRIS *v., pres.*) to hold tightly, shut tightly, tighten, close, etc. Cf. also SARRALY *adv.*[1]) — And the formast of his menȝe .. richt sarray [*E.* sarraly] to-gidder raid; BARB. VIII 296 (C). Bot thai war mony sturdie men and stout And held thame sarre togidder in ane rout; HAY *Alex.* 1612.

Sarrazen(e, -ezen(e, varr. of SARACEN(E *n.* and *adj.* **Sarrély,** var. of SARRALY *adv.*[1] **Sarrie,** var. of SARY *adj.*

Sarris, *n. pl.* (F. *serres* pl. pieces of wood used to bind or reinforce interior structural parts of a ship (1538 in Wartburg), f. *serrer* v. (see SARRAY *adv.*).) — For j^{m}iij^{c}xxv fut of gret sarris to the schip, .. for vj^{m}j^{c}xl fut of sarris pullan and plank to the said schip and xiiij pece of barrotis; 1508 *Treas. Acc.* IV 46.

Sarris, Saris, *v., pres.* (F. *serrer* to hold tightly, grip fast, squeeze (13th c. in Larousse). Cf. also SARRAY *adv.*, SARRALY *adv.*[1]. Perhaps, however, merely a further example, with irregular spelling form, of SAR(E *v.* a, or, confusion with SAR(E *v.* may explain the Ruddim. spelling.) — The silly dow .. he clippis at the last And lowkyt in his punsys sarris [*Ruddim.* saris] fast Thristand his tallans sa throu hir entralis [L. *columbam comprensamque tenet pedibusque eviscerat uncis*]; DOUG. XI xiii 170.

Sarsaparilla, *n.* [e.m.E. *sarcaparilla* (1577), *sarsa parilla* (1597), etc., F. *salseparille* (*c*1560 in Larousse), Sp. *zarzaparrilla*.] The dried root of sarsaparilla, prob. used as a medicine. — Sewyn vnce and ane half sarsaparilla price of the vnce iiij s. viij d.; 1590 *Edinb. Test.* XXI 311.

Sars(e)net, varr. of SARCINETT *n.* **Sartan,** var. of CERTANE *adj.*

Sartan(e, -aine. [Varr. of CERTANE *n.* and *adj.* Cf. e.m.E. *sartayne* (*a*1533).]

A. *adj.* **1.** Certain, sure. = CERTANE *adj.* 2.

Quhair I might haue bene sex houris before thame, gif I had bene sartane of thair way; 1615 *Highland P.* III 217. But when people goe soe far off, they can hardly be sartaine of their diett; 1685 *Seafield Corr.* 12.

2. Not specially designated or enumerated. = CERTANE *adj.* 3.

[He] gayf grund sessyng .. of that sartan land and junt fefment; 1456 *Peebles B. Rec.* I 117. [He] stud vp in the curt and clamyt Wil Peblys of a sartan som of gold and siluer; *Ib.* 118. Allowit to the clark for sartane extraordeneris debursit be him; 1597 *Household Bks. Jas. VI and Anne* 28 Feb.

B. *noun.* **a.** An unspecified number (*of* persons). = CERTANE *n.* 2 a.

With consent of the balye and a sartan of the dussan; 1457 *Peebles B. Rec.* I 125.

b. An unspecified quantity (*of* something). = CERTANE *n.* 2 b.

That Mechal Forest has ressyngit .. a sartan of his land liand [etc.]; 1456-7 *Peebles B. Rec.* I 117.

Sartefecation. Var. of CERTIFICATIOUN *n.* 2. — And to requyr thame to concur with hym to the effet forsaid with sartefecation to thame; *c*1554 *Corr. M. Lorraine* 379.

Sarttyfie. Var. of CERTIFY *v.* 1 b. to assure by way of warning. — And yf your honour do not shortly appoynt your offecer for Este Tyvedale to met I moste sarttyfie the Lordis of the Counsell; 1576 *Douglas Corr.* 218.

Sarueice, -ves, -vice, -vyse, -wis, -wyce, varr. of SERVIC(E *n.* **Sarvand, -vant, -vend, -wand, -wantt,** varr. of SERVAND(E *n.* **Sarvator, -vettour, -vitour, -watour,** varr. of SERVITOUR(E *n.*[1] **Sarve, Sarwe,** varr. of SERV(E *v.*[1] and *v.*[2] **Sarving,** var. of SERVING *vbl. n.* **Sarvyet,** var. of SERVIETTE *n.*

Sary, Sory, Sorry, *adj.* Also: **sarye, -ie; sairie, -y; sarrie, sorye, -(i)e, -ey; sorri(e.** [North. ME *sari* (Cursor M.), *sary* (*a*1400-50), ME and e.m.E. *sorie* (*c*1200), *sori* (Cursor M.), *sory* (Piers Plowman), *sorry* (*c*1555), OE *sárig*.]

1. Of persons, etc.: Full of distress or sorrow; distressed, vexed, unhappy; regretful; sorrowful, sad. Also *transf.* Variously const.

(1) *predic.* The lave .. Off thar tynsaill sary and wa; BARB. XII 141. That mycht all his mengne se Sary and wa; *Ib.* XIX 41. *Ib.* XX

298. Sary ware thai thane ilkane; *Leg. S.* xxiii 123. [Th]ane that tyrand can rodly rare, As man that had bene wondit sare, Na he sicht nocht priwely, As men dois quhile that are sary; *Ib.* l 1025. *Ib.* xxv 247, etc. His battale than men mycht haue sene Sary and wraith, abaisit and mad, And Porrus battall blyth and glaid; *Alex.* II 9841. That bathe scho sek was and sary, And all lyk for to be dede, Bot gyff scho sonnare gat remede; WYNT. V 1458. Wpe sho ras, and .. Owte off the tempyll gretand yhede, .. dulefull and sary; *Ib.* 1966. The modyr than, that .. saw [*pr.* swa] hyr barne deyd, wes sary; *Ib.* VII 1278. The stait of man dois change and vary Now sownd now seik now blyth now sary; DUNB. (OUP) 178/10. Lat us in hairt nevir moir be sary; *Ib.* 185/37. Thouch I made yow sarie in a pistile; NISBET *2 Cor.* vii 8. Id. *Matth.* xxvi 37, *Luke* xviii 24.

(*b*) Thocht he had ire Was na ferly, And Gaudifeir Was mair sory Than man that war .. disherist aganes richt; *Alex.* I 2770. Sory and wraith war thare menȝe; *Ib.* II 4389. Art thou contrite and sorie in thy spreit; HENR. *Fab.* 698. Welcum, oure indeficient adjutorie .. That never saw Scot yit indigent nor sory Bot thou did hym suport; DUNB. (OUP) 109/27. LYND. *Test. Pap.* 369. King Corbrede sorie and trist .. departit to Scotland; *Boece* 135b. This ilk captane vpoun the wall that stude, Richt sorie wes and murnit in his mude; STEWART 52636. Scho was angrye and sorye for a seassone; KNOX II 358. Quhen syt and sorrow semis superlatiue Sagis sould seme leist sussious and sorie; *Maitl. Q.* 144/84. The king was werie sorrie and cryit on all his serwandis .. to ryde to the feild and stanche slaughter; PITSC. I 319/7. Quhan the governour got this letter .. that the erle of Angus .. sould haue sufferit he was vondrous sorie; *Ib.* II 50/34. Quhaireuer I ryde or go, ȝe sall nocht sorie be; *Montg. Suppl.* ii 13.

attrib. The drerie damiesall Distres And eik hir sorie sister Hivines; DUNB. (OUP) 147/19. Sum stuid besyde ane wondderous sorie man; 1573 SEMPILL *Sat. P.* xxxix 211.

ellipt. Als lang leivis the mirry man As the sory; LYND. *Sat.* 107.

transf. I latt no lovaris pas with sorry [*Ch.* ane sair] hairt; LYND. *Sat.* 290 (B). My saule is sarie [L. *tristis*] and sad evin to the dede; HAMILTON *Cat.* 152. With wonder sorie hart With few hors men [he] fra his camp did depart; ROLLAND *Seven S.* 87. When my hopes proue naught with sory mind, I sigh & say vnkind, vnkind; CRAIG II 100.

(2) He [*sc.* King Edward] wes off his [*sc.* Bruce's] eschap sary; BARB. II 65. [Clifford] That for his tynsaill wes sary; *Ib.* V 450. He worth sary for to be, For the gret aith that mad had he; *Leg. S.* xxxvi 521. Quhen deide wes he .. few for that cas wes sary; WYNT. V 290. Fortoun .. For quhois fals change suld none be sary; DUNB. (OUP) 184/9.

(*b*) The king and all his barnie Sall of thir tydingis sory be; *Alex.* I 2136. For Sampsoun war thay all sory [: tenderly]; *Ib.* 2180. The king Acastrus richt gretly Of his sonnes was maid sory That was slane; *Troy-bk.* II 2680. For he saw afoir the errouris tocum, for the quhilkis he also afoir wes sorie; WINȜET II 56/9. And the said Christen being sorye thairat, the said Patrik said, 'Christen I promit I sal nevir have uthir wyf bot yow'; 1569 *St. A. Kirk S.* 330. In respect that hir prayer stopit and that sche wes sorie for itt; 1590-1 *Crim. Trials* I II 236. Donalde now being the kingis frinde, was sair sorie for his deith; DALR. I 242/16. Quhair ye wrytt the counsell cannot understand the benefett thay sall have [etc.] .. I am sorie thairoff; 1614 *Aberd. Council Lett.* I 128. And that thairby he kyitchis [*sic*] to be sorie and grewit for the same; 1617 *Aberd. B. Rec.* II 353. SPALDING I 97, 294.

(3) For nane wes in that company That wald haiff bene halff sa sary For till haiff sene his broder ded; BARB. IX 44. I am verray sary at my herte to se samony that suld be noble men gevin to mak lesingis; 1543 *Corr. M. Lorraine* 36.

(*b*) The godlie luffis thair pastours .. And will be sorie to se thame want or wrangt; LAUDER *Minor P.* iii 78. I am sorie to know from Scotland that [etc.]; 1588 *Cal. Sc. P.* IX 649.

(4) The pure pepill .. was wonder sorie That thay war sa destitute of sic glorie; ROLLAND *Seven S.* 5016. We ar rycht sorie that this is treu for the maist part and mair; WINȜET I 44/3. I am the more sorie in deid that suche as ye are hath interteniment in the palaices of princes; FERG. *Answer* in *Tracts* 48. I am sorey that ony sic attemptat suld fall out; 1576 *Douglas Corr.* 206. I ame sorie that ye uas so nyce uith the Lorde Murley, for he is my olde freinde; 1599 *Maxwell Mem.* II 9. We ar sore we can nothe expeid our selfis but your lordschipis miscontentment; 1602 *Misc. Spald. C.* II 289. Yet I soulde be sorri he soulde be to dow yow any onkindnes; 1606 *Laing MSS* I 107. Wee ar heartily griued and sorey, that [etc.]; 1638 Balfour *Ann.* II 267.

(5) Quhen Gaderanis saw sa hard ane pais Be held thame, thai war all sary; *Alex.* I 999. *Ib.* 1972. Quhen that scho herd off this cas Suppos in hart scho sary wes; WYNT. II 48. And quhen he herd that stedfastly The caus that made thame sa sary; *Ib.* V 2730. My hart was wounder sarye Quhen comoun weill so sopit was in sorrow; LYND. *Dreme* 997. Ȝit quhen scho sichis, my hairt is sairy; Id. *Sat.* 4375 (B). — Bot I think and my moder wer levand I wald be richt sorey and she wer in sa hard estate; 1586 *Laing MSS* I 57.

2. Expressive of or displaying distress or sorrow.

(*a*) With sary cher he held him still; BARB. XIX 83. Than answerd scho with sary cheire; *Seven S.* 1238. Than .. the stewart .. Past furth and maid a sary mane; *Ib.* 1256.

(*b*) The empreour .. Hard the loud schout and sorie pieteous call Of the emprice; ROLLAND *Seven S.* 1258. His sory sang was 'Oche and wallaway!'; 1567 *Sat. P.* iii 4. Quhen sall my hart ceis of this sorie sang? *G. Ball.* 89.

3. Involving or accompanied by, invoking or causing distress or sorrow.

(*a*) Hop .. can refresch a sary thing And mak with wyt recomfortinge; *Ratis R.* 571. Thir hid nettis for to persaue .. This sarie calf wyislie to vnderstand; HENR. *Fab.* 1938. Drunkynnes .. May be reput of sorow and of syne A sary smyt tyle hyme that leffys thar in; *Bernardus* 75. Ane othir aukwart a sarye straik tuk thar; *Wall.* II 109 (McD). [The serpents] with thar cruell byt and stangis fell .. tuke mony sary morcell [L. *miseros morsu artus*]; DOUG. II iv 26. Tha sary bukis makis mention How wemen betryit in speciall Adame Dauid Sampsone and Salamon; *Bann. MS* 271b/18.

(*b*) This warldis vane prosperitie Quhilk hes thir sory properteis thre That is to say gottyn with grete laboure Kepit with drede and tynt is with doloure; HENR. *Orph.* 550. Leif I thaim thus, blamand thar sory [*1570* sary] chance; *Wall.* III 239. I standand bundin in ane sorie plite; DOUG. *Pal. Hon.* 932. That sory [*M.* sary] sicht me sudaynly affrayit; DUNB. *G. Targe* 207. Furth of the flude of Stix that sory place; DOUG. III iv 13. Ane of the sory furyus sisteris thre .. quhilk causis all myschefe; *Ib.* VII v 101. *Ib.* VI viii 116. The causis all of his sory banishement; 1567 *Sat. P.* iii 206. Couatus men .. Sall nocht at lenth eschaip ane sorye end; LAUDER *Minor P.* i 599. Hammiltoun he me huif, I ȝow declair, Ane sorie surname; 1571 SEMPILL *Sat. P.* xxviii 40. In the begynnyn of the seik[nes] at the vring is reid & thik .. sall the seiknes seikerlie be sorie; 16.. *Herbarius Latinus Annot.* (Bot.).

proverb. Manie mirrie evening maks manie sorie morning; CARMICHAEL *Prov.* No. 1148.

4. Of a person or animal: **a.** That has suffered misfortune, hapless, pitiable.

(*a*) 'That sary Benet,' [he] sayd, 'am I'; WYNT. VI 1117. Feil woful turmentis of wrachit catyvis sary [: tary]; DOUG. VI Prol. 50. 'Palynurus,' quod scho, 'thou sary syre'; *Ib.* V 159. Doutles richt sone this sarie fox war slane; STEWART 58111.

(*b*) For brawle doggis at the dure All settis on the sory hound That lyis euir at the grund; *Colk. Sow* I 136. For sory swyne for thair golfing affraid; *Ib.* II 179. Sen the deces Of my sory husband; DOUG. IV i 42. Lyke as the littill emmet haith hir gall The sorie banestikkill haith hir fin we sie; *Bann. MS* 211b/2.

b. Habitually grieving, plaintive, miserable.

Bot, wit ȝe weill, he wes ane sory man; For falt of gude he wantit all his will; *K. Hart* 511. In place of a sarie las that never leuche, God gave me .. a pleasand boy; MELVILL 323.

5. Wretched, worthless, of little value, pitiful, mean. **a.** Of places or things, material or non-material. **b.** Of persons or animals. **c.** *proverb.* **d.** *adv.*

a. That sary lyff contenwyd he; WYNT. VIII 5577. The eldest sister sperit at hir gest Giff that scho thocht be ressone difference Betuix that chalmer and hir sarie nest; HENR. *Fab.* 277. With ery curage, febill strenthis sary; DOUG. *Pal. Hon.* 103. Bot soft and soupill as the silk is his sary lume; DUNB. *Tua Mar. W.* 96. Ȝour sary ioys beyn bot ianglyng and iapys; DOUG. IV Prol. 20. Scho thinkis it nocht eneuch .. fortobe wrokin .. Apon thar sary levingis and remanys; *Ib.* V xiii 50. Incaise Harbargary had bene the name of ane village or sary toun; *c*1554 *Misc. Bann. C.* II 29. Not ane sarie house, as this eirthlie bodie was, bot ane fair buylding; ROLLOCK *Wks.* I 305. Faith hes bot ane sarie licht, bot the licht be presence is mervellous; *Ib.* 325. Is sarie base wordis changed in glorious wordis for nathing? *Ib.* 397.

b. Vnhelsum meit is of a sarie mous; HENR. *Fab.* 1493. And settin down lyk sarye crockis; DUNB. (OUP) 105/18. At hame remanis the sarie lairdis That giffis no giftis nor ȝit rewardis; *Maitl. F.* 193/3. They seing this sarrie thing, said to themselves [etc.]; MONTG. *Flyt.* 386 (H). Quhair, sevin ȝeiris, it sowkit, sweillit, singit and sarie [*H.* sarrie]; *Ib.* 494 (T).

c. He is a sairie begger that may not gae by ane mans dore; FERG. *Prov.* No. 334. It is a sairie brewing, that is not good in the newing; *Ib.* No. 474. Shee is a sairie mouse that has but ane hole; *Ib.* No. 762. Sairie be your meil poke and ay your neive in the nook of it; *Ib.* No. 775. *Ib.* No. 369, 512. Sarie; CARMICHAEL *Prov.* No. 192. *Ib.* No. 689.

d. Sarie shitten .. ill shapen shit; *a*1600 T. MAITLAND in *Watson's Coll.* II 54.

Saryzine, -zyn(e, Sarzan(e, -zen(e, varr. of SARACEN(E *n.* and *adj.*

Sas, *n.* [e.m.E. *sasse* (1642), F. *sas* (16th c. in Larousse), late MDu. *sas* 'statio navium et cataracta' (Kilian).] A river- or canal-lock, a sluice. — Wee shifted boattes 3 times that night because of ther slousses or sases .. by the which the water is made navigable; 1646 J. HOPE *Diary* (1958) 183.

Sasaphas, *n.* Only *attrib.* (Appar. put for e.m.E. *sas(s)afras* (1577), Sp. *sasafras*, the name of the tree or of its bark or wood.) — My sasaphas coope edged about the mouth with silwer and my table men of whalebone; 1654 *6th Rep. Hist. MSS* App. 683/1.

Sas(c)er, varr. of SALSAR *n.* and *v.* **Sase,** var. of SES(E *v.*[1] **Saseing, Sasine, -ing,** varr. of SESIN(E *n.* **Saserit,** var. of SALSARIT *ppl. adj.* **Sasiabilitie,** var. of SATIABILITY *n.* **Sasinett, Sasnett,** varr. of SARCINETT *n.* **Sasis,** var. of SAUCIS *n. pl.* **Sasone, -oun(e,** varr. of SESON *v.* and SESO(U)N(E *n.* **Sassar,** var. of SALSAR *n.*

Sasse. ? Var. of, or ? erron. for, FAS *n.* 2, or *sussie* SOUCY *n.*[1] — The said Johne sayd .. that he [*supra* ane of the balyes] tuk na sasse of hyme, for he wald put all the towne in his sleyf for ane punt ayll in ane penne purse; 1558 *Inverness Rec.* I 19.

Sasseit, *v.* [Appar. f. L. *sēcēdere*. Cf. 18th c. Eng. *secede* (1702).] *tr.* To secede from, disassociate oneself from (here, a statute). — The provest, Thomas Wilson [etc.] .. sasseited tharfra; 1664 *Kirkcaldy B. Rec.* MS 18 April.

Sasser, var. of SALSAR *n.* **Sassering,** var. of SALSARING *vbl. n.* **Sassone,** var. of SESO(U)N(E *n.*

Sasteing, Say-sting, -styng, *n.* [North. ME and e.m.E. *saa stange* (Cursor M.), *sastange* (Cath. Angl.), *saystang* (1593), Sw. *såstång*, Dan. *saastang*; SA *n.* and STING *n.*] A large pole for carrying a 'say' (SA *n.*) or tub of this sort. — He bar a sasteing in a boustous poille; *Wall.* II 33. Ane cumyeone, ane bukat, say & say styng; 1538 *Aberd. B. Rec.* MS XVI (Jam.). The said Ewin .. straike at me wyth ane say sting; 1566-7 *Inverness Rec.* I 146.

Saster, Sasur, varr. of SALSAR *n.* **Sasyne,** var. of SESIN(E *n.*

Sat, var. of SALT *n.*[1] and *adj.*, SET *v.*, SETE *n.* and SET(T *n.*, p.t. and p.p. of SET *v.* and p.t. of SIT *v.* and SITE *v.*[2] **Sataine,** var. of SATIN(E *n.* **Satand,** pres. p. of SIT *v.*

Satan(e, Sathan(e, Sathanas, -es. Also: **Sathana; Sythynnis.** [ME and e.m.E. *sathanas* (*c*1160), *satanas* (*c*1220), *sathan(e, satan, sathone, sathonas* (all Cursor M.), *Satan* (*c*1425), L. (Vulgate) *Satān, Satanās*, f. the Gk.]

Satan, the Devil. Also *transf.*

(*a*) Satane bad hyme mak bred ȝare; *Leg. S.* ix 155. The principale of dewilis al Selfe Satan; *Ib.* xxxii 262. Att Satanis command he stayit; 1590 *Crim. Trials* I II 210. The Lord .. restraining Satan and ceasing himself to let out his hand against me; *a*1699 *Sel. Biog.* II 105.

(*b*) Thairfoir hy ȝow to the pott of hell With Sathan our Abirone to dwell; ROWLL *Cursing* 148 (B). Off Sathanis senȝie syne sic ane vnsall menȝie; DUNB. (STS) xiv 79. A swenyng swyth did me assaile Off sonis of Sathanis seid; *Ib.* (OUP) 161/4. Lytchore haldis men fast in Sathanys net; ABELL 7*b. We persaving how Sathan in his membris the Antechristes of oure tyme, crewellie dois raige [etc.]; 1557 *Facs. Nat. MSS* III xl. That Fillok .. That semis ane sanct new send vs fra Sathan [: dane, agane, fane]; ROLLAND *Seven S.* 8018. *G. Ball.* 10, etc. Efter diuers rebuikis giffin him being ane ȝoung boy for gifing the first fruitis of his bodye to Sathaneis wse; 1619 *Perth Kirk S.* MS 10 June. Betaking him self to the seruice of Sathan the ennimie of his saluatioun .. he enterit in pactioun and covenant with the devill; 1630 *Justiciary Cases* I 144. 1662 *Soc. Ant.* XXII 219, etc.

(*c*) The erle Ferandis modyr was Nygramansour and Sathanas Scho rasyt; BARB. IV 242. With dedigne He can ansuer ma [to Peter] And sad 'Ga agane, Sathana [: ma]!' *Leg. S.* iii 354. He .. Our set all the Sathanas syde Sarazenis mycht; *Howlat* 509 (A). Off wikkit birth and sone of Sathanas [: was]; HAY *Alex.* 17320. The Deuillis net Quhilk Sathanas for all sinfull hes set; HENR. *Fab.* 2445. Sathanes; *Contempl. Sinn.* 974 (Arund.).

(*d*) Quhen scho went to se To far landis in hir verginite With Sythynnis syne scho was corruppit sone; *Bk. Chess* 2096.

transf. Quhen that glowrand gaist [*sc.* her husband] grippis me .. Than think I hiddowus Mahowne hes me in armes. Thair ma na sanyne me save fra that auld sathane, For thocht I croce me .. He wil .. clap me to his breist [etc.]; DUNB. *Tua Mar. W.* 102.

Sate, var. of SET *v.*, SETE *n.* and p.t. and p.p. of SIT *v.* **Sateable,** var. of SUTABLE *adj.* **Satefé, -ie, -y,** varr. of SATISFY *v.* **Satefeing,** var. of SATISFYING *vbl. n.* **Satein(e, -en(e, -eyn,** varr. of SATIN(E *n.* **Satele,** var. of SATTIL(L *v.* **Satelling,** var. of SATTILLING *vbl. n.* **Saterday, -daye, Saterurisday,** varr. of SET(T)ERDAY *n.* **Saterik,** var. of SATIRIC *adj.* and *n.* **Satesfé, -y,** varr. of SATISFY *v.* **Satfat, -filt,** varr. of SALTFAT(T *n.*

Sath, var. of SAITH *n.* **Sathan, -ana, -anas, -ane, -anes,** varr. of SATAN(E. **Sathing,** var. of SATIN(E *n.* and SETHING *vbl. n.* **Sathurday,** var. of SET(T)ERDAY *n.* **Sathyne,** var. of SATIN(E *n.*

(Satiability,) Sasiabilitie, *n.* [Cf. med. L. *satiabilis* adj. (*c*1350 in Latham), e.m.E. *satiable* adj. (1570).] Capability of being satiated. — Withouttin rancour, perfyte cheritie, Withouttin hunger, sasiabilitie; LYND. *Dreme* 586.

Satiat, Saciat, *v.* [e.m.E. *saciate* (*c*1532), *satiate* (1611), L. *satiāt-* p.p. stem of *satiāre* to satiate.] *tr.* To fill, satisfy (with food). Hence gen., to gratify to the full (a person or his desires). — Nothir may the est nor the west pertis of the warld saciat thame [*sc.* the Romans]; BELL. *Boece* I 149. To that fyne that .. thay may saciat .. thair botumles couatise of spoyle and thift; BUCH. *Wr.* 35. Onlie blude man satiat the bludethristie; 1594 *Misc. Bann. C.* III 169. Resolveing .. to satiat his filthie .. lust and unchaist appetyte with the said Lady Katharene Grahame; 1633 *Justiciary Cases* I 215.

Satiat, Saciat, *p.p.* Also: **-ate, saceat.** [e.m.E. *sacyate* (1526), L. *satiāt-*, p.p. stem of *satiāre*; SATIAT *v.*] Satiated, glutted. Const. *with, of, in, be.*

The Inglismen .. Of Scottis blude quhilk mich[t] not satiat be; STEWART 52268. The godly man .. in reading, talking, or meditation .. is never satiate; 1548 KNOX III 10. In mennis sicht presenting ane angell, And sine inwart satiat with subteltie; ROLLAND *Seven S.* 8026. Ȝit culd thay neuer get power ouer this sempil man, nor haif thair hartis satiat of him; *Lyndesay Pref.* (STS) 398. The said Johnne, not beand satiat heirwith .. come .. and .. removed hir tennentis thairof; 1579 *Reg. Privy C.* III 128. 1582 *Crim. Trials* I II 103. And that he not only wes satiat thairwith bot lykwayis kest hir doun the said tolbuith stair; 1582-3 *Perth B. Ct.* 9 Feb. Not having satiat thair hartis with this enorme forme of dealing; 1584 *Reg. Privy C.* III 683. Let thy heart fixe thy love as lang as thou will upon the creatures, thou sall never be satiate... But gif ains thou love God [etc.]; R. BRUCE *Serm.* 155. The said Mathou tuik ane heavie seiknes .. with ane extreme drouth, that he culd navayes be satiat; 1597 *Misc. Spald. C.* I 168. Ane man will drink in that spreit, and desire to be satiat with that water; ROLLOCK *Wks.* I 447. Tables removed, efter Homer's fassion well satiat, he honord me so much as to schow me his bookes and papers; 1614 DRUMMOND in Insh *Colonial Schemes* 44. Thay quha, in thair ȝouth-heid, war trwbillit with famine.., can nocht.. by no plentie, be satiat; 1616 *Crim. Trials* III 586. FORBES *Rec.* 448. As that the thrist and crueltie of the procurers therof might be satiat and slockened therewith; CALDERWOOD IV 36.

(*b*) Quha euer enteris in it he how hungry that euer he be he salbe fulfillit and saciat; *Prester John* 310b. And with his lust couth nocht be saciat; BELL. *Boece* I xiii. Quhen his lust was saciat; *Ib.* 199. *Ib.* 83. Be this torment saciat was nocht the kingis ire; *Boece* 339b. To sla thame & be saciat of blud; *Ib.* 579. The pepill .. mycht nocht be saciate to woundere sa grete formosite of hir .. gudlie portratoure; *Ib.* 610b. Saceat; *Compl.* 96/6. *Ib.* 81/14. ROLLAND *Ct. Venus* I 211. Wes neuer none, nor neuer salbe, Saciate, that sycht tyll that he se; LYND. *Mon.* 5049. And he him self als appeiring drunk .. ȝit culd nocht be saciat; 1624 *Perth Kirk S.* MS 7 June. Oft surfetting, yit never saciat [: that]; GARDEN *Kings* 11.

(Satiety,) Satietie, Saciety, -ie, *n.* [e.m.E. *sacietie* (1533), *satiety* (Spenser), OF *sazieted, sacietet* (*c*1120 and 12th c. in Larousse), MF *satiété* (1530 in Larousse), L. *satietāt-*, f. *sātis* enough.] **a.** The condition of being fully gratified; full attainment of a desired object or goal. **b.** A sufficiency or abundance. — **a.** *Satias*, sacietie; *Despauter* (1579). He .. hes not gottin that joyfull countenance of Christ, in the quhilk countenance is onlie saciety of plesures; ROLLOCK *Wks.* I 330. In whose countenance is sacietie of joyes; *Rollock's Thess.* Dedication. — **b.** Hath thou not a full satietie of all the vanities under the sunne? 1569 KNOX VI 566.

Satifé, -ie, -y(e, varr. of SATISFY *v.* **Satifeing, -ieing,** varr. of SATISFYING *vbl. n.* **Satiffy,** see SATISFY *v.* 4 b. **Satill, -illit,** var. and p.p. of SATTIL(L *v.*

Satin(e, *n.* Also: **satt-, sath-; sait(t)-, set(t)-, seat(t)-; saut-, salt-** and **-(e)ing(e, -en(e, -ein(e, -(e)yn(e, -aine, -on, -oune.** [ME and e.m.E. *satyn(e* (*c*1369), *sat(t)en* (*c*1400), *sathan* (*c*1460), *sateyn* (1506), etc., OF *zatin, satin* (1351 and 1387 in Larousse).]

1. Satin.

The forms in *salt-* occur only in 1580 *Aberd. B. Rec.* II 36.

For many further examples, see BIRGE *n.*, BROCHIT *ppl. adj.*[1], CRAM(M)ASY *adj.* and *n.* and CRAM(M)ESYN(E *adj.* and *n.*; DAMAS *n.* and DAMES *n.*; FIGURÉ; PIRN *n.* 1 e, and PIRNIT *p.p.* and *ppl. adj.* (*n.*); and the Indexes to the various vols. of *Treas. Acc.*

(1) In empcione tresdecim vlnarum de satyne; 1373 *Exch. R.* II 440. Vnum vestimentum de rede satyne; 1436 *Reg. Episc. Aberd.* II 147. For ij½ elne of sating .. to be a doublat to the king; 1473 *Treas. Acc.* I 15. Satyne; *Ib.* 74. Twa tippatis of silk ane blak satyn lynyt with grece; 1483 *Acta Aud.* 112*/1. [A cope] of red satyn with I and K of gold; *c*1500 *Coll. St. Salvator* 153. Black satin; 1501-2 *Cal. Doc.* IV 336. [3 ells] satyn; 1507-8 *Rentale Dunkeld.* (SHS) 249. Grene satin; 1508 *Treas. Acc.* IV 117. A chesabill of auld grene satene; 1510 *Stirlings of Keir* 293. Ane pece of tanne satene of Rouans [*ed.* Romans]; 1516 *Acta Conc. Publ. Aff.* 70. Satyne; 1526 *Treas. Acc.* V 274. Gray sateyn; 1531 *Ib.* 416. Quhite railȝeit sating; 1538 *Ib.* VII 109. For thre quarteris of reid Darnyk sating; 1539 *Ib.* 155. Sating; 1548-9 *Ib.* IX 279. Ane nycht goun of blak sating railya; 1548 Thomson *A Collection of Inventories* (1815) 78. His hors was bairdit full richelie Couerit with satyne cramesie; LYND. *Meldrum* 386. Of satine greine all of ane leveray; *Clar.* II 734. Ane paill of reid saten with ane kaip of clayth of gold; 1559 *Edinb. B. Rec.* III 43. Half elle of incarnatt satin; 1567 *Misc. Maitl. C.* III 186. Muttoun gluiffis pasmentit lynit with satyne & taffeteis price of the pair iij s.; 1575 *Edinb. Test.* III 362. Claith of gould welwete sataine and dameis; PITSC. I 368/6. Crammasie sating; 1596 *Montgomery Mem.* II 235. All vther satyne of the lyke making, out of grayne; 1612 *Bk. Rates* (Halyb.) 327. Ane quarter [etc.] .. of rid satein to cover the kingis buik, xl s. the ell; 1615-16 *Glasg. Univ. Mun.* III 567. Ane pair of breikis and doublit of minnome sateine; 1620 *Edinb. Test.* L 284b. And mort claithe with the armes and mort heidis .. all on sateine; 1643 *Funeral Acc.* in *Berw. Doc.* 1 May. 1644 *Acts* VI I 76/2.

(*b*) For fowre elne of blak sattin to be [etc.]; 1490 *Treas. Acc.* I 143. A stek sattyn cost [etc.]; 1495 HALYB. 22. For .. armasyn sattin to be ane hogton for the king; 1525 *Treas. Acc.* V 253. Ane govne of gray satting; 1546 *Elgin Rec.* I 88. For vii ells of fyne grein Jenua sattin; 1635 *Black Bk. Taymouth* xxii. Isabella sattin; 1648 *Thanes of Cawdor* 307. Any cloathing .. of velvet sattin or vther stuffs of silke; 1672 *Acts* VIII 72/1.

(*c*) Sathyne; 1505 *Glasgow Dioc. Reg.* II 114. Ane doublat of gray sathing; 1519-20 *Stirling B. Rec.* I 2.

(*d*) Ane purs of saiting price thairof x s.; 1588 *St. A. Test.* II 97. Ane coverlet of ane burd of reid weluot upon quhyt saitting; 1603 *Reg. Privy C.* VI 576. Half ell half quarter saitten to be ane picadell; 1613 *Edinb. B. Rec.* VI 362. Ane gowne .. with lapis of saittinge; 1627 *Brechin Test.* IV 291b.

(*e*) Welwit .. lyle worsat setyne of silk sewing silk [etc.]; 1556 *Dundee B. Ct.* III fol. 63 (1 Aug.). Ane goune of flourence setoune; 1633 *Stirlings of Keir* 447.

(*f*) Ane tuo legged purse with ane bitt of seattine; 1658 *Brechin Test.* VI 163.

(*g*) Thrie steming cloikis .. ane lynt witht salting; 1580 *Aberd. B. Rec.* II 36.

(2) To be teachit and learnt in making of woolens, sautings, silks, taffities [etc.]; 1587 *Aberd. Trades* 335. Violat welvets and satynes, dropit with gold and golden letters; 1591 *Misc. Spald. C.* III 160. Certane remanes of novatois duratois perpetuanois water philp and chinais bischopis sateines; 1620 *Edinb. Test.* L 352. Fowr ellis of florit satenes; 1641 *Ib.* LIX 278b.

2. *attrib.* **a.** Made of satin.

Ȝon fair las with the sating goun; LYND. *Sat.* 4412. x dosane of satene buttonis; 1561-2 *Treas. Acc.* XI 107. Twa saltene dowblaittis, .. ane pair of satteing breikis; 1580 *Aberd. B. Rec.* II 36. For the ordering of his yellow satting cloithis; 1590 *Thanes of Cawdor* 197. Grein silk to sew on the freingȝeis and saittein pasmentis; 1593 *Edinb. D. Guild Acc.* 502. To mak my quhyt setting goun; 1603 *Montgomery Mem.* II 245. Ane broun saittene dowblet; 1606 *Crim. Trials* II 508. Sex pair plane kawzit sating heidis at xxxvj s. the pair; 1615 *Edinb. Test.* XLVIII 214b. For ȝour leddisschipis purpore seaton goune; 1620 *Tailor's Acc. Bk.* A 99b. Seatein; *Ib.* Seatin; *Ib.* Seatoune; *Ib.* 105. Sex pair of satene ribbennet gloves; 1643 *Edinb. Test.* LX 235. Ane codwar with ane seattine bonnett; 1658 *Brechin Test.* VI 163. Satten; 1662 LAMONT *Diary* 146. That plain sattin ribbons may be worne vpon apparell in the same manner as taffetie ribbons; 1673 *Acts* VIII 212/1.

b. Designating a particular type of silk or cloth made of silk.

The former may be merely tautological.

With blak veluott satyne dammes [etc.]; 1536 (*c*1580) *Edinb. B. Rec.* II 77. Satyne [*Ch.* sating] damers; LYND. *Sat.* 3771 (B). Thre vnces sating silk .. blak satyne silk to be howpis to his hienes clock; 1582 *Treas. Acc.* MS 36.

Satinisgobe, *n.* [e.m.E. *satinisco* (1615), *sattinisco* (1619), *satanisco* (1639); pseudo-Spanish formation f. SATIN(E *n.*] An inferior sort of satin. — [Convicted of] stealing .. out of Alexander Young's loome two webs, whereof there was ane of satinisgobe and ane other of dornick; 1679 *6th Rep. Hist. MSS* App. 688/1.

Satir, -yr, *n.* [e.m.E. *satyre* (1509), *satire* (1596), OF *satre*, MF *satire* (*c*1290 and *c*1355 in Larousse), L. *satira, satyra*, alt. f. *satura* a satire.] A satirical poem, a satire. Also *attrib.*, of satire, satirical. — Ane pleasant satyre of the thrie estaitis [*B.* The play maid be Dauid Lynsayis] in commendatiovn of vertew and vituperatioun of vyce, as follovvis; Lynd. *Sat.* heading (Ch.). Ane satyre of the thrie estaitis; *Ib.* frontispiece. — *attrib.* Thair was the satir [*L.* satyr] poet, Juuenall; Doug. *Pal. Hon.* 907.

Satirday, var. of Set(t)erday *n.*

Satiric, *adj.* and *n.* Also: **satyricke, satyrik, satirique, saterik, satorik.** [ME and e.m.E. *satiric* n., *Satiricus* prop. n. (Trevisa), *satyryke* adj. (1509), *satirick* adj. (1581), *satyricque* n. (1589), F. *satirique* (late 14th c. in Larousse), vulgar L. *satiricus*, f. *satira* a satire.] **a.** *adj.* Containing satire. **b.** That writes or composes satires. **c.** Given to using satire; satirical. **d.** ?*noun.* Satirical utterance or composition. — **a.** The foirsaidis obiectis causit me ken him, Satirique all and full of venum; *Rob Stene* 23. — **b.** Euripides, Iuuenal, Perseus, Horasse, nor nane of the satiric poiettis; *Compl.* 66/5. Bid the satyricke find-fault poet, take him; Craig iv 32. — **c.** [The king and Mr. John Lindsay his secretary knew of it, who is a subtle] satorik [man]; 1596 *Cal. Sc. P.* XII 279. I think some satyrik lawier sall [etc.]; 1626 *Ancram & Loth. Corr.* I 42. — **d.** Nor have I used minȝearde nor effeminate, tantting invective, .. vane saterik or lowse wowsting and wantting speiches; Bisset I 77/22.

Satisfactio(u)n(e, *n.* Also: **satisfacton, -faccio(u)n(e, -faxion, -faccon; sattisfactione, -faccioune; setesfacten.** [ME and e.m.E. *satisfaccion, -iun* (Cursor M.), *satisfacioun* (Ayenb.), *satisfaccoun,* (1477), *satisfaction* (1531), F. *satisfaction* (*c*1155 in Larousse), L. *satisfactiōn-*.]

For many further examples (in senses 2 and 3) see *St. A. Kirk S.* Index, s.v. *Satisfaction*, and Ireland *Asl. MS* I 22-32.

I. With reference to obligations.

1. a. Compensation for financial loss; repayment of a debt; settlement of money, etc., owed. **b.** Restitution or reparation made for harm done, goods stolen, physical injury committed.

Also const. *mak, (tak, give, get).*

a, b. (1) Off the sattisfaccioune [L. *emendatione*] of scathis don in burghe; *Acts* I 363/2. The sade Duncane become plege and souerte for satisfaccione of party; 1496-7 *Acta Conc.* II 56. 1501 *Ib.* III 26. Oliver his wyffe and he lyis cursit and aggregit quhilk I traist thai wald nocht do without the hoipe of sum way by my satisfactioun [*sc.* of a debt]... Beseiking your lordschip that ye will nocht gif ane absolutioun without I be satisfiit [etc.]; 1546 *Corr. M. Lorraine* 162. Commissioun for satisfactioun of the clergie for thair lyverentis .. prelattis .. quhais rentis be the said annexatioun [of ecclesiastical lands] salbe diminissit sall .. ressaue recompence and satisfactioun thairfoir; 1587 *Acts* III 438/1. & that in full contentatione & sythment & satisfactione of all & haill his portione naturall .. be the deceis of his wmquhill fader; 1588 *Prot. Bk. J. Inglis* 10 July. Tua hunderethe markis .. quhilk scho acceptis in satisfactioun .. of all contractis infeftmentis [etc.]; 1623 *Brechin Test.* IV 130b. I declar my selff to be awand to .. my laufull dochtir .. ten thowsand pundis .. in satisfactioune to her of all that scho can clame of me .. and in satisfactioune to hir of quhatsumevir portioune naturall, bairnes pairt of geir [etc.]; 1647 *Maxwell Mem.* I 338. In full .. satisfactione of all provisione, bairnes part of geir .., or portione naturall; 1681 *Echt-Forbes Chart.* 155. The satisfaction of an old security has been always sustained .. from circumstances proven; Stair *Inst.* iv xlv § 23. *Ib.* i xviii § 5.

(2) Qvhillez hale gre satisfactioun and payment to the forsaidez religiouse .. be .. made; 1398-9 *Liber Melros* II 489. He [*sc.* a sick man] suld mak satisfaccione of al he has tane wrangwysly or at he aw; *Cr. Deyng* (STS) 264. James Scot was ordanit .. to .. paye and caus to be maid satisfactioun as borowus for Robert Scot; 1509 *Selkirk B. Ct.* (ed.) 8. The inquest .. culd nocht fynd to compel the said Wilkesoun to mak satisfaction nor pament of the said hors; 1520 *Ib.* 56. To mak oppin satisfactioun to the said James afore the kirk; 1562 *Edinb. Hammermen* 252b. That for thair trespas thair fredomeis be cryed doune .. quhill thai mak sik mendis and satisfactioune thairfoir; 1605 *Glasgow B. Rec.* I 229. Anent the satling of that feid Mr. James seamis to be very penitent and wald glaidlie .. mak satisfactioun and assyithement; 1607 *Lett. & St. P. Jas. VI* 111. Mr. James Gordone has removed therefrom [*sc.* from the ministerial charge], and his wyfe offered to move upon satisfaction made for the manse; 1649 *Moray Synod* 95. — And nocht gaynstandand that this said knycht had tane satisfactioun and assithement of the barouns gudis; Hay I 166/21. Trew men can gett no satisfactioun Becaus our king is nocht of eild; Kennedy *Bann. MS* 93a/31. Howsoever my men .. deny any wrang .. commitit be thame yit .. I have send thame to be tryet .. that as ye find thame giltie, thay may give satisfactioun; 1621 *Aberd. Council Lett.* I 194.

c. *specif.* Compensatory payment or indemnification made for manslaughter or mutilation.

In collocation with Kinbut(e *n.*, Assithment *n.* and Contentatioun(e *n.*

In assithment and satisfaction of the said mutilation; 1557 *Prot. Bk. Gilbert Grote* 17. For full assyithment, haill satisfactioun and compleit kinbut paying for the slauchter of .. Johne Mure, .. to .. pay to Janet Mure £233. 6 s. 8 d.; 1574 *Misc. Abbotsf. C.* 37. In full contentatione & sattisfactione of the allegit mutillatione & bluid schedding of the said Allexander Dunbar; 1588-9 *Prot. Bk. J. Inglis* 7 March. Gif the hurt fall out to be a maymeing or mutilatioun, that then the pairtie offending sall .. performe the order .. for satisfactioun of the pairtie maymed; 1605 *Reg. Privy C.* VII 707. In name of kinbute assythement and full satisfactioun [for the cruel and unnatural slaughter of the said Thomas Johnestoun]; 1623 *Douglas of Morton* 704. I was .. inforced .. to submit myselfe to arbitrement satisfaction and reconciliation; Lithgow Trav. i 7. He .. was fynet and haid found cautioun for satisfactioun of his bluidfyne; 1657 *Rothesay B. Rec.* 269.

d. Restitutory payment of expenses. **e.** Payment for a service rendered.

d, e. For satisfactioun and payment of are done dede and service; Hay I 138/32. To content and pay to Williame Grant xx s. money .. in satisfactioun of the mending of the said Thomas wyfes face being hurt with ane chandilar; 1570 *Canongate Ct. Bk.* 279.The bearer is cummit to you on his awin charge .. therefore ye mon have respect to him. The last bearer that ye send, cam heir behind the hand, and hath gottin na satisfaction as ȝit; 1592 *Crim. Trials* I ii 324. Concerning the feyis and satisfactioun to be gevin to the constables and clerkis for thair paynes; 1617 *Acts* IV 540/2. Ye sall .. gett acquittances upon sic as ye deburss for discharge of ȝour diligence and satisfactioun of the counsall at ȝour returne; 1618 *Aberd. Council Lett.* I 164. He making satisfactioun to the javellour for his expenssis; 1620 *Crim. Trials* III 483. To giv satisfactioun to the said maister Alexander Lyndsay .. both for his debursingis amd for his awin pains; 1636 *Glasgow B. Rec.* II 42. Sex punds Scots in satisfactioun for his paines in building the steiple staire; 1670 *Kirkcudbr. B. Rec.* MS 23 July.

f. Payment of a tax.

The maister of the Inglis schip .. entiris his vessel for the pettie customes, and agries to give four hundredwecht of guid tobacco in satisfactioun of claims under that head; 1638 *Dumbarton B. Rec.* 56.

g. Appar. a receipt (for money received).

Efter your husbands buriall we sall then give order for receaving these moneyes from ȝour ladyship and to give yow all just satisfaction in securitie for your liberation thairof; 1630 *Aberd. Council Lett.* I 318.

2. a. In the pre-Reformation church: The performance by a penitent of penal and meritorious acts, as part of the sacrament of penance, an instance of this.

(1) And than the ded .. pacyently tholyt makis satisfaccioune, nocht alanerly of the vaneall synys, bot als .. for mony dedly synis; *Cr. Deyng* (STS) 198. Allmous deid is a part of satisfactioun; Ireland *Asl. MS* 33/6. *Ib.* 27/21. Satisfaccioune; Id. *Mir.* I 47/20. Grant me of thi sonne to haue compassioun .. And for my synnis do satisfactioun; *Remembr. Passion* 658. O Sanct James and Sanct Josaph .. grant me full satisfactioun for my synnis or I de; *Rois Garlandis* 520. Off undone pennence and satisfactioun sure; Dunb. (OUP) 18/93. Scantlie dois he sufficient satisfactioun; Abell 17a. [That] throuch his patience, humilite and satisfactioun, and

recompensatione done to God and religioun and to ws .. he serue .. to be deliverit .. of the said pennance; 1534-5 *Prot. Bk. Sir J. Cristisone* 35. And this lawly and meik submissioun in thi confessioun .. is ane part of satisfactioun; HAMILTON *Cat.* 223. That satisfactioun quhilk is heir spoken of is nocht sa to be takin that ane penitent synnar may .. deserve remissioun of synnis be ony paine .. tholit be him or to mak to God ony just .. recompence equivalent to the syn .. for in this maner it is nocht in the powar of man to satisfie for that satisfactioun hes only our salviour Christ wrocht and fulfillit be his bitter passioun. .. Bot satisfactioun quhilk is callit the thrid part of pennance is diffinit thus. .. Satisfactioun of pennance is to cut away the occasionis of synne; *Ib.* Qvhy hald ȝe satisfactioun Papistrie? WINȜET I 104/23. A gryt misdeid .. necessarlie requiris gryt satisfaction; KING *Cat.* 91b.

(2) And ȝit thai [*sc.* St. Cyprian and others] speik .. of a satisfactioun requirit of our obœdience; WINȜET I 105/8. Satisfactioneis of men for theyr synnes; 1580 *Facs. Nat. MSS* III lxx.

b. In the post-Reformation church: The performance of (public) penance for one's offences against church discipline, a particular instance of this.

Lord I haif done full littill reuerence Vnto thy sacramentis .. And of thy halye supper for my sin recompence Gif I for my sin bewaill and mak satisfactioun [*Bann. MS 18a/44* The haly supper for my syn recompence And of my gilt the holy satisfactioun]; Dunb. *Bann. MS* I p. 9/44. Thai ar decernit to mak public satisfaccion this nixt Sundaye in the essemble; 1561 *St. A. Kirk S.* 142. Quhen satisfactioun is dewlie maid be him for the .. crymes [of fornication] .. mariage than to proceid; 1582 *Ib.* 491. Thai haif oftimes bein chargit to enter on the penitent stuill and mak satisfactioun for thair offences; 1587 *Ib.* 602. 1588 *Ib.* 630, etc. George Richeson .. was ordainit to mak publik satisfactioune the nixt Sabbath befoir the congregation; 1616 Ritchie *Ch. S. Baldred* 144. Except he preveened the sentence by satisfaction; 1638 BAILLIE I 164. CALDERWOOD II 303. The haill elders .. declairit they ressauit full satisfactione of him [*sc.* the Earl of Crawford] and thanks the Lord for him; 1651 *Cupar Kirk S.* 7 Feb. in Lamont *Diary* 21 n. John and Alaster McGueines have closed their satisfaction and their repentance is received; 1654 *Moray Synod* 119. They might enter for the third time and have ther satisfactione accomplished; 1666 *Bonckle Kirk S.* 8. The minister .. graunted the bill .. appointing them to enter .. ther satisfactione for the third time, and to pay four pounds of penalty; *Ib.* Thir marked with a cross are to give satisfactione to the church; 1684 *Dumfr. & Galloway Soc.* 3 Ser. XXXVI 162.

3. The atonement made for human sin by Christ in his passion.

(1) For originale syn he him allane maid satisfacton be his passioun; IRLAND *Asl. MS* 23/18. He offerit His precius lif for ws and maid perfit satisfaccioun quhar nouthir we nor vthir creatur mycht or had power to mak satisfaccioun; Id. *Mir.* II 29/19. His deid and passioun .. was necessar for our satisfaccioun and deliccioun of our synnis; *Ib.* 39/20. *Remembr. Passion* 589. Dunb. *Bann. MS* 18a/44 (see 2b). GAU 4/29. To mak satisfactioun and mendis for al our synnis; HAMILTON *Cat.* 154. *Ib.* 223 (see 2 a above). The doctrine of the kirk of God .. hes declarit this mistery [of the blessed sacrament] .. as ma, aucht, and suld be to the contentacione, contenment and satisfaccione of al guid Cristiane men; Q. KENNEDY *Breif Tract.* (ed.) 112/8. To the effect that we mycht obtene euerlastand lyf, satisfactioun for syn, redemptioun fra the .. serwitude of the devill; Id. *Compendious Ressonyng* (ed.) 158/28.

(2) To .. bring us agane in the favour of his father eternal makand to Him ane just and perfit satisfactioun for al our offencis; HAMILTON *Cat.* 154. The sacrifice of the mes was institute .. nocht as ane new satisfactioun, or redemptioun, bot as ane new sacrifice; Q. KENNEDY *Compendious Ressonyng* (ed.) 158/31. His satisfactions of men for their sinnes; 1638 *Nat. Covenant* in *Facs. Nat. MSS* III xcvii.

II. With reference to desires or feelings.

4. Gratification or pleasure occasioned by some event, or a particular instance of this; contenting (a person) by the complete fulfilment of a desire or supply of a want.

For satisfactioun of ȝour desyr, ȝe sall consider that the markispoinct of oure ressoning is [etc.]; Q. KENNEDY *Compendious Ressonyng* (ed.) 169/28. Quiche loiss, altho great to me, hes nocht .. geven these quho ar so heid set aganes me full satisfactioune; 1641 *Bk. Carlaverock* II 139. When that shall be done by his majesty that will give satisfaccon to his subiects, no man living shall joy more thereat; 1646 *Lennoxlove MS* (The Duke of Hamilton) C1/2094. The satisfaxion I shall reseaue in sieng you is a uerie great uan; 1653 *Wemyss Corr.* 97. Dier Madam, It is a verey gret setesfacten to me to heir from your la[dyship]; 1664 SRO *Seafield MS* GD248/556.

Satisfactorie, *adj.* [e.m.E. *satisfactory* adj. (1547), noun (1530).] = SATISFACTOUR *adj.* — The wicked opinion of it, to wit, that it is satisfactorie for sinnes, in this argument also is included; FERG. *Answer* in *Tracts* 47.

Satisfactour, *adj.* [MF *satisfactoire* (1495 in Larousse), med. L. *satisfactorius* 'atoning (theol.)' (*c*1225 in Latham), f. L. *satisfacere* to satisfy.] Serving to make satisfaction or atonement for sin (cf. SATISFACTIO(U)N(E *n.* 2). — Bot thocht his glorius passioun be satisfactour sufficiently for all mortale syn .. neuertheles [etc.]; IRLAND *Asl. MS* 23/20. That ony werk or sufferance of pane be sufficiently satisfactour ar fyve condiconis requyrit; *Ib.* 24/23. — *pl.* And thow dois ony werkis nocht penale na dolorus thai suld be callit mair meritouris na satisfactouris; *Ib.* 25/27.

Satisfaxion, var. of SATISFACTIO(U)N(E *n.*

Satisfear, *n.* [e.m.E. *satisfier* (*a*1565); SATISFY *v.*] One who satisfies a creditor by paying (*to* him) the due payment. — Satisfearis and payaris to him of the foirsaid summes; BISSET I 249/26.

Satisfeing, -fieing, varr. of SATISFYING *vbl. n.* and *ppl. adj.* **Satisf(f)é, -ei, -ie,** varr. of SATISFY *v.* **Satisfiet,** var. of SATISFYIT *ppl. adj.*

Satisfy, Satify, *v.* Also: **satys-, sattis-, sat(t)es-; sati-, saty-, sate-** and **-fie, -fye, -ffie, -f(f)é, -fei; seteisffé; setifé.** [Late ME and e.m.E. *satisfye* (Lydgate), *satisfied* p.p. (1433), *satysfy* (1455-6), *satisfy* (1529), also *satefie* (*c*1475), *satify* (*c*1555), OF *satisfier* (Hatz.-Darm.), also *satefier, sattifier* (1294 in Larousse), L. *satisfacere.*]

I. With reference to debt or obligation.

1. *tr.* **a.** To pay (what is owed or due) (*to* another), to discharge (a debt), to pay (a fee, expenses or the like); to fulfil (an obligation) (*to* another); to comply with (a request or demand).

(1) Thy air .. Haif he thy gude, he takis bot small tent To execute, to do, to satisfie Thy letter will, thy det, and legacie; HENR. *Fab.* 836 (Ch.). The haill kirk annuellis and kirk rentis of this burgh is nocht habill to satisfie and pay the ministeris stipendis; 1580-1 *Edinb. B. Rec.* IV 196. Bring .. sic .. allabast as salbe sett doun .. quhilk we sall not faill to se defrayed and satisfeit to your gude contentment; 1594 *M. Works Acc.* (ed.) I 314. A decreit of removeing is sufficientlie satisfied if the partie find caution to remove; HOPE *Major Pract.* II 98. Be the practique of Scotland, the maills and dewties of the lands should be first satisfied; *Ib.* 116. To exact for ilk staine's weying of inbrocht guidis .. four pennies .. and the pryces to be satisfeit be him quha sall requeir the service; 1648 *Inverness Rec.* II 198. Jeane Millar aristed .. ay and whill shee satisfie the jayl fies; 1672 *Bk. Old Edinb. C.* VI 112.

(*b*) The .. commissioneris of burrowis .. promitis to satife and fulfill the said act; 1601 *Conv. Burghs* II 107.

(2) The said Jhone .. sall .. cancell and destroy this present contract as a full discherg of ather of thair pairtis honestlie satisfeit and performit to vtheris; 1594 *Facs. Nat. MS* III lxxv. Duncane Paterson .. oblist him to satisfie to Jhone Tailyefeir, massoun, the

soume of ij[c] pundis money for beiting of the brig of Inneralloun; 1596 *Stirling Chart.* 221.

(*b*) Quha being suorne deponit that his said vmquhill father satefeit sez firlottis mele to the said Dauid; 1582-3 *Perth B. Ct.* 18 Jan.

b. To compensate for (a financial or other loss).

(*a*) By that to restoir and satisfie at the leist the haill goodis and geir to everie man quhilk he had reft and taine away; PITSC. I 129/15.

(*b*) Caidmour hes lying waist without teling .. quhilkis hillis wuld haif sawin till us xviij[xx] bollis of aittis [etc.] .. extending to the sovme of [blank] quhilk sowme we desyre to be satifyit and mendit; 1562 *Peebles B. Rec.* I 282.

c. With a person as object in senses a and b above. Also const. *for, of, anent(is* what is owed, etc., and *with* the compensation offered.

(1) [Andrew Hamilton] contentit, applesit, satisfiet and assithit plesandly the tenants and fermorers [by giving them other mailings]; 1555 *Miss Hamilton MSS* in *8th Rep. Hist. MSS* App. 309/1. For to inhabill thameselffis to satisfie the saidis creditouris be ingaddering of the sowme; 1598 *Acts* IV 167/1. David Watsonn becumes catioun for them that they sall mack their repentance and they [*sc.* the Session] askit them to seteisffe him undir the doubill of the paine; 1620 *Dumbarton B. Rec.* App. 1 2. Satisffie; 1623 *Aberd. Council Lett.* I 208. [The moveable goods of the prisoner were ordained to be escheat and the] pairties damnified .. to bee satisfied out of the first and readiest thereof; 1659 *Aberd. Sheriff Ct.* III 59. Sir Andrew Dick pursues James Baillie compriser of certain lands .. pertaining to Sir Andrew, to .. see it found, that James was satisfied by his intromissions within the legal, and produced a rental .. where-with James intrometted .. he having medled at least with some of the duties from the tennents; 1662 *Decis. Lords* G 41.

(*b*) That sindry at this præsent .. ar nocht eschameit nocht to satifie thair inferiouris .. in thair deuitie of dettis; WINȜET I 107/24. We grant ws satiffeit and payit be them in numerat money foirsaid; 1586 *Perth B. Ct.* 16 Dec.

(2) Lett her l. be satisfeit anentis thai x chalderis victual; 1565 *Reg. Morton* I 14. That kyndlie possessouris .. sould nocht be removit .. without thai wer satisfiit for thair kyndnes; 1566 *Reg. Privy C.* I 467. To .. aduise with him how Dauid Orme may be satisfeit anent the byganis; 1570 *Reg. Morton* I 55. The said Mr. Johne Spottiswod said the Act wald caus him be satisfeit for the foure akeris be the Bischope of Sanctandorois; 1573 *Reg. Privy C.* II 314. He delt thair landis and goodis to thair creditouris and complenaris quhill they war satisfieit of the thingis taine fre thame quhairof the misdoaris war convickit; PITSC. I 83/19. He levis .. to satisfie the laird of the grund for thair tak fourty pund; 1579 *Edinb. Test.* VII 25. [Persons convicted are to] repair with diligence to my lord thesaurar .. and compone for thair eschetis, quhairthrow the parteis may be satisfeit and thay orderlie relaxt; 1582 *Reg. Privy C.* III 503. Untill the pettie customer be satisfied of the dewties thereof; 1626 *Dundee Chart.* 157. Togedder with fourtie lib. for my hous maill .. off the quhilkis I hald me .. satisffiet; 1630 *Aberd. Council Lett.* I 312.

(*b*) And declarit that the said James Logane satyfy thaim for thair part of the pris libellat; 1558 *Admir. Ct. Bk.* (St.S) 84. Swa are thaye payit and setifeit of thar fathyre arff; 1562-3 *Rec. Earld. Orkney* 116. Quha that enteris nocht to satifie the gude toun and fermoraris of thair dewteis to be callit the nixt law day; 1573 *Edinb. B. Rec.* IV 4.

(3) And the air is bundin .. to satisfie the woman with ane uther manse; BALFOUR *Pract.* 109.

d. To remunerate; to reimburse; to pay (a person) (in respect *of* wages or expenses) (*for* work done or services rendered).

(1) Labourand everie man according to his craft .. unto sic tyme that all the artailyearie .. be .. put in .. ordour and the samyn being done be thame thai to be considderit and satisfiit extraordinarlie; 1565 *Reg. Privy C.* I 403. The said Georg .. to repent for the vrang and to satisfie the leich; 1612 *Inverness Rec.* II 99. *Ib.* (see 2 d). The edifice of the kirk is helpit and the skletter satisfied with halff a dolor and his expenses; 1650 *Boharm Kirk S.* 2 June.

(2) He hes .. acceptit upoun him to satisfe the furniseris of his majesteis house according to thair .. comptis and als to satisfie all feis accustomat to be payit be comptrollaris .. and his majestie being of gude mynd that the said Sir George be satisfeit of the saidis debursmentis; 1599 *Reg. Privy C.* VI 40. The counsell .. dischairgit Hew Monro .. of his chairge over the brucht and ordaines that he be satisfeit of his bygane pey; 1644 *Inverness Rec.* II 183.

(3) Clerk, to continew .. compleitting of the said buik .. thae promeist to se him satisfeit for his paynis; 1581 *Conv. Burghs* I 112. Jhonn Jhonestoun to pay to .. Mr. Adam Colt, ane of the regents .. the Alhallowmes terme nixtt to cum of his stipend .. he satisfeand .. the intrant regent in his place for the monethis service to be maid before the said terme; 1590 *Edinb. B. Rec.* V 25. Sche haifing done pleasour to the gudwyffe of Gallowschelis, for the quhilk sche did nocht satisfie hir; 1590-1 *Crim. Trials* I II 236. The toune .. did obleis him to satisfie him ane dollor ilk day for ilk day he suld remaine in the toune's imployment; 1646 *Inverness Rec.* II 191. Thai ordayne the said William to sattisfie the pairtie for the chirurgians paynes; 1659 *Dumfries Council Min.* 16 May. The said earle is to satisfie the said Johne to wite for eury hors shoing [etc.]; 1663 *Galloway P.* 2 July.

2. To compensate for, atone for (an offence).

(*a*) And to satisfie my foirsaid symple dyte In recompance of it now will I wryte Of the secund penny; *Colk. Sow* II 5. Ȝour bludie boist na syith can satisfie; *G. Ball.* 104.

(*b*) The bittir deid salbe ane sacrefice The grete offence of man to satifye; BELL. *Bann. MS* I p. 4/64.

b. *To satisfy the kirk,* also, *the discipline* (of the church) (*for, of* an offence), to atone, by public repentance or in some other fashion, for the offence.

(1) He sal satisfye the kirk at sic tyme as he may be lasered; 1579 *St. A. Kirk S.* 439. The sessioun .. ordanis thame .. to begin thair repentance upon Sonday nixttocum and thairefter to continew quhill the kirk be satisfeit; 1581 *Ib.* 465. Scho sall satisfie the kirk for hir relapse in fornicatioun; 1605 *Dundonald Par. Rec.* 88. Thai salbe ansuerabill to satisfe the kirk for payment of the bluids and satisfeing of the parte efter tryall; 1606 *Melrose Reg. Rec.* I 8. That if the lyk be fund in him he sall satisfie according to the law both magistrat & kirk; 1615 *Fraserburgh Kirk S.* 28 March. Ordaning that scho sall satisfie the kirk as ane incestuous persone; 1626 *Justiciary Cases* I 49. The Laird of Banf and his complishes have not yit satisfied the kirk for the slauchter of James Ogilvie; 1630 *Banff Ann.* II 28.

(2) Confessing their sin were appoynted to satisfie the discipline of Lochell in sackcloath; 1672 *Alford Rec.* 185. The elders .. brings in .. delinquents .. and constraines them, nill they will they, to satisfie discipline; 1677 *Inverness Presb.* 79.

c. To discharge by atonement.

That Georg Muchet should satisfie his penultie and repentance; 1655 *Stirling Ant.* II 18.

d. To make, chiefly financial, reparation to (a person) (*for, of* an injury or wrong inflicted on or perpetrated against him).

1538 *Breadalbane Doc.* No. 48 may, rather, belong in sense 1.

(*a*) The said James Grant .. sall satesfe Johne Grant .. of all skaythis quhilk salbe .. done be hyme; 1580 *Prot. Bk. W. Cumming* 248. The judges .. ordanis him to satisfie the pairtie and the leich as salbe modifiet be the judges; 1612 *Inverness Rec.* II 99. They ar many vther men heir who thinkis that the performyng of thir offeris war ane ware ease way to satisfe his majeste for the contempt; 1614 *Denmylne MSS* in *Highland P.* III 165. The said pannell stoll from John Campbell .. ane mart kow for which he likewayes satisfied the owner; 1677 *Argyll Justic. Rec.* 85.

(*b*) Swa that gif ony falt standis in him he to satefy the pairty sustenand the scaith; 1500 *Edinb. B. Rec.* I 80. The saidis Johne Campbell .. hes nocht satefeit the said Archibald .. tuyching ane cleme at the said Archibald hed to the landis of Stronmonessaik; 1538 *Breadalbane Doc.* No. 48. That thai .. sall assythe satefy and pleise Robert Cathcart .. for the mutilatioune committit be thame vpoune him; 1548 *Crim. Trials* I I 336.

3. *intr.* To make satisfaction, reparation, or atonement (*as* an offender, or *for* the offence).

(1) Thai ar ordanit to be impressonit and to satisfie; 1585 *St. A. Kirk S.* 558. That Edward Steivin & Gilbert Watt be holdin in memorie wntill thay satisfie; 1630 *Fraserburgh Kirk S.* II 20 Jan. Shee .. hade begune her repentance alrady, and was content to satisfie to the full; 1631 *Strathbogie Presb.* 5. 1635 Ritchie *Ch. S. Baldred* 241. 1646 McKay *Hist. Kilmarnock* 317. Raisers off neidfyre in the presbytrie of Strathboggie ar ordained to satisfie three days in sackcloath; 1649 *Moray Synod* 96. Compeired Jhone Williamsone & .. was injoyned to satisfie in repentance; 1652 *Cullen Kirk S.* 25 July. I am glaid he is weill and come home and abll to satisfie for himself; 1653 Hossack *Kirkwall* 254. James Jafray .. to satisfy in the joges two dayes; 1670 *Soc. Ant.* X 655.

(2) It wes .. ordanit that the said Marioun suld satifie as adulteres in the paroche kirkis of the parochinis quhair scho faultit; 1588 *Haddington Presb.* 10 July. Whosoever should bleach cloath on the Sabbath day should satisfie as a fornicator; 1686 *Boharm Kirk S.* 10 Oct.

(3) That he sould satisfie for his fornicatioun .. as the kirk sall inioyn to him; 1602 *Dundonald Par. Rec.* 1. Scho reportit that scho hade satisfeit in Rathen for hir fornicatioun with the said man; 1628 *Fraserburgh Kirk S.* II 12 March. [He] is appoynted to satisfie for subscryving the act; 1650 *Dumfries Kirk S.* 4 July. John Robisone .. being seited befor the sessione becaus he had not satisfied for his fall in fornicatione; 1651 *Cramond Kirk S.* I. John Stirling .. was ordained .. to satisfie publicklie before the congregacioun .. for takeing away a young woman; 1669 *Dunblane Synod* 71. James Rob and Alexander Rob ordained to satisfie for their troubling the church; 1678 *Boharm Kirk S.* 17 March.

(*b*) The allegit spulȝearis and troublaris of the said Doule .. may be eselie compelit to satefie for the samyn; 1574 *Edinb. B. Rec.* IV 20.

II. With reference to feelings or needs.

4. *tr.* To fulfil (a person, his attributes, etc.) in respect of (*of*) his needs, desires, etc., to cause a person to be fulfilled or contented or satisfied in some respect; also, in the passive, to be fulfilled; to be content; to be mollified or pacified, freq. *of* or *with* what fulfils or satisfies; also, to find it sufficient, to require no more than *to* do something.

(1) Had I rewarde amang the laif, It wald me sumthing satisfie; Dunb. (OUP) 130/75. Hys twa last bukis sall swage thar fantasy Les than na resson may thame satysfy [*Ruddim.* satify]; Doug. *Direct.* 70. Thare is no warldly thyng at all May satysfie ane mannis saull; Lynd. *Mon.* 5043. This taill could nocht his mynd all satisfie; *Clar.* iv 106. God .. Quhilk feidis vs with thy sweit word, Sen, all that ever tuke lyfe of thé, Thow satisfyis aboundantlie; *G. Ball.* 21. To bawdische bourdis ȝit man I oft gif ear To satisfie ane flesch[l]e cumpanie; Arbuthnot *Maitl. F.* 51/59. Singing and iusting and rinning of great horse at the listis witht all wther kyndlie games that might satisfie the quene; Pitsc. II 17/8. Heirin we lippin to be satisfeit, our requeist being swa resonabill; 1583 *Edinb. B. Rec.* IV 290. The poore ministers .. compleaned .. that neither were they able to live upon the stipends allowed nor gett payment of that small portioun which was allowed. So faine would the comptroller have played a good varlett and satisfeid the queene, or elles have made up his owne profite; Calderwood II 172. But the incensed multitude wer as yet so litle satisfeed .. that [etc.]; J. Gordon *Hist.* I 11.

(*b*) Nocht as diligent of thar seruice for he was evir als glorius omnipotent & satifyet as he is now; *Asl. MS* I 299/10. Quho had all riches unto Ynd And wer not satefeit in mynd With povertie I hald him schent; Dunb. (OUP) 187/7. Of this oure .. desyre, ye will [etc.] .. and that ye .. induce the remanent of oure said burrois till satifie ws in this behalf to the effect aboue written; 1541 (*c*1580) *Edinb. B. Rec.* II 109. Bot thay salbe all satyfeit and content; Lynd. *Mon.* 6181. Thow opinnis furth thy hand full gratiouslie And satifyis [*v.rr.* satisfyis] all flesche aboundantlie; *G. Ball.* 20. Our pretence is not to satifie and delite the delicat earis of curius men; Q. Kennedy *Tractive* 101.

(2) Sum wald tak all this warldis breid And ȝit not satisfeit [*M.* satisfeid] of thair neid; Dunb. (OUP) 211/32. Iuno, .. Not satisfyit [*Sm., Ruddim.* satyfyit, L. *saturata*] of hir ald fury ne wrok; Doug. v xi 11. Than to be satefeit of his admiratione, he send for that ald vyif, and inquirit hyr [etc.]; *Compl.* 134/8. Sen that ȝe ma me satisfie of all my panis; *Bann. MS* 238b/29.

(3) Thai chasit and slew ravenus beistis; and nocht satifyit [*v.r.* satyfeit] with sic besynes, slew .. thevis; Bell. *Livy* I 18/23. The Samnetes nocht beand satesfit nor contentit of this ansuer; *Compl.* 99/32. And wes nocht satifyit nor content Off his awin regione nor his rent; Lynd. *Mon.* 2009. Our auld enemeis quho can never be satisfeit with thair awin, bot gif they haue thair nichtbouris to; Pitsc. II 35/17. Swa that he culd newir be satisfeit of drink; 1596-7 *Misc. Spald. C.* I 85. Of the quhilkis uitenseillis, abulyements .. above uritten we half us weill satisfyit; 1609 *Wemyss of Bogie MSS.*

(4) Scarsly kowth he satisfyit be Forto behald thir armour bryght and schene; Doug. viii x 52. Saxons nocht satifyit to haue coverit al regions towart the ryver of Tyne with dede carions [etc.]; *Boece* 261b. The heretors & elders present, are satisfied to holde the schoole of Leny in blench; 1692 *Cramond Kirk S.* III 15 Aug.

b. With desire, cruelty, curiosity, etc., as object.

(*a*) Henr. *Fab.* 836 (Ch.) (see 1 (1) above). To conques gudis to satisfy [*pr.* satiffy] thair foule desyris; Bell. *Boece* (M) II 391/20. Id. *Livy* I 28/22. We wyll schaw gude wyll to satisfie thare curiositie sumpart; Q. Kennedy *Tractive* 229. *Magnum .. implevit genitoris amorem*, satisffeit the desyre of Aeneas that thoucht long efter him; Buch. *Comm. on Virgil Æn.* I 716. The defender .. must presently find caution to satisfie [*1699* satifie] the kings will; Mackenzie *Laws & C.* II xxx 4 (1678) 560.

(*b*) He altyrris hys style sa mony way .. To satyfy ilk wightis fantasy; Doug. v Prol. 36. For the Romanis wald nocht dedenye thair majeste, to satefy the desire of barbar pepill in ony othir sort; Bell. *Boece* I 123. Satefie; *Boece* 327. Quhil be sum meretis thare ire war satifyit; Id. *Livy* I 4/21. For quhen thir pepill may nocht be gottin to satifye his cruelte, he behuffit finalie to rage on him self; *Ib.* 115/12. He requires the bischopric of Moraue to be gyuen him be the quene. The quene gentillie ansueris that how sune scho cumis to Scotland, to satifie his askeng; Dalr. II 454/22.

c. To satisfy sexually, to satiate.

I .. will bauldlie say With wyfe or maid to play the man .. But for to satefie ane whore Fiftene sies I can not induire; *Maitl. Q.* 254/41.

d. To alleviate or cure (a disease).

Quhen sair seiknes, that nane can satisfie, Commix[i]t is with mad melancolie; Stewart 14182.

5. a. To resolve or settle (an issue). **b.** To comply with, to fulfil the demands of (an injunction, petition, etc.) **c.** *intr.* To give an answer (*to* a question).

a. Controuersie and stryfe .. laid war al to sleip, satifiet and put to rest; Dalr. II 342/10. Looke that the grounde of youre uarres be satisfeid in youre peaxe; James VI *Basil. Doron* 103/4.

b. I wald thou satisfiit my petition quhilk is necessar; Fowler II 44/4. Haueing satisfied the injunctiones of the presbytery for ther malignancy; 1649 *Kingarth Par. Rec.* 6.

c. This is an apt quæstioun .. to the quhilk nochttheles it becumis to satifie nocht of my awin ingyne bot be auctoritie of the diuine law; Winȝet II 32/6.

d. *tr.* To fulfil the requirements and pay the membership dues of (a craft, trade).

To pay and satisfie the craft conforme to the dewitie of ane brotheris second sonne; 1613 *St. A. Baxter Bks.* 80. Dauid Bull comperand personally satisfeit the said craft for his fredome and of all that thay have to lay to his charge; 1625 *Dunferm. Hammermen*

MS p. 20. Robert Robertsone .. is bound prentes .. for the space of three ȝeares and ane ȝeaire for met and fie and has sattesfied the craft; 1656 *Ib.* p. 60. Johne Stevinsone hath satisfied the craft of his freedum muney and all other thing that can lay to his charg befor thir witneses [etc.]; 1657 *Ib.* p. 61. And hath served the samen tyme for that whilk the said Robert Lawsone hath satisfied our trade; 1667 *Ib.* p. 71.

6. To assure or convince, to free from doubt or uncertainty.

(1) The angell satisffiis Josephs mynd; NISBET I 6. His father was sumthing displiceit, bot at the last was appeisit and satisfeit; MAITLAND *Ho. Seytoun* 28. I am sufficientlie satisfeit .. with the effect and substance of the mes; Q. KENNEDY *Compendious Ressonyng* (ed.) 181/3. *Ib.* 181/5. I do intend .. to send Monsieur Dosell to the quene youre maistres .. quho sall declair that unto hir from me that, I trust, sall satisfie hir; KNOX II 170. My .. will is that ȝe .. cause hir to convein the counsall .. quhair the matter may be tryed and ȝe may be purged thairof and than I will be sattisfied; PITSC. II 211/11. Ather to satisfei hir majeste heir with ressounis or els [etc.]; 1586 Rait & Cameron *King James's Secret* 91. Albeit thais ressones may satisfie thame quha [etc.]; KING *Cat.* in *Cath. Tr.* 205/4. There are with you more worthy and learned than I am, Messrs. Dickson, Blair, and Hamilton, who can more fully satisfy you; 1637 RUTHERFORD *Lett.* (1894) 467. 1643 BAILLIE II 70, 71.

(*b*) The nerrest, and onelie way, .. to be satifiit and establischeit in all materis .. concerning faith; Q. KENNEDY *Tractive* 99. To be satifiit in his doute; *Ib.* 174. Satifeit; Id. *Breif Tract.* (ed.) 112/2. Sene we have hed just occasione .. tilbe resolut and satifyet with the first hede .. of our disputacione; *Ib.* 134/34. Satifyet; *Ib.* 131/13. Plesandly to satifie thame, according to ȝour doctrine, of the expres word of God; WINȜET I 67/20.

(2) Na thing that I can do may satesfy my lord governour to gif me my pece; 1548 *Corr. M. Lorraine* 233.

(3) Whereby I hope your lordship .. will be satisfied that I need not come; 1664 STAIR in Brunton & Haig 365.

Satisfying, Satisfeing, Satifieing, -feing, *vbl. n.* Also: **satyfeyne, satefeing.** [e.m.E. *satisfiynge* (1560), *satisfying* (1647); SATISFY *v.*] The action of SATISFY *v.*, in senses 1-4.

(1) For satisfeing of Alexander Brog barbour for mending of the said Alexander heid; 1569-70 *Canongate Ct. Bk.* 143. For satifeing of the baillies for hir vnlaw; 1579 *Inverness Rec.* I 272. For decorment of the streit and satifieing [*pr.* satisieing] of the wholle heretores concerned; 1674 *Edinb. B. Rec.* X 200.

(*b*) Satefeing of Nicolas Vddart of his thousand merkis debursit [etc.]; 1574 *Conv. Burghs* I 28.

(2) For payment of the bluids and satisfeing of the parte efter tryall; 1606 *Melrose Reg. Rec.* I 8. For satisfeing of the kirk the bluid to the baillie and the parte; *Ib.* 9.

(3) If the sinners had been actually put-at for satisfying in their own persons; DURHAM *Comm. Rev.* 296.

(4) For satyfeyne of gret men, sic as my lord Angus [etc.]; 1551 *Corr. M. Lorraine* 352. For satisfeing of thame that mycht not abyde his avancement; 1567 Anderson *Collect. Mary* I 91.

Satisfying, -fieing, *ppl. adj.* [e.m.E. *satisfying* (Shakespeare); SATISFY *v.*] That satisfies, satisfactory, convincing. — His anssr was he had not uttensills &c.: quhich is not satisfying; 1695 *Ayr Presb.* 23 Oct. Thomas Wells gives noe satisfieing reason of his withdrawing; 1695 *Cramond Kirk S.* IV 12 July.

(Satisfyit,) Satisfiet, *ppl. adj.* [p.p. of SATISFY *v.* Cf. mod. Eng. *satisfied* (1816).] Of a debt or obligation: Discharged, paid in full. — For ane exhibitioun of ane satisfiet band he haid of his father; 1659 *Rothesay B. Rec.* 26.

Satisieing, see SATISFYING *vbl. n.* **Satlat, Satl'd, -led, -let, -lid, -lit(t,** p.t. and p.p. of SATTIL(L *v.* and varr. of SATTILLIT *ppl. adj.* **Satle,** var. of SATTILL *n.* and SATTIL(L *v.* **Satlement,** var. of SATTILMENT *n.* **Satling, -lyng,** varr. of SATTILLING *vbl. n.* **Satly,** var. of SALTLY *adv.* **Satorik,** var. of SATIRIC *adj.* and *n.*

Satrap, *n.* (ME and e.m.E. *satrap(e* (Wyclif), OF *satrape* (*c*1265 in Larousse), L. *satrapa* (also, *satrapes, satraps*) 'a governor of a province, a viceroy (among the Persians)' (Lewis & Short), f. the Gk.) — Dare king of kingis .. till princis and primatis Till satrapiz counsallaris .. Greting [etc.]; HAY *Alex.* 4501.

Satt(e, varr., p.t. and p.p. of SET *v.*, SETE *n.* and SIT *v.* **Satteing,** var. of SATIN(E *n.* **Satteled, Sattel(l)it,** p.p. of SATTIL(L *v.* and varr. of SATTILLIT *ppl. adj.* **Sattel(l,** varr. of SATTIL(L *v.* **Sattel(l)ing,** varr. of SATTILLING *vbl. n.* **Satten,** var. of SATIN(E *n.* **Satterday, Satternsday,** varr. of SET(T)ERDAY *n.* **Sattesfy,** var. of SATISFY *v.* **Sattie,** see SITTY-SATTY. **Sattier,** var. of SATYR *n.* **Sattiled, -it,** p.p. of SATTIL(L *v.*

(Sattilitnes,) Settlidnes, *n.* [e.m.E. *settlednesse* (1571); SATTILLIT *ppl. adj.*] The state of being ordered, resolved or arranged on a permanent or durable basis. — I stayed him until the parliament was past, and matters put to sum settlidnes; 1585 *Lett. Jas. VI to Eliz.* 25.

(Sattill,) Satle, Setill, *n.* [SATTIL(L *v.* 14. Cf. e.m.E. *settle* (*a*1660).] **a.** The action or fact of settling a person's affairs, or of rehabilitating him or her. **b.** Resolution, solution. — **a.** To giue satisfactioune to my creditouris and relieue my cationeris quhairinto my setill and resaiueing of my remistioune standis; 1618 *Haddington Corr.* 137. — **b.** [They] did still tend towardss the best satle and accomodation of all thir difficultyes; 1639 *Hamilton P.* (Camden Soc.) 103.

Sattil(l, Sattle, Settill, Settle, *v.* Also: **sattyl, -el(l, satill, -le, -ele; saitle; setill, -ell(e, -le, settel(l.** P.t. and p.p. **sat(t)il(l)it, satlit, settillit, setlit,** etc. [ME and e.m.E *satle* (Cursor M.), *sedle(n* (*c*1315), *sattele(n, set(t)le(n* (14th c.), *satill* (? *a*1400), *sattle* (*c*1440), *satell* (1525), OE **sætlan, setlan.*]

I. To descend, sink down.

1. *intr.* **a.** Of persons: To fall to the ground, to fall *down*, to collapse. **b.** Of the sea: To become low, or lower, to subside. **c.** Of the earth of a grave: To settle.

a. He failied force and settled down; *Sir Eger* 1570. The proclamator thairof, or he com to the mides of the proclamatioun settillit doune spicheles, and with meikle adoe ane uthir wes enterit into his rowme; MELVILL 632. He fainted and setled downe, and before they cowld reach .. the bookebinders howse .. ther was no breath in him; 1665 LAMONT *Diary* 177.

b. The sey .. sall ryse on heycht .. lyke ane wall ewin straycht vpstand, Syne sattell doun agane so law That [etc.]; LYND. *Mon.* 5466.

c. Quhen the graf is sattillit that .. the samyn laris be laid agane; 1555 *Stirling B. Rec.* I 66. To evin and fill vp the graives as thai sall happin to sattill and fall down; 1583-4 *Edinb. B. Rec.* IV 317.

2. Of troops: To fall back, give ground, retire. Also *tr.*, To cause to retire.

(*a*) Turnus a litil .. Begouth frawart the bargane to withdraw, And sattil [*Sm.* sattyl] towartis the ryveris syde a law; DOUG. IX xiii 28. The Scottismen syne hapnit at the last .. Sum thing to sattill and to pas abak; STEWART 56815. LESLIE 199 (see SATTILLING *vbl. n.* 2). Ane companie of fresche men .. come so fercelie wpoun the Earle of Huntlieis wangaird quhill they war compellit to satill a littill abak; PITSC. I 97/12. Bot thay causit the Inglismen to sattill frome thame; *Ib.* 275/28. He persauit .. the rycht wing of the regentis vantgard put bak and sattill lyk to fle; MELVILLE *Mem.* 201.

(*b*) The Earle of Angus battell quho sieand thame .. command wpoun him settlit and reteirit done to the Earle of Huntlieis battell to haue had help of thame; PITSC. II 100/4. Then Monsieur Dosuell .. luikit to my lord governouris host quho .. had settillit a lyttill abak; *Ib.* 158/16.

tr. So cruell counter .. Quhilk satlit hes the Sutheroun far abak Beȝond the place quhair that tha first began; STEWART 56849.

3. To arrange (things or objects) in order, or in a desired position.

That however these pewes were seatled there should left [*sic*] .. roume for ane entrie to the dask; 1676 *Cullen Kirk S.* 27 Nov.

b. To put (a ship) in order, to set her to rights (prior to sailing). Also *fig*.

For the chelder helpet me to satele her [*sc.* the ship] xv s.; 1595 *Skipper's Acc. (Morton)* 6b. For met to the chelder that helpet to sattell her; *Ib.* 8b.

fig. Bot thole my joyes .. To settell thame within an portt of rest; JAMES VI *Poems* II 165/6.

II. To come or bring to order after disturbance in a pre-existing state of affairs.

4. *intr.* To become pacified: **a.** Of a commotion or mêlée. **b.** Of persons.

a. Grit sture arrayse incontinent, for sum Hamiltounes suspectit that this cruell turnes was done be the Douglassis. .. In the end, the sture sumquhat settiled; 16.. ANDERSON *Hist.* in *Crim. Trials* I i 135.

b. He .. wald settill of his ire sa sune as they became penitent of thair offences; PITSC. I 111/13. Thairfor, the king satteling, aggreagit thair cryme verie hilie; MELVILL 307.

5. *tr.* To impose order on, regulate (a situation, matters, disorder, etc.). **b.** To bring (a place or community) to (*in*) a settled state, to pacify (a country). **c.** To calm, appease, pacify (a person or persons).

Some examples may belong in senses 10 or 15.

His hines myndis therefter sa to sattill the present stait of materis in this realme as [etc.]; 1583 *Cal. Sc. P.* VI 579. Thir fascheries war skarse weill sattelit, when out braks a graitter contradiction; MELVILL 123. Efter lang vexation and mikle adoe the peiple's insurrection was sattelit; *Ib.* 308. To haue satled the confusioun and disordoure of the estate; 1602 Cochran-Patrick *Coinage* I 205. Iniurious wordis or licht braulis wer then easilie to be satlit; 1609 *Acts* IV 434/2. My Lord of Argyll .. hes settelled thingis in Ilay and the greatest he found thair adherent with the rebels he hath caused execut; 1615 *Denmylne MSS* in *Highland P.* III 295. The Marques of Huntly and sum weill set freindis satlit this feid and Frendracht ordaint to pay to the ladie relict [etc.]; SPALDING I 14. Leaving all thinges so soundlie satled both in kirk and pollicie as thay thame selffis could devyss; *Ib.* II 86. *Ib.* 152.

b. Gif we mai sattill the contrai in quyetnes and purge it of sik iniquetie; 1597-8 *Warrender P.* (SHS) II 355. The Yllis, quhilk his majestie with so grite paines and travellis hes satled under his majesties obedience; 1614 *Bk. Islay* 183. The cuntrey being now reasonable weele satled, without appeirance of ony new commotioun or trouble; 1614 *Crim. Trials* III 286. 1625 *Moray Synod* 10. For all belived .. That it [*sc.* Parliament] should satle this distracted natione; 1639 *Bk. Pasquils* 83.

(*b*) True, there was swearing and cursing and brawling .. whereat we were grieved; but we hoped, if our camp had been a little settled, to have gotten some way for these misorders; 1640 *Misc. Maitl. C.* III 478. And conteinous yit the pryce of the aill in respect the marcats ar not settillit thrugh want of watter to the mylnis; 1656 *Glasgow B. Rec.* II 348.

c. [He] appeirit crabit at the beginning, but is satlit sensyne; 1597 Moysie xxiii. We satlit them in end and maid them to drink and schake handis; 1628 *Sutherland Bk.* II 147.

(*b*) John Galie .. and John Ker .. struglit and straik utheris .. and .. Patrik Ker came to setle thame; 1671 *Rothesay B. Rec.* 217.

d. To cause to diminish, to reduce the severity or virulence of (plague).

That the Lord would satle the plague of pestilance at London and other places ther abowt; 1665 *Alyth Par. Ch.* 105.

III. To establish, put in place, a (new) order, state of affairs, etc.

6. a. *passive.* To be placed, located. Also *fig.*

Thomas .. clam vp in the thikest of the branches .. and he scarslie satlet, thear enters tua Englishmen on hors; 1611 *Reg. Panmure* I xxxiv.

fig. Thus, finding the warrand of God sattelit in my hart; MELVILL 172.

b. *reflex.* To establish one's residence, domicile oneself, in a place.

(*a*) Leslie .. resoluit to cum hame .. and sattill him self besyde his chief; SPALDING I 130. That they should not suffer any stranger weiveris to satle himself [*sic*] within the said jurisdiction; 1692 *Glasgow B. Rec.* IV 56.

(*b*) [That] all .. Egiptianis .. settill thame selffis at certane dwelling places with maisteris; 1573 *Reg. Privy C.* II 210.

c. *intr.* To take up one's abode.

(*a*) Thus come and satled heir, Our worthy wyse divyne; GARDEN *Elphinstoun* 1879. For at the laist itt will be seuyne yeirs befor they gett the way off briding, excepe thay satill at home; 1636 *Fam. Innes* 225.

(*b*) George, Earl of Huntly, resolving to setle at Invernes; FRASER *Polichron.* 108. The [l]ose likewayes of a Sc[ots ship] .. filled with our countriemen who designed to setell with us; 1685 *Dunlop P.* III 10.

7. *tr.* To install, establish, confirm (a person) (*in* an office, *to* a calling or *with* an employer). Also with the office, etc. in apposition.

(1) Maister Gudman wes satlit heir in Sanctandrous; *Wode's Psalter* (ed.) 166. 1573 *Reg. Privy C.* II 210 (see 6b above). As also quhairthrow the ministerie suld not haif occasioun to calumniat me thay wer content presentlie to offer and being peciable sattilit in ther awin estaitis to performe quhatsumevir .. thai culd for satisfactioun of the kirk; 1594 Warrender *Illustr. Sc. Hist.* 49. The .. provist .. bailleis [etc.] .. hes thocht guid to sattill thair commowne clerk in the said office; 1606 *Glasgow B. Rec.* I 248. Ane proper young man .. bot verie suspitious for not being sattillit to ony gude calling; 1606 *Reg. Privy C.* VII 715. [Mr. John Mersir] .. to byd at hame, wait upon his buik and calling, seeing in his stipend he is sattilled; 1617 *Ellon Presb.* 100. That he be satlit and contractit with ane maister and he to be his prenteis; 1642 *Elgin Rec.* I 273. I have followed the service of this croune, whair I am now satlid; 1653 *Scotland and the Commonwealth* 223.

(*b*) They had not settled him [*sc.* the schoolmaster] till they tried him; 1682 J. M. Beale *Fife Schools* 207. To provyde and setle a scholemester in a fitt and commodious place; 1698 *Strathendrick* 30.

(2) The former, being nott by law setled secured infeft and seased in his own estate, that he .. was setled principall baliefe; FRASER *Polichron.* 183. Robert Miller .. is (God willing) within a short tyme to be setled minister of the said new erected paroch; 1697 *Glasgow B. Rec.* IV 245. The session .. did unanimously agree to settle him a pensienor of the hospital dureing life; 1712 *Inverness Kirk S.* 235.

b. *passive.* To be established (*in* a state or situation).

The'are .. to giue place compeld, .. To their successours, or them selfs Be satled in their seats. This day one rules .. Tomorrow he is dead; GARDEN *Garden* 88. Ther ar sindrie sowmes promeist to gait them all setled and sustenit whill God provyd better; 1662 *Glasgow Trades House* 377.

c. To be established in the matrimonial state *with* a spouse.

In beulding and reparing ȝour hous of Burnbank which should reather be ȝour exercis when ȝow ar satled with a halfmarrow then now; *c*1650 *Red Bk. Grandtully* II 144.

d. To be established *in* legal possession of property, lands, etc.

The landis and leving of Howpaslet war .. acquyret be vmquhile Sir James Douglas .. and he thairby setlet in possessioun thairof; 1616 *Crim. Trials* III 383. FRASER *Polichron.* 183 (see (2) above).

8. a. To grant, secure (an income, etc.) *to* or *upoun* a person.

Your majestie would likwayes graciously setle upon the petitioner such ane annuity .. as [etc.]; 1661 *Laing MSS* 286. Anent ane sufficient locall modified stipend and mans and gleib setled to the said Mr. Alexander; 1672 *Kirkcudbr. B. Rec.* MS 13 Jan.

b. To impose (a charge) *upon* (a community).

That .. your lordship [*sc.* General Monk] wald tak some course to setle a localitie upoun some place; 1653 *Stirling B. Rec.* I 208.

9. To put in place and ensure the stability or permanence of (a state of affairs); to establish ((good) order, government).

(*a*) Crevand .. the said deacon [etc.] .. to satle ane good ordour amongest thame heirefter so that the poor be not oppressit; 1638 *Glasgow Trades House* 191. Most humblie entreatting your majestie wold .. sattill ane firme and durabill peace against all innovasiounis be sea and land; 1640 *Anal. Scot.* I 389. As to the government of the Church his maiestie will make it his care to satle & secure the same in such a frame as shall be most agreeable to the word of God; 1661 *Acts* VII 88/1.

(*b*) My lord regentis grace cuming .. is .. to settill and estableis universall peace; 1569 *Reg. Privy C.* I 667. To settill a perfyt amytye and sincere intelligence betuene these two crownes; 1588 *Laing MSS* 75. Being myndfull to settill sum guid ordour with the ministrie of our kirkis of the pryorie of Sanctandrois; 1603 *Reg. Morton* I 190. The goverment of the Church was setled by law; 1666 *Laing MSS* 353. Ane .. watch of the inhabitants [to] be setled; 1672 *Aberd. B. Rec.* IV 279. To make and setle such rules and methods .. as shall be found necessarie; 1697 *Glasgow B. Rec.* IV 234.

b. To establish (an enterprise, institution, etc.).

The said brugh of Dundy lenis to the said brugh of Edinburgh the said Henre de Turk quhill the wark be satillit; 1601 *Conv. Burghs* II 117. Anent our resolutioun for changing of the staple port at Camphere .. we desyre expreslie that the stapill be transported from Camphear and setled at Middelburght; 1629 *Aberd. Council Lett.* I 305. And that the ky mercat being appoyntit to stand .. within the lands of Aikenbar .. that thay had appoyntit sex persounis to attend thair at the bar to settill the said ky mercat and not to suffer the kye to go to Bonyll; 1640 *Dumbarton B. Rec.* 61. That care be taken to setle ane scool vith a compitent provision and a qualified scoolmester in evrie congregatione and that for this effect the Act of Parliament concerning the provision for ane scool be put in execution; 1649 *Alyth Kirk S.* in *Sc. N. & Q.* XII 75/2. If a packquet boat were setled betueen John of Groats house .. and Kirkwall; 1711 *Conv. Burghs* IV 513.

c. To establish or confirm (health or strength).

And quhen thai wer baith a lytill recoverit and thair strenthis not ȝit fullie settillit they returnit to thair auld pastyme agane; Buch. *Detect.* (1727) 12.

d. To confirm (a person, his mind) *in* a state, belief, etc.; to make (him or her) steadfast or resolute.

(1) *Fingitque premendo, componit animum vi*, settellis; Buch. *Comm. on Virgil Æn.* VI 80.

(2) Giff he fand him selff satteled in his auld fathe, he wald continow a Jesuist; Melvill 73. Otheris lymmaris wickedlie disposit and not yitt fullie satled in obedience; 1622 *Denmylne MSS* in *Highland P.* III 314. Till shoe be some better satled in her mynd; 1643 *Moray Synod* 66.

IV. To fix (what is uncertain), to decide (a question).

10. a. To appoint or decide upon.

(*a*) [The Commissioners chose Robert Glencors] as a person who knew best to sattle post stages betuix Carlill and Portpatrik; 1642 *Reg. Privy C.* 2 Ser. VII 327.

(*b*) The wardanis setlis a day, tha meit [etc.]; Dalr. II 158/28.

b. To fix by mutual agreement.

(*a*) At our last meting I wes dismissit and na thing sattellit betuix ȝour maister & me; 1608 *Breadalbane Lett.* 4 July. Thir matteris being this wayis satlitt; *Hist. Kennedy* 14. Understanding that the most pairt of the rentis belonging to the churche of this burgh ar not satled; 1636 *Edinb. B. Rec.* VII 180. That materis may be satled in peace; 1638 Spalding I 101. To .. meit .. thes quha war appoyntit to satle the .. tyndis, and to .. give in the names of thes quha hes to gried; 1656 *Glasgow B. Rec.* II 352. This expres particular provisioun and conditioun allwayes sattiled and aggreit vpon; 1660 *Montgomery Mem.* II 324.

(*b*) Quhill we had settillit our plat; 1608 *Crim. Trials* II 257. The Lairde of Lundy .. to be heere .. airlie in the morning at whitche tyme this mater wilbe setled; 1615 *Denmylne MSS* in *Highland P.* III 246. *Ib.* 255. That thais maters may be setlit to the effect the poore ones may be intertenit; 1630 *Aberd. Council Lett.* I 321. That the mater might be hard and setled anent his cleame to the office of ischearie; 1641 *Acts* V 332/1.

c. *passive*. To be determined *to* do something, or (*in* a resolution).

If we perceive him satled sa To satisfie him with the slae; Montg. *Ch. & Slae* 1222 (Wr.). In quhiche peremptorie resolution he is so satled that [etc.]; 1590 *Cal. Sc. P.* X 279.

11. *To settle one's estate* or *affairs*, to put one's affairs in order; to arrange for the disposal of one's property, etc., before one's death. **b.** *specif.* Of the king: To put in order the affairs of the kingdom. Cf. 5b above.

(*a*) The cair your maiestie hes to sie my Lord off Murray's estait sattlit; 1602 *Wemyss Corr.* 55. Being .. of intencioun to sattill thair wardlie effeiris; 1618 *Edinb. Test.* L 125b. Before they could gayne the oppertunitie .. to satle thair estaites; 1631 *Misc. Spald. C.* II 227.

(*b*) I James Cathkin .. being cairfull to have my wordlie affaires settiled and set appairt; 1631 *Misc. Bann. C.* II 252.

b. Our souerane lord .. being now myndit to satle his estait in perfite .. concord; 1585 *Acts* III 385/2.

(*b*) Be lykeliehode he [*sc.* the king of Spain] will inclyne to quyetnes, settill his state, assuire his dominions, [etc.]; 1590 *Warrender P.* (SHS) II 137.

12. To resolve or bring to a settlement (a dispute or disagreement); to decide on (a question).

There is some overlap with sense 5 above.

(*a*) That all controverse and question of law be saitlit & takin away betuix .. thame; 1611 *Brechin Test.* II 133. Quhill the eylistis and jaris newly arrysin betuix the nytboras and indwellaris in Striuling war satlit and removit; 1611 *Sc. Hist. Rev.* XIX 109. That he sould gett all theis turnis satled: quhairin .. laweris haid nather pouar nor moyan to do it; 1615 *Black Bk. Taymouth* 48. Hoping .. ye may satle the questionne amongis thame; 1619 *Aberd. Council Lett.* I 176. And the laird haifing satlit the particulare all to the witnessis expenssis, quhilk they culd not agree vpoun; 1620 *Grant Chart.* 332. The differences latlie arysen betwixt .. counsell & the gildrie .. of .. Innernes quhairby the Commissioners of Pearth [etc.] .. wer appoynted to com to this burghe and sattle all enormities & tack away animosities; 1676 *Inverness Rec.* II 269.

(*b*) Bot, in end, a glib of the college gear settlet that [*sc.* a dispute]; Melvill 123. I dowt nocht bott ȝowr moist sacreid maiestie will swa settill thatt turne, as heirefter thay be na cawisee of gruge on ather syd; 1607 *Lett. & St. P. Jas. VI* 107. Quhill the saidis differences had bene finallie setled and composed; 1631 *Justiciary Cases* I 167. They thoght guid to setle first the contraversie now standing amongst the cordoners; 1654 *Glasgow Trades House* 314. The birlamen, .. quhair thair is any controversie either in merches setting or any vther thing that is in oddis .. shall .. setle the said bussines; 1662 *Forbes Baron Ct.* 230. Setelle; 1676 *Conv. Burghs* III 663.

13. *intr.* To arrange matters in dispute, to come to terms or agreement (*with* a person) (*for (anent)*, *to* do, *in* doing something, *that* something should be done).

(1) I sall rather be willing to sattill freindlie nor be rigour; 1609 *Breadalbane Lett.* 17 Oct. That they saitle not with him .. vnles he gif his obedience to his maiestie; 1622 *Grant Chart.* 220. [The commissioners are] to treat and satle with thame anent the best way of lifting of this excyse; 1660 *Aberd. Council Lett.* IV 48. Severall persones .. quho efter ther goods ar stowine and findis the theiff therof does satle & compone with him; 1663 *Old Ross-shire* II 53.

(*b*) Being willing to settle freindlie with the said Old Toune for the said ryot; 1663 *Aberd. B. Rec.* IV 205. To setill and aggrie with them both for thair feu dutie of the fraternitie house; 1664 *S. Leith*

Rec. 120/1. His brother .. whos just concerns I never oueind furder then to deal with himself to tak it away and did settel as I thought to his content; 1667 *Laing MSS* 373.

(2) As they haue formerlie agried & satled with Andro Ross and Robert Nicolsone to rear wp the back of the tounes hous at the water syd; 1677 *Inverness Rec.* II 272.

(3) And if they can setell with them in proportionating of the said mark upon the boll; 1675 *Kirkcaldy B. Rec.* MS 26 April.

(4) I did agrie and sattill with my collers .. that they shall have for every leade of gritte colle .. 22 penes Scotts; 1654 WEMYSS in *Sc. Diaries* 125.

b. To settle accounts by payment.

Becaus that he hes satled with all the creditors at quhaes instance he wes warded; 1672 *Inverness Rec.* II 255.

14. *tr.* To satisfy (a person) in respect of his wishes or demands.

Thay stude content thinkand that thay haid bene satlit in that point; 1593 *Acts* IV 34/2. We intend, God villing, to sattill your lordschipis baith; 1602 *Misc. Spald. C.* II 290.

b. To reconcile (opposing parties in a dispute).

(*a*) That yow use your best endeavours .. to sattle the parteis in peace and quyetnes; 1629 *Reg. Privy C.* 2 Ser. III 24.

(*b*) Johne Maxwell of Slognaw is desyred to setle pairties betuixt and this day fourtnight and gif not to report; 1679 *Kirkcudbr. Sheriff Ct. Processes* No. 303 (29 April). In obedience to your comands I have takine paines with both pairties bot can not settill them thairfor it must be decerned be the judg; *Ib.* (undated letter). For my necessar expence in going to Orialand to settle Sir Charles Hay and the tennentis; 1684-5 *Dumfr. & Galloway Soc.* 3 Ser. XXXVI 168.

c. *passive.* Of two or more persons: To be reconciled, to be brought to or have come to mutual agreement.

Till your forder order and aduertisement houe you and hie are setlit; 1653 Hossack *Kirkwall* 254.

Sattilling, Settilling, *vbl. n.* Also: **sat(t)ling(e, satteling, sat(t)elling, satlyng; saitling; set(t)ling, settelling, set(t)leing.** [Late ME *satlynge* (once, Prompt. Parv.), e.m.E. *setling* (1553), *settling* (1609); SATTIL(L *v.*]

1. The action of SATTIL(L *v.*, in senses 2, 5, 6 c, 7, 7 c, 7 d, 8, 9, 9 b, 10, 11, 12 and 13.

Some quots. from (8) may belong in (2).

(1) That thay war constrayned to retire thame selffis .. satling to the gret battell to get help of thame; bot the multitude of the Scottismen, beleiffing be thair satling that all had bene losed [etc.]; LESLIE 199.

(2) The panes and travellis tane be his majestie in satling of the same marche undir obedience; 1603 *Reg. Privy C.* VI 831. Seing thair is some cours now in handis for satling of the Clan Gregour; 1612 *Ib.* IX 308. To gif thair advise anent the satling of the disordouris of the middle shyris; 1629 *Ib.* 2 Ser. III 21. For reformatioun of sum escaipes and satlyng of the countrie; SPALDING I 100.

(*b*) And meanyng .. to remove and reforme the same [abuses and corruptioun] as well for the health as seurtye and better settilling thairof; 1590 *Cal. Sc. P.* X 317.

(3) I wold fain have pleassed myself with (I fear vain hopes) that betwext intervalls of settelling you would have com home; 1687 *Dunlop P.* III 37. Lay no stresse at all in your settelling again upon any help from Scottland, for I assure you .. you will get non; *Ib.*

(4) For electioune and sattlinge of thair commowne clark of the said bruch be voit of counsale; 1606 *Glasgow B. Rec.* I 248. It .. being .. now withholden from the ministery at that kirk, hinders thair satling, and maid all my predecessouris non residentis, .. quhilk I hope shall now be gratiouslie amendit to the perpetuall satling of a ministery at that kirk; *Rep. Parishes* 8. For setling of a minister there; 1652 *Soc. Ant.* I 161. For remeid and setling them in ther distinct profesiouns; 1674 *Dunferm. Hammermen* MS 88.

(5) To signifie .. the satisfactione I haue in my sones nou satling of himself but also my hartly consent to your allyance and his maching uith your family; 1649 *Wemyss Corr.* 94.

(6) In the satelling of him and his dearrest in the securitie of their levingis; 1572 Bann. *Trans.* 474.

(7) There is liberty grantit to Mr. George Clark to gang south for sattling of his provision and getting entrance to his benefice; 1616 Cramond *Ch. Aberdour* 15. To attend upone the effairis of the kirk levingis anent the satling of the ministeris stipendis; 1617 *Edinb. B. Rec.* VI 171. For establishing and sattling ane stipend for a scholmaister; 1658 *Inverurie* 325.

(*b*) For the perpetuall setling and mortification thairof [*sc.* a sum of money] in all tyme cuming to the proper uses whairunto the same is destinat be my said umquhill father; 1630 *Aberd. Council Lett.* I 324. For plantatione of kirks in order to the setling ane locall stipend; 1673 *Alford Rec.* 205. Mortified monyes for building of the library house, and setleing a yeirlie sellarie upon a bibliothecarius; 1691 *Misc. Bann. C.* III 236.

(8) Till the countrie be in sum better satling; 1641 *Acts* V App. 641/1. Til they restore the richt aiere nauer shal they haue paise or any satling; 1659 *Blairs P.* 102.

(*b*) That a monarch's suretie no way stood In victories in warrie broyles and blood, But in the loue of subiects trust and true, Thence, said the saige, did setling sure ensue; CRAIG I 10. And for setling of perfyite ordour quhairby all the measouris that ar now vseit may be reducit to the conformitie; 1621 *Acts* IV 632/1. The Southerland men .. drew to some setleing for a short space; GORDON *Geneal. Hist.* 184. These thrie propositions for setling religion the militia and Ireland; 1646 *Acts* VI I 576/2. [A meeting of the National Synod] for settling the Church in its intrinsics; 1674 in *Dunkeld Presb.* I 242. Of overtures anent setling this presb. into two meetings; 1696 *Ayr Presb.* 13 May.

(9) For the settling of schooles in their several parishes; 1666 J. M. Beale *Fife Schools* 157.

(10) For satteling anent the money for outriging of the commissioner; 1642 *Banff Ann.* I 87. Anent the sattilling the localitie of Sir Charles Erskynes troup; 1650 *Acts* VI II 582/1. That some [*sc.* elders] .. sould have maid relatioun to the proveist what every mans voyce was anent the setling of the seats of the kirk specially who .. were for him or against him; 1654 *Dumfries Kirk S.* 19 Oct. For settling a measure to be made for knockit beir; 1659 *Edinb. B. Rec.* IX 174. Anent the satelling and establishing of the pryce; 1702 *Culross* II 44.

(11) As for my effairis wardlie, and satling thairof; 1628 Pont *Cunningham* App. 208. For satling of my warldlie affaires; 1634 *Edinb. Test.* LVI 321. For the sattelling of your royall effairs in this kingdome; 1641 *Wemyss Corr.* 68.

(12) And for secluding and satling downe of all contraverseis quarell and debaittis; 1580 *Inverness Rec.* I 278. I .. coft his right for ease of the tenents and setling of jarres; 1642 *Orkney Rentals* III 19. Now, materis in the way of satling, his majestie desyrit .. that sum of the Scottis nobles .. sould cum and confer with him at his camp; SPALDING I 217. The committe of Estaites .. travellis betuixt thame for agrieans; bot no satling; *Ib.* II 88.

(13) For my awin chargis .. in settling with the ladie, & resaving of the rents from the tenents; 1646 *Caldwell P.* 110. At the setleing with the coilȝiars; 1673 *Dunferm. Coal Acc.* 20b. Upon settleing with Smith in the teerms thatt the rest of the silk stocken men are upon; 1682 *New Mills Manuf.* 33. In a constant wse and possessione of setling with the rector .. in the .. suppleing of vacant places within the colledge; 1682 *Scot. Ant.* XVI 3.

2. *concr.* **a.** The sediment or lees ? chiefly or only of wine. Also *fig.* and *attrib.*

Ane medecyne for the maw .. With sueit satlingis and sowrokis; HENR. *Practysis* 29. Nans Smyth .. being greatlie rebuikit for going with ane tub and ane barrow and ane wther with hir put satlingis thairin; 1620 *Perth Kirk S.* MS 5 Nov. For draff and satlings to my wame, six placks I am addebted; *Mare of Colinton* 511.

fig. My heart .. Whose dregges hath done the whole world delude, Which is but draffe and satlinges full of sinne; NORVELL *Meroure* 31a.

attrib. Saitling [*H.* Erse] slaiker, glaid glaiker, rum raiker for releif; POLWART *Flyt.* 785 (T).

b. A settlement, an appointment to an office or a ministerial charge, a post of this kind. Cf. 1 (4).

Doctour Johne Gordoun, the deposit minister at Elgyne, haueing gottin sum satling in Ingland, returnit to Elgyne, quhair he sauld his plenishing and bookis [etc.]; SPALDING II 175.

c. A payment (made in settlement of a debt.) Cf. 1 (9) and (12).

vij li., quhairof scho ressavit certane setlingis extending to [£4 7 s. 8 d.]; *c*1596 *Glasgow Prot.* V Pref. xiv.

(Sattillit,) Sattel(l)it, Settillit, *ppl. adj.* Also: **sat(t)lit, set(t)lit; satl(e)d, -id, -at; satteled; sattilt; settlid, set(t)led, setlaid.** [e.m.E. *setled* (1556), *settled* (1568); SATTIL(L *v.*]

1. Fixed permanently in place or position.

That all .. friemen of the said craft .. resident out with this burgh repair to the samen and thair mak thair satled dwellingis and abodes; 1483 *E. Loth. Antiq. Soc.* II 105. Donald Dow was minister .. and .. lived in Finask, there being no setled mans at Kirkhill; FRASER *Polichron.* 184.

2. Of a mental state, etc.: Decided; fixed, determined; firmly adhered to. Also *transf.*, of a person, const. *in* (a state of belief).

(*a*) With grave and sattellit judgement; BISSET I 111/2. A .. man .. Whose judgment sound, sure satl'd, sharpe, & sage; GARDEN *Worthies* 77. Quhen I haif callit my mynd to ane satlit posture; 1631 *Red Bk. Menteith* II 129.

(*b*) It uas my settlid mynde That [etc.]; JAMES VI *Poems* II 149/37. With ane settillit purpois .. to bareiff him of his lyffe; 1624 *Crim. Trials* III 572.

transf. Whow could he .. think that anie sattlit man in the treuthe could .. be moved to alter his judgment? MELVILL 461.

3. Of an arrangement, state of affairs, course of action, etc.: Fixed, established or regulated; characterized by stability or permanence, in senses 5, 9 and 10 of SATTIL(L *v.*

(*a*) Sic satlid ordour as the saidis craveris of reformation may lesumlye vse; 1590 *Conv. Burghs* I 335. Satlat; 1599 Murray Lyon *Hist. Lodge Edinb.* 14. Quhill the said cuntrie .. may be brocht to ane satlit and peceable estait; 1600 *Reg. Privy C.* VI 154. [That they might be compelled to betake themselves to some] lauchfull and sattelit trade; 1605 *Ib.* VII 713. That thair shall be ane satlit forme of the best and maist commoun and approvin grammer .. collectit establischit and prentit; 1607 *Acts* IV 374/1. Bot .. many bludy battailes wilbe fochten first or it tak an satteled effect; MELVILLE *Mem.* 203. [They] sould .. haif .. haldin these landis vnder a perfyte and satled obedience; 1615 *Denmylne MSS* in *Highland P.* III 257. I .. willing to put my warldlie effairis & bissines to sum satlit ordour; 1625 *Edinb. Test.* LIII 173b. That sum satlit course sould be taiken for instructing the people in wearfair; 1626 *Edinb. B. Rec.* VI 301. That no servinge person be harboured within the paroche that tack not themself to satled service with ane master; 1636 *Boharm Kirk S.* 24 April.

(*b*) Gif it war found .. that the saidis .. counsall had transgressit the said auld constitutiounis and setlaid forme of gouernement; 1590 *Conv. Burghs* I 333. Sall the Kirk of Jesus Chryst be les regardit .. in a setlit and constitut esteat; 1601 MELVILL 493. All policie and cair of the liberteis of this burgh hes bene .. neglectit, .. for the want of ane soleid and setlit ordour amongis thame selfis; 1605 *Glasgow Chart.* I 1 dcviii. A most Chrystiane king, liveing our kirk and realme in a settlit religioune and policie; 1606 MELVILL 628. The laird of Lundy .. hes vndirtane .. to bring the same to some setled perfectioun; 1612 *Denmylne MSS* in *Highland P.* III 123. If it pleas God to send a setled peace; 1629 *Aberd. Council Lett.* I 305. 1645 *Old Ross-shire* II 13.

b. Of an institution: Established on a permanent footing.

Seing sic contrair opinionis amonges the clergie within a reformed satled kirk; SPALDING I 97.

4. Of a Presbyterian minister: Established in an office or charge.

(*a*) Maister Gudman .. wes the first satlit minister in Sanctandrous; *Wode's Psalter* (ed.) 165. Ane supplication .. anent the provyding of that kirk with ane sattilt minister; 1613 Cramond *Ch. Aberdour* 11.

(*b*) Through the want of a settled minister the people have been deprived of preaching; 1689 *Cramond Kirk S.* III 6 Dec.

5. Of an office, income, institution, etc.: Guaranteed to the holder; settled, established in senses 8 a and 9 b of SATTIL(L *v.*

Becaus ther is not a satled provisioune for his maintenance, the minister and elders ar ordeined to stent the paroch; 1634 Cramond *Ch. Rathven* 13. Mr. Willm. Lighton had a schole at Dun, but not a satled provision; 1650 *Brechin Presb.* 30. The moderator did ask .. if they had a setled schooll, they ansred they ever had one till now, which the school master hes deserted becaus he had no settled benefice; 1665 *Lanark Presb.* 105.

Sattillit, p.p. of SATTIL(L *v.*

(Sattillitly,) Sattletlie, *adv.* Also: **sattledlie; setledly.** [e.m.E. *setledly* (1602); SATTILLIT *ppl. adj.*] In a rational or reasonable manner. — This re-establishing of bishops, I skarse understand wherunto it sould be referred, whether to their estat in papistrie, or the idle life of tulcans that they have occupied since the reformatioun of religioun; or neither, but some thrid and mixed conditioun patched up of both, not yitt sattledlie considered; 1585 Calderwood IV 330. Quhilk maid men sattletlie, gravlie, and quietlie to bring out thair reasones; MELVILL 413. At the entree I could not setledly dyte foor lynes; 1638 JOHNSTON *Diary* I 308.

(Sattilment,) Satlement, Setl(e)ment, *n.* Also: **setll-, setelment.** [e.m.E. *settlement* (1626), *setlement* (1651); SATTIL(L *v.*] The action of SATTIL(L *v.*, in various senses.

1. Establishing, instituting, setting up in sense 9 b of SATTIL(L *v.*

The most fitt place for setelment of the said staple port is Campheir; 1674 *Edinb. B. Rec.* X 174. If it be fitting to doe the affairs that are intrinsecall to the satlement of a church by halves; 1674 *Laing MSS* I 398.

2. Installing, investing (a person *in* an office) in sense 7 of SATTIL(L *v.*

Non .. compierd to object any thing against his setlment in the said office of the ministrie; 1687 *Dunkeld Presb.* I 480.

3. Fixing, arranging, putting in order in senses 5, 8 a and 9 of SATTIL(L *v.*

(*a*) As for any affairs that may trouble yow .. I intend to come in my selfe in order to the satlement therof; 1693 *Lett.* (Annandale Papers, R. C. Reid transcript) 31 Aug.

(*b*) I know this tyme bypast .. ther wes alarums and extraordinarie consultations to keep you bussied which I would not interupt with such inconsiderable things as I had to say until the great setlement of this new aggreement for peac gives me libertie, in the first place, to render yow exceeding thankes for your kindness to my Lord Drummond and his brother; 1667 *Laing MSS* 362. Havinge now obliged myself in the setelment of my sone Francis patremony more liberally then I was weel able [etc.]; 1698 *Old Ross-shire* II 25.

b. An arrangement, agreement (*with* a person, *for* doing something) in sense 13 of SATTIL(L *v.*

To .. mak ane setllment with the stoking wevers for working the pair of the new fashoned stript hoes; 1683 *New Mills Manuf.* 58.

Sattilt, var. of SATTILLIT *ppl. adj.* **Sattin(e,** varr. of SETTING(E *n.* **Sattin(g,** varr. of SATIN(E *n.* **Sattirday,** var. of SET(T)ERDAY *n.*

Sattire, var. of SATYR *n.* **Sattisfaccioune, -factione,** varr. of SATISFACTIO(U)N(E *n.* **Sattisfie,** var. of SATISFY *v.* **Sattle,** var. of SATTIL(L *v.* **Sattledlie, -letlie,** varr. of SATTILLITLY *adv.* **Sattling(e,** varr. of SATTILLING *vbl. n.* **Sattlit,** var. of SATTILLIT *ppl. adj.* and p.p. of SATTIL(L *v.* **Satturday,** var. of SET(T)ERDAY *n.*

Sattwall, Setwal(l, *n.* Also: **satwell; settuall, set(t)well; (seteuale).** [ME and e.m.E. *ge-, ȝedewal, zeduale* (Ancr. R.), *sed(e)wale* (*a*1310), *settewale* (Chaucer), *setwale* (*c*1440), *setwall* (1530), *setwal* (1610), AF *zedewale* (OED), OF *citoal* (1190 in Greimas), med. L. *cytowalla* (*c*1242 in Latham), ultim. f. the Arabic.] Setwall, the root or plant.

1. The root of the East Indian plant *Curcuma Zedoaria*, used as a drug.

(*a*) viij vnce of satwell price of the vnce v s.; 1537 *Acta Conc. & Sess.* MS IX 157. Foure vnce of sattwall price x s.; 1582 *Edinb. Test.* X 302.

(*b*) Of karkys or balys of .. gynger, setwell [med. L. *seteuale*], almondis [etc.]; *Acts* I 305/2. A li. settwell; 1496 HALYB. 57. Thrie vnce of setwall price viij s.; 1575 *Edinb. Test.* III 387b.

2. The plant valerian.

Valeriana, settuall; *Herbarius Latinus Annot.* cxlix (Bot.). [Phu,] Anglice, Setwal or capons taile; *Dioscoridis Annot.* 8.

Satty, see SITTY-SATTY. **Sattyl,** var. of SATTIL(L *v.* **Sattyn,** var. of SATIN(E *n.* **Sattyre,** var. of SATYR *n.* **Saturday(e,** varr. of SET(T)ERDAY *n.*

Saturitie, *n.* (e.m.E. *saturyte* (1533), *saturitie* (1548), *saturity* (1643), L. *saturitāt-*, fullness, satiety.) — It [*sc.* avarice] sowis soir without saturitie; STEWART 13862.

Saturnday(e, varr. of SET(T)ERDAY *n.*

Saturn(e, Saturnus, *n.* [ME *Saturnus* (Trevisa) late ME and e.m.E. *Saturne* (*c*1400), *Satourn* (Lydgate), *Saturn* (1651), OE *Saturnus*, L. *Sāturnus* the god Saturn.]

1. The god Saturn in mythology.

(*a*) Saturne .. With lippis bla and cheikis leine and thin; HENR. *Test. Cress.* 151. *Ib.* 172, etc. Saturn and Jubiter his son goddis of the aire; *Asl. MS* I 324/17. Thare saw I crabbit Saturn ald and haire; DUNB. *G. Targe* 114. And the pepill Sycany hecht alsswa By quham the land of Saturn war and wys Hes left [etc.]; DOUG. VIII vi 39. Than Jupiter prayit to god Saturn In liknes of ane tod he wald him turn; *King Berdok* 39.

(*b*) A lorde callit Saturnus .. the quhilk Saturne techit the folk of that contree to mak housis and labour cornis [etc.]; HAY I 38/11. That tyme quhen levit the king Saturnus; *Want of Wyse Men* 9 (Ch. & M.). Sum held he had bene Dedalus .. And sum Saturnus kuke; DUNB. (OUP) 163/68. *Ib.* 111/75.

2. The planet Saturn.

(*a*) Than Saturne auld wes enterit in Capricorne; HENR. *Fab.* 635. Throw the speir [= sphere] of Saturn ald, Quhilk fader is of all thir sternis cald; Id. *Orph.* 189 (Ch. & M.). Ane sterne callit Saturn quhilk is hie abufe al the laif of the planetis and for that cause it aperis verray litil to mennis sycht. It makkis reuolutione in thretty ȝeir [etc.]; *Compl.* 53/11. Quhen Saturne is warme and Venus wicht; *Bann. MS* 266b/5.

(*b*) & the mone .. In heldyne was of Martis house & in the bowndis of Saturnus; *Leg. S.* xxi 398. The vij planet is Saturnus the planet of hunger deid & seiknes; *Porteous Noblenes* 170/22 (Asl.). *Bk. Chess* 1, 4. Undir Saturnus fyrie regioun; DUNB. (OUP) 160/31.

Saturneday(e, Saturne his day, Saturnsday(e, varr. of SET(T)ERDAY *n.* **Saturudaye,** see SET(T)ERDAY *n.* **Satwell,** var. of SATTWALL *n.* **Satyfé, -fy,** varr. of SATISFY *v.* **Satyfeyne,** var. of SATISFYING *vbl. n.* **Satyn(e,** varr. of SATIN(E *n.*

Satyr, Sattyre, -ire, Sattier, *n.* [ME (*pl.*) *satury* (Chaucer), *satyri* (Trevisa), *satiri* (Gower), e.m.E. *satyr(e* (Caxton), *satire* (1569), OF *satire* (1372 in Larousse), L. *satyrus*, pl. *-i*.] A satyr, the mythical creature or a representation of it. — At libertie alas shall fauns and satyrs lope; CRAIG I 21. Bastien deuysed a nomber of men formed lyk sattyres, with lang tailes and whippis in ther handis, runnyng befoir the meit; MELVILLE *Mem.* 171. The sattiers .. pat ther handis behind them to ther tailes, quhilkis they waggit with ther handis; *Ib.* Sa schone as they saw the sattires waging ther tailes or romples; *Ib.* 172. Those satyrs and doleful unknown creaturs of islands and deserts; KIRK *Secr. Commonw.* (1964) 364.

Satyr, var. of SATIR *n.* **Satyricke, -ik,** varr. of SATIRIC *adj.* and *n.* **Satysfie, -y,** varr. of SATISFY *v.*

Sau, var. of SAW *n.*[2], *v.*[1], *v.*[3] and p.t. of SE *v.*[1] **Sau-,** var. of SAW *n.*[3] **Sauce,** var. of SALS *n.* **Sauch,** var. of SALS *n.*, SAUCHT *adj.* and p.t. of SE *v.*[1]

Sauch(e, Saugh(e, *n.* Also: **salch, salk; sawch; saucht, sawtht; sach;** pl. also **saquhis.** [North. ME and e.m.E. *salfe* (1368), *salghe* (Cath. Angl.), *saugh* (1641, also, once, e.m.E. 1548 in TURNER *Names Herbes*), also in the later Sc. and north. Eng. dials., OE (Angl.) *salh*.]

1. Sallow or willow trees or bushes, collectively; also, *specif.* as timber.

(*a*) Salch et alryn; 1398-1400 *Aberd. B. Rec.* (SHS) 84.

(*b*) In .. plantatioun of treys, that is to say, eysses, osaris, and sauch; 1472 *Reg. Cupar A.* I 163. Sawch; 1473 *Ib.* 188. *Siler*, sauch; CARMICHAEL *Etym.* 4. Lie hawthorne, sauche, aller et esp; 1587 *Reg. Great S.* 455/2. John McKnab verefeit he fensit the hegis and tymbir sauche and esche; 1607 *Kirkcudbr. B. Rec.* II 20. He raif up sum sauch and haisseill; 1621 *Black Bk. Taymouth* 372.

(*c*) Nor dikyne of na maner of wod .. except it be of thorne, saucht, or erde; 1499 *Exch. R.* XI 395. Peces of tymmer intromettit wyth be him .. viz. saucht, asche and holyne; 1579 *Inverness Rec.* I 273. Johne Neilson .. hes .. cuttit .. tuaye horse laidis of the greyne young saplyne, saucht, birk and of the vther young growtht; 1586 *Inverness Rec.* I 307. Sawtht; *Herbarius Latinus Annot.* cxxxvi (Adv.).

(*d*) Bushes of elder, saugh and birks; 1633 *Sibbald's Orkn. & Shetl.* 5. The .. species of tries are declared to be as follows esh, elm, allar and saugh; 1694 *Stitchill Baron Ct.* 114.

(*e*) Send me as many willow stakes for planting as you can and a deall of privet sach; and .. grafts of pearlie [etc.]; *c*1687 *Cromarty Corr.* I 56. [The said William is to cut no growing timber .. excepting] thorns sach and hessell; 1696 *Kirkcudbr. Sheriff Ct. Deeds* II 752.

b. With defining term: (A particular variety of) willow.

I must desire of you some sets of those saughs I like best of yours, the black saugh, willow, osier, and laurell saugh [etc.]; 1702 *Hist. Carnegies* 377.

c. *pl.* Sallow or willow trees; some instances may refer to different varieties of the tree.

To George Cambel, gardiner of Strivelin, to by sauchis and to set thaim, iiij Franch crounis; 1501 *Treas. Acc.* II 83. In .. plantatioun of treis, eschis, osaris, and saquhis; 1540 *Reg. Cupar A.* II 7. Schaw [= sow] .. in the feldis folloving .. The thortir feld, all aittis .. The neder salkis, aittis; 1552 *Elgin Rec.* I 116. That evirie of the saidis partye plant sauchis and allar treis endlang the sydis of the said .. fowsie; 1586 *Prot. Bk. J. Merser* 4b. [To pay to William Carnecroce of Hilslope 20 merks as the agreed price of] the bark of certaine saughes; 1665 *Melrose Reg. Rec.* II 135. To the Lord Raiths gardiners at setting sauchs in the meadows; 1692 *Foulis Acc. Bk.* 146. 1695 *Ib.* 176. 1702 *Hist. Carnegies* 377 (see b above).

2. A rope made of twisted sallow withies.

Filling of tauch, rak sauch, cry crauch, thow art our sett; DUNB. *Flyt.* 245. For this foule deid ȝour seid man rak ane sauch; 1570 SEMPILL *Sat. P.* xii 56. For graithing the mekill thorne .. sawchis to be bandis to the samyne and wikaris to bynd the samin; 1595 *Edinb. D. Guild Acc.* 565.

3. *Attrib.* and *comb.* with *bus* and *tre, timmer, yaird.*

(1) To ane gray stane beneth ane sauch bus; 1556 *Peebles B. Rec.* I 230. — The wrangus taking fra him .. of ane sauch tre; 1511 *Alloway Baron Ct.* 6 May. Thar targettis bow thai of the lycht sauch tre; DOUG. VII x 73. Sawch; *Ib.* VIII i 54. On the sauch treis our

harpis we hang; *G. Ball.* 114. To pay xviij[d] to the said James for ane sauch trie borrowit ane yeir syne; 1569-70 *Canongate Ct. Bk.* 167. *Salix*, a sawch tree; DUNCAN *App. Etym.* With foure hundreth grit growand sauch treis; 1607 *Crim. Trials* II 522. *Silur*, a sauch tree; *Despauter* (1617) 8. An old ruinous castle with planting of sauch or willow trees for the most part about it; SYMSON *Descr. Galloway* 81.

(2) [To cut no timber] except ramell saugh timber; 1673 *Kirkcudbr. Sheriff Ct. Deeds* I 416.

(3) [Part of the hemp yard next to the] saugh yaird with als maney of the saughes [as will be needful for his use]; 1671 *Kirkcudbr. Sheriff Ct. Deeds* I 344. The sach yeards and the wat rigs that lyes west of the mansione hous; 1673 *Ib.* 454.

Sauchen, -in, *adj.* Also: **saiching.** [SAUCH(E *n.*] **a.** Of (the) sallow or willow; *specif.* in describing the tree. **b.** Made of sallow or willow. **c.** In a place-name. **d.** *fig.* Like the willow, flexible and tough rather than rigid. — **a.** Southeist doun be the dyke of Gartchan to ane sauchin bus; 1570 *Strathblane Par.* 322. For cutting of ane aischin trie and ane saiching tree; 1592 *Breadalbane Ct. Bk.* 38b. — **b.** Seven dozen of nests of sauchen boxes; 1681 *Blackness Customs* 18b. — **c.** Sauchinford; 1688 Wodrow *Hist.* (1721) II App. 201. — **d.** *fig.* A sauchin maister will overcom ane aikin servand; CARMICHAEL *Prov.* No. 194.

Sauchit, p.p. of SAUCHT *v.* **Sauchnyn, -nyng,** varr. of SAUCHTYNG *vbl. n.*

Saucht, *n.* [ME *sæhte* (1154), *saht(e* (Layamon), *seihte* (Ancr. R.), *sauht* (*c*1275), *saght* (Cursor M.). Also in the later Sc. and north. Eng. dial., late OE *seht, seaht* masc., ON **sæht*, **saht* fem. (OIcel. *sætt, sátt*).] Freedom from strife, vexation or trouble; peace. Chiefly in the phrase, *to sit in saucht.* — For he may nevir sit in saucht Onles he be hir sklaif; *Bann. MS* 160b/43. Mak swir the dwirs giff ȝe wald sit in saucht; SEMPILL *Warning* iii. The feind wald faine man be ȝour wyfe, Can neuer sit in saucht; *Philotus* 1144. — *proverb.* Better saucht with little aucht, nor care with many kow; FERG. *Prov.* No. 164.

Saucht, *adj.* Also: **sawcht, sauch.** [ME *saht* (*c*1175), *sahhte* (Orm), *sauhte* (*a*1272), *saght* (Cursor M.), *sauȝt* (*a*1320), *sawht* (Gower), OE *seht, sæht,* ON **sahtr* (OIcel. *sáttr*).]

1. In agreement, reconciled.

Qwhen the king ves thus with him [*sc.* Randolph] saucht [*E.* sauch; *rh.* betaucht]; BARB. X 300 (C). And how thaire modyre made thaim saucht; WYNT. IV 1178. Thir kyngis twa togyddyr met, .. Thare thai departyd nowcht all sawcht [: awcht]; *Ib.* VII 1370. *Ib.* V 510. Now lat ws change scheildis, sen we beyn sawcht [: taucht]; DOUG. II vii 48.

2. At peace, free from molestation, in security, not engaged in aggression or molestation.

And all his liegis louit him richt weill, .. In all his tyme salfe thai war and saucht [: aucht]; STEWART 2005. We trowit from thence thay suld haif sittin saucht [: faucht], And suld haif tyrit of all thair tyrannie; 1570 *Sat. P.* xiii 179.

(Saucht,) *v.* P.t. **saucht, sacht.** P.p. **sauchit.** [ME *seyte* (1297), *saȝt* (Cursor M.), *sauȝte* (*c*1320), *saughte* (*c*1400), p.t. (pl.) *(ȝe)sehtodan* (1154), p.p. *(i)sauȝt* (*c*1250), OE *sehtan.*] *tr.* and *intr.* To reconcile; to become reconciled. — *intr.* Thai forthocht that thai faucht, Kissit samyn and saucht [*B.* sacht]; *Howlat* 844 (A). — *tr.* Johne Meik is deprywit of his fredome in respect of his daylie toilȝeing and can nocht be sauchit; 1578 *Kirkcudbr. B. Rec.* I 80.

Saucht, var. of SAUCH(E *n.* **Sauchté,** var. of SAFETY *n.*

Sauchtine, *v.* [ME *sahtnien* (*c*1175), *sæhtnien* (Layamon), *sauȝte(n* (*c*1320), *saughte(n* (Piers Plowman); SAUCHT *adj.* Cf. SAUCHT *v.*] *tr.* To reconcile. — Dere lady, I thé pray, that .. thu succure me & sauchtine me & thi sowne; *Leg. S.* xviii 1485.

Sauchtyng, Sauch(t)nyn(g, *vbl. n.* Also: **sawch(t)-, saught-** and **-ing, -enynge.** [ME *saghting, sahutinyng* (both in Cursor M.), *sauȝtening* (*c*1320), *sauȝtynge* (*a*1400); SAUCHT *v.*, SAUCHTINE *v.*] Reconciliation; pacification; agreement.

(*a*) Wald thu .. ask forgiffnes of thi syne, Yheit mycht thu sauchtyng with hym wine; *Leg. S.* l 921.

(*b*) & that mane suld eftyre syne Recouire his blyse & sauchnyn wyne; *Leg. S.* xvi 248. & with ȝoure God wil sauchtnyng mak; *Ib.* xxi 742. Sauchtnyn; *Ib.* xviii 4. Qwhyll at the last the hale barne Off Brettayne knyt thame in sawchtnyng [*C.* sauchtenynge, *W.* sauchtnyng]; WYNT. V 3177. Quham euir he hyt, thair sawchnyng was wnsoft; *Wall.* X 332. Nane syk trety of sawchnyng [*Sm.* sauchnyng, *Ruddim.* sauchning] nor cunnandis; DOUG. X xiv 176.

(*c*) It semys saughtnyng thai seik, I se be thair feir; *Gol. & Gaw.* 1264.

Sauciat, *adj.* [e.m.E. (once) *sawciate* p.p. (1509), L. *sauciātus*, p.p. of *sauciāre* to wound, hurt.] Wounded, hurt. — Godlie youth, Thy sauciat soule hurt with a holy love; GARDEN *Worthies* 88.

Saucie, var. of SALSY *adj.*

(Saucis,) Sasis, Sauses, *n. pl.* [OF *saucisse* (*c*1268 in Larousse). Cf. e.m.E. *sawsyge* (15th c.), *sausage* (1553), *sawsege* (1573), *sassage* (1641), etc., ? ONF *saussiche.*] Sausages. — xij rodekeins, xij quhit sasis; 1598 *Stirling Ant.* III 310. We have eated puddings heir also that we call sauses which they make most usualy of suine; 1665-7 LAUDER *Jrnl.* 77.

Saud, var. of SOUD *n.*

Saudall, *n.* (L. *sodālis* a mate, fellow, companion.) — Euin as the bird into the breir, Dois cry vpon his saudall deir; BUREL *Pilgr.* II 396.

Saudre(e, see SANDRE(E *n.*[2] **Saueing,** var. of SAWIN(G *vbl. n.*[2] **Sauell,** var. of SAUL(L *n.* **Sauen,** p.p. of SAW *v.*[1] **Sauer,** var. of SAVO(U)R *n.* and *v.* and SAWAR *n.*[1] **Saues,** pl. of SAW *n.*[1] **Sauf(e,** varr. of SAW *n.*[1] and *v.*[2] **Saufer, -ir,** varr. of SAPHIR(E *n.* and *adj.* **Sauf(f)-,** varr. of SAFE-, SAV(E)-. **Sauf(f)(e,** varr. of SAFE, SAV(E. **Sauf(f)roun, Saufferoun,** varr. of SAFFRO(U)N(E *n.*

Saug, see SANG *n.* **Sauge,** var. of SAGE *adj.* **Saugh,** p.t. of SE *v.*[1] **Saugh(e,** varr. of SAUCH(E *n.* **Saught,** p.t. of SEK(E *v.* **Saughtnyng,** var. of SAUCHTYNG *vbl. n.*

Sauill, var. of SAUL(L *n.* **Sauill-haill,** var. of SAUL(L)-HELE *n.* **Sauin(e,** p.p. of SAW *v.*[1] **Sauin(g,** varr. of SAWING *vbl. n.*[1] and SAWIN(G *vbl. n.*[2] **Sauir,** var. of SAVO(U)R *n.* **Sauister,** var. of SAWSTAR *n.*

Saukleslie, var. of SAKELESLIE *adv.*

Saul, var. of SAIL(L *n.*[1] **Saulce,** var. of SALS *n.*

Sauld, *ppl. adj.* [e.m.E. *sold* (1535), p.p. of SELL *v.*] That has or have been sold. — Thair is to be deducit the rest restand upoun the comptar at the fute of the sauld victuellis; 1590-1 *Exch. R.* XXII 162.

Sauld, var. of SALT *adj.* **Sauld(e,** p.t. and p.p. of SEL(L *v.* **Saul(e,** varr. of SAIL(L *n.*[1] and SAUL(L *n.* **Saul(e)-,** varr. of SAUL(L)-. **Saules-mes,** var. of SAUL(L)-MES *n.* **Saulf(e, Saulff,** varr. of SAFE *adj.*, *prep. and conj.* and SAV(E *v.* **Saulf(e)-, Saulff-,** varr. of SAFE- and SAV(E)-. **Saulie,** var. of SAUL(L)IE *n.*

Saul(l, Soul(l, *n.* Also: **saul(l)e, sawl(l, sawle; sauell, -ill, sawill, -ell; sal(l; soulc, sowll, sowlc, sowcl(l; sollc; soylc.** [North. and north-midl. ME and e.m.E. *sawle* (*a*1175), *saule* (*c*1175), *sawel, sawll, saul* (all Cursor M.), midl. and south. ME and e.m.E. *sowle* (*c*1175), *soule* (13th c.), OE *sáwol, sáwel, sáwl.*]

1. The spiritual part of a person, seen as the source of knowledge, intellect, perception, spirituality and the like; a person's spirit seen as distinct from the body, freq, in collocation with *body, flesh* and *banis,* etc.

(1) We .. committis to thé Saull and lyf ladye; *Howlat* 737 (A). Mak for thé intercessouris the saull of Jhesu, the blissit moder of God [etc.]; IRLAND *Asl. MS* 47/8. The first peticione The saul sais o heuinlie fader [etc.] .. The secund peticione The saul ansuaris [etc.]; GAU 97/19, 98/7. There be nane of theme sa rude bot he beleuis hyme selff to haue ane immortall saul; Q. KENNEDY *Breif Tract.* (ed.) 130/30. And how to liwe .. when as my wofull hart Is

soundred from his spreit and soule, her lyiflie vitall part; FOWLER I 56/240. They that are in this way misseled up in their saull, of all men in the earth they are maist miserable; for als long as thou may keep in thy mind a spunk of this knowledge and spirituall light [etc.]; R. BRUCE *Serm.* 107. My sensis and my saull I saw Debait a deadly strife; HUME 12/21. My saule violented and urged God by this argumenting prayer; 1634 JOHNSTON *Diary* I 200.

proverb. Thai .. Skantlie durst say thair saull wes thair awin; STEWART 23395. *Ib.* 48694. I dar not say my saule is myne awin; CARMICHAEL *Prov.* No. 998.

(2) Hou scho demanyt hir flesche, Til saule & body to-gydir ves; *Leg. S.* Prol. 82. The saule of the man has twa habitudis and fallouschippis with the body that it is vynt and iunyt with; IRLAND *Mir.* III 25/35. To the Feind saule flesch and banis Thay gaif thame; DUNB. (OUP) 169/77. He hais gevin us our body and our saule; HAMILTON *Cat.* 142. Be this men suld leif all thare kyn And with thare wyffis mak dwellyn .. For God hes ordanit thame trewlye To be two saulis in one bodye; LYND. *Mon.* 784. Q. KENNEDY *Breif Tract.* (ed.) 119/10. Neuir sal the body be turnit in the saul, nor the saul in the body; WINȜET II 41/17. Baith saule and body to defend; *G. Ball.* 10, etc. How far the .. saull .. dois surpas The mortall .. corps, a lowrd and brukill mas; HUME 19/85. The bodie .. is ane lead layit on the back of the saull; ROLLOCK *Wks.* I 312. A spait of mischiefs, which .. are readie .. to overflow not onlie our bodies and estates, bot our verie soules; 1640 BAILLIE I 264.

b. *specif.* In regard to its health, well-being, comfort, nourishment or its retention or loss or in regard to its relationship with God. Also in collocation with *body*.

(1) His testament than has he maid .. For hele of his saule; BARB. XX 167. That his travale Micht eftir till his saull [*E.* saule hele] avale; *Ib.* 346 (C). Als he slew Petir and Paule, Till eke the tynsale of his sawle; *Leg. S.* ii 828. & vthir women mony ma Thar husbandis left for heile of saule; *Ib.* 1089. Sancte Paule That tacht men to kepe the saule; *Ib.* xxxi 96. God .. Set his saull in solace; *Howlat* 466 (A). *Ib.* 985 (B). For the vertues of the body ar all subjectis to the vertu of the saule; HAY I 117/33. Science .. is riches that euer sall indure .. To mannis saull it is eternall meit; HENR. *Fab.* 140. Than sall ressoun ryse, rax and ring And for thy saull thair is na better thing; *Ib.* 1124. The saule is Godis dochtir deir, And eik his handewerk; Id. *Bludy Serk* 113. Na vthire bodely creature may resaue .. the hie diuinite as dois the saule of mane; IRLAND *Mir.* I 59/3. The .. sevyne sacramentis .. are callit the .. medicyne of the saull; Id. *Asl. MS* 8/15. So suld his saull with gret intelligens Of knychtlie werteu be enarmed richt; *Bk. Chess* 855. To .. clenge my saull fra lipper syne inwert; KENNEDY *Asl. MS* II 275/68. To gar men noit how weill he the saule lude; Id. *Pass. Christ* 287. Thair wyl nolder monk freir nor preist answeir for my saule bot my self; Q. KENNEDY *Tractive* 125. Id. *Breif Tract.* (ed.) 136/22. Seuin deidlie sinnis Quhilk that the sillie saull persewis; ROLLAND *Seven S.* 1443. Ȝe may sing merilie, For all our saulis helth; *G. Ball.* 156. *Ib.* 150, 193. Saulis to saif frome suffering sorowis seir; SCOTT i 70. R. BRUCE *Serm.* 111. Faith .. is a sure .. aprehension of the promises of God aplying thame to youre saull & thairfore maie it iustlie be callid the goldin chaine that linkis the faithfull saull to Chryste; JAMES VI *Basil. Doron* 37/9. MONTG. *Devot. P.* iii 46. I uill go uerray neir the hazart of my saull, befor [etc.]; 1610 *Lett. & St. P. Jas. VI* 183.

(*b*) Most gladly do they covet the foode of their soules; 1556 KNOX IV 135. 1565 *Ib.* VI 253.

(*c*) To think his grace can his own soyle forget And to storme-foorth his princely wrath on babies; 1639 *Fugitive Poetry* II xvii 4/122.

(2) A hait fyry power .. Beyn in thar sedis quhilk we saulys call, Sa far as that thir noysum bodeis cald Nocht tareis thame tharfra; DOUG. VI xii 18. Siknessis ather of the saule or the boddie; JAMES VI *Basil. Doron* 131/2.

c. *specif.* With regard to the survival after death, entry into heaven, etc. of a person's spirit. Also in collocation with *body, banis, cors*.

The *c*1614 Melvill lxiv quot. in 1 (*c*) below may be somewhat garbled and/or may belong elsewhere.

(1) The Magdelane .. his saule bare vpe in hewine; *Leg. S.* xvii 341. For to find a preste chanoun or secular for the saul of the Excellent Prince Robert; 1402 *Scone Indenture* (Reg. H.). A messe for the said Johne Steuartis saule hys elderis saulys and all Crystyn saulys; 1454 *Glasgow Chart.* I II 44. For suppos all the saulis that euer was or salbe war in hell [etc.]; HAY I 28/3. The balyheis .. has grantit thwa mark that langit Sant Jamys autar to Sir Jon Heltson to swpple our Lady seruis and the said Ser Jon sal mak serwis for the sal that left tha thwa mark to be don at that autar; 1457 *Peebles B. Rec.* I 121. God resawe My petows spreit and sawle; *Wall.* II 175. The qwhilk Marione Thomsone lewit to the morne mes for hir sall; 1489 *Dunferm. B. Rec.* 17. Sa deliuer thow the sely saulis that ar in the panis of purgatory; *Rois Garlandis* 229. *Remembr. Passion* 557, etc. I leiff my saull for evirmare .. In to my lordis wyne cellar; DUNB. (OUP) 115/18. Heir sal he be heisit vp .. and an craw or ane ke salbe castin vp, as it war his saull; LYND. *Sat.* 4242 heading (Ch.). The saulis departit may be helpit be celebratioun of the mes; HAMILTON *Cat.* 282. The immortalitie of the saulis .. and thir wordis of the saulis crying vnder the altar; 1560 *Conf. Faith* in *Acts* II 531/1. FOWLER I 23, etc. I delyuer .. my saull to God; 1590 *Edinb. Test.* XXI 361. Our Kirk sall be infected with the error of the locall descent of Christis saull to hell; 1631 *Misc. Bann. C.* I 241.

(*b*) [To] big ane kirk, and feft preistis to pray For all thair soullis; STEWART 38978. Fye upoun that knave Death that will come whitther we will or not .. and the seally sowll, I fear, shalbe so feable, that it can neather cary with it gold [etc.]; KNOX II 389. *Elysium,* a pleasant fielde quhair gude mens soules dwells after their death; DUNCAN *App. Etym.*

(*c*) Efter the said schir Jhonis deces to ramane with the principall patrovn [etc.] .. of the said altar .. the said chaplane prayand perpetuelle for the said Jhone and Jonat thair kyne and frendis .. and for all Crissin sauillis; 1526 *Cart. S. Nich. Aberd.* I 155. Sax chaplannis to syng and do sufforhe for the sawill of the said wumquhill Jhone; 1530-1 *Maxwell Mem.* I 261. He endit his dayes quhilk .. he holy spent in the service of God .. and he eattis the fruittis of his labouris in ane pairt of the sawell amonges the number of the righteous; *c*1614 Melvill lxiv. Any that ar shot be vs, their sowell will goe to hevin; 1662 *Crim. Trials* III 604.

(2) Quhen saule and cors disseueryt er; BARB. XX 196. To the Lord .. I bequyth my saull, My banis to ly in the Kirk of Sanct Paull; ROLLAND *Seven S.* 2421. The synfull flesche .. Sall .. be closit in to clay .. The silly sawill sall pas a wilsome way; *Bann. MS* 50a/21. I leif my saull to God and my body to the erthe; 1580 *Edinb. Test.* VIII 62. Sum will say, saull to God, and banis to the borrow mure; ROLLOCK *Wks.* I 312. Till the day of .. redemption .. quhen, in bodie and sauell, he sall schine in the heavens; *c*1614 Melvill lxiv. Banes to the fyre, and saull to the devil; 1632 Dalyell *Darker Superst.* 99. I .. doe comit my sowel to God and my bodie to Christian bwiriall; 1687 *Kirkcudbr. Test.* (Reg. H.) 31 Oct.

d. *specif.* As capable of experiencing emotions, feeling joy, grief, guilt, etc. Also in collocation with *body*.

The quhilk joy Petir and Paule Browkis now in body and sawle; *Leg. S.* ii 878. Than rynnis the blude throwout the corps .. be the gaynesse and the blythenes of the saule; HAY II 122/11. Thelya .. quhilk can our sawlis bring To profound wit and gret agilite; HENR. *Orph.* 54 (Asl.). Apon my wrechit sawle haue sum mercy That gyltles sufferis sik dyseys; DOUG. II ii 158. For till my saull it is full sweit, Quhen sin besettis me; *G. Ball.* 125. Ceis than my saull and truble me no moir And for this warldlie truble mak na mone; ARBUTHNOT *Maitl. F.* 60/155. My saul displesantlye doth prove .. a heaven and hell in love; FOWLER I 148/13. Hope .. Which .. brings disparing saules [*v.r.* hartes] some ease in love; *Ib.* 181/14. Lift vp mine hart, my lips disclose My tendered tung vntie Then sall my singing saull reioyce; HUME 15/131. Schau then if loue .. Such sings of sorow in thy saule compellis; MURE *Early Misc. P.* ii 18. O my soule! . . I charge thee, that thou loupe for joy with me; D. LINDESEY *Heavenly Chariot* 23.

e. *comb.* In many, freq. nonce, collocations.

Saull preist, a priest employed to pray for the souls of the dead; *Soul cowper,* a 'trader' in souls; *Soul swarfe,* a 'swooning' of the soul, ? a spiritual crisis; *Soul confirmation,* ? the settling of religious doubts; *Soul sleper,* a psychopannychite, 'one who believes that the soul sleeps between death and the day of judgement' (OED).

Siluer in gret quantite He ordanyt for his saule weill; BARB. XX 169. Thane all the folk of that cyte . . Hyme as fadire & saule-hyrd Resauit; *Leg. S.* xxvii 1457. The grete chargis and dewiteis that folowis the said ordre, with saule perile, and worschip oft in were; HAY II 39/6. Be the pilgramage compleit I pas for saull prow; *Gol. & Gaw.* 269. For mannis saule thir arlis offerit he, Quhill he be his ded the saule price laid doun; KENNEDY *Pass. Christ* 151. Ane good measor is ane good soul freind; Ferg. *Prov.* MS No. 157. The chaplanreis callit the saull preistis and all vtheris chaplanreis fundit of auld; 1606 *Acts* IV 293/2. Christ promiseth soul-ease; RUTHERFORD *Tryal Faith* 93. That death, that soul-hell in the want of Christ; *Ib.* 260. We then know that these soul-cowpers and traffickers show not the way of salvation; 1650 Id. *Lett.* (1894) 653. Albeit he faint, and have soul-swarfes now and then, yet shall spiritual life be kept in him; DICKSON *Psalms* (1653) 277. O! says the poor body, I can never come at soul-confirmation in the matters of God, I have ay a hink in my heart about the Covenant, and . . about the work of Reformation, and fain would I be soul-confirmed, if I could win at it; M. BRUCE *Soul-Confirmation* 14. There was never . . one minister censured for error, save only Mr. John Hepburn, a soul-sleper; KIRKTON *Hist.* 28.

f. With *pres.p.* and *p.p.*, forming objective and instrumental combs.

(1) *pres.p.* A sanctuarie from soule-thralling snares; DRUMMOND II 33/27. The heartsomenesse of that soul-refreshing exercise of praise; DURHAM *Clavis Cantici* 51. Against a toleratione of soulmurthering and deluding sects; 1659 *Cramond Kirk S.* 15 Oct.

p.p. With interchange of manie soulesooke kisses; CRAIG II 60. The soule-sunke sorrow of godlesse epicures and hypocrites; LITHGOW *Trav.* X 435. The soul confirmed man leaves ever the devil at two moe; M. BRUCE *Soul-Confirmation* 13. *Ib.* 14 (see e above).

(2) Snakie hatred, soule conceav'd disdaine; MURE *Early Misc. P.* xxi 25.

2. a. In asseverations or exclamations, freq. *be, for, on* (one's) *saul.* **b.** In legal contexts, *to swear in* (*be*) (one's) *saul, to tak on (the court of)* (one's) *saul,* also *to declare upon* (one's) *saul.*

a. (1) The bischop sad than, be his sawle, [etc.]; *Leg. S.* ii 361. Schir, be my saull, ȝe neid not be effraid; HENR. *Fab.* 436. *Ib.* 211, etc. Now, on my saule, ȝe ar bot lurdanis all; *K. Hart* 524. Now, be my saul evin on a case; LYND. *Sat.* 646 (Ch.). Be my faderis saull Schir; PITSC. I 198/15. For the very saul of you, touch not a hair of his tail; *Presb. Eloq.* (1693) 78.

(2) My saull cum neuer in haly steid, I sall set sum in hard affray; *Alex.* I 2492. On ȝour saule beid, That ȝe be sicker of this sedull I send ȝow; HENR. *Practysis* 21. He wald wish God to sinck his saule in hell if he evir had to do with Margaret Scott; 1600 *St. A. Kirk S.* 935.

b. (1) Gif the principale party be absent the advocat sal suere in the saule of hym eftir as is contenit in thir meteris; 1429 *Acts* II 19/1. Litiscontestacione to mak, the aith of suthfastnes in my saule til swere [etc.]; 1500 *Acta Conc.* II 473. For us & in our name to defend & persew, litiscontestatioun to mak, the aith of suthfastnes in our saule to sweir [etc.]; 1518 *Lag Chart.* 54. His absence to excuse litiscontestatioun to mak the aith of suithfastnes in his saule to sweir; 1570 *Prot. Bk. J. Robeson* 82b. All schirefs sould compeir ȝearlie in the cheker, or ane sufficient depute for him, haueand power to sweare for him, and in his saull; SKENE *Reg. Maj.* II 57.

(2) And that uith the qualitie of sweiring the samyn be thair saule and conscience; 1629 *Justiciary Cases* I 118.

(3) He denyit the wax and tuk it on the court of his sawl he was awand bot v s.; 1447 *Ayr B. Ct.* 17 Nov. That wmquhile Eleizeus Foret . . tuke one his saule before witnes that he causit the . . evidentis . . to be abstract; 1516 *Fife Sheriff Ct.* 30. And this I tak one my saull; 1535 *Selkirk B. Ct.* (ed.) 158. I Johne Campbell of Ardkinglas testifies afoir God and takis it on my saull that [etc.]; 1594 Innes *Sketches* 532. Nather is the testimoniall producet gevin out be the subscryveris thairof upone thair saule and conscience that the persewer is mutilat etc.; 1631 *Justiciary Cases* I 166.

(4) The deponar declaris upon his saull and conscience that [etc.]; 1594 *Warrender P.* (SHS) II 247. 1603 *Reg. Privy C.* VI 577.

3. The vital, sensitive or rational principle in human beings.

Man has bathe saule vegitative that he lyfis by, sentitive that he felis by, intellective that he understandis by [etc.]; HAY II 157/9.

4. An individual endowed with a soul, a person.

Also pl. without inflexion.

Sometimes with admixture of sense 1 b, as also in *cair of saulles,* and *cure of saulis,* CURE *n.*[1] 1 d.

(*a*) Apostil syne chesit Paule, To helpe and succour many sawle; *Leg. S.* Prol. 165. In to the Katryne . . The saulis had sonkyn throu the syn of thé War not [etc.]; KENNEDY *Flyt.* 455. Tha sawlys valȝeant [L. *egregias animas*] Quhilk, with habundans of thar blud besprent, Has conquyst ws this realm; DOUG. XI i 57. Sufferage and prayerris to be done . . for the saullis that lyis in the said kirk and kirkyaird; 1532 (*c*1580) *Edinb. B. Rec.* II 59. The feind a sawll I trow will ken me; LYND. *Sat.* 770. Ane parisoun Quhare thare bene two thousand saulis to gyde; Id. *Trag. Card.* 412. Thow art ane vnclene land . . The prophetis that ar in it . . haue deuorit the saulis; WINȜET I 14/8. Quhilk . . Dumbar, North Berwick [etc.] . . with the depauperat saullis that this day dwell thairin can testifie; KNOX I 404. Bot, sillie saulis, thay ar sa daft, Thay ken nathing; 1572 *Sat. P.* XXXI 112. A lord, ane earle [etc.] . . , Bot simple sauls, vnskilfull, moyenles, The pure [etc.]; HUME 74/203. Nather could they find tham selves able to discharge so grait a cair of saulles; MELVILL 4. Off purpois . . to convert saules to the Romane religioun; 1613 *Crim. Trials* III 253.

(*b*) That . . mony sowles perished in your iles and highlandes; 1548 *Cal. Sc. P.* I 144. And to the puir solles almous daylie conforme to the diet comptis hard be auditouris; 1561 *E. Loth. Antiq. Soc.* V 20. The saikles soule Philoxenus was slaine; CRAIG II 65. Below the middle part there was but one body and aboue the middle there was two liuing soules; LITHGOW *Trav.* II 52. [£1 0 s. 8 d. given to] Andrew Folie, a distressed poore soule in Arnadge; 1635 *Ellon Par.* 78. He will not be entreated, but let a poor soul stand still and knock, and never let-on him that He heareth; 1637 RUTHERFORD *Lett.* (1894) 355. That Adam Taite . . had most wickedly traduced Mr. Walter Greig and laboured to weaken his ministrie, calleing him a betrayer of sowles, a scruffer of scriptures, and one unworthie of his place; 1658 *Cupar Presb.* 183. Be the Christian soule quhat he will in himself, he is Christ's match and fair one; 1659 HAY *Diary* 30.

b. As a term of endearment.

In his hall wes nocht admittit bardis na fenȝeit fulis bot ane certane of pure barnis callit his saulis & ane of thame wes his play fule; ABELL 113b. Thow, fair saule, incace thow len thy eares Vnto the mornfull accents of my mone; FOWLER I 184/5.

5. *transf.* Said of a person who endows another with life or action or who animates, controls or manipulates another.

'I am fare, and almychty, And haly Israel am I, And I am Goddis saule,' he sade; *Leg. S.* i 235. Ane gredie, godles man, to quhom the Secretar is the saule; BANN. *Memor.* 15. The Secretar, saull to the Erle of Atholl; *Ib.* 17. Lordis Home, Seatoun and the Secretar, soule to all the godles band; *Ib.* 19.

Saul(l)es(s, *adj.* [e.m.E. *soulless* (1553), *soule-less* (1594), OE *sáwel-léas.*] Ignoble, base. — Sic saules thingis [*sc.* women] hes nouther sin nor schame; ROLLAND *Seven S.* 8008. Or than to God ȝe do grit injure And sall accus ȝow saules of my ded; *Bann. MS* 223a/54. The Caluinistis .. ar lyk saules men of law in peruerting Christis latter vil and testament; HAMILTON *Cath. Tr.* 52b. A saulles suinger, seuintie tymes mensuorne; MONTG. *Sonn.* xxiv 3. Thir saulless lovnis plunderit .. quhair euer thay cam; SPALDING I 205.

Saul(l)-hele, *n.* Also: **saule-, sawle-, sauill-** and **-heil(l, -heile, -heyle, -hell, -hail(l.** [ME and e.m.E. *saul-hele* (Cursor M.), *sawle-heale* (*c*1440), *soule-hele* (Piers Plowman), *sowle-heele* (1418), *soule-heale* (*c*1550), OE (*fore uncerra*) *sáula héla*. SAUL(L *n.* and HELE *n.*² 2.] The health of the soul; spiritual well-being; salvation. Freq. with possess. adj.

(1) That his trawaill Mycht till his saule hele awaill; BARB. xx 356. He .. askyt .. Quhow he suld do for his sawle-hele; *Leg. S.* xii 226. 'Ȝet,' quod scho, 'be myn saule-heile!' *Ib.* xxx 153. He .. said he wald .. For his saull heill that viage Tak; WYNT. VI 924 (W). Thai men that .. for thar saule hele visitis in haly steddis; *Acts* I 11/2. HAY I 203/13. It is mar spedfull .. to the manis saul-heill to her mes with clen hart; *Vert. Mess* (STS) 10. For the saule heil of the said maister Jhone; 1486 *Bk. Carlaverock* II 447. The said Jhone movit of devocioun and for his sauill haill has infeft ws .. in all and haill his tenement; 1526 *Cart. S. Nich. Aberd.* I 154. Prosperite causis oft tymis a man to forȝet God and his saull heill; ABELL 10b. Put nocht thi saul heill in perell for wnsewir redusin of othiris; *Ib.* 87a. Saule hail; HAMILTON *Cat.* 225. Saull heill; ROLLAND *Seven S.* 2430.

(2) That he .. fore saule-heile come in that land; *Leg. S.* xiii 94. Saul-heyle; *Ib.* xxi 565. Sawle hele; *Ib.* xxvii 706, etc. For athe aganis saule hele is nocht tobe kepit; HAY I 149/15. A hous of almus for tyl harbry in it pur foulk for saull heile; 1464 *Peebles B. Rec.* I 151. Anvellis .. gewyn be ald Willyam of Bothwell .. to Sant James awter for saul hell; 1475 *Ib.* 175. Men of suffisans .. Seikand saull hele without dissimilans; *Contempl. Sinn.* 108 (Asl.).

(Saull-helth,) Saule-helth, *n.* [ME and e.m.E. *soule helthe* (Gower), *sawle healethe* (*a*1450), *soulehealth* (1574); SAUL(L *n.* and HELTH *n.*] = SAUL(L)-HELE *n.*(1). — That thay intromet with my gudis & geir .. as I haif left the same for my saule helth; 1575 *Edinb. Test.* IV 47.

Saul(l)ie, *n.* Also: **salie.** [SAUL(L *n.* Also in the later dial.] A hired mourner. — That no duillweidis be givin to herauldis, trumpetoris or saullies except by the earlis and lordis and thair wyffes; 1621 *Acts* IV 626/1. Tuentie four saulies, 2 and 2, in order, the foremost of them carying one hes staffe a gumpheon; 1625 Balfour *Her. Tracts* 101. *Ib.* 106, etc. Twenty four hoodies or saulies tua & tua; 1636 *Marquis Huntly's Funeral Roll* MS (Nat. Mus. Antiq.). Salies; 1654 Rogers *Social Life* I 161. And Saint Chrisostom condemns feign'd mourners, whom we call saulies, because they pray for the souls of the dead; MACKENZIE *Observ.* 365.

Saull-knell, Saule-knell, *n.* [ME and e.m.E. *soul-cnul* (*a*1300), *soule-knylle* (*c*1400), *sowle-knell* (1515); SAUL(L *n.* and KNELL *n.*] A peal of a bell rung to mark a person's death. — For mony ane dois haist thair awin saule knellis; LYND. *Answ. Flyting* 40. Let not be rung for me, that day, saull knellis; Id. *Test. Meldrum* 181.

(Saull-massing,) Soule-massing, *vbl. n.* [e.m.E. *soule massing* (*c*1555); SAUL(L *n.* and MASSING(E *vbl. n.*] The action or practice of saying masses for the dead. — The exorbitance of superstitious exiquies are found infinite (as in their .. holy watering, letanie praying, soule-massing [etc.]); BIRNIE *Kirk-b.* xi.

Saul(l)-mes, Saules-mes, Soul(e)-mass(e, *n.* Also: **saule-, sawll-, sawle-, sawill-, sal-** and **-mese, -mess(e, -mas, -mis.** [Late ME and e.m.E. *sowlemasse* (*c*1450), *sowlemas* (1461), *sawmos* (1533), *soul-masse* (1661); SAUL(L *n.* and MES *n.*¹] A mass for the soul of a dead person. Also *attrib.*

For many further examples, see the Indexes to the various vols. of *Treas. Acc.*

(*a*) He doande yerely ane obbyt of Sir Johne of Kindelochis saule mes; 1466 *Cart. S. Nich. Aberd.* II 330. 1478 *Grey Friars* II 14. To the king to offir at the qwenis sawle mes; 1488 *Treas. Acc.* I 89. 1490 *Ib.* 170, etc. Sawll mes; 1495 *Old Dundee* I 568. Atour he sal nocht ring Laurence at the saule messe nor menyngis; 1503 *Aberd. B. Rec.* I 72. To the bellman for passing throw the tovne to the saullmes & derige; 1505 *Edinb. Hammermen* 53. To the regentis of the sayd patigoge and to the studentis .. to compeir to the sayd derige and saule mes; 1509 *Reg. Episc. Glasg.* 522. Saule mis; 1511-12 *Treas. Acc.* IV 182. Quhen thair decessis ane craftisman within this burght, the segster to ring bot to his sawill mes; 1525 *Aberd. B. Rec.* I xli. My wife .. to caus ane saule mes and dirge to be done for me about the sammyn day twelmont; 1547 *Glasgow Test.* I 17b. Wther funerall expensis .. wyth saull mes and dirigie; 1562 *Inverness Rec.* I 89. Saull mese; PITSC. I 310/6. Thaireftir [they] dansit and sang the saule mes and dairgie for the ministeris deid meir; 1579 *Reg. Privy C.* III 215. In and to the haill anniuersareis, sawle messis, diergeis, lady messis syluer; 1583 *Edinb. B. Rec.* IV 282. Ther wes ane solemne saule mess with a dergie soung afternoone; BIRREL *Diary* 7. For saying of ane saul-mess for the Lady Auchindoun; 1607 *Crim. Trials* II 530.

(*b*) To feft ane solemne salmes and dirige yeirly, the day of my obite, in the Cathedrall Kyrk of Murray; 1540 *6th Rep. Hist. MSS* App. 671/1.

(*c*) [The] dekin .. sale haue the samyn office at the saules messis; 1491 *Cart. S. Nich. Aberd.* I 256. The secund day of November, to the kingis offerand at the saules mes, xiiij s.; 1504 *Treas. Acc.* II 266.

(*d*) And if thou be to ly at the altar, how wantst thou a priest to say thy soule masse? BIRNIE *Kirk-b.* xi. As being a meer popish institution appointed for singing of soul-masses; 1685 Fountainhall *Decis.* I 353.

attrib. Anent the dergis, sawle mas feists, and vthers; 1504 (*c*1580) *Edinb. B. Rec.* I 99.

b. *comb.* *Saul mes day*, All Souls' Day, 2nd November. Also *Saul mes evin*, the eve of All Souls.

(1) The halidayis that hapnis within the samyn [*supra* acht dais], sic as Alhalow day, Saul mes day [etc.]; 1507 *Edinb. Chart.* 191. I leif to Sanct Mongowis bell to pas throwe the townne one Salmes daye .. to gar praye for mye faderis saule [etc.] .. and all Chrisstyne saulis aucht peneis of annuall; 1509 *Liber Coll. Glasg.* 209. I leif to the blak freris of Glasgve acht schillyngis of annuall .. to syng derige on salmes day efter nowne and mes on the morn; 1509 *Reg. Episc. Glasg.* 522.

(2) Saulmes Evyn; 1535-6 *Ayr B. Acc.* 75. To George Blair to ryng the bellis on Salmes eyn; 1536-7 *Ib.* 25. For rynging of the bellis all nycht on Salmes evin; 1552 *Edinb. B. Rec.* II 338.

Saulpeter, var. of SALPETER *n.* **Sault,** var. of SALT *n.*¹, *n.*² and *adj.*

Saumbu(e, *n.* [ME *sambu* (14th c.), *saumbu(e* (*a*1450), etc., OF *sambue* (*c*1175 in Larousse).] A saddle-cloth. — Onerat se de ij saumbues pro sororibus .. regis; 1331 *Exch. R.* I 389. Pro duabus saumbuys [£20. 13 s. 4d.] Et pro vna cella [= sella] palfredi pro domino custode; *Ib.* 402.

Saumond, var. of SALMOND *n.* **Saundre,** var. of SANDRE(E *n.*¹ **Saune,** p.p. of SAW *v.*¹ **Saup,** var. of SAP *n.* **Sauraly,** var. of SEVERAL(L)Y *adv.* **Saur(e,** varr. of SAVO(U)R *n.* and *v.* **Sause,** var. of SALS *n.* **Sausered,** var. of SALSARIT *ppl. adj.* **Sauses,** var. of SAUCIS *n. pl.* **Saus(s)ar, -er, -our,** varr. of SALSAR *n.*

(Sauster,) Sawster, *n.* [ME and e.m.E.(? chiefly north.) *salsister* (Durh. 1347, Cath. Angl.), *sawsestyr* (Durh. 1434), *sawcyster* (Prompt. Parv.), *sawstyre* (15th c.), ? altered f. F. *saucisse* SAUCIS *n.*] A sausage. — *Farcimen*, 'moretum', a pudding, a sawster; DUNCAN *App. Etym.*

Sauster, var. of SALSAR *n.* and SAWSTAR *n.* **Saut,** var. of SALT *n.*¹ **Sauté, -ie,** varr. of SAFETY *n.* **Sauting,** var. of SATIN(E *n.*

Sautouer, Sawtry, *n.* [OF] A saltire or St. Andrew's Cross. — A chefe, a bend, a fes, a pole, a cheueron, a wnde, a geron, a croix and a sautouer [F. *sautouer*]; *Loutfut MS* (ed.) 12a. And giff ȝe will wit quhat is a sautouer [F. *sautouer*] luk in this buk the armes of Boudeuille; *Ib.* 12b. There bene in armes callit two pynyons. One is quhen the feild is a sawtry Saynt Andrewes crosse; *Ib.* MS 61b.

Sauuour, var. of SAVO(U)R *n.* **Sauvage,** var. of SAVAGE *adj.* (*n.*). **Sauvant,** var. of SAV(E)AND *pres. p.*, *prep.* and *conj.* **Sauve-condyt,** var. of SAFE-CONDUCT *n.* **Sauvely,** var. of SAFELY *adv.* **Sauyng,** var. of SAWING *vbl. n.*[1].

Sav, var. of SAV(E *v.* and SAW *n.*[1]

Savage, *adj.* (*n.*). Also: **savadge, savaige, sawage, sauvage, sawwage, saffage; sov-, sowage.** [ME and e.m.E. *sauuage* (*a*1300), *sauage* (*a*1330), *salvage* (1542), OF *salvage, sauvage* (late 11th c. and *c*1175 in Larousse).]

Forms in *-o-* may represent an editorial misreading.

A. *adj.* **1.** Of persons, their behaviour, etc. **a.** Fierce, cruel, ferocious.

(*a*) Thus that king that wes sawage Held this folkis in hard thirlage; WYNT. II 567 (W). Bot Orestes cachit in furyus rage For cryme of his moderis slauchtir and savage In lufe hait byrnyng for his spows byreft .. Set on this Pirrus .. And .. Smate of his hed; DOUG. III V 95. Contempnand all humane societie and honest forme of traid .. ; takand ȝour selffes to ane maist wyld, barbarous, savage, inhumane and crowell form of lyff [*sc.* piracy]; 1610 *Crim. Trials* III 103.

(*b*) The pairtis of ȝour majesties kingdome committed to my spirituall oursicht being so barbarous that without seuer animaduersione thay can not be cohibite from thair wonted souage behauiour; 1615 *Denmylne MSS* in *Highland P.* III 189.

b. Of fighting men: Fierce, bold, indomitable, valiant.

There is some overlap with a above; however, this sense expresses approval not evident in a.

Quot. *Wall.* IX 457 may belong in a.

Ȝonge he was and to Sothroun rycht sauage; *Wall.* I 339. Ȝong Wallace fulfillit of hie curage; In prys of armys desirous and sauage; *Ib.* II 2. Gude expensis till him thai maid to pas Maister Jhone Blayr was offt in that message A worthy clerk, bath wys and rycht sawage; *Ib.* V 534. Schyr Wilȝham off Leis graithit thaim that nycht With v thowsand welle garnest and sawage Apon the Scottis thai thocht to mak scrymmage; *Ib.* VIII 813. Than Longaweill that ay was full sawage With Wallace past as ane to that scrymmage; *Ib.* IX 457.

c. Expressing disapproval: Reckless, ungovernable.

A mane .. of wyne that has vsage Ande habundance and syne is nocht saffage Throw mychtines and confort of the wyne At temperance bydis and sobyr syne; *Bernardus* 300.

2. Of animals: Wild, untamed, hence, fierce, ferocious.

The rauand sauuage volffis of strait montanis ande vyild forrestis that deuoris nolt ande scheip for ther pray; *Compl.* 2/17. Wylderines quhairby the wyld beistis sik as hart hynd dois rais wolfis toddis foxes beiris bairis sangleiris wyld swyne with uthir savaige and distroying bestiall incressed and multiplied; BISSET I 28/34.

3. Untouched by civilisation, 'natural'. **a.** *Savage man*: The conventional naked male figure of heraldry. **b.** Pertaining to or characteristic of uncivilised persons, their dress, etc.

a. The tuo rough sauadge men (with a batton or club in either of their hands) who doe support and uphold the armes are peculiar to the house of Southerland; GORDON *Geneal. Hist.* xiv.

b. BUREL *Queen's Entry* 216 (see B *noun* 1 below).

4. a. Of a place: Wild, uncultivated. **b.** Of the inhabitants of a place: Uncivilised, untamed, barbaric.

a. I went throw ane wood sauage; BUREL *Pilgr.* 1 1.

b. Duryng his tyme so justice did preuaill The sauage [PITSC. I 277/9, sawwage] Iles trymblit for terrour; Eskdale, Euisdale, Liddisdale and Annerdale Durste nocht rebell; LYND. *Test. Pap.* 494. The souage daillis he dantonit and maid thrall To serue thair king he gart thame gif thair band; 1570 *Sat. P.* xix 29. To repair .. our said fortalice .. the west pairt of Sutherland in trobollis tyme beand ane barbare and sowage rowme; 1571 *Old Ross-shire* I 365.

B. *noun.* **1.** A wild or uncivilised person; passing into one who behaves in a fierce, cruel or barbarous fashion.

Hir grace past to ane vther pairt Quhair sche beheld sum .. Acoutert in an sauadge sort .. Thir are the Moirs .. Quha dois inhabit in the Ynds .. Thir savagis I you assure Wer well decord; BUREL *Queen's Entry* 235. There are two kings in Arabia .. beeing vntamed sauages and mis-regarders of ciuility who .. corroborate the malignity of their dispositions with bloody and inhumane interprises; LITHGOW *Trav.* V 228. Some scattering Arabs sold vs water .. two of which sauages our captayne hyred to guide vs; *Ib.* VI 292. Most Christians like to savadges go eat; 1680 *Dunferm. Ann.* 350.

2. A 'savage man', as in A 3 a above.

For his crest a burning hill supported be two savadges or naked men, proper; 1673 *Grant Chart.* 248.

Savagelie, *adv.* [ME *savagelych* (*a*1400); SAVAGE *adj.*] In a savage or ferocious manner. — Raigeing .. nocht only aganis .. the haly, catholik Kirk, bot maist sauagelie aganis thame selfis; WINȜET II 7/31.

(Savagenes,) Sevegnes, *n.* [ME and e.m.E. *savagenes* (14th c.), *savedgenes* (1600); SAVAGE *adj.*] The state or condition of being savage, barbarous or uncivilised. — Ane .. desyer to have removit frome amange his subjectis of the cuntrey of Scotland all sic cuustomis, faschiouns and behaviouris as did in ony wayis smell of barbaritie and sevegnes, and to have reducit thair haill cariage and behaviour to that comeliness and decencie that is requirit in a civill and weill governit pepill; 1604 *Reg. Privy C.* VI 594.

Savaige, var. of SAVAGE *adj.* (*n.*). **Savand, -ant,** varr. of SAV(E)AND *pres. p.*, *prep.* and *conj.*

Savar, Safare, Salver, *n.* Also: **save(a)r, -or, saf-, sa(i)ffer.** [ME and e.m.E. *sauuer, savere* (both Cursor M.), *savour* (*c*1330), *saver* (*c*1410); SAV(E *v.* Cf. SALVIO(U)R *n.*]

1. One who delivers, protects, preserves, keeps safe, in senses of SAV(E *v.* **a.** Applied to God, etc. as a saviour of souls.

Safare of sawlis, ta me sone; *Leg. S.* iv 171. & lofing til his savor ȝeld; *Ib.* xl 699. Christ Jesus, saiffer of men; LYND. *Mon.* 4532. O Mary, .. gentilest saiffer of synnaris; *Arundel MS* 281/86. That thou [*sc.* the Virgin] be to me helper and saver in my letter dayis .. be help and supple off .. oure saluioure Jesu Christ; *Ib.* 282/121. Throwgh Jesus Crist our Lord and onlie saluer; 1561 *St. A. Kirk S.* 129. Thou quho namd is Paracletus .. Be my instructor, saiffer, and conwoy; J. STEWART 257 § 228.

b. In other applications.

[The persons to whom the jewels were] laide in wed [..] savaris [of the said jewels]; 1573 *Cal. Sc. P.* IV 621. And yit .. they ar .. the savears of his honour and privileages of his croun, and we the empearers thairof; 1596 MELVILL 382.

c. One who recovers and restores lost property. Cf. SAFER *n.*

The pannell declairit that he nawayis staw the purse .. bot .. fand the purse vpone the flure of the chalmer; quhilk he retenit, and spendit of the money that was thairin ten merkis allanerlie, in recompance to the saiffer, and restoirit the said purse; 1618 *Crim. Trials* III 441.

2. ? A protective covering of some sort.

For ladies hoods and muffs, and for spectacles: ? a protective bag, case, or the like for keeping them in. Also ? for the hub of a wheel.

(1) For ane elne of quhite sating .. xl s. and ij elnis of blak sating price of the elne xxxiiij s. to be thair saferis to thair hudis and thair mufellis; 1539 *Treas. Acc.* VII 167. For safer of a mufe 2lb. 18.; 1694 *Household Bk. Gr. Baillie* 257.

(2) To the boy in Rosses for safer of my spectacles and silver caise; 1689 *Foulis Acc. Bk.* 106.

(3) For saffer of the ring of the nave of on of the cart wheills; 1699 *Foulis Acc. Bk.* 259.

Save, *n.* (Appar. ME *saue* (Chaucer), *save* (? *a*1450), appar. not recorded thereafter, L. *salvia* the herb sage.) — Yow sall prepare yowre self .. by taking cooling broath [*blank*] saue [*ed.* sane] and oximell; 1653 *Laing MSS* I 277.

Sav(e, Saf(e, Saif, Sauf, Salve, Saff, *v.* Also: **saw(e, saive, saif(f)(e, sayf(f(e; sauf(f)(e, sawf(f)(e, salf(f, saulf(f)(e, salwe; seff.** P.t. and p.p. also **saift, -d, sauft, sawft(e, (sawth).** [ME and e.m.E. *sauue(n* (Ancr. R.), *saluin* (*a*1225), *save(n* (*c*1250), *sauf* (Cursor M.), *safe* (Rolle), *sawe* (*c*1425), *salve* (1432), OF *salver, sauver* (1050 and late 12th c., in Larousse), late L. *salvāre*.]

1. *tr.* To bring or deliver out of danger; to rescue from (impending) misadventure or misfortune; to avert danger from. Also, rarely, in the passive without agent, to be saved, to escape.

There is considerable ambiguity with sense 2.

(1) Hyre barnis .. ware fletand One a burd, & sawit ware; *Leg. S.* xxi 83. Sir, ȝe sawit ws now; *Ib.* xxvi 201, etc. Gif the duk .. be tane, quhethir gif [any] suld save him, have merci of him and ransoun him; HAY I 120/13. Quharfor he chesis quham he lufis best, and helpis to save him; *Ib.* 194/26. Thai held him lost, ȝit God him sawth [*1570* sauit] agayne; *Wall.* II 419. At he wald haf sawit him and he micht haue wone to him; 1486 *Bk. Carlaverock* 447. Thou sal thole iniuris & spulȝe, ande ther sal be na man that can saue thé; *Compl.* 24/15. If he wald haue sauit his sones [*sc.* from the gallows], he wald not haue stopit for ane woman; MAITLAND *Geneal. Setoun* 20. *Erepte*, save, sane; BUCH. *Comm. on Virgil Æn.* III 476. John Robertson .. peressitt at the Saynd ness, in the watter, sawand Alexander Myll; CULLEN *Chron. Aberd.* 46.

(*b*) And to hir [*sc.* Venus'] court siclik I am annext Wnder all pane, and refuge nane I haif, Without ȝour grace [*sc.* Vesta] with sum salue wald me saif; ROLLAND *Ct. Venus* II 946. Vnto the tyme the prowest and wtheris nychtbouris saiffit the said James; 1563-4 *Inverness Rec.* I 109. Hou Andro Wood of Largo saivit the Erle of Glencairne; PITSC. I 317 h. of ch. Quho raid suoftlie throw the feild to saif all freindis that thay might comprehend on lyffe; *Ib.* 319/29. Bot inmedeatlie Androw Wode .. tuik him and saiffit him and brocht him away on lyue; *Ib.* 320/6.

(*c*) Thu ma lichtly .. Sauf hyme & do thi-self na il; *Leg. S.* xxx 250. All the folk off Tessaly Had nere bene drownyt. .. Wpone ane hille was callyd Parnas Thare thai ware that sawfyt was; WYNT. II 460. Salve ws, lattir wardis of Troy, that we ne spill; DOUG. III ii 39. All drowkyt and forwrocht, Thai salwyt [*Sm.* salffit, *Ruddim.* saiffit] war, and warpyt to the cost; *Ib.* X vi 45. It wes aggreed that he suld keip vessellis to sauf and conduct theme towardis there landis frome the said harbarie; BISSET II 234/9.

(2) And of his gud the most part all He kest out .. To saif his folk; *Troy-bk.* II 2390. All that schyltrum thai slw down, And sawffyd off Berwyk swa the town; WYNT. VIII 1770. For to sawffe his comwnate Than wyth the wardane tretyde he; *Ib.* 5551. Schir .. Ȝe wald haue reuth and take this ambassade For thocht ȝe had neuer vther seruice made Till Alexander ȝe seruit ȝoure warisoun To saif his men or thai be dongin doun; HAY *Alex.* 3467. Quhou Eneas the trayson dyd persave, And quhat debait he maid the town to save; DOUG. II vi heading. On horsbak in this Tarchon baldly draw, Wilfull his pepill to support and saw; *Ib.* XI xiv 10. Maid proclamationis .. That ilke man .. Sulde cum, and saif thare awin cuntre; LYND. *Mon.* 2030. Ane woman than .. Saifit the toun; *Maitl. Q.* 94/160. Hester .. Saifit the peopill all; *Ib.* 94/162. That now or never is the tyme to gaine honour .. and to saive or lose thair countrie; 1640 *Kirkcudbr. Min. Bk.* 51.

(3) Bot quha sa redis Virgill suythfastly Sall fynd Neptune salf [*Ruddim.* sayffit] Eneas navy; DOUG. I Prol. 162. The maryneris .. with lang bolmys of tre .. Inforcis of to schowyn the schyp to save; *Ib.* V iv 95. Awalke! .. and put to ȝour hande stoutlie to saif Petiris schip; WINȜET I 6/21. Ane .. wall .. within quhilk schippis and boittis are drawin and salvit; *Descr. Isles* 437. 1602 *Conv. Burghs* II 139 (see b below). Meattis it behuifis to cast out the guidis to saif the schip; BISSET II 242/34.

(4) Ȝour hous .. laitlie onlie on the Lord it lenit, .. Quha send ȝour self to sawe it and sustenit; *Maitl. F.* 424/175.

b. To extricate (goods or possessions) from (the threat of) destruction or loss; to salvage.

There is some overlap with (3) above.

Gyf a schyp be brokyn .. the men ar haldyn to saufe als mykyl of the gudes as thai may; *Ship Laws* c. 21 (B). Gif be aventure of sey ony parte of spuleȝe .. be salfit and drevin bakwart apoun thare coist; *Boece* 87b. Giff ony skipper be schipbrokin than sall he and his cumpany be haldin to saiff and succour the schip and guidis with the reidship of the schip [etc.]; 1602 *Conv. Burghs* II 139. BISSET II 241/24. The bischops seruandis saiffit his bookis and vther insicht and plenishing and hid thame in nichtbouris houssis; SPALDING I 194. My lords monny was saved [from the fire]; *c*1685 *Annandale Corr.* 314.

2. To keep safe, to preserve from danger, to protect. Freq. said of God.

Gol. & Gaw. 793 is included here ((1) (*b*)) as, appar., an alliterative use of a phonetic doublet. This seems to make rather better sense in the context than that *salf* is an instance of SAW *v.*[2] to heal, remedy. Cf., however, the example in *Piers Plowman*, *sauen .. and saluen .. of alle synnes* cited by OED (*salve* v.[1] 2 b).

(1) Now ga we furth .. Quhar he that maid off nocht all thing Lede ws & saiff [*C.* sauf] ws; BARB. VIII 263. Bot thane for the takine of the rud He mad on hyme .. thocht he vntrewful was, Ȝet wes he sawit nocht-the-les; *Leg. S.* xxxii 22. And God .. Mot saif our king, and gif him hart and hand All sic wolfis to banes of the land; HENR. *Fab.* 2775. The king grantit the gait to schir Gawane, And prayt to the grete God to grant him his grace, Him to saue and to salf; *Gol. & Gaw.* 793. Bot fader ȝit I dreid that we Be tane .. He said, Na, son, God will ws saif, Becaus that we sic mister haif; *Seven S.* 865. And for girth come he hidder than, And we haf savit him as we may; *Ib.* 1947. Quhow Grekis maid thare deuotioun haill To Mars, to saif thame in battaill; LYND. *Mon.* 2352. Without that ȝe [*sc.* the seventh sage] find sum help .. Conforme to the sayingis ȝe said .. That was ilk ane to saif ȝour day about; ROLLAND *Seven S.* 6168. He sall saif thé mychtfullie, And na mischance sall cum to thé; *G. Ball.* 108. Thow may me saif, thow may me spill Baith lyfe and deide lyis in thy will; *Ib.* 170. Fair weill leill lady .. Fair weill that may me saif and spill; *Bann. MS* 225b/22. And Jonas, in the quhellis bellie, Thow safit thre dayis; *Montg. Suppl.* xxxi 35. Let na man do evill for counsell of greit men, or thair maisteris, thinking thay sall save thame; BUCH. *Detect.* (1727) 160. Becaus he pat downe the puire vitches and saiffit gretter in his awin cumpanie daylie with him; PITSC. II 218/6.

(*b*) Der God, .. Sawff hym and scheld him fra his fayis! BARB. II 145. To sawffe hym and hys swnnys thre .. He gert, of Goddis byddyng, mak Ane ark; WYNT. I 379. Salf; *Gol. & Gaw.* 793 (see (1) above). Sum be myrknes of the nycht war salfit; *Boece* 49b. Aidane thus deliuerit be reskewe of his son was horssit and salfit be flicht; *Ib.* 317b. Quha haillit hir and teichit hir all thingis, and speikis and wairnis hir of thair cuming and saulfis hir; 1588 *Crim. Trials* I II 163.

(2) Now se I na remeid To saif the selie beistis that I keip; HENR. *Fab.* 2467. Gevin to Arche Pennecuik to saif the quenis hors [*sc.* when being taken onboard ship], vj frs. vij s.; 1538 *Treas. Acc.* VIII 159. Preistis, pray na mair To Sanct Anthone to saif thy sow; *G. Ball.* 197.

(3) His douchtir .. to the dragone Suld be gyffine, to sauff the ton; *Leg. S.* xxxiii 106. The Scottis men .. Made ordynawns thare land to save [: gave]; WYNT. VIII 3739. Thow has convoyit .. this victorios armye without ony harme resauit. .. Be thy manhede .. oure army is salfit but dammage; *Boece* 87b. That my citee was sauffit be thi helpe; BELL. *Livy* I 34/31. The first .. said .. this day the cietie I sall saue; .. The secund said the nixt day fallis me Fra enemeis for to keip the cietie; ROLLAND *Seven S.* 6135. God saifd our ship, and ruled our noble ruther, And helpt vs throu; MONTG. *Misc. P.* xlviii 161.

(4) Gif God savis my neuoys weill And my suord and my mais of steill, I sall me venge; *Alex.* II 1243. I say nocht na he sulde .. abandon all his gudis, lyf, and body to sauf the Cristyn faith; HAY I 131/5. We tak on hand .. To saif baith myrrour & ymage; *Seven S.* 1792. That nowthir tre nor other thing .. war sauffit or left vndistroyit; BELL. *Livy* II 181/16. 17 salt hyds, which were laid upon the highest lofting of the steeple and bells for saving the said lofting and bells [*sc.* from fire]; 1671 *Orkney Rentals* App. 64.

b. In phrases of benediction, sometimes also by way of greeting, as *God save ȝou, God save the King.* Cf. 7 f.

(*a*) God sawe thé, gif his willis be; *Alex.* I 584. And God .. Mot saif our king; HENR. *Fab.* 2775. 'Quhat set yow thus,' scho said, 'so God yow saiff'; *Wall.* VIII 1324. The hyndmest [said] .. 'God saue ȝow, schir, of Scotland salbe king!'; STEWART 39718. Freyr Allane said I pray to God ȝow saue Heir is aneuche forsuth of Goddis gude; *Freiris Berw.* 316. God saif him & the haly gaist And keip the man fra manting mekle; *Sym & Bruder* 124. [He] saluist thame .. God saif ȝow seid of Salomon; ROLLAND *Ct. Venus* II 38. God saif our king and send him lytill ado; 1573 SEMPILL *Sat. P.* xxxix 379.

(*b*) Him that now regnis, quham God sauff; *a*1540 *Misc. Bann. C.* III 41.

c. To guarantee immunity or safe-conduct to, to warrant the safety of.

To sayff the erll Wallace .. Set on his clok a takyn for to se, The lyoun in wax that suld his condet be; *Wall.* XI 905. Sa be this maner of brevis all schipis and guidis be saiffed [*v.r.* assured] of the richt of the said prince; BISSET II 233/24. *Ib.* 235 n. 1.

3. With non-personal agent: To serve as a protection, defence or means of deliverance to.

(*a*) Alssua .. suppos he pas to do sum pugny of were .. that savis him nocht; HAY I 116/33. His helme was crasit and in the harnes him clave Quhill all the gold of Grece mycht nocht him save; Id. *Alex.* 3266. Couert of treis sawit him full weille; *Wall.* II 71. His dowblet wes maid of ledder And saift [*M.* sauft] him; *Christis Kirk* 97 (B). Sa it become .. That the four toris sauit the childis face; ROLLAND *Seven S.* 1804. Thy boist .. sall not saif thee I say; *Ib.* 6409. Quhen gallow pin or cutting knyfe Suld stranglit thé, and saift my deir; 1570 *Sat. P.* xi 55.

(*b*) That he [*sc.* Cyrus] suld hang hym on a tre, Quhare nevyre rayne wytht mekyll wete Na nakyn swn mycht sauff with het; WYNT. IV 424. Bot .. gif he war suilke a mysdoar that gyrth of haly kyrk aw nocht to sauffe hym; *Acts* I 38/2. Gif a man be tane apon ane otheris sauf condyt, traistand in it that it suld sauf him and lede him seurly; HAY I 181/8. We .. revokis all .. respittis to the effect that the samin sall be of nane avale .. to sauf ony maner of person to be accusit .. for recent slauchteris .. committit be thaim; 1533-4 *Acta Conc. Publ. Aff.* 417. The said conduct .. is sufficient anewcht to salf the said Jhone that he be na presoner; 1543 *Acts* II 431/1.

4. a. To spare (a person), to allow, freq. out of clemency, to escape impending punishment, death, etc. Also *absol.*

(*a*) The kyng Pryame .. The which thow aught .. To saue in thyne protectioune; *Troy-bk.* II 848 (C). That nane that was in that cittie .. sould sauit be; *Alex.* I 26. Quhilkis for na gude thai mycht do thare suld be savit, bot thai suld tyne thair hede; HAY I 117/2. He is ane iuge to nane suspect, To pvneis synffull man and saif; HENR. *Abbay Walk* 36. Wallace commaundede thai suld na wermen saiff; *Wall.* IV 256. Wemen .. And ȝong childer, forsuth thai sawyt no ma; *Ib.* 492. [He] slew thaim all .. and saiffit neuer one; *Bk. Chess* 1037. Haue mercie, luif .. Quhat have I wrocht .. That ȝe suld murdir me .. Leif creuelte and saif ȝour man for schame; DUNB. (OUP) 39/13. That da tha sauit nother ȝoung or ald; STEWART 32258. I ordand ȝou to slay doune al the Romans, and nocht to saif ane of them; *Compl.* 100/28. The laird of Virmestoune .. passit to the regentis ludging and thair gave him fair vordis to come furth promeisand to saue him, quha instantlie come furth .. beleivand to have beine sawit bot it stuid nocht in the handis of thair backaris to save thame at that tyme; PITSC. II 263/17, 19, 20. Quhen thow did all the warld distroy for Adame sine; And sauitt Lott; *Montg. Suppl.* xxxi 31. [He] wes set at libertie by the Earle of Catteynes who saiffed him not out of any favor towards him .. bot onlie [etc.]; GORDON *Geneal. Hist.* 185.

absol. In batall is mor worschipe .. To wyn & saif than for to sla all dovne; *Bk. Chess* 1033.

(*b*) He gert that woman sawffyt be That sauffyt hys spyis .. Quhen thai come that towne to spy; WYNT. II 1083. He chesit of thai twa Men erare to saufe than sla; *Ib.* V 4088 (W). Auchten schyppys gret thai wan: Off thai thai sawffyd nevyre a man; *Ib.* VIII 1758 (R). Sauff; *Ib.* II 1140. *Ib.* VI 1080. That sawfte suld be nayne Off Sotheroun blud, quhar thai mycht be ourtayn; *Wall.* VII 1039. Thai sawft na Sotheroun for thair gret riches; *Ib.* VIII 525. To sauff thi king or spill; *Ib.* XI 1378. Salf; *Seven S.* 612. Bot and thow salf me and thoill to pas hame to my cuntre haill and feire thaire salbe perpetuall remembrance of the clemence .. of the Impriour of Rome; ABELL 22b. Barclay wes hangit in Strieueling, and the vther wes sauffit vpoun cautioun; *Diurn. Occurr.* 181.

b. To spare (a thing), to leave undestroyed.

The admirall .. set Craigfergus into fyre, And saifit nouther barne nor byre; LYND. *Meldrum* 96. That na maner of .. persones .. mak murburne .. bot yeirly .. and that .. upone grass only, saiffand hedder and young wood; 1621 *Black Bk. Taymouth* 352. They brunt and wasted the toun of Wick bot they saiffed the church; GORDON *Geneal. Hist.* 196.

5. *reflex.* To extricate oneself from danger, to act to assure one's safety, to keep oneself safe.

(*a*) Amend thy mis thy self to saif; KENNEDY *Bann. MS* 268b/44. [They] Gaif him counsall na langar for till byde .. And saif him self vnto ane better tyme; STEWART 19616. Thairfoir, deir hart, ȝour self ȝe saue, Howbeit thay tak me; LYND. *Meldrum* 1246. Saif; Id. *Mon.* 3343. Thay .. saifet thame selfes in wodis, mosse, and mure; DALR. I 170/5. Gif .. he had nocht .. fled in Irland .. to safe him selfe; *Ib.* 183/9.

(*b*) For this tempeste sal lest .. Ay till we herbry this ded cors; For-thi, to sawfe ws, cast it owt! *Leg. S.* XVI 479. Gyf that ȝe .. That now may sauf ȝow, gif ȝe will, And out of dangere lithtly fare; *Troy-bk.* I 184. A schyp .. hapnys to .. ryve thai sauf thaim [*A.* sau thaim, *H2* saulfe thaimself] at that tyme the beste wyse at thai may; *Ship Laws* c. 22 (B). Nother party wist .. To salf hym self quhar away to fle; DOUG. X xiii 8.

6. To deliver, preserve or spare (one's own or another's *life*, also *banys, blude, liffis savite*).

(1) That na man acces suld hafe To leche hire na hir lyf safe; *Leg. S.* xlii 190. Gif me some Of ȝour mete .. That I tharewith my lyve may saif [: haif]; *Troy-bk.* II 2540. The Bauderane .. tane With men that wald him sone haue slane Na war Cassanius .. That safit his lyfe as man of pris; *Alex.* II 1904. I say, treuly he awe to save his lyf and defend him; HAY I 200/29. Fra thow be bund thow may mak na defence To saif thy lyfe nor ȝit thy libertie; HENR. *Fab.* 2929. To saiff his lyff thre ȝer he duelt in But; *Wall.* VII 936. The grewhound savit his sonnis lyf; *Seven S.* 483, etc. The fidenatis slang away thare wappynnys, and askit thare live to be saiffit; BELL. *Livy* II 99/23. Thay war fane .. To saue thair lyfis, for to cut thair mane mast; ROLLAND *Seven S.* 569. [He] said that gif thaj wald let in my lord Bothwill he wald saif all thair lyvis and let thame gang hame; *Diurn. Occurr.* 101. Quhen Donald had tint the victorie, he thocht best to saif his lyfe be flychte; DALR. I 182/26.

(*b*) Ilk man to sauff his lyff, Fled; BARB. X 703. To saufe his lyf I hicht lely; *Leg. S.* xl 1143. That kyngys fyve Crape in a cove to sawffe thare lyve; WYNT. II 1194. All man is behaldyn to sauf his awin lyf; HAY I 198/34. I say he may defend him lefully and sauf his lyf again his juge; *Ib.* 201/21. *Ib.* 183/20 (see 7 c below). Bid him salff oure livis and honoure; Id. *Alex.* 3442. Be quhais defence his lyf was saufeit; LESLIE 200. [They] had not faillit to have slane thame, wer not Nicolas and Gilliechallum Rossis, brethir to the said George, saulffit thair lyveis; 1596 *Reg. Privy C.* V 293.

(2) Ane is our liffis savite That suld on na vis savit [*E.* sawft] be Gif thai had vs at thair liking; BARB. IV 537 (C). To saif the blude of man Nane vther way [*sc.* than by single combat] as now counsall I

can; HAY *Alex.* 1793. And we agane .. Besowth my fader to salue his wery banys, And not be wilful to perys; DOUG. II x 161.

b. With inanimate agent or instrument.

That bataillyne with-owtyn dout Sawyt [*C.* saffit] thar lywys; BARB. IV 137. Ane woman .. leit meill fall doun throw the loftis of the toure, be quhilkis his life wes certane dayis savit; BELL. *Boece* II 476. The said Haggerstoune .. shot in his corslate of pruife, which sauet his lyfe; BANN. *Memor.* 133. My lyf with the lyfe of the skipper and companie miraculusly saived by the kok bot; 1640–1 *Ruthven Corr.* 72.

7. To keep intact or unsullied, to maintain, preserve, safeguard (honour, chastity, conscience, etc.). **b.** To maintain or uphold (a right, privilege, or the like). **c.** To keep, not to break (an oath). **d.** To maintain, preserve, not to allow to lapse or be lost or destroyed. Also const. *to* (= for) a person. **e.** To reserve or retain (*to* oneself) from destruction, alienation, etc.: cf. SAV(E)AND *pres. p.* 1. **f.** In expression of benediction, by way of compliment: (May God) preserve (the attribute of the person). Cf. 2 b.

(*a*) Wyrk yhe then apon swylk wys That ȝour honour be sawyt ay; BARB. II 341. Be besy concord fore to draw .. gif that thow may, Saifand thi consciens alway; *Ratis R.* 356. HAY *Alex.* 3442 (see 6 (1) (*b*) above). HENR. *Fab.* 2929 (see 6 (1) above). And grant the frekis on fold farar to fall, Baith thair honouris to saif; *Gol. & Gaw.* 1008. I pray God better his honour saiff Na to be lychtleit sua; DUNB. (OUP) 99/44. A lufsummar leid .. That couth be secrete and sure and ay saif my honour; Id. *Tua Mar. W.* 284. Fy on hir that can nought feyne her fame for to saif; *Ib.* 461. Gyf thaire be ony service at ȝoure grace wyll command me or ony of myne alegeance saiffyt, we sall be redy at command; 1525 *Douglas Corr.* 99. I alone, the quhilk abone all men Suld saif ȝour schame and als ȝour honestie; ROLLAND *Seven S.* 5947. Timothe, says he, saue that thing geuin thé in keiping, eschewand the prophane nouelteis of woices; WINȜET II 61/13. Sa my private and publick impugnatioun did saife my conscience sufficientlie; 1567 Anderson *Collect. Mary* II 282. To saif hir pure virginitie; MONTG. *Misc. P.* xvii 78.

(*b*) I sall thi kyndnes quyte, And sauf thyn honoure; *Gol. & Gaw.* 1102. Thai suld .. nocht fle bot be vassalage sauf thare honoure; *Boece* 115b. That be deid thai mycht saufe thare wirginite; ABELL 13b. I did gritly mervell that he .. shuld not haif considerit of these materris befor thay wer brocht in suche termes that culd not be calde back agayn, princely honour salveit; 1586 *Cal. Sc. P.* VIII 383. Sir James Hackett and Colonell Scotte [were] cleired by the comittee, ȝet that did litle salue ther honor amongest honest men; BALFOUR *Ann.* IV 89. A slender paper used by him for salving his conscience at the swearing of the Test; 1681 LAUDER *Observes* 54.

b. Till ilke kyrk [the] rycht suld stand .. Before all sawffyde [*C.* sauffyt, *W.* sauf, *A.* sauff] the dygnyte; WYNT. VII 740. Till all prestys he dyd reverens And sawffyd thare statys wyth diligens; *Ib.* 3578. The Erle Jhon dyde besynes .. To sawffe the rycht off [his] cwntre; *Ib.* VIII 4470. To sauff my state I haiff maid hir my advocate; *Ib.* IX Prol. 49. To advyse with lawyers how their priviledges might be salved if such a patent should pass; 1681 *Edinb. Surgeons* II p. 238.

c. Quhen he cowertly To sauf his aith did felny; *Leg. S.* xxxvi 548. The said knycht is in a difference: to enter to sauf his ath, or to absent him to sauf his lyf; HAY I 183/19.

d. (1) Be him .. God disponit to saufe the natur of man all othir distroyit be ane flude; ABELL 3a.

(2) Gif goddis lykyt lenth my life langar space Thai wald haue salwyt to me this litil place; DOUG. II x 140. The realme of Latynis and Troianis was sauffit to this childe Ascanius be prudent tutorie of Lavinia his moder; BELL. *Livy* I 15/20.

e. Our souerane lord according to the libertie and priuilege saulffit to him selff .. hes [etc.]; 1587 *Acts* III 436/1.

f. 'Aa, souerane lord, saif ȝour mercie!' quod he. 'My purpois wes with him for to haif plaid'; HENR. *Fab.* 1078.

8. To deliver (a person, his or her soul, etc.) from sin and its consequences; to bring to spiritual salvation.

active He .. til [his] mayster consale gaf To do sa and hymself sawe; *Leg. S.* iv 66. He callit Paule To safe & succur syndry sawle; *Ib.* xv 14. Thy blud that sauit ws all; *Fifteen Ois* 304. God of his iustice mon correct And of his mercy petie haif He is ane iuge .. To pvneis synffull man and saif; HENR. *Abbay Walk* 36. To saif thi saull now schawe thi sapiens; *Contempl. Sinn.* 580 (Asl.). It is mekle better finaly to saif the man na to tyn him; IRLAND *Mir.* II 126/15. Lang tyme eftir that he [*sc.* Trajan] deit .. Sanct Gregor .. prayit for him and at his prayer God of his mercy savit Traiane; *Ib.* III 128/14. He that saulis saifis and dammis; ROWLL *Cursing* 17 (B). Saife [*BD* sayffe]; DUNB. (OUP) 21/167. *Ib.* 13/25. Vthir sawlis to saive it settis thame weill, Syne sendis thair awin saule to the deill; LYND. *Sat.* 4535 (B). The king is not saif be his greit armie .. And mony thousand hors may na man saif; *G. Ball.* 94. Na sanct can saif ȝour saule; *Ib.* 193.

(*b*) Of hyme that sal ws sauf on tre; *Leg. S.* xxxvi 80. Sawfe; *Ib.* xvi 4, etc. Jhesu Cryst [was] of Mary borne Tyll sawffe oure lyff that wes forlorne; WYNT. I 494. Justice .. Haldis pepill in pes & gude prosperiteys Salfis thaire saulis makis thaire saluacioun; *Regim. Princ.* 158 (Fairf.). Salue; NISBET *Jude* 23.

(*c*) And vthir sempill peple trowit as the kyrk trowit that tym and that saffit thaim; IRLAND *Mir.* II 13/4. He saffit us nocht be the mediation of Moyses, bot be the bitter .. passion of his awin naturall sone; HAMILTON *Cat.* 38.

passive & mony sonnis of Israel .. To God, thar lord, turne sal he, & sagat thru hyme sawit be; *Leg. S.* xxxvi 66. He .. that treuly kepys this informacione .. he beis sauit; *Cr. Deyng* (STS) 268. Saiffit or damnit, knew we in speciall, We suld na prayer .. for thame expreme; *Contempl. Sinn.* 669 (Harl.). Sauet; NISBET *John* iii 17. Saiffit; HAMILTON *Cat.* 127. Qvhat haif ȝe for ȝou, that the infantis of the faythful ar saifit alrady but baptim; WINȜET I 81/18. I .. luikis to be sauit be na mene bot be .. Jesus Christ; 1593 *Edinb. Test.* XXV 121b. In tham that ar saiffed and in tham that pearishe [2 Cor. ii 15]; 1600 Melvill 475. 1615 *Edinb. Test.* XLVIII 256.

reflex. This way heffing seffit my lyff .. the Lerde of Fentre thinking on guid ressoun .. to seff him selff sufferit allenerle for his relligioun; 1593 *Innes Rev.* III 103.

b. In asseverative expressions, esp. *sa God* or *Christ me save.*

(1) Sa God him self me saiff [: haff *v.*; *C.* saff] Ik hald me rycht weill payit; BARB. XX 220. The Bauderane said, 'Sa God me saif, The ressoun quhy fane wald I haif'; *Alex.* II 2381. Schyre, (sayd he,) sa God me sawe, The kynryk yharne I noucht till have; WYNT. VIII 243. So me Crist saif; *Howlat* 120 (A). Sa Christ me saue; *Rauf C.* 500. Yhe sall it haiff, als God me saiff in saille; *Wall.* v 675. My sempillnes .. Wait off na way, sa God me saiff; DUNB. (OUP) 71/22. Sa God me save; *Ib.* 121/98. Saife; LYND. *Sat.* 4461 (Ch.).

(2) Marge Gelletlie .. purgit hir be hir great aith Lat neuer hir saull be saifit if euer scho knew him; 1619 *Perth Kirk S.* MS 7 June.

9. In prec. senses, with participial or adjectival complement: To rescue, keep, preserve, etc. *lefand, on plunderit*, etc.

(*a*) For thare hele .. God had hym sawyt than lefand; WYNT. II 408. To saiff the lordis injustified [PITSC. I 183/13, from iustifieing] in the tyme of the kingis furie; Pitsc. (1814) 201. The toun wes saiffit on plunderit by payment of this soume; SPALDING I 213.

(*b*) Fere leware he had .. A man off his to sauff wnslayne Than off hys fays to sla .. a thowsand; WYNT. V 1445. So hail and feir mot salf me Jupiter; DOUG. IX iv 98. Ȝe may sauff the bairnis of Fergus vndefraudit of thair kyndlie heretage; BELL. *Boece* (M) I 53. I pray to God him speid And sauf him sound in till his leill travale; *Freiris Berw.* 65 (M). He wald haif sauffit the sone of ane theif, being his tennent, unjustifiit, allegeand [him] .. to be his awin; 1564 *Reg. Privy C.* I 306.

10. a. To keep undisposed of, to retain intact in safe keeping (goods). **b.** *absol.* To save up (money, goods), accumulate by saving.

a. That sic schip .. nor na thing within hir salbe adiudged wraik bot the guidis sall be saifed and keiped be the sicht of the schiref coronall; BISSET II 213/1.

b. Gett and saif and thow sall haif, Len and grant and thow sall want; *Bann. MS* 147a/4.

11. a. To safeguard or deliver a person from (disgrace, shame).

e.g. by behaving discreetly. Const. possess. of the person and noun as direct object.

It moste be to me leful rewth to haue Of my frendis myschance, hys lak to save, Quhilk in hys awyn defalt tynt not the gre; DOUG. V vi 112.

b. (To take the action needed) to avoid or prevent (an undesired outcome).

That Wm. Fraser .. would be excommunicat the next sabboith day according to ane process led and deducit againis him befoir the Presbytry unles he saveit the same; 1629 Cranna *Fraserburgh* 184. To .. saive the schedding of much innocent blood; 1639 *Rec. Kirk Scotl.* 215.

12. To give protection from, fend off, keep out (something material).

With inanimate agent.

The plesant plane-trie will the leavs vnfauld With fairest schaddow to save the sone in symmer; 1584 SEMPILL *Sat. P.* xlv Pref. 46. Thre daillis to saiff the weit abowe the sclaittis that na weit cum within the scoole; 1595 *Edinb. D. Guild Acc.* 582.

13. In prec. senses, const. *fra (from)* and *for* **a.** A person, **b.** An impending misfortune, undesirable consequence, etc.

a. I a maid; and standis in mony stour, Fra Inglismen to saiff my womanheid; *Wall.* v 691. Than think I hiddowus Mahowne hes me in armes Thair ma na sanyne me save fra that auld Sathane; DUNB. *Tua Mar. W.* 102. Fra sic a syre God ȝow saif [*M.* sauf] my sueit sisteris deir; *Ib.* 145. Quhou Perseus sauit Andromada fra the cruel monstir; *Compl.* 63/14.

b. (1) Giff God ws sawys fra persawing Off thaim; BARB. X 576. The ded he tholyt .. To safe ws fra the wethirwyne; *Leg. S.* xvi 246. Saf; *Ib.* xliii 120. Tharfor women suld kepit be, That thai mycht na licht women se, .. It savis thame mony a tym fra ill; *Gud Wyf & D.* 226. He commandit the round tempil .. to be cassin down... Yit the inhabitantis saiffit the samin fra uter eversioun; BELL. *Boece* II 378. He is callit .. saluiour for he saiffis ws fra our sinnis; GAU 36/1. LYND. *Mon.* 290. This fell serpent he hes ouircum .. And sauit the childe fra perrell; ROLLAND *Seven S.* 1809. ȝe haue saift this day fra schame and skaith ȝour awin honour & als my douchters baith; *Ib.* 9572. To procur in the caus of Elizabeth Payt .. to save hyr caus fra paresing becaus sche is seik and maye nocht compeir; 1565 *St. A. Kirk S.* 251. Defend vs from temptatioun .., Saif vs from schame and from dispair, From unbeleue and Lollardis lair; *G. Ball.* 13. To .. saif vs fra exces and drunkinnes; *Ib.* 19. SKEYNE *Descr. Pest* 17. From forfalting he sauit his lands; 1570 *Sat. P.* xvi 56. He saift thé from the gallous; *Ib.* xviii 9. As they might saiff thair contrie fre skaith and boundage; PITSC. II 103/28. Saiffe; *Ib.* 239/21. Of the office of ane Christiane magistrat .. To hauld hand to thame alsweill concerning thair awin personis, saifing thame from iniurie and oppin violence; 1581 *Bk. Univ. Kirk* II 503. Sawitt [CULLEN *Chron. Aberd.* 61, saifitt]; *Montg. Suppl.* xxxiii 36. A verie high mountaine .. which saiffs and gairds the shipps from the wind which doth come on the east; *c*1630 *Macfarlane's Geog. Coll.* II 187. Monro causit big wp .. ane court de guard, for saifing of his soldiouris fra weit and cauld on the nicht; SPALDING I 278.

(*b*) To sauf thame fra his venyme; *Leg. S.* xxxiii 78. And we sall sawffyd be fra blame; WYNT. IV 799. All that ryfft togyddyre crape And sa was sauffyde the cyte Fra stynk than, and mortalite; *Ib.* 1223. Devotly Prayer he mad .. For till relewe hym .. And tyll sauff hym fra that deffame; *Ib.* v 4203. [They] suore to sauf thame and kepe thame sekirly fra all harme; HAY I 29/29. The wemen sone he sauffyt fra the dede; *Wall.* v 1127. Ambrois .. in ane littir tursit be twa grete hors to sauf his person fra troubill; *Boece* 286b. Traisting throw strenth of the place .. to sauf thare flamynis, thare preistis [etc.] .. fra occisioun & fire of inemyis; BELL. *Livy* II 209/4. Our lord Jesus me sauf from sic ane cace And keip me out of perrell and fra schame; *Freiris Berw.* 88 (M).

(*c*) Quhou Neptune his navy salvyt [*Sm.* saifit, *Ruddim.* sauit] fra wraik; DOUG. I iii heading. Be my laid star virgyne moder .. from hell, salue al go not to wraik; *Ib.* III Prol. 45. God salf ws all from sik a syre as this! *Ib.* XI Prol. 72. Because he tuke the saidis gudis be ane pretendit titill .. the quhilk at that time not beand declarit null, aucht and sould salve him fra all crime of spuilȝie, albeit the samin may not salve him fra wrang; BALFOUR *Pract.* 470. Salve; PITSC. II 55/25.

(*d*) The pylgram, that sa Saffyt hym fra his felone fa; *Leg. S.* iii 1124.

reflex. He counsalit hir to seike remeid, To saif hir-self frome felloun deid; *Troy-bk.* II 2004. Now Crist, .. Cum of the croce & saife thé fra the ded; KENNEDY *Pass. Christ* 966. Hys hed coverit, to salf hym fro the dynt; DOUG. XI xiii 68. As maist cowart .. bestis saiffis thaimself for dangeir; BELL. *Boece* I 152. King Druskene .. with ane fewe cumpanye salfit him self fra dangere; *Boece* 365.

(*b*) Sa ma thu thé sauf fra hel; *Leg. S.* xxx 257. Gyf thow will sauf thé fra schowris Of gret dises, of new amouris Be weil ware [etc.]; *Ratis R.* 1038.

(2) Alswa betwene the chapellis guteryt with hewyn stane to cast the watir owte and to save the werc for the watir; 1387 *Edinb. Chart.* 35.

14. To clear oneself from (the imputation of legal default, or its consequences) by offering an acceptable excuse.

At the ferd court of the mut he sal at his entra warand his essonȝes and gif he may set for hym certane resonis to sauf his distres thai salbe sufferit; *Burgh Laws* c. 102 (A). Sa that .. the tennande sal be neuermar herde thairof [*sc.* if he fails to appear on the fourth day] bot gif that he may sauf the dais of his defalt be the lorde the kyng warande & be the kyngis breff; *Reg. Maj.* c. 5. And gif the ferd day the party deffendand aperis & cumis & can salf his defaltis he sal nocht be amercyt bot he sal ansuer til his party of the querel; *Quon. Attach.* c. 4.

15. *To sauff* occurs in BARB. and WYNT. in senses of SAV(E)AND *prep.* and of SAFE *prep.* **a.** *prep.* Excepting only, except for, except. = SAV(E)AND *prep.* 2 a. Only in WYNT. **b.** With due respect to, without offence to. = SAV(E)AND *prep.* 2 c.

a. Quhare nowthire man na wywe To sauff geawntis thare he fand; WYNT. III 541. For nane thare governale than had To sauff barnys off yhowthad, Threllys [etc.]; *Ib.* IV 1598. But barnys tha bredyre deyde ilkane, To sawff Robert oure kyng allane; *Ib.* VIII 1326. *Ib.* IV 1168, 2438, V 420.

b. 'Schyr,' said he, .. 'To sauff ȝour presence it [is] nocht swa'; BARB. III 173.

Save, var. of SAFE *adj., prep.* and *conj.* and SAW *v.*[2]

Sav(e)and, Sav(e)ing, Sa(i)fand, Saufand, Sauffing, Saffand, *pres. p., prep.* and *conj.* Also: **savyng, sawend, -yn, sevand, saifeand, saifane, saiffand, saif(f)ing, sailfand, sayffande; sauff-, saulf-, salf(f)-, salv-, sawuand(e, saulf(f)ing, salving; sa(u)vant.** [ME and e.m.E. *sauynge* (Chaucer), *savinge, sauing* (both Tindale), (late) north. ME *sauing* (*c*1400), *safeyng* (*c*1440), *sauand* (*a*1400-50), absol. use of pres. p. of SAV(E *v.*]

A. *pres. p.* Reserving or retaining *to* a person, freq. the original possessor (a right, possession, expense, etc.); also, once, *fra* harm, an imposition. Cf. SAV(E *v.* 7 e.

(1) Sauand to me my seruys aught and custum thar of; 1400 *Maxwell Mem.* I 139. Safand to the said Johne Stevart the price of

all distroyit byggyngis that may be recouerit be the law; 1460 *Stewart Mem.* 64. Ve herfor .. has .. confermyt the saide chawmer [etc.] .. sayffande to ws and our successouris suffrage of dyvyne seruice of the chapellains of the saide alter alanerly; 1493-4 *Reg. Episc. Aberd.* I 334. 1528 *Acts* II 329/2. Saiffand to thaim and every ane of thaim thair just and lauchfull defenses; 1532-3 *Dunferm. Reg. Ct.* 94. Saifeand; SKENE *Reg. Maj.* I 13. Saiffand and reserveand allwayes to our said soverane lord .. rightes service and few dewties aucht and wont to his hienes; 1641 *Acts* V 438/1.

(*b*) That the forsayde schyr Jon sall gyf to the forsaide Doncane .. al hys land of Kylmanyne saufand til the forsayde schyr Jon hys tenandrys [etc.]; 1389 Slater *Early Sc. Texts* No. 16. Saufand to the forsaide religeous men courtis of lyfe & lym [etc.]; 1402 MacRae *Early Sc. Texts* No. 9. Giff that he has mony dochtiris betwix thaim salbe departit the heretagis .. saufand the cheff tenement till the eldest sister; *Reg. Maj.* c. 59. The said Cristiane has set and to ferm latyn .. to the said Alex. [etc.] .. al and sindry hir landis of Stratoun [etc.] .. saulfande anerly to the said lady bludwite and merchete; 1447 (1451) *Reg. Great S.* 106/1. Saufand to the saidis partijs lauchfull excepcions again the persouns that salbe chosin vppoun the saide gret assise; 1478 *Acta Aud.* 83/2. Saulfand to the procuratouris abonewritin .. al excepcions of law .. til be proponit and allegit be thame agane the sade day; 1498 *Acta Conc.* II 200. A precept of confirmatioun of the crownarschip of Carrik .. salffand to the kingis hienes service auch and wount; 1499 *Reg. Privy S.* I 50/1. Johne Sym offerit .. to .. mak the twa pannallis .. and offerit to find souerteis for doing thairof vpoun his awne expens, sauffand to him the said tymmer and sowme of xl li.; 1556 *Edinb. B. Rec.* II 254. Saulfand alwayis to oure soverane lord .. the sufferage of devoit orisoun .. aucht and wont to his hienes; 1575 *Reg. Privy S.* VII 27/2.

(*c*) To be haldin .. to the said Johne and Agnes .. saulfing alanerly to ws and oure successouris the cariage of samekl of the saidis landis as the said Johne and Agnes occupiis and manuris nocht; 1477 *Exch. R.* VIII 403 n.

(2) Bot sauand my senyeoury fra subiectioun And my lordscip vn-lamyt withoutin legiance All that I can to yone king .. I sall preif all my pane to do hym plesance; *Gol. & Gaw.* 441.

B. *prep.* **1.** Reserving or withholding only; except for; excepting, except. = SAFE *prep.* I.

(*a*) His straikis ar .. of the best that leuand is In warld, saiffand the king allane; *Alex.* I 1363. I sall infefft the said Dauy .. in ten pundis wortht of land qwhar euer him likis best .. safand anerly the chemys of the Bord; 1425-6 *Antiq. Aberd. & B.* III 202. Als mekil land liand on the south half thar oxgang safand a gate liang betuix of fourty fute brede; 1437 *Reg. Dunferm.* 285. All .. Left the castall .. Saifand a barne; *Seven S.* 427. Saiffand the moving of the kingis knycht; *Bk. Chess* 2068. To resing in the handis of James Abbot of Dunfermlyn his landis of Kildun .. saiffand & reservand the franktenement of the said landis for al the dais of his lif; 1505 *Reg. Dunferm.* 374. Quhilk salbe maid betweix the dait heirof and the nixt processioun saiffand the octavis of Corpus Christi day; 1532 *E. Loth. Antiq. Soc.* II 99. And to deliuer thair schip agane with hir pertinentis safand awentour of see to the saidis Thomas and Robert; 1540 *Aberd. B. Rec.* I 173. Besekand yowr grace, sawend greteir besynis, advertys me gyf yowr grace ples command me ony wthyr service; 1548 *Corr. M. Lorraine* 231. Ordanis James .. to pay for the hale xvij stane of lint .. saiffand that is the rafues of thre stane allanerlye; 1563 *Dumfries B. Ct.* fol. 182. The sea cummis almaist betwix thame, saifand ane small grip of the lenth of twa .. pair of buttis; *Descr. Isles* 429. Quhilkis all ȝeildit to, saifand Glenurquhay; 1615 *Black Bk. Taymouth* 48.

(*b*) To my best hound .. Quhilk I lude best, without ony compair, Saifing my wife [etc.]; ROLLAND *Seven S.* 1946. The said Jhonne had na vtheris guddis saifing his prenting irnis; 1563 *Edinb. B. Rec.* III 170. They did sequester all my servandes from me, saiffing two; 1571 *Misc. Bann. C.* III 121. GORDON *Geneal. Hist.* 417. Not to suffer said loadinges .. to be wait or damnified .. during said voadges, saiffing perells or men of war be sea; 1660 *Old Ross-shire* I 186.

(*c*) Bot thole the sayd Robert & hys airis joyis thaim frely .. saufand the ten markys worth of land at is callit Murraf & Hedhouse [etc.] .. and saufand the successyoun of lyne gyf it hapnis to faylȝe of the forsayd Robert; 1400 Slater *Early Sc. Texts* No. 48. Al my landys of Kilmany saufand the landis the qwhilkis Alisander of Ceton haldis of me; 1400 *Charter* (Honess Doc. No. 1, Reg. H.) MS. Na man may mak knychtis bot he be knycht before, sauffand the pape; HAY II 20/37. Thai sal .. vphald the dykyng of the sayd hill to the sayd tym, sawuand hedromys of croftis; 1472 *Peebles B. Rec.* I 168. Becaus thar is accquittance schewin .. of the remanent of the hale some salfand a mark; 1483-4 *Acta Aud.* *131/2. Savffand; 1491 *Ib.* 151/2. The thre first bukis he has ourhippyt quyte Salfand a litill twychyng Polidorus; DOUG. I Prol. 155. Salvand [*Ruddim.* Saufand]; *Ib.* IX Prol. 96. All ourecoverit sauffand thare hedis; BELL. *Livy* I 18/9. Saufand beaute I can prise na mare Of hyr; *Dunb. App.* vii 13. Saulfand and exceptand allanerlie one half aikre of the foirsaidis landis; 1592 *Waus Corr.* 491.

(*d*) Becaus na thing was than restand amang thame sauffing thare mvnycioun and armoure; BELL. *Livy* I 13/8. Sauffing suche as ar attaynted; 1569 *Reg. Privy C.* II 4. And without ony provocatioun maid .. saulffing assisting to the proclamatioun .. he wes assaultit; 1570 Bann. *Memor.* 359. That thay wan litill honour .. sauffing that thay brint the kirk of Annan; LESLIE 202. Sa that na place thairof salbe .. garnist saulffing the castellis and fortressis that .. hes bene accustumat to be fortefeit; 1572 *Reg. Privy C.* II 159. Saulffing onlie a small tyme of abstinence frome crueltie and unlauchfull hostilitie; 1592 *Douglas Corr.* 38. Thay haveing nathing saulffing thair buikis in thair handis; 1594 *Reg. Privy C.* V 162. Saulffing that particulair of pistollettis; 1611 *Lett. & St. P. Jas. VI* 199.

(*e*) The said schire Jone of Wemys .. sal for this fredome .. be next in the hoppyr qwhays corn sa euire be thare in, saffand the said Jone Malvyle; 1427 *Wemyss Chart.* 53.

(*f*) The enhabitans of al Teuydale sauuant the forest of Jeddeworth; 1386 Slater *Early Sc. Texts* No. 9. Ther sal be non entercomynyng betwene the roialmes sauant the manere exprimyt; *Ib.*

2. Reserving, without prejudice to.

(*a*) [He] gayf grund sessyng .. to Sym Conno [etc.] .. safand ilk manis rych; 1456 *Peebles B. Rec.* I 117. To perten to the said Lard of Innes saifand all ryt of bundit chartir gif ony be; 1493 *Reg. Episc. Morav.* 250. Saifand allwais the priuilege of fre burgessis of this burght; 1541 *Perth Guildry* MS p. 513 (31 Oct.). Saifeand alwise the service to the other over-lord for the lands haldin of him; SKENE *Reg. Maj.* 142. Saueand alwaie the fredom and liberties in sic cases, giuen to lords be kings of Scotland; *Ib.* II 21b.

(*b*) Sawyn all mennis rychtis; 1488 *Dunferm. B. Rec.* I 1. Saiffing all other mennis rigchtis; 1551 *Dundee B. Ct.* II fol. 71 (30 June).

(*c*) Thai that hingit hym sal mak fine wyth the kyng saufand thaim lyf and memberis and disherisone; *Acts* I 5/2. Sauffande; *Ib.* 26/2 (see SAFE *prep.* and *conj.* 1 b). Puttand erd and stane in thar handis as maner is, salfand ilke mannis rycht; 1490 *Lanark B. Rec.* 8. Sauffand the persecutioun of the caus befor the juge spirituall; 1545 *Reg. Privy S.* III 193/1. And saufand every mannis rycht obtenit heirto before; 1559-60 *Ib.* V I 171/1. Promitting that he suld doe ony thing possible to be done salfand his lyif; *Diurn. Occurr.* 65.

(*d*) I gaiff .. sesing [etc.] .. saffand all menis rychtis; 1490 *Reliq. Antiq. Sc.* 49. And gif ony .. debait happinis amangis ony of the saidis craftis, thai till meyne the samyn amangis tham self .. saffand the priuilegis pertenand to the toun; 1520 *Edinb. B. Rec.* I 200.

3. With deference to, with due respect to.

(*a*) Thairfor (saiffand thair honour) guid it war to interdit [etc.]; *Art of Music* 9b. Saifane ȝowir lordshippe's better awys and opinione; 1586-7 *Cal. Sc. P.* IX 328.

(*b*) I .. hir demandit, sauyng hir plesance, Quhat wes hir name; LYND. *Dreme* 152. Than said the sone, Sauing ȝour graces will, To that sentence na way grant will I till; ROLLAND *Seven S.* 10625. And ȝow (saifing ȝour dew honoris we speik); WINȜET I 9/18.

(*c*) Salffand the plessour of the kingis hienes ȝour soverane, and of your grace, I set it at litill; 1527 *Douglas Corr.* 122.

4. Leaving out of consideration, setting aside, not to mention.

[They] denyit the thrid and secund exceptionis peremptouris saufand the ineptitudis and irrelivantis thairof; 1559 *Admir. Ct. Bk.* (St. S.) 124. Ȝit considering thair injurious and intollerable proceadings afoir this tyme be secreit surmysis, salving thair untrew sklanderis agains our said soveranis honor, and now publiclie, ȝea and desperatelie braiging to cloik thairby thair treasonable doings, we cannot [etc.]; 1568 Anderson *Collect. Mary* IV II 160.

5. Avoiding, eschewing (a commitment or duty).

Thair is mony inconstant sluggish & idle persones uha saveing service gives thameselves to traffict and play the merchant; 1615 *Kirkwall Sheriff Ct. Act Bk.* 5b.

C. *conj.* **1.** Introducing a finite clause, with or without *that*, stating an exception: Except that.

(*a*) Jon sal resyng al the landis .. wyth al the fredomys .. savand the for sayd Jon sal joys the frank tenement for the tym of his lyve; 1395 MacRae *Early Sc. Texts* No. 4. The sade Wilȝam beand alegyt and fre of al aucht and wunt seruys savand gyfe in tyme of harueyst we gader our tendis that he supple and help; 1472-3 *Reg. Cupar A.* I 168. This bataile nedis na borrois of thair entrais .. saiffand that euery partii sale enterchangis selis; *Loutfut MS* 110b. All the pepill of his tovne Askit at God for to depryve him dovne Saiffand thair was ane ald wyf .. That for him prayit; *Bk. Chess* 1544. He sal laboure .. to mak tenentis vndir hym, saifand Donalde Makferchar sal remane with his tak for his lesing [*pr.* lefing]; 14.. *Reg. Cupar A.* I 232. All bair heided, saiffand he hade on his croune; *Clar.* IV 2616.

(*b*) Saufand that his spousyt wif aucht till haf .. the innerhalf of the hous; *Burgh Laws* c. 85 (A). [He shall] gife oure his lyferent of the foirsaidis landis .. saulfand that gife the benefice .. beis evictit .. he sall have full and fre regres; 1550 *Lochwinnoch Par.* 22.

2. Introducing an adv. phrase: Except.

Saiffing allenarly at Witsounday; 1443 (*c*1580) *Edinb. B. Rec.* I 7. That the saide merchandis of Irwine .. aucht to cum within .. Are nouthir on the merkate day na yit nane uthir day of the wolk to tap na gudis of merchandise .. sauffand anerly on the fare day; 1443 *Ayr Chart.* 28. We sal tak the said erlis .. part in all .. quarellis .. contrar ony personis, sevand in our soviane lord the kyngis service, 1527 *Caldwell P.* 59. That na nebur .. persow ane vther affor ony vthir juge sailfand affor the prowest and balyeis; 1564 *Peebles B. Rec.* I 297. As for the graithe of the stabill .. he .. denyis to haif had further intromission with ony of the same saulffing in ane coffer left [etc.]; 1573 *Inv. Q. Mary* clvi. Thair salbe na passage ower Forth be the ferrie boittis at ony pairt, saulfing onlie betuix Leith and Brint Island; 1584 *Reg. Privy C.* III 696.

Savear, var. of SAVAR *n.* **Save-conduct, -conduit(e, -condute,** varr. of SAFE-CONDUCT *n.*

Save dira. Appar. repr. (? erron.) F. *sans dire* (without saying). — Save dira adew his leive he tuik; 1584 SEMPILL *Sat. P.* xlv 662.

Savegaird, -garde, -gaurd, varr. of SAFEGUARD *n.*

Sav(e)ing, Saufing, *vbl. n.* Also: **savyng, sawing, saif(f)ing, saiveing, saufyng, sauffing, saulffi(i)ng.** [ME and e.m.E. *saueing* (Cursor M.), *safynge* (Rolle), *sauynge* (*c*1460), *saving* (1477), *sawing* (1551); SAV(E *v.*] Saving, delivering, preserving, salvation, etc., in senses 1, 2, 6, 8, 9, 10 and 13 of the verb.

(1) The samin [*sc.* shipwrecked goods] salbe recovered be thame [*sc.* the owners] they refoundand the uncostis maid in saiffing thairof; BISSET II 214/3.

(*b*) Of saufyng of gudis of a brokyn schyp; *Ship Laws* c. 21 (B). Quhen a prisonare is tane, merci is aucht him and that his maister aw to do him grace .. And gif him thinkis he will have sum courtasy for his saufing and keping and costis making [etc.]; HAY I 159/14. That na maner of persoun of the inhabitantes of Edinburgh now remaning in Leith pay stent or furneis money to the saidis rebellis .. now abiding in Edinburgh under cullour of saulffiing thair geir within the toun or utherwise, under the pane [etc.]; 1571 *Reg. Privy C.* II 90. Saulffing; *Ib.* 102.

(2) [To] big pelis .. for saifing of thare selfis men tennentis and gudis; 1535 *Acts* II 346/1. The laird of Wormestoune was slaine in defence and saveing of my lord regent; PITSC. II 263/27. For saueing of the botis scheris iiii s. 8 d.; *c*1598 *Skipper's Acc. (Morton)* 14a. Thair hes bein na haineing or saiveing for the young grouthe; 1605 *Elphinstone Chart.* 147. To bige ane ewin dyick .. and that for saiffing [*ed.* saissing] of his awin cornes; 1629 *Inverness Rec.* II 167. Sufficient herdes for keiping of their bestiall and saveing of their corne; 1658 *Stitchill Baron Ct.* 15.

(3) And it salbe comptit to the .. prouffit and saufing bathe of mony saulis and lyfis of thy peple; HAY II 102/24. The saifing off thair toun from distruction ther bulwark from the water ther estaits ontakin from them and the saifing off ther lyfs; 1644 *Aberd. Council Lett.* II 370.

(4) Think on Traiane .. Haid singulare sauffing [*Asl.* savyng, *Arund.* saving] throw iustice and piete; *Contempl. Sinn.* 218 (Harl.). HAY II 102/24 (see (3) above). For the health and saifing of ȝour awin saulis; WINȜET I 39/15. The destruction of vyces and sawing of men; *a*1599 James VI *Basil. Doron* App. 309.

(5) Be the plesour off God, the sawing off Crissting blude oneschade, and velhte off bothe the ralmis; 1545 *Douglas Corr.* 158. 1644 *Aberd. Council Lett.* II 370 (see (3) above).

(6) Wee sall graunt .. conuenient place .. to ressaive the said teyndis [*sc.* sheaves] for keping and sauffing of the samynn; 1540 *Antiq. Aberd. & B.* II 252.

(7) For saiffing of the jestis therein fra rotting; 1590 *Culross* I 131. 1644 *Aberd. Council Lett.* II 370 (see (3) above).

Saveing, var. of SAV(E)AND *pres. p., prep.* and *conj.* **Savely,** var. of SAFELY *adv.*

(Savenap,) Saufnap, *n.* [ME *sauvenap* (1312), *salvenap* (*c*1330), *savenap* (1337), *saufenap* (1391), *sawnape* (*c*1400), *sanap* (14-15th c.), last recorded *a*1440, OF **sauvenape* (OED), f. *sauver* SAV(E *v.* and *nap(p)e* table-cloth (see NAPKIN *n.*).] 'A strip of cloth placed over the outer part of the table-cloth to preserve it from being soiled' (OED). — In saufnapis, xvj vlne; 1331 *Exch. R.* I 387.

Saveour, var. of SALVIO(U)R *n.* **Saver,** var. of SAVAR *n.* **Saveraill,** var. of SEVERAL *adj., adv.* and *n.* **Savering,** var. of SAVOURING *vbl. n.*

(Save-skalbert,) Salf-skalbert, *n.* [SAV(E *v.* and SCAUBERT *n.*] ? A protection for a scabbard. — Deliverit to Thomas Softlaw to be skalbertis .., Item deliverit to him to be salf skalbertis ane blak cordwall skyn; 1541 *Treas. Acc.* VIII 26.

Savet(i)e, -ye, varr. of SAFETY *n.* **Savice,** see SAFETY *n.* **Savin,** var. of SEVIN *num.* and *n.*

Saving, *ppl. adj.* [ME and e.m.E. *sauuand* (Cursor M.), *sauynge* (1535), *sauing* (1581).] (Spiritually) saving, imparting salvation, redeeming. — Thy sauing heill and rychteousnes; *G. Ball.* 136. My soule reioice mott in thy sauing grace; JAMES VI *Poems* II 24/14. An honest and saving mean of Christ for mortification and faith's growth; 1637 RUTHERFORD *Lett.* (1894) 504.

Saving, var. of SAV(E)AND *pres. p., prep.* and *conj.* and SAV(E)ING *vbl. n.*

Saving-tree, Sawing-tree, *n.* [ME and e.m.E. *savin* (*a*1387), *saveine* (Gower), *savine* (1590), OE *safin*, OF *savine* (*c*1130 in Larousse), L. *sabīna.*] Savin, the shrub (*Juniperus sabina*). — Two handfullis of the cropes of the sawing tree; 16.. *Nat. Lib. MS 22.2.11*, last p. Elspeth Jamison .. is suspect to be with child and indeavouring to put back the birth and for that end had been seeking the saving tree from the gardiners and did leap over the bastealie this day fourteen dayes; 1709 *Wemyss Kirk S.* 11 Sept.

Saviour, var. of SALVIO(U)R *n.* **Savirles,** var. of SAVOURLES *adj.*

(Savit,) Saifed, *ppl. adj.* [SAV(E *v.* 1 b; cf. ME (Cursor M.) and e.m.E. *saued* (1509) spiritually saved, delivered from damnation.] Preserved, salvaged (goods). — The maister may lay to pledge sum of the saifed guidis; BISSET II 241/28.

Savit(i)e, varr. of SAFETY *n.* **Savoer, -oir, -or,** varr. of SAVO(U)R *n.* and *v.* **Savor,** var. of SAVAR *n.* **Savorand,** var. of SAVO(U)RAND *ppl. adj.*

(Savorie,) Suvorie, *adj.* [ME and e.m.E. *sauure* (*a*1225), *savory* (Chaucer), *sauoury* (*c*1400), OF *savouré*; SAVO(U)R *n.* and *v.*] = SAVOR(O)US *adj.* b. Only *fig.* — Yea, shining lights and suvorie salt, In word and deed they may be that; ANDERSON *Winter Night* 23.

Savoring, var. of SAVOURING *vbl. n.* **Savorit,** var. of SAVO(U)RIT *ppl. adj.*

Savor(o)us, *adj.* Also: **savourous, sawouris, sawris.** [ME and e.m.E. *savorous* (Chaucer), *sauourous* (Caxton); SAVO(U)R *n.*] **a.** Sweet-smelling. **b.** Pleasant to the taste.

a. Vnder the cerkill solar thir sauorus seidis War nurist be dame Natur; *Howlat* 31 (A). With precious unguementis eftir the sesoun savourous; HAY II 122/4. And syne pas till a faire bed, wele grathit with savourous clathis and clene; *Ib.* 144/13. Ane place .. Moir sauorus sueit than sempill style may schaw; J. STEWART 19/146. Sueit sauorus smell in sop all soupit scheine; *Ib.* 54/52. *Ib.* 171/3. Hir breathing savorous, Hir rosie lippis most eminent; MONTG. *Misc. P.* xxxv 42.

b. Mony sawouris [*B.* sauorous] sals with sewaris he send; *Howlat* 705 (A). Servit thai war of mony dyuers meis, Full sawris sweit; *K. Hart* 420. Gif salt be vnsauorous in quhat thing sal ye mak it sauorous [*W.* sauori, *P.* sauery]? NISBET *Mark* ix 49.

(Savorouslie,) Savoruslie, *adv.* [e.m.E. *saverously* (Caxton); SAVO(U)R *n.*] With a good savour, pleasantly, sweetly. — Ane mydow fair, Quhair flouris fresche maist sauoruslie did smell; J. STEWART 84/163.

Savoryng, var. of SAVOURING *vbl. n.*

Savo(u)r, *n.* Also: **savoure, savoir, sawo(u)r(e, sawowre, saweoure, sawure; sauer, -ir, sawer; sauuour; saur, sawr; sair.** [ME and e.m.E. *sauur* (Ancr. R.), *sauueur* (Cursor M.), *sauer* (Manning), *safour, sauour* (both *c*1315), *sauor* (14th c.), *saueure, saueoure* (both 1422), *sauoure* (Caxton), OF *savor, saveur* (*c*1130 and 1256 in Larousse), L. *sapor.*]

1. Flavour, taste.

Claret wyne .. chosin be the odour colour and savour; HAY II 140/33. Ole of the olive causis the watter to haue bettir sauour and gust; IRLAND *Mir.* III 16/20.

2. (A) smell, perfume, aroma.

(1) Of the cors com swet sawoure; *Leg. S.* ii 275. With that thai feld a sweet sawor; *Ib.* l 673. Saweoure; *Ib.* iii 850. Sawure; *Ib.* xvi 130. Sum [waters] may be knawin be thair weicht and hewe and be thair savoure as bryntstane; HAY II 138/25. Rosis, lilleis .. and all manere of colouris, sauouris and precius odouris; IRLAND *Mir.* I 58/22. As the balme .. als lang as it is includit in the boist makis na gret sauour; *Ib.* 145/26. Quhat sweit savour and smel from hir spryngis; DOUG. V xi 95. Pommes de pin .. quhen thai ar new rype thai haue guid sauour; *Lindsay MS* 39. *Nidor*, a brint sauour; CARMICHAEL *Etym.* 7. Ambre .. is .. of a goldne colour and of a sueit sauor; DALR. I 48/1. He let a fart farneyeir and is gon to sie quhat savour it hes the yeir; Ferg. *Prov.* MS No. 772. With a strong senting savor [*sc.* of sulphur dioxide]; 1646 J. HOPE *Diary* (1958) 178.

(*b*) And [in] the end [the Phoenix] passis til a gud tre and of gud sauuour [F. ondeur]; *Loutfut MS* 22b.

(*c*) My wyne ȝard .. Full of gude sawr and suetenes; *Gray MS* iv 38.

(*d*) For sair [*v.rr.* savour, sawour; *Asl.* stynk] syne he did spitt; DUNB. (OUP) 156/177. Within few dayes na man durst com neir [the corpse] for evill sair; *Diurn. Occurr.* 334.

(2) Persawand prowd sawoure thare Of sottyne flesche; *Leg. S.* vii 705. I seke the sawowre of that ros; WYNT. I Prol. 126. Men that liffis with the sawoure of quhyte apillis and deis quhen thai want that sawour; *Asl. MS* I 167/18, 19. And blist be thi haly neis that wes worthy to feill the hevinly sauour of thi barne Jesus; *Rois Garlandis* 361. For till eschew the dishonesty and euil disposit savour thairof [*sc.* of tallow]; 1505 (*c*1580) *Edinb. B. Rec.* I 107. Or beir my bodie ad tabernam Quhair I may feill the savour of drink; DUNB. *Maitl. F.* 156/43. We smel with our neyse the savoir of breid and wyne; HAMILTON *Cat.* 206. Sawer; ROLLAND *Seven S.* 5668. Sic as caill and growand treis, moist heuie sauer of lynt, hemp [etc.]; SKEYNE *Descr. Pest* 6. Be the sauir of ȝour end [= breath]; 1573 DAVIDSON *Sat. P.* xlii 63. That thai rynd nor melt na talloun or craklingis quhair the savour thairof may cum to the hie streitis; 1599 *Edinb. B. Rec.* V 258.

(*b*) Ye are like the pipers dog the [*sic*] smels the saur of a feast a far of; CARMICHAEL *Prov.* No. 1719.

b. In *fig.* context.

Our prelates .. apprehendit .. Paull Craw .. & maid ane sacrifice of him .. And findand the sawour of this sacrifice fragrant and smelland, thay tuke .. mony ma quha .. war cuttit of be the fyre; *Lyndesay Pref.* (STS) 400. The book .. intituled, 'A Declaration of the Just Causses of his Majestie's Proceedings against these Ministers ..', a birth .. fitlie .. representing the father, and a fruit carryeing the kyndlie savour of the trie; FORBES *Rec.* 522.

3. *fig.* In various uses deriving from 1 or 2 above. **a.** A positive or sustaining feeling. **b.** A characteristic or indicator *of* a quality, state, etc.

a. Thinkand one his dolorus passioune .. thar is causit in thi hert ane merwalus sueitnes and sauour that liftis thé aboue all waurldly thingis to comfort and strenthe the curage; IRLAND *Mir.* I 45/30.

b. Of the balme that .. betakynnis the odour and sueit sauour and smelling of gud fame .. of the persoune; IRLAND *Mir.* III 16/2. The gospell .. sall .. be unto thee ane savour of life unto life, or ane savour of deith unto deith; ROLLOCK *Wks.* I 395.

Savo(u)r, *v.* Also: **savoer, sauer, sawor, -er, saur(e, sawr(e, sar(e, sair.** [ME and e.m.E. *sauur(e, sauir* (both Cursor M.), *sauyr(e* (Rolle), *sauer* (Piers Plowman), *sauour* (Chaucer), *sauowr(e* (*c*1430), etc., OF *sav(o)urer* (*c*1112 and *c*1225 in Larousse), late L. *saporare* to savour.]

1. *intr.* To give off a (particular) smell or scent; to smell *of* something.

A pupyll .. lywand .. But othir met than of ayre Savorand [*C.* saworande] off ane appyll fayre; WYNT. I 728. Wyth spycery welle savorand, And off kynd welle flevorand; *Ib.* VIII 1477. Fy, quod the feynd, thow sairis [*M.* saris] of blek; DUNB. (OUP) 168/34.

proverb. Its kyndly the pock savour [*1641* sare; *Carmichael No. 872* saure] of the herring; Ferg. *Prov.* MS No. 800.

b. *proverb.* ? To taste or ? a further example of a above.

Manie wats not quhairof the wel sauris quhill it fall drie; CARMICHAEL *Prov.* No. 1140.

c. *fig.* To show traces *of*, have (some of) the characteristics *of*.

A man cummyn of noble vertueus .. spekand of all gud sauorand of all suetnes; *Loutfut MS* 36b. Thy reasonis sawres of reik; MONTG. *Flyt.* 684 (T). Utring wordes wich savor of disaffection; 1651 *Rec. Kirk Scotl.* 641. Yit it savoering of sorcerie or superstition sche [etc.]; 1655 *Old-lore Misc.* V III 134.

2. *tr.* To flavour (*with* a spice).

Powderit with prymros sawrand [*M.* savorand] all with clowis; DUNB. *Flyt.* 192 (B).

3. *intr.* To have the capacity to smell smells, to be able to smell things.

Thair nois thirlis can nouther sauer nor smell; *G. Ball.* 110.

4. *tr.* To perceive, apprehend.

Sathana, .. thu miskennis, and saweris nocht The thinge he God will haf in thocht; *Leg. S.* iii 355. That we in hartis may sauour Thy mercy and thy fauour; *G. Ball.* 146.

Savourable, *adj.* [e.m.E. *sauorable* (1502), OF *savorable* (1204 in Greimas).] Capable of being perceived by taste or smell. — That all thing God and nature has maid is till him sensible and savourable be way of reasoun; HAY II 111/25.

Savo(u)rand, Sawerand, Sawrand, Sairing, *ppl. adj.* [e.m.E. *sauering* (1578), *savouring* (1598); SAVO(U)R *v.*] **a.** Having a (good, sweet) smell. **b.** Sweet-smelling. — **a.** Flouris Well sawerand [*C.* savourit] of ser colouris; BARB. XVI 70. Sauorand; *Alex.* II 5022. And in thai bathis suld be gude precious herbis and wele savourand in thair sesone; HAY II 134/1. Sueit sairing flouris; DALR. I 44/11. — **b.** Thir gressis and gers, Outhir sawrand or sour; HENR. *Practysis* 88.

Savoure, var. of SAVO(U)R *n.*

(Savouring,) Savoring, -yng, Savering, *vbl. n.* [ME and e.m.E. *sauorynge* (Chaucer), *sauering* (Wyclif), *saweringe* (*a*1450); SAVO(U)R *v.*] **a.** Tasting. **b.** *fig.* Something that conveys the 'flavour' or gives a 'taste' of (something else). — **a.** Se that suppose we feil suggestioun to syn other be .. heiring smelling savering [etc.]; HAMILTON *Cat.* 116. — **b.** I wald into my rurall wlgar gros Wryte sum savoryng [*Sm.* -ing] of thyne Eneados; DOUG. I Prol. 44.

Savo(u)rit, Sawrit, Saird, *ppl. adj.* [e.m.E. *savored* (1590); SAVO(U)R *v.*] Having savour, scent or smell, with defining *adv.*; -smelling. — Savourit; BARB. XVI 70 (C) (see SAVO(U)RAND *ppl. adj.* a). To pas .. By gowsty placis, welch savorit, must and hair; DOUG. VI vii 79. Mary that is ane evill saird mes [*Ch.* il sauorit dische]; LYND. *Sat.* in *Bann. MS* 187a/334. The sweit smell and the suaue odour .. Sa weill sawrit and viuificatiue; ROLLAND *Ct. Venus* II 852. Ipothecar .. In his buith .. half ane dram of must euill sauorit price iiij s.; 1575 *Edinb. Test.* III 319b.

(Savourles,) Savirles, *adj.* Also: **sawr(e)les, sairles.** [ME and e.m.E. *sauourles* (Trevisa), *savourlesse* (1627); SAVO(U)R *n.*] *fig.* Insipid, jejune, lacking in interest, dull. — Moyses .. wrait the law .. In thare awin vulgare language of Hebrew That all the bairnis of Israell .. Mycht knaw the law .. Had he done wryt in Latyne .. It had to thame bene bot ane sawrles [*v.r.* sauirles] jest; LYND. *Mon.* 564. Thocht ȝe think pryde of paramouris .. set on syde sic sawreles game; ROLLAND *Seven S.* 4808. Sic preichouris .. Quhen thay haue run ouir with ane reill Thair sairles sermone; 1573 DAVIDSON *Sat. P.* xlii 537.

Savourous, var. of SAVOR(O)US *adj.*

Savoy, *n.* [? F. *Savoie* the district of S. E. France.] Appar. some type of cloth. — One Chinay Tollercloth of stamped Savoy; 1653 *Soc. Ant.* XXIII 303.

Savyng, var. of SAV(E)AND *pres. p.*, *prep.* and *conj.* and SAV(E)ING *vbl. n.*

Saw, Salve, *n.*[1] Also: **sav; salf, saufe.** Pl. also **sawuis, saues.** [ME and e.m.E. *sallfe* (Orm), *salve* (Ancr. R.), *sau(e* (once, Chaucer), *salffe* (15th c.), *saulve* (1530), *salve, sawf* (1663), OE *salf, sealf.*]

1. (A) healing ointment, unguent, balm.

pl. Bot mekly vald scho wesch thar fet & with soft sawis thare saris bet; *Leg. S.* xlvi 102. All hir vondis .. Enonte with sawuis soft & swet; *Ib.* I 687. Lechis war noght to lait with sawis sa sle; *Gol. & Gaw.* 883. Is nowdir .. Seiknes nor sairnes .. Bot I can .. leiche thame fra lame & lesure With sawis thame sound mak; HENR. *Practysis* 21. To Jhonne Robesoun barboure to by sawes to the warkman that my lord Borthuikkis man hurt, xliiij s.; 1548 *Treas. Acc.* IX 176. *Clar.* V 1510. Sindrie kyndis .. of sawis droggis & oylis; 1576 *Edinb. Test.* IV 154b. Emplasteris and sawis; 1578 Creswell *Royal Coll. Surgeons Edinb.* 14. 1579 *Edinb. B. Rec.* IV 124. 1584 SEMPILL *Sat. P.* xlv 386. MONTG. *Flyt.* 224 (T). That scho saw the guid nychtbouris mak thair sawis with panis and fyris; 1588 *Crim. Trials* I II 163. Dilaitit of .. rubbing him with sawis maid of dyuerse grene herbis; 1607 *Ib.* II 535. Certanis vllis & sawis in pigis; 1612 *Brechin Test.* II 253.

(*b*) [They] laid soft salves to thair woundis reid; *Clar.* I 560. Pretious salves and balmes; *Ib.* 1162. [Like surgeons] in applicatioun of thair salues and plaisteris; FERG. *Serm. iii Malachi* in *Tracts* 62. In his buith of medicein and salves with instrumentis of chirurgerie; 1632 *Edinb. Test.* LV 269. That he had goode salves .. whair-with he wold cure hir dochtir; 1649 *Crim. Trials* III 599.

sing. (1) Thom gaif hir .. ane thing lyke the rute of ane beit and baid hir .. seith and mak ane saw of it; 1576 *Crim. Trials* I II 53. Sa son as sche rubbit the saw vpoun the patient .. and it drank in, the chyld wald mend; *Ib.* Scho maid a saw of reid nettellis and butter and wodbind and laid it to the patientis lunȝie; 1597 *Ib.* II 28. She coft poysone .. to mak a saw to cure a bairne; 1650 *Elgin Rec.* II 273.

attrib. I thought upon what would be the best forme of a salve box; 1687 *Bk. Old Edinb. C.* XXVII 137.

(2) xxx s. money as for expenses depurssit be hir for saw and heilling of hir heid; 1596 *Paisley B. Rec.* 175. Lyke as scho tuk in hand to haill the young Laird of Bargany .. with cure of saw maid of quick-siluer and rubbit it on the patient; 1597 *Crim. Trials* II 28. Jonet Morisone .. bund vp her head and gave a peice saw to rub to her breist; 1660 *Rothesay Par. Rec.* 25. A whyte iron box with saw in it; 1685 *Soc. Ant.* LVIII 355. For searching the horse cap^ns^ foot and puting in saw quhen he was pricked; 1691 *Foulis Acc. Bk.* 137.

(*b*) Precyus invnctment salve [*Ruddim.* saufe] or fragrant pome; DOUG. XII Prol. 146. Let vs go in our garding and seik Herbes to be salf that is of maist vertew; ROLLAND *Seven S.* 5660. Ane sponefull of salue maid of meikill wort; 1633 *Orkney Bp. Ct.* 87. The forsaid salve of sousely leafes; 1650 *Maxwell Mem.* I 351.

b. With defining terms.

[She] bad his vyf vasch him with vinegir and grein sav; 1604 *Dundonald Par. Rec.* 54. Ane orange-coloured saw; 1650 *Culross* I 238. For eye sav 1 d.; 1687 *Sc. Ant.* VIII 157. To Mrs. Hair for 3 unce neit saw for the young coatch hors; 1699 *Foulis Acc. Bk.* 249.

2. *fig.* A remedy, cure.

(*a*) Pennaunce .. is the werray saw and hevinly medycyn that is maist proffitable to the man; IRLAND *Mir.* III 67/25. Till all his sair he soucht na saw bot ane The quhilk wes ded; KENNEDY *Pass. Christ* 257. With dispensatiounis, sawis for euerie sair; 1571 SEMPILL *Sat. P.* xxviii 106. Italye .. exspecteth quhat person may aryse .. to applye medecins and saues to her auld cancred disease and yssew; FOWLER II 162/12.

(*b*) Haile moder of our makere .. Haile succour & salf for the synnis sevyne; *Howlat* 720 (A). HENR. *Test. Cress.* 411 (see SAW *v.*[2]). And refuge nane I haif Without ȝour grace with sum salue wald me saif; ROLLAND *Ct. Venus* II 946. And how that be ane other knott agane he hes him bound Whiche onlye is the salve [It. *rimedio*] that may in contrare lowe be found; FOWLER I 50/86. To seik for salue of her that gaue the sair; *Philotus* 446. There is never a stripe that his back keeped but it will be an healing salve to thy soul; WELSH *Forty-eight Serm.* 498. No salue butt death could cure her inward sore; MURE *Dido & Æneas* III 38. ADAMSON *Muses Thr.* 157.

proverb. A salve for al sairs; CARMICHAEL *Prov.* No. 172.

Saw, Say, *n.*[2] Also: **sawe, sau, sall, sa(a.** [ME and e.m.E. *saȝe* (*c*1175), *sawe* (Layamon), *sagh, sau(ue* (both Cursor M.), *saw* (1566), *say* (1602), *saye* (1614), OE *sagu*; SAY *v.*[1]]

1. Something said (or recorded in writing); spoken (or written) material, also, as distinguished from action, *in deid and saw*; speech; a statement; a report, account or tale. Also *to say at saw*, to record, put into words; *but a say*, mere words, a verbal quibble. Also, *transf.*, something sung.

(1) Aymo recordis in his saw That Paule ay fra the cok craw .. oysit [etc.]; *Leg. S.* ii 53. For that he .. Resisted than hys sawes [L. *verbis eius*] ageyne; *Troy-bk.* II 108 (C). He was weill taucht in deid and saw; *Alex.* I 1285. The haile caus Was said in to schort sawis; *Howlat* 246 (A). He [*sc.* God] sais expressely He will nocht the dede of the synfull man. .. Than is nocht to presume that He will do again His sawe; HAY I 257/2. Schir Spynagrose with speche spak to the king 'Yone lord [etc.] .. ' 'Hevinly God' said the heynd 'how happynis this thing? Herd thair euer ony sage sa selcouth ane saw!' *Gol. & Gaw.* 266. A churll thai had that felloune byrdyngis bar [etc.]. .. Quhen Wallas herd spek of that mery saw He [etc.]; *Wall.* II 36. The master said I ken thai wordis Of his stepmoder that is the saw; *Seven S.* 1864. Quhen that the sound of his saw sinkis in my eris Than ay renewis my noy; DUNB. *Tua Mar. W.* 115. To myne eris come this saw [L. *vox*]: 'Ene, quhy [etc.]'; DOUG. III i 76. Tell me withoutin mony sawis That ȝe speak not the maner and the caus; ROLLAND *Seven S.* 997. Lord .. gar thame mend it Quhome thai offendit In deid or saw; MAITLAND *Maitl. F.* 428/42.

(*b*) 3it so 3e sall not from thair [*sc.* detractors'] sayingis saue 3ow Bot gif thay sie 3e sussie of thair sais [: caus, knawis, wais (= walls)]; J. MAITLAND *Maitl. Q.* 134/39.

(*c*) 3our deid is not lyk 3our say; 1571 *Sat. P.* xxix 30.

(2) Mony kynrik to his clame .. Maneris full menskfull .. Selcouth war the sevint part to say at saw; *Gol. & Gaw.* 409.

(3) Some say they do submit to episcopal government in synods because they join with the Presbyterians the same persons that were with them before. This is but a say for it is not the persons but the form of government that gives the denomination to a government; FRASER *Lawfulness Separ.* 30.

(4) *transf.* Be nocht sa lerge unto thir sawis sung; DUNB. (OUP) 38/19.

b. In legal contexts: A claim, assertion or agreement; a statement (of a witness, etc. before a court); a sentence (of a court). Also in allusive use. Passing into **c.** A person's word, indicating commitment or a promise.

b. (1) Peraventure the ayr walde eftir the fadris disses that suilk gyftis war of na vale. .. Suilk saw of the ayre sall nocht be herde be this resoun that the fadre gafe his landis sua in his lege pouste; *Acts* I 42/2. Eftir the said proves sawis and deposicionis Richard of Rutherford forspekar of the said Lankyn strekit a borch in Alexander of Kintoris hand; 1433 *Aberd. B. Rec.* I 389. The man is blist that dois [baith] dome and law Quhilk quhen he giffis a sentence suthfastly Gerris execute .. Not anirly to sele a nakit saw For to Godwart is bot derisioune A bair sentence but execucioune; *Regim. Princ.* 243 (Marchm.). Quhat burges deuly put upone the inquiest that revelis menes sawis said thairinto .. sal tyne his fredom; 1531-2 *Selkirk B. Ct.* (ed.) 118.

(*b*) He brings .. no prooff safe his simple or naked saa or dixit; 1616 *Misc. Hist. Soc.* II 207.

(*c*) Gif it happyn .. that he .. be nocht assethit of his payement fullily at the say conteignit in owre obligacons; 1407 Slater *Early Sc. Texts* No. 65. The tennent aucht .. till haif tane lauchfull witnes .. before ane juge .. for his nakit sall [BALFOUR *Pract.* 208, say] is nocht aneuch in that mater; *Harl. MS 4700* fol. 282a.

(2) Thocht my sawis haif no seillis I sall leir hir to fan; *Bann. MS* 256a/27.

c. Ane lord and he be leill .. His sau is ay als sickker as his seill; HENR. *Fab.* 2282. Thocht ane off thaim be fals in till his saw; *Wall.* x 229. Lat ay 3our deid follow 3our saw; *Bann. MS* 52a/93. Sueit hairt sen I 3our freind only wes ay I windir quhy so fremmitly 3our say Frome me away 3e do attray; *Ib.* 239a/2.

d. A prophecy.

And said till hyr thir thre wers .. *Rex ruet in bello* [etc.]. .. This wes the spek he maid perfay And is in Inglis toung to say, 'The king sall fall in the fechting [etc.] .. ' This is the sentence off this saw That the Latyn gan hyr schaw; BARB. IV 260. Thi [*sc.* the Sybil's] secret sawis and thi propheceis Endite of my kyn and genology I sal gar kepe; DOUG. VI i 140.

2. A sententious saying; a piece of advice cast in a traditional form, a maxim, a proverb.

(*a*) Fore it is sad in elderys saw 'Ful harde is hungyre in hale maw'; *Leg. S.* vii 653. Knawlage .. of the prophetis the sawys; *Ib.* xii 376. It is said in commone sawis That mastry mawis the medow doune ay; WYNT. V 1498 (W). For-thi haldis clerkis be thare sawe That custwme is the tothir lawe; *Ib.* VIII 657 (R). My deir sone, .. enprent thire sawis of thire haly and wys men in thi mynd; MYLL *Spect.* 281/19. As it is said in mony suithfull saw, Far ethar is to hald nor for to draw; STEWART 5837. Now fynd I trew the saw quhilk Dauid said: [etc.]; LYND. *Trag. Card.* 236. ROLLAND *Ct. Venus* I 491. Giffe sawis be suth; SCOTT i 193.

(*b*) & gyf 3e wil kene this thinge Clethe 3ow [etc.] .. quhen the paianis hard this sa [: ay] Thai sad [etc.]; *Leg. S.* xv 84.

3. ? Language or idiom, a particular language or ? a further example of I above.

Set I myne intent .. Fra that I sene hade storis sere In cronnyklys quhare thai wryttyne were Thare matere .. to drawe Off Latyne in tyll Ynglys sawe; WYNT. I Prol. 30.

Saw, *n.*[3] Also: **sawe, sau-.** [ME and e.m.E. *sagh* (Cursor M.), *sawe* (*c*1340), *sa3e, sae* (both *a*1450), *saw* (1495), OE *saga.*]

1. A saw, the tool.

Also with defining words, indicating special types or purposes, etc., as in *dismembering (leg) saw; hand saw; iron saw; tenent saw; tug saw* and *whip saw.*

For further examples with defining words, see ARM *n.* 4; ARMIT *adj.*[2]; BAKKIT *adj.*; BOW-SAW *n.*, and RUG-SAW *n.*

Thre sawis ane armyt a tugsaw a handsaw [etc.]; 1501 *Acta Conc.* II 477. For ane saw send to the werkmen; 1532 *Treas. Acc.* VI 155. Ane aiche & ane saw; 1574 *Edinb. Test.* III 86. Ane saw and furmour plane [etc.] .. pertaining to ane wrycht; 1600-1 Whitelaw *Sc. Arms Makers* 162. Tenent sawes .. whip sawes .. leg sawes; 1612 *Bk. Rates* (Halyb.) 326. [Paul Martin's essay to be] a dismembering saw for the leg, a trepan [etc.]; 1691 *Arch. Scot.* I 180.

2. *Attrib.* and *comb.* in *saw-bled* (= blade); *saw-tre* (= some part of a mill); *saw-man, -wricht.*

(1) Four saw bleddis estimat to iij lib.; 1597 *Edinb. Test.* XXX 290b.

(2) The expenses of the Lytill Myl in rasying hir saw tree and hir queheles and mending hir foirschot; 1558-9 *Edinb. Old Acc.* I 294. *Ib.* 289, 295.

(3) To the saw wrichtis of Lochmaban that tha wantit of thair wages, xxviij s.; 1506 *Treas. Acc.* III 339. 1607 *Edinb. B. Rec.* VI 334. John Blak the saw-wright; 1657 *Dumfries Kirk S.* 12 Feb. Dobie, Robert, sawman, his child; 1666 *Greyfriars Interments* 181. Gairns, John, .. sauman in Todrick's Wynd; *Ib.* 239.

Saw, Sow, *v.*[1] Also: **sawe, sau, schaw(e, schau; sowe.** P.t. **sew(e, seu, schew; sawit, -yt; sowd.** P.p. **sawin(e, -yn(e, -ing, -en, sauin(e, -en, sawn(e, sawnne, saune; sown, soune; saw; sewyt.** [Early and north. ME and e.m.E. *sawe(n* (*c*1200), *sau* (Cursor M.), midl. and south. ME and e.m.E. *sowe(n* (*c*1200). ME and e.m.E. p.t. *seow* (*c*1175), *sew, siew* (both *c*1200), *seu3* (*c*1250), *seu* (Cursor M.), also *sceued* (Cursor M.), *sowide* (Wyclif), *sowed* (1535); p.p. *sawenn* (Orm), *sauun* (Cursor M.), *sawen* (Rolle); *isowe* (*a*1250), *sowen* (1523), also *i-sowed* (Piers Plowman), OE *sáwan.*] To sow.

1. *tr.* To scatter or deposit (seed) on or in the ground so that it may grow; to plant (a crop) in this way. Also *fig.*

pres. He socht to saw thare bere & ry; *Leg. S.* xxvii 988. To schau his seiyde & wine his corne; *Ib.* xl 133. That all thar tenandis plant woddis and treis mak heggis and sawe brvme; 1457 *Acts* II 51/1. Lauboraris .. Sum sawand seidis fast frome place to place; HENR. *Fab.* 1723. *Ib.* 1744. For sedis to saw in the garding in Strivelin vj s.; 1501 *Treas. Acc.* II 82. 3e sal sau the cornis on 3our feildis bot 3our enemeis sal eit it; *Compl.* 24/24. That na persoun this 3eir schaw vther seidis in the feldis folloving; 1552 *Elgin Rec.* I 116. That the soorest waye to make them [*supra* firr tries] prosper is to saw them in the seid; 1637 Innes *Sketches* 519. He sawing the cropt therof upon the ground the said yeir; 1662 *Irvine Mun.* II 88.

p.t. At that corn sal .. remayn stil witht the husband of the biris that sew it; 1434-5 *Chart. Cupar A.* II 33. Trytolanum that first sew cornes on erth; *Asl. MS* I 325/22. That the sade James laubórit telit and sew the aitis and cornis that grew on the twa merk of the Stanyburne; 1500 *Acta Conc.* II 461. Certane neichbours .. sew ane pest of wald seyd in the eist end of thair said croft; 1553 *Blackfriars Perth* 234. Ane crop of quheit quhilk the said Michaell sew on twa riggis of land of the landis of Restalrig; 1569 *Canongate Ct. Bk.* 75.

(*b*) They sowd 19 bolls corn on it and had about elevin scoir threeves growing; 1707 CUNNINGHAM *Diary* 26.

p.p. Quhil the tym the corne be sawyn agayne; 1402 *Reg. Panmure* II 184. And ony thyng that is byggyt or sawyn anyrly pertenys to the ayre; *Bute MS* c. 109. Saune; *Acts* I 44/2. Sawin in the said Grange thretty and thre bollis of quhete; item, sawin xviij bollis of pese; 1439 *Cop. St. A.* 177. Certane corne sawine within the commone; 1450 *Reg. Episc. Brechin.* I 146. And thare was the first quhete that ever was sawin with man; HAY I 38/27. A part of his seid being sawing and the remanent redy til saw; 1497-8 *Acta Conc.* II 123. The proffittis of .. vj score of bollis [of aitis] that he micht haf sawin upone the sammyn landis [erit and telit] be him of the sade last yeire; *Ib.* xxx bollis of alkin corn sawin on the ground; 1508 *Reg. Privy S.* I 240/2. Sawnne; NISBET *Luke* xix 21. Sawn; 1548-9 *Reg. Privy S.* IV 20/2. ȝit sall it efter bring furth frute as liuely seid sawin in gude ground; 1560 *Conf. Faith* in *Acts* II 532/2. We .. confes that dornell cockell and caffe may be sawin .. in the middis of the quheit; *Ib.* 534/1. Thrie bollis beir to be sawin .. sex bollis beir by the seid, price of the boll xx s.; 1570 *Edinb. Test.* II 122. Sawyn; 1581 *Lanark B. Rec.* 81. xvj bollis aittis reddy to the happer albeit thair was than sawin on the ground .. xxiiij bollis aittis; 1592 *Edinb. Test.* XXIV 168. Sawen; CLELAND 17.

(*b*) Ye said 'I would he had sown my lint seed'; 1662 *Soc. Ant.* XXII 220. Ther was soune hotspurs in the east border .. last weeke and this day .. next the mid. row of gooseberrie bushes; 1689 *Foulis Acc. Bk.* 117.

b. With *sede*, etc., *fig.* and in *fig.* context.

(1) Cristis lare techit he And sew sa Goddis sede thane At he to Criste mony wane; *Leg. S.* i 77. And schawand Goddis sede scho drev Towarte Arle; *Ib.* xvii 13. In Goddis ȝard to set plucht To schau his seiyde & wine his corne; *Ib.* xl 133. Thare he saw sawyne il seide; *Ib.* 203. Schaw; *Ib.* vi 119, etc. Quha dreidis al vindis .. sall neuir weill saw his corne; *Wisd. Sol.* (STS) 478. The seid of lufe was sawin in my face; HENR. *Test. Cress.* 137. Quha lykis to .. saw the seid of werteu & sapiens; *Contempl. Sinn.* 269 (Asl.). In fragil flesch ȝour [*sc.* Venus's] fykkil seyd is saw; DOUG. IV Prol. 8. Quho hes this seid [*sc.* of sin] amang ws sawin; LYND. *Mon.* 431. Vpon that guid seid of the quheit quhilk the Sone of Man hes sawin in his croft; WINȜET II 56/17. And nocht sa mekle bastard seid Throw out this cuntrie sawin; *G. Ball.* 189. It is necessarie that seid be sawin for the tyme to come .. that the kirk be not left barren and waist to the posteritie; 1580 *Cath. Tr.* 37/1. Sauing the seid of arrogance throu opinion of knalege and treu religion; HAMILTON *Cath. Tr.* in *Cath. Tr.* (STS) 78/2. Sewe; DALR. I 325/11. About this tyme was sawen The seminar and seeds Of .. futur warre; GARDEN *Elphinstoun* 1307. Shall coloured knaues so malapertlie lie And shamelesse sowe there poysoned smitting seede; JAMES VI *Poems* II 111/20.

(2) Love .. sawes his breirs and thornes within my hart; FOWLER I 156/13.

2. To scatter seed on (land, etc.); to plant or supply (land) with seed. **b.** *fig.* To strew (land) with (*of*) something. **c.** *transf.* To strew (something, material or non-material) *with, full of* (something) in the manner of sowing land with seed.

pres. Lat tham .. tel and saw and dyk and delff the erde; *Bernardus* 391. Quhar thou thi ryggis telys forto saw [L. *serentem*]; DOUG. VI xiv 96. Scho tholit .. Alexander Dundas .. to ryfe out teill and saw the marcheis; 1540 *Wemyss Chart.* 157. The kirk of God .. is the ground that he takis plesour to labour he earis it he sawes it and he scheiris it; 15.. *Misc. Spald. C.* IV 105. And causit delff the north half of the said ȝearde saw the samin and sett the samin with bow caill; 1585 *Perth B. Ct.* 31 May. SKENE *Reg. Maj.* II 13b. The haill husbandmen .. sall saw the twalff pair of thair haill laboring in pese; 1640 *Rec. Old Aberd.* I 69. [£11 5 s. Scots] for ceartane beir .. to saw my croft; 1663 *Kirkcudbr. Sheriff Ct. Deeds* I 262. With sixtein load of corne and tuo bolls of beir to sau the land; 1688 *Kirkcudbr. Test.* (Reg. H.) 25 Feb.

p.t. To the takin Alex. Makfersan bailȝie to the bischop .. sew the Larde of Echtis falde .. uppoun the samyn grond; 14.. *Reg. Episc. Aberd.* I 250. [He alleged] that he nor his forbearis thir xl yeris sew nocht the sadis landis; 1501-2 *Acta Conc.* III 139. Quhilk rig he possessit sew schuir and led away; 1609 *Melrose Reg. Rec.* I 79.

p.p. Al the wast land that was lynt or corn sawin; 1458 *Peebles B. Rec.* I 128. He [shall pay] .. the penny male .. for the land that was sawin with his sede; 1483 *Acta Conc.* II cxxxiii. Saune; 1516 *Fife Sheriff Ct.* 33. The landis foyrsaidis now savin with aittis; 1534 *Dunferm. Reg. Ct.* 111. Wilȝam Dempstar [etc.] .. has rewyn out telit and sawyn ane part tharof [*sc.* of the common]; 1535 *Reg. Episc. Brechin.* II 186. Ve haue sauen oure feildis to the behufe of oure enemeis; *Compl.* 29/5. My berne .. hes of hir awin .. Ane calf ane hog ane futebraid sawin; *Jok & Jynny* 19. I ordane my land to be sawin with my awin seid; 1580 *Edinb. Test.* VIII 62b. [They] have intrusit thameselffis in hir haill landis foirsaidis sawne and occupyit the same; 1586 *Reg. Privy C.* IV 65.

b. Off dede and hurt men .. thik the feild was sawin; HAY *Alex.* 3235.

c. Hys tawbart and array sewyt [L. *consertum*] with breris; DOUG. III ix 13. Argumentis sawin full of sophistrye; LYND. *Mon.* 672. I send unto yow ane sepulture of hard stane collourit with blak sawin with teiris and bones; BUCH. *Detect.* (1727) 145.

d. Of seed: To be sufficient for (a certain area of land).

That is alsmekill land as a celdr of atis will schawe; 14.. *Reg. Episc. Aberd.* I 250. The peiss sows farder than anie grain .., for example fourten pecks of peiss, of beir and wheit a boll, and fyve furlets of oates will sowe an aker; *Skene Agric. MS* (ed.) 68.

e. Of an area of land: To be sufficient to take (a certain quantity of seed).

With the thrid of the yerd and als meykill of the croyft to schaw ane pecce of linyget; 1566 *Crail B. Ct.* MS 7 Aug.

f. Of dung: To fertilize that area of land for the sowing of which (a specified quantity of seed) is requisite.

That quhar William McCoren sauld him of muk that wald haif sawin thre bollis beyr .. nochttheles the said Patre hes not gottin bot alse mekle muk as will saw xiiij peckis beyr; 1566-7 *Inverness Rec.* I 146.

3. *intr.* or *absol.* To scatter or plant seed on or in the ground. Also *fig.* and in *fig.* context.

And men he kend tyll ere and sawe; WYNT. I 1620. With power to tele and saw nochtwithstanding .. statutis of the forest made in the contrare; 1509 *Reg. Privy S.* I 285/1. Caus .. Mechanikis wirk, husbandis to saw and scheir; SCOTT i 174. Saw thin and maw thin; FERG. *Prov.* No. 751. Let us nocht say with the slugart .. nor for feir of wound and wather leave af to saw and schear; MELVILL 287. Unhappie is the man .. That teils the sand and sawis in the aire; BOYD *Sonet.* I sau I sett—no flour nor fruit I find; MONTG. *Misc. P.* xxxi 17.

b. Of seed: To be capable of being sown (over a certain area).

Skene Agric. MS (ed.) 68 (see 2d above).

c. *transf.* With *frute, seid*, to beget or as if to beget (a child).

The frute thou beris .. The Haly Gaist it in my body sew; KENNEDY *Pass. Christ* 1099. He sew na seid in to hir Sandelandis; *a*1568 SEMPILL *Sat. P.* xlvii 40.

4. *tr., intr.* or *absol.* In collocation with *reap (raipe)* (REPE *v.*), *sheer* (SCHER(E *v.* 10 e, q.v. for further examples) and, once, *ryise* (RIS(E *v.* 4) in *fig.* or *transf.* uses of the above senses.

Mark quhow the apostle S. Paul vritis on this maner 'It quhilk is savine ane naturall body sall ryise ane spirituall bodye.' S. Paul callis our bodeis efter the resurreccione spirituall bodeis nocht changit in sprites bot clede with spirituall giftes; Q. KENNEDY *Breif Tract.* (ed.) 128/17. Edinburgh since that day had reaped as thai sew; KNOX II 289. Quhat so euer a man saues the same sal he raipe, for quha sawes in his flesh he sal sheer corruption of the flesh, bot quha sa sawes in the spirit shal sheer lyf ewerlasting; KING *Cat.* 185.

5. *tr., fig.* To disseminate or spread; to propagate or foment (beliefs, states of affairs, etc.).

pres. He [*sc.* the Devil] sett him .. to saw errouris and herisies in the faith; HAY I 13/35. Sawand poysoun and mony wickit thocht In mannis saull; HENR. *Fab.* 1900. O wicket tung, sawand dissentioun; Id. *Hasty Credence* 41. I departit in Cathtnes to saw discord; 1543 *Corr. M. Lorraine* 13. Sawis; WINȜET I 77 marg. Thair culd na better way be fundin nor to saw iolasie betuix vs; 1568 *Haddington Corr.* 271. Efter that Sathan .. To rais his kingdome tentation did sau Into the hairtis of men; 1581 *Sat. P.* xliv 83. The fasones of evill fame quhilk ever reports of all things to the warst and oftentymes sawes abrod .. evill newes for guid; MELVILL 326.

p.t. Vlixes .. dangerus rumour amang the commonys hedis Skalit and sew of me in diuers stedis; DOUG. II ii 70. Causys of stryfe and batale I wald thou sew [*1553* schew]; *Ib.* VII v 140. God .. Sewe in their sinnat ciuile seditione Which sprang and grew to their confusione; NORVELL *Meroure* 19a. Knox held steue the pleuch And Methuen seu adulterie so teuch; 1581 *Sat. P.* xliv 125. But as the diuill is ready bent goode uorkes to hinder ay So seu he in this nauie stryfe; JAMES VI *Poems* I 219/155.

p.p. Quhen the apostolis had al-quhare In that land sawyne Goddis lare; *Leg. S.* xi 404. That thair gude renoune be wyde sawin and publist our all realmes; HAY II 89/7. With meiknes ressave the word that is plantit or sawin amongis yow; HAMILTON *Cat.* 26. Sawin; ROLLAND *Seven S.* 3381. *G. Ball.* 203. I feir that men .. Hes sawin suspitioun; MAITLAND *Maitl. F.* 304/25. 1573 DAVIDSON *Sat. P.* xlii 898.

b. To start up, establish (a school).

Mr. William Forbes .. is ordayned to have a care to sowe a school there to propagate the gospell; 1648 *Inverurie* 303.

6. To scatter, mainly, on the ground, as one scatters seed; to sprinkle or spread about.

And all trunschionis sawin on the grene; HAY *Alex.* 1542. Ane sow to quhome men .. In hir draf troich wald saw the precious stanis; HENR. *Fab.* 147. Sawand this calf thir small birdis to sla; *Ib.* 1896. Armouris swerdis [etc.] .. I sall do saw and strow our all the feildis; DOUG. VII ix 28. The Dewill .. commandit [them] to saw the peces of glas langis the mwre; 1590-1 *Crim. Trials* I II 238. Siluer lyk seid he sew vpon the ground; DALR. I 308/7.

b. To distribute or disperse. Also *reflex.*

Bot the prophetis sa thyn ar sawyn That nane in erd now is knawin; BARB. IV 686. And quhen thar meble ilke day Ves nere delt clenely away Thane sawyt thai .. Thare gudis w[n]mowable sone; *Leg. S.* xliv 89. And in hys almus he sew sylver Till all pure folk that had myster; WYNT. VI 1897. Feill off that kyn in Scotland than was sawyn; *Wall.* XI 1226. To the ȝet come rydand on ane day Wirschip of Weir, quhilk sawis honouris hie; *K. Hart* 818. Yett are thay catcht in girnis that thay did sau; JAMES VI *Poems* II 18/5.

reflex. Crystis discipilis .. sawit thame in placis seyre; *Leg. S.* xvi 187.

7. To shed, spill (blood).

Wemen .. sall nocht .. draw abak quhair mekill blude is sawin; STEWART 9601.

(Saw,) Sawe, Salve, *v.*[2] Also: **save, saif, sauf.** [ME and e.m.E. *sallferm* (Orm), *salue(n* (Ancr. R.), *sauuin* (*a*1240), *saue* (*c*1400), *sauf* (Lydgate), OE *salfian, sealfian.*

Some examples may properly derive from SAV(E *v.*]

tr. **a.** To anoint with salve. **b.** To heal, cure (a person, or his hurt). Also *absol.* **c.** *fig.* To remedy, make good, smooth over, put to rights (something amiss, a difficulty, source of disharmony, or the like).

a. For that tyme men in custum had The gret heit of the sone that bad Bathis til oyse & vneyment To sawe thame [L. *unguentis et balneis utebantur*] that thai var nocht schent; *Leg. S.* xvi 118.

b. A quyk wel sprang & sic dring gef That al seknes it can sawe [L. *omnes languidos sanat*]; *Leg. S.* xxxiii 320. Thair is na salue may saif or sound [*Anderson* save or sound, *Ch.* saif thé of] thy sair; HENR. *Test. Cress.* 411. Rycht couertly thai kepe him in that caiff Him for to sawe so secretlye thai mycht; *Wall.* II 278. Gif thair may be found ony defence Help or remeid that may ȝour seiknes saue [: haue]; ROLLAND *Seven S.* 9892.

(*b*) Albeit thow sich ane thousand syth It will nocht sauf thé of thy sair Nor ȝit remeid thé of thy cair; *Maitl. F.* 252/12.

(*c*) Bot ye [*sc.* the queen] my onlie medicinar remaines And easilie .. May salue my sores and mitigatt my paines; JAMES VI *Poems* II 69/8.

absol. Thy skirming Geuis to vs all recomforting Efter thy dint na sawe may saue; *Alex.* I 1079.

c. [It imports me greatly that the intelligence began] betwix ws [may continue]. .. I made yow some overture at Londone how to salve all matiers; 1560 *Cal. Sc. P.* I 511. Onles speciall remeid was providit in expres termes to salve that mater, that it culd not faill to be in tyme predudiciall; 1586 Rait & Cameron *King James's Secret* 77.

Saw, *v.*[3] Also: **sawe, sau, schaw.** P.t. **sew(e.** P.p. **sawin, -yn, sewin, sowne, shoin.** [ME and e.m.E. *sahe(n* (*a*1225), *sagh* (Cursor M.), *sau* (*a*1325), *saw(e* (*c*1340). Cf. MLG, MDu. *sagen*, ON *saga* to saw; SAW *n.*[3]]

1. *tr.* **a.** To cut (up) (timber, a tree) with a saw; to manufacture (boards, etc.) by sawing. **b.** To cut down (a tree) by sawing. **c.** *intr.*

a., b. *pres.* A wricht he tuk .. And ordand him to saw the burd in twa; *Wall.* VII 1151. And for the sawing of the said tymmer being in the wod my lord sal ger schaw thame on his expens; 1511 *Aberd. Sheriff Ct.* I 103. As weil thame that sawis as thame that awis the tymber; 1620 *Orkney Rentals* App. 51. James Seatone haith promised to cause saw dealls .. to mak dead kysts therof; 1645 *S. Leith Rec.* 58/1. Sawe; 1679 Meikle *Old Session Bk.* 209. To cutt heugh saw and dight .. als many couples and jests as would serve anc housc; 1693 *Melrose Reg. Rec.* III 115.

p.t. To the sawaris that sewe the said rauchteris; 1513 *Treas. Acc.* IV 488. To the sawers that sew the treis; 1599 *Edinb. D. Guild Acc.* 701.

p.p. A grene tre fra the rwte wes sawyn [*W.* sewin]; WYNT. VII 545. 1506 *Glasgow Dioc. Reg.* II 152. 1531-2 *M. Works Acc.* (ed.) I 73. Greit treis of asche sawin to be couppillis; 1567 *Edinb. Test.* I 49b. The trie being sawin in it was seine .. a meruellous multitude of wormes; DALR. I 61/11. Quhilk trie is lyeing at the ȝett sowne in tuo peices; *c*1643 *Rothes P.* The remanent part of the loading of good merchant deals being good wood four shoin being free of rotten and riven; 1688 Smout *Sc. Trade* 294.

c. To othir tua sawaris at sew with thaim; 1496 *Treas. Acc.* I 281.

2. *tr., transf.* To cut as with a saw.

Uith bluntid knyfe beginn To sau his iointis & leithis & for to hakke his runklid skinn; JAMES VI *Poems* I 161/454.

Saw, var. of SAV(E *v.*, S(C)HAW *v.*, SEW(E *v.* and p.t. of SE *v.*[1]
Sawage, var. of SAVAGE *adj.* (*n.*).

Sawar, Sawyer, *n.*[1] Also: **sawer, sauer; sair;** SAWSTAR, (SAIRER). [ME and e.m.E. *sawier* (1350), *sagher* (1379), *sawer* (1457), *sawyer* (1497), *saer* (1536); SAW *v.*[3]] A sawyer.

For many further examples, see the Indexes to the various vols. of *Treas. Acc.*, s.v. *Sawer, Sawyer*; and *M. Works Acc.* (ed.) I and II Indexes, s.v. *Sawyers.*

See also Black *Surnames*, s.v. *Sawyer.*

Sertane sawaris feyt .. for the sawen of the tymmyr to the rowbarge; 1494 *Treas. Acc.* I 246. 1502-3 *Ib.* II 355. To the sawaris for complete lautht to the said kechin; 1532 *M. Works Acc.* (ed.) I 101. Saweris; 1553 *Edinb. B. Rec.* II 183. To the sawars for sawing of vj treis to be barcatts trests and neidillis; 1554-5 *Edinb. Old Acc.* I 148. 1600 *Glasgow Wrights* 39. Sair; 1670 *Greyfriars Interments* 66. Knowes, William, sawyer, a child; 1675 *Ib.* 368. Saver; 1696 *Ib.* 57.

Sawar, -er(e, *n.*[2] [ME and e.m.E. *sawere* (*c*1175), *sowere* (*c*1200), *sower* (1380), *sawer* (*c*1450), OE *sawere*; SAW *v.*[1]] A sower.

1. One who sows seed. Also in *fig.* context.

Sum sawand seidis fast frome place to place The harrowis hoppand in the saweris trace; HENR. *Fab.* 1724. Sawere; NISBET *Matth.* xiii 18.

fig. Cankcarit corruptars of the Creid, Homlok sawares amangest good seid; *a*1574 *Three Reformers* 13.

2. *transf.* One who disseminates, promotes or propagates (discord, sedition, etc.).

Remembering how he .. had removit the sawaris and movaris of all divisioun; *Boece* 84b. Seditious lymmaris sawaris [L. *semina*] of discorde; BELL. *Livy* I 271/7. Ane scismatike and sawer of discord; STEWART 17584. *Bk. Disc.* 235. WINȜET II 6/30. *a*1568 SEMPILL *Sat. P.* xlvii 2. MAITLAND *Maitl. F.* 24/16. Gif I knew who wer the authoris or sawaris of sic vild brutes; 1571 Bann. *Memor.* 142. Sum .. vncannie sawers sew .. causes of contentioun; DALR. II 58/24. They ar not rycht Bot saweris of idolatrie; *Montg. Suppl.* xxxiii 102.

Sawce, var. of SALS *n.*

Sawcer, *n.* (? MF *saucier* (15th c. in Larousse) 'saucer', sauce-maker, also med. L. *salsarius, sausarius* (1201 and 1415 in Latham). — Jo. Scot the sawcer of Edinburgh; 1666 M. P. Brown *Suppl. Decis.* II 424.

Sawcer, var. of SALSAR *n.* **Sawch(e,** varr. of SAUCH(E *n.* and SALS *n.* **Sawchnyng, Sawchtnyng,** varr. of SAUCHTYNG *vbl. n.* **Sawcht,** var. of SAUCHT *adj.* **Sawcie,** var. of SALSY *adj.* **Sawd,** var. of SOLD *v.*

Sawe, var. of SAFE *adj.*, SAV(E *v.* and SAW, p.t. of SE *v.*[1] and see S(C)HAW *n.*[1] **Sawen,** p.p. of SAW *v.*[1] and var. of SAWIN *ppl. adj.*[1] **Sawend,** var. of SAV(E)AND *pres. p., prep.* and *conj.* **Sawen(g,** varr. of SAWING *vbl. n.*[2] **Saweour(e,** varr. of SALVIO(U)R *n.* and SAVO(U)R *n.* **Saweraill, -ale,** varr. of SEVERAL *adj., adv.* and *n.* **Sawerand** var. of SAVO(U)RAND *ppl. adj.* **Sawer(e,** varr. of SAVO(U)R *n.* and *v.*, SAWAR *n.*[1] and *n.*[2] **Sawetê, -ie,** varr. of SAFETY *n.*

Sawfa, Sawfeir, varr. of SAFER *n.* **Sawf(f)-, Sawf(f)(e,** varr. of SAFE-, SAFE *adj., prep.* and *conj.* and SAV(E *v.* **Sawft(e,** p.t. and p.p. of SAV(E *v.* **Sawieour,** var. of SALVIO(U)R *n.* **Sawill,** var. of SAUL(L *n.* **Sawill-mes,** var. of SAUL(L)-MES *n.*

Sawin, *ppl. adj.*[1] Also: **-ine, -yn, -en.** [e.m.E. *sawen* (1536), *sawne* (1634); SAW *v.*[3]] Sawn.

Of sawyn beche the ribbis forgyt was; DOUG. II i 6. viii score of sawin burdis of fir; 1517 *Misc. Spald. C.* II 77. To the skaffalting .. lx sawin dalis; 1531-2 *M. Works Acc.* (ed.) I 80. The subjectis of Denmark and Norroway ar dischargeit of .. selling of all aikin tymmer and sawine daill to our subjectis; 1564 *Warrender P.* (SHS) I 39. For ane sawin wanscot; 1589 *Edinb. B. Rec.* V 327. Of sawen planckis fyftein; 1631 *Buccleuch Household Bk.* 26 Sept.

Sawin, -yn, *ppl. adj.*[2] Also: **-ine.** [e.m.E. *sowen* (1578); SAW *v.*[1]] Sown, in senses 1 and 2 of the verb.

1. Of seed, etc.: That has been sown.

Vpon his sawyn corn; 1508 *Prestwick B. Rec.* 40. If any horssis sall be fund .. wpon any manes sawin corn; 1649 *Aberd. B. Rec.* IV 96.

b. With limiting term preceding.

Caile and leikis .. cummyne of new sawine seide; *Leg. S.* xl 406.

2. Of land: Sown with seed; seeded, planted.

Vpoun sawin ground; 1588-9 *Crail B. Ct.* MS 27 Feb. Nane of thame sall .. feid or pasture ony of thair horsis within sawin land; 1631 *Monteith Stewartry Ct.* 7 Oct. 1652 *Peebles B. Rec.* II 5. 1669 *Glasgow Merchants House* 130.

Sawin, p.p. of SAW *v.*[1], *v.*[3] and S(C)HAW *v.* and var. of S(C)HAVIN *ppl. adj.*, and SEVIN *num.* and *n.* **Sawine,** p.p. of SAW *v.*[1] and see SAIM *n.*

Sawing, Sowing, *vbl. n.*[1] Also: **sa(u)-, schaw-, schau-, sou-, sowe-** and **-inge, -yng(e, -in(e, -yne; sawn; sowen.** [ME and e.m.E. *sauing* (Cursor M.), *sowynge* (Piers Plowman), *sowyng* (*c*1440), *soweynge* (1529), *sowing* (1577); SAW *v.*[1]]

1. The action or fact of sowing (seed, a crop); an instance of this. **b.** The period of the year during which this takes place.

(1) The whilk forestar sall frely haff of land to the sawing of sex ferlotis of bere; 1388 *Antiq. Aberd. & B.* III 294. Anent plantacione of woddis and heggis and sawing of brovme; 1457 *Acts* II 51/1. In sauyng of quhet pes ry and benys; 1473 *Reg. Cupar A.* I 164. For the allegit wrangus .. riving out of ane parte of the saydis .. landis and sauing of cornis thairupoun; 1548 *Antiq. Aberd. & B.* III 359. That Johane Annan aucht keip nychtborheid in schawin of ry in the crofts betuyx [etc.]; 1550-1 *Elgin Rec.* I 107. 1574-5 *Ib.* 150. For dressing laubouring sawing [*ed.* saiving] and setting of heirbs seidis and flouris; 1579 *Edinb. B. Rec.* IV 111. He mareit the said Elspett .. eftir the sawing of the beir seid and knawis nathing of the abstractit multurs the yeir preceiding; 1603 *Aberd. Sheriff Ct.* II 42. And the other twa laid at the sawing of the nixt cropt; 1659 *Kirkcudbr. Sheriff Ct. Deeds* I 10. — Gryte frostis and snaw .. no plewis going an litle seid sawing; SPALDING I 262.

(2) He sais that al thing has a tyme in this warld .. as .. tyme of seting and of sawinge; *Wisd. Sol.* (STS) 168. The said land .. has bene broukit .. in tymes bigane outher in ering & sawing or in pasture; 1473 *Acta Aud.* 27/1. With ympyn and sauyng with plantation of frote treis [etc.]; 1473 *Reg. Cupar A.* I 189. The tyme of sawing is callit rycht tedious And eik of schering rycht sueit is the sesoun; *Contempl. Sinn.* 281 (Harl.). It is na les convenient to haif sumtyme of vacance at the tymes of the sawing as of the scheiring; 1585 *Acts* III 376/1. Tomas Lege passit to the saidis landis & output the said Archbald & his pleucht & input the saidis James & his tenentis thairto be sawing & harrowing; 1589-90 *Prot. Bk. J. Inglis* 18 March. That landward men .. sall take land fra their maisters and sall labour it be teiling and sawing; SKENE *Reg. Maj.* II 13b.

(3) Ane auchtane pairt of Ester Innerarite .. quhilk Jhone Baron .. brukyt afoir and resignit to him at this sawing and sway entres at Witsunday in xlviij° ȝeris; 1549 *Reg. Cupar A.* II 237.

b. With all malis [etc.] .. dawerkis at teling sawing and schering in hervist; 1539 *Reg. Privy S.* II 470/2.

c. *transf.* Of procreation.

Woman .. Suld man haf hyre make to be .. For the sawyng of thare sede; *Leg. S.* xxxii 337.

d. *attrib.* and *comb.*

With *seasoun, sive* (= sieve), *tyme.* Also *sawin(g schete*, a sheet, tied on one shoulder, and forming a bag to hold seed for sowing.

(1) Ane baikin clath, ane sawn schet, ane vondoclath; 1534 *Selkirk B. Ct.* (ed.) 144. [3 canvases .. 2 sowing sheets 6 harrows and 6 sleds; 1560 *Edinb. Test.* in Sanderson *Rural Soc.* 173.] Twa sawin scheittis estimat baytht to xxvj s. viij d.; 1636 *Edinb. Test.* LVII 259b. — For a mand 2 s. 8 d.; sowen sive, 3 s.; 1678 *Sc. Ant.* IX 107.

(2) Twa tymes of vacance the ane in the sawing and the vther in the scheiring seasoun; 1587 *Acts* III 447/2. — For a hors labour in sawing tyme and heruist; 1595 *Breadalbane Ct. Bk.* 92.

2. The action of sowing or planting (land) with seed.

Sal ger John Voket .. ger sede be [? bocht] for the schawynge of the forsaid crofftis; 1461 *Aberd. B. Rec.* MS V 1 p. 422 (11 May). To pruf the crop & proffit of the sawing of the landis; 1478 *Acta Aud.* 67/2. The actione .. perseuit .. for the wranguis occupatione bruking sawyng & intromitting of his landis; 1519 *Fife Sheriff Ct.* 156. 1542 *Ecclesia Antiqua* 309. Teling [etc.] .. manuring & schawing of the saidis landis; 1558 *Inverness Sheriff Ct.* 68b. Cornis for sawing of the ground [etc.]; BALFOUR *Pract.* 495. In teillinge schavinge and harroving of certain ruids lyand [etc.]; 1603 *Inverness Rec.* II 7. 1612-13 *Misc. Spald. C.* V 91. 1619 *Crim. Trials* III 469. For the teiling and sawing of ane yard lyand in the Clachan of Campsy; 1627 *Rep. Hist. MSS, Var. Coll.* V 126.

3. a. A quantity of seed for sowing or having been sown; the amount of seed required to produce a particular crop. **b.** The crop resulting from the sowing of a particular (specified) quantity of seed. **c.** An extent of land defined by the quantity of seed required to sow it.

There is some ambiguity between the senses and with sense 1 above.

a. That Henry Petkarne sall restor .. a boll of bere twa bolle of atis sawin & a swyne price ij s. quhilkis gudis .. he tuke as his eschete; 1483 *Acta Aud.* 122*/1. To cause .. the said William .. to mak me thankfull payment of xiij bollis aittis for my sowde ij bollis aittis sawing pruiff corne and all wther dewytie; 1567 *Inverness Rec.* I 157. Thair will grow commonlie 20, 18 or at the leist 16 bolls beir yeirlie eftir ilk bolls sawing; *Descr. Isles* 429. Lismoir .. will grow alsmekle eftir ane boll sawing as in the Lewis; *Ib.* 435. The said Marioun to haif ȝeirlie ane pect of lingett sawing and the said Jonet to haif half ane pect sawing ȝeirlie; 1595 *Edinb. Test.* XXVIII 343b. The sawing of kaill seid in xij beddis estimat to iiij lb.; 1606 *Ib.* XLII 18b. The sawing of thrie bollis sax pekis corne measour; 1622 *Dumfries Test.* IA 23b. The sawing of four bollis corne great and small the growthe quhairof estimat to ten bollis corne meassour; 1627 *Ib.* 141. Twa key .. ane old meir .. the saving of sex pecks beir and ane small corne measour of Kirkcudbright; 1668 *Kirkcudbr. Test.* (Reg. H.) 15 June. That every inhabitant labourer within the toune .. saw halfe a fortpeit peck of peize upon ilk boll sawing of corne and beir; 1678 *Rothesay B. Rec.* 356. Ane peack of beair souing estimet 40 s. Scots; 1682 *Kirkcudbr. Test.* (Reg. H.) 17 Aug.

b. For the wrangus cutting doune and distruccione of twa bollis of bere sawing growand upon the Tempilland of Menstre; 1500 *Acta Conc.* II 435. Allane Fyndelay .. twk .. furthe of the said half merk land the sawing of twa bollis of aittis; 1548 *Lanark Sheriff Ct.* MS 7. Foure bollis aittis schawing vpoun the ground; 1569 *Edinb. Test.* I 344. Sawin of infeild aittis xx bollis estimat to the thrid corne .. price of the boll with the fodder xxxiij s. iiij d.; 1582 *Ib.* XI 332. The sawing of four bolls beir deduceand twa makingis of malt; 1667 *Kirkcudbr. Test.* (Reg. H.) 8 Feb.

(*b*) Sowyne: (Item, j peis sowyne); 1597 *Stirling Palace Larder Bk.* MS. Thrie loads corne sowing .. pryce of the load 10 merkis; 1665 *Kirkcudbr. Test.* (Reg. H.) 31 Jan.

c. (1) The landis .. of Ballinbrech .. extending ȝerely .. to thre chalderis of aitis sawing estymeit to the thrid corne .. twa chalderis of quhete & ber sawing estymett to the ferd corne; 1490 *Acta Conc.* I 156/1. The samyn tennentis occupijs ane fald of vj bollis atis sawing to the Lard of Echt; 14.. *Reg. Episc. Aberd.* I 250. George of Gordoun .. occupeis a celdre of atis sawyne pertenand to Dunmetht of the bischoppis land; *Ib.* The sade Kirkland of Parkle wilbe waistit be the hurting of the sade oxin extending til a chalder of aitis sawing; 1499-1500 *Acta Conc.* II 387. & ane rud of land extendand to ane firlot bere sawin callit Hedrige ane croift in Drownaris extendand in the haill to xv libs. & twa firlotis of ane sawin; 1588 *Prot. Bk. J. Inglis* fol. 38a (15 Nov.) Gif the said half cottere will nocht extend to the said sawing .. the said Duncane sall haf samekill of that cotterie callit Maltis fald; 1629 *Macleod P.* (Reg. H.) 17 June. Thrie aicker of land saing worth £10 the aicker; 1682 *Kirkcudbr. Test.* (Reg. H.) 17 Aug.

(*b*) Twa buttis of land .. quhilkis will contene ane forlot of atis soving; 1555 *Prot. Bk. T. Dalrymple* 2. An bear butt of thrie pecks sowing; 1675 *Kirkcudbr. Sheriff Ct. Deeds* I 513. And that other rigg or butt of land .. lyand in the field called the Gallowbank .. extending to ane halfe boll of beir soweing [in Rosemarkie]; 1681 *Acts* VIII 295/2.

(2) Thess four to have two bollis sowing off land; 1650 *Frasers of Philorth* II 286.

(3) The said Johne .. to the ȝard croftis for samony sawine as the said croft wes wount to have; 1550 *Breadalbane Doc.* 17 May. The number of all the bollis beir sowings within the brughe of Tayne; 1612 *Old Ross-shire* I 30. Ane coat hous and ane yeard with four bolls sawing; 1642 *Aberd. Sheriff Ct.* III 41.

4. *fig.* Dissemination or spreading (of doctrines, sedition, etc.).

That Phigellus and Hermogines, by sawing of strange doctrine resisted his preiching; BISSET II 304/23. Tending to the sawing of seditione and discoird betuixt his maiestie and the foresaid nobleman; 1631 *Justiciary Cases* I 176.

Sawin(g, *vbl. n.*[2] Also: **sau-, shau-** and **-eing, -ine, -en(g.** [Late ME and e.m.E. *sawyng* (c1440), *sawing* (1598); SAW *v.*[3]] The action of SAW *v.*[3]

For the sawing of the pik barrell; 1494 *Treas. Acc.* I 252. Sawen; *Ib.* 246 (see SAWAR *n.*[1]). Sawin; 1496 *Ib.* 323. Sawine; 1506 *Glasgow Dioc. Reg.* II 152. For the treis makyn and savin, xviii d.; 1508 *Aberd. B. Rec.* I xxi. For .. ane half estland burd sawing for mwildis to the maister masoun; 1529 *M. Works Acc.* (ed.) I 4. Sawin; 1531-2 *Ib.* 106. For iiii ruidis ii ellis lath sawing ilk ruid xii s.; 1535-6 *Ib.* 176. Saweng; 1558 *Ib.* 296. 1573 *Treas. Acc.* XII 344. For the sauing of the trey to pitt about the asiltrie; 1576 *M. Works Acc.* (ed.) I 300. For saueing of vi scheris to our peges nos; 1599-1600 *Skipper's Acc.* (*Morton*) 54b. For sawin of thrie great spares in cliftis and four dailles; 1633 *M. Works Acc.* (ed.) II 369. The sauster for sawing the trie in four standerts; 1660-1 *Aberd. Shore Wk. Acc.* 468. For working the timber work with the shauing of the dayles and sylling of the house; 1666-7 *Rec. Old Aberd.* I 216.

Sawing(e, p.p. of SAW *v.*[1] and varr. of SAV(E)ING *vbl. n.*, and SAWING, *vbl. n.*[1] **Sawing-tree,** var. of SAVING-TREE *n.* **Sawintene,** var. of SEVINTEN(E *num.* **Sawister,** var. of SAWSTAR *n.* **Sawit,** p.t. of SAW *v.*[1] **Sawité,** var. of SAFETY *n.* **Sawld(e,** p.t. and p.p. of SEL(L *v.* **Sawl(e, Sawll, Sawle-,** and **Sawll-,** varr. of SAUL(L *n.*, SAUL(L)- and p.p. of SEL(L *v.* **Sawme,** var. of SAIM *n.*

Saw-miln, *n.* Also: **-milne, -mylne.** [e.m.E. *sawe-mylle* (1553); SAW *n.*[3] and MILN *n.*] A sawmill. — Cum molendino granario et sawmylne; 1661 *Retours* I Inq. Spec. Inverness (87). Iff any persone desyr to have the libertie of ane saw milne abow the samen he sall be licencied to suplicat for the samen; 1671 *Inverness Rec.* II 246. That there be an saw miln erected in some considerable port or harbour; 1695 *Acts* IX 490/2.

Sawmound, var. of SALMOND *n.* **Sawmpyll,** var. of SAMPIL(L *n.* **Sawn(n)(e,** varr. of SAND *n.*, SAWING *vbl. n.*[1] and p.p. of SAW *v.*[1] **Sawo(u)r(e, Sawowre,** varr. of SAVO(U)R *n.* and *v.* **Sawouris,** var. of SAVOR(O)US *adj.* **Sawpheir,** var. of SAFER *n.* **Sawr(e, Sawr(e)-,** varr. of SAVO(U)R *n.* and *v.*, SAVO(U)R-, and SEWAR *n.* **Sawris,** var. of SAVOR(O)US *adj.* **Saws,** var. of SALS *n.* **Sawsar, -er,** varr. of SALSAR *n.* and *v.* **Sawsie,** var. of SALSY *adj.*

Saw-silver, *n.* [Appar. irreg. reduced var. of SAFER-SILVER *n.*, ? perh. with substitution for the first element of SAW *n.*[2] (saying, statement).] Chiefly or only in the Western Borders. = SAFER *n.* — [George Bell in Annand denied the robbery of four cows and calf] afferming giff the said complener could speir thame on him albeit it coist him ane hundreth pundis of saw silver that he sould pay to him the valour of the saidis guidis and saw silver. [Since then] the said complener hes speirit the same on him and hes thairupoun debursit ane hundreth pundis of saw silver; 1598 *Reg. Privy C.* V 469. [Rob Haliday .. servant of the said Laird of Johnnestoun was decerned to make redress .. of 35 sheep stolen .. price of each 40 s. and £20 of] saw silver; 1600 *Ib.* VI 116. For the speiring and trying of the quhilkis guidis [pursuer gave 130 merks of] saw silver; 1602 *Ib.* 469. That the Commissionaris [of the Borders] haiff power upoun ony trew manis complaint for stouth [etc.] .. to puneis the malefactour and mak redres to the compliner of his guidis and saw silver, for be this doing men wilbe cairfull to speir thair geir and the kingis majestie shall have greater commoditie be the nomber of escheatis that shall fall; 1606 *Ib.* VII 721. 1607 *Ib.* 728.

Sawstar, Sawister, *n.* Also: **sawster, sauster; sauister.** [SAW *v.*[3] with ending as in BAXTER, BREWSTER, LITSTAR, etc.] A sawyer. = SAWAR *n.*[1] — Robert Murry sawster; 1582 *Perth B. Ct.* 16 Oct. [For ship-building] thre tymmermen twa sausteris; 1589 *St. A. B. Ct.* 19 Sept. William Trumbule sawstar; *Ib.* 24 Oct. Adem Donaldson sawister; CULLEN *Chron. Aberd.* 45. To the sauisteris for breking of timmer and sauing of lathe; 1599 *M. Works Acc.* (ed.) I 322. Sawisteris; *Ib.* 321. Patrik Lamb sawster; 1600 *Crim. Trials* II 195. To the sausters for sawing of xxxiiii draught of lath; 1611 *M. Works Acc.* (ed.) I 334.

Sawster, var. of SAUSTER *n.* **Sawsy,** var. of SALSY *adj.* **Sawt,** var. of SALT *n.*[1] and *n.*[2] **Sawtable,** var. of SALTABLE *adj.* **Sawth,** p.t. and p.p. of SAV(E *v.* **Sawthe,** var. of SALT *n.*[1] and see SALS *n.* **Sawtht,** var. of SAUCH(E *n.* **Sawtry,** var. of SAUTOUER *n.*

Sawuand(e, varr. of SAV(E)AND *pres. p., prep.* and *conj.* **Sawuis,** pl. of SAW *n.*[1] **Sawure,** var. of SAVO(U)R *n.* **Sawwage,** var. of SAVAGE *adj.* (*n.*). **Sawwyoure,** var. of SALVIO(U)R *n.* **Sawye,** var. of SA *adv.* and *conj.* **Sawyer,** var. of SAWAR *n.*[1] **Sawyn(e, -yng,** varr. of SAV(E)AND *pres. p., prep.* and *conj.*, SAWIN *ppl. adj.*[1] and *ppl. adj.*[2] and SAWING *vbl. n.*[1] and p.p. of SAW *v.*[1] and *v.*[3] **Sawyt,** p.t. of SAW *v.*[1] **Sawyté,** var. of SAFETY *n.*

Sax(e, pl. of SEK *n.*[1] and varr. of SEX *num.* and *n.*[3]

Saxeane, *adj.* [L. *saxeus* made of stone, f. *saxum* stone.] Made of stone. — Ane closter . . Triangill maid, with craftie wark saxeane; ROLLAND *Ct. Venus* II 488.

Sax-earing, var. of SEX-AIRRING *adj.*

Saxeere-stane. (Nonce use after Atkinson, see below, some sort of mineral. Cf. also SPAR *n.*[2]) — [All the last gold which I brought of Scotland and presented unto the king's majesties owne hands in certaine sundry places mingled with the sappar, callamineere, saxere & sallinere stones, was gotten at Short-clough brayes; ATKINSON *Gold Mynes in Scotland* (Bann. Cl.) 33. If you please to desire for gold . . then seeke it, as in other nations, out of these dry meneralls or menerall stones (viz.!) the sapparr, the calamineere, the saxere and the salinere stones; *Ib.* 67. *Ib.* 15.] — Stevin Aitkinsoun, Englishman . . sall mak ane new searche, tryall and discoverie of the mynes, seames and minerallis in Crafurde Mure with the saxeere the calumyneere and the salyneere stanes; 1616 *Reg. Privy C.* X 531.

Sax-fauld, var. of SEX-FAULD *adv.* **Saxhering,** var. of SEX-AIRRING *adj.* **Saxlye,** var. of SEXTLIE *adv.*

Saxon, *n.* (*adj.*). Also: **Saxone, Saxoun, (Saxion).** [ME and e.m.E. *Saxoyn* (1297), *-o(u)n* (1388), *Saxson(e* (c1420), F. *Saxon* (1512 in Larousse); OF *Saisne* (1080 in Larousse), L. *Saxon-*, of WGerm. origin.]

A. *noun.* A member of a Germanic people, an inhabitant of Saxony, a number of whom invaded and conquered parts of southern Britain; an Englishman presumed to be descended from this people.

The Dowglas so dewe Wan wichtly of weir . . Fra sonnis of the Saxonis; *Howlat* 577 (A). Our ald ennemys cummyn of Saxonys blud; *Wall.* I 7. *Ib.* XI 1124. And sa remanit the Saxonis in the south part and we in the north vnto the tyme that the Danys dantit the Saxonis; *Asl. MS* I 192/15, 17. The quhilk fals Saxonis blud war evire ȝit aganis the Cristin faith; *Ib.* 199/1. Heldolus, quhilk pullit ane dagger fra ane Saxoun; BELL. *Boece* II 48. *Ib.* 35. Saxouns; *Boece* 270, etc. King Arthuris banis wes eirdit xv fute deip that he sould nocht be found be the Saxionis; ABELL xvi a. Sergestes and Engestes . . tua Saxons that cam vitht aleuin thousand Saxons fra thair auen cuntra to support . . the kyng of Grit Bertanȝe; *Compl.* 86/5, 6. The same peple plane ignorant of the Ingles toung, oft names the Inglise men Saxons; DALR. I 3/30. This Carancie . . seruet . . to Emperour Diocletiane. . . He was amrel of ane nauie contrare the Frenches and Saxounis; *Ib.* 187/5. Herduntus . . desconfeitt the Inglismen and Saxonis in ane battall at Lochmabane; *Chron. Kings* 47.

B. *adj.* Of or belonging to the Saxons, as prec. **b.** *specif.* *(The) Saxone langage, toung*, English (as opposed to Gaelic).

Syne Saxon and the Scottys blude Togyddyr is in yhon frely fwde; WYNT. VII 583. The Saxone herald thair remaning; STEWART 27309.

b. Adanus . . In Saxone langage wes nocht richt perfyte; STEWART 29388. The king [*sc.* Oswald] quhilk leirit quhen that he wes ȝoung The Saxone langage and the Scottis toung; *Ib.* 29392. The Ingles men evin as the mair politick Scottis vses that ald Saxone toung; DALR. I 85/19. As the Saxon languadge prevailed (which banished the patronimicks) the surname of Innes arose from the lands; 1698 *Fam. Innes* 6.

Saxon(e, *n.* [ME and e.m.E. *Saxon, Saxoyne*, OF *Saxoine* (OED), late L. *Saxonia.*] Saxony, the country of the Saxons. The STEWART quot. may belong in SAXON *n.* — Germanea superior hie Ducheland the impyre Swewia Bawar and Saxon; *Asl. MS* I 306/12. In tha boundis the blude of Saxone Thair duelling maid first into Albione; STEWART 24448. The duc of Saxon; *Compl.* 168/2.

Sax-part, var. of SEX-PART *n.* **Saxpens,** var. of SEXPENCE *n.* **Saxsum,** var. of SEX SUM *n.* **Saxtane, -ein(e, -en(e,** varr. of SEXTEN(E *num.* and *n.* **Saxt(e,** var. of SEX *num.* and *n.*[3] and SEXT *adj.* (*n.*[1]). **Saxté,** var. of SEXTY *num.* **Saxteint, -tenit,** varr. of SEXTEND(E *adj.* **Saxtelie,** var. of SEXTLIE *adv.*

Saxter-aithe, *n.* [A calque of ON *séttar-eiðr* 'an oath of six', i.e. of six compurgators; *sax* SEX *num.* B.] In Shetland: The 'sixfold oath' (see notes below, and quots.).

'While the sixfold or twelvefold oath might be imposed in the first instance in cases of certain serious crimes, the general rule was that the "larycht aith" was the "first offender's" oath, the sixfold being normally prescribed for persons who had "failyeit the larycht aith of befoir," and the twelvefold for those who had previously failed to "quit" themselves with the six-fold oath' Gordon Donaldson (ed.) *Shetland Sheriff Ct.* iv-v, and cf. LAWRICHT-AITH(E *n.*

For a list of crimes for which the *saxter aith* was imposed, see W. Douglas Simpson (ed.) *The Viking Congress, 1950* (1954) 105, 155.

Donald Smyth is dempt to quyt himselff . . with the saxter aithe in caice he failyie the larycht aithe; 1602 *Shetland Sheriff Ct.* (ed.) 8. Walter Turvous is dempt to quyt himselff and his boitsmen of the drawing of Airthour and Gutheromis at the Nes lynis . . with the saxter aithe becaus he failyeit the larycht aithe of befoir; *Ib.* 12. Jonat Archbald is dempt to quite hirselff with the saxter aithe for the turning of ane siff and riddill for ane pair scheiris; *Ib.* 29. *Ib.* 20, 30, etc.

Saxtlie, var. of SEXTLIE *adv.*

Say, Sey, *n.*[1] Also: **saye, sae, se(a.** [ME and e.m.E. *say* (1297), *saie* (Wyclif), *seeay* (1519), *sey* (1636), OF *saie* (c1212 in Larousse).]

1. A fine cloth similar to serge; a piece or length of such cloth. Also *pl.*

(*a*) Ane stik of red say; 1490 *Acta Conc.* I 158/1. A goun . . lynit with say; 1499 HALYB. 161. Thre stremaris for the schip, of the say send hame be Jerome Friscobald; 1506 *Treas. Acc.* III 89, etc. vi ruffs and courtings of say; 1517 *Lanark & R.* 195. Ane pund and ane half of fyne say; 1566 *Treas. Acc.* XI 501. Thre elnis ane half of densyre say; 1572 *Edinb. Test.* II 257.

(*b*) A doble curlet of sey; 1493 *Acta Conc.* I 315/1. For iiij steik of sey rede blew grene and ȝallow; 1501 *Treas. Acc.* II 26. For ane blew steik of sey to the banar for the schip; 1504-6 *Ib.* III 89. Ane pair of courtingis of bew sey; 1580 *Coll. Aberd. & B.* 391. 1587 *Aberd. Chart.* 350. Of champit reid sey; 1598 *Black Bk. Taymouth* 323.

(*c*) Thre gren courtingis of sae with frontellis of gren tafiteis; 1562 *6th Rep. Hist. MSS* App. 647/2.

pl. The chawmer hingin witht ald sayis palyt reid blew and ȝallow; 1518 *Reg. Episc. Aberd.* II 175. Serge, sewin silk, sewin gold and siluer sayis; 1554 *Edinb. B. Rec.* II 201. Sayes of Flanders making cullored; 1612 *Bk. Rates* (Halyb.) 326. 1661 *Acts* VII 253/2.

(*b*) Na superior cleth, seys [etc.]; 1472 *Lennox Mun.* 90. Seyis, worsetts, linings and single sergis; 1587 *Aberd. Trades* 335. Thair is sevin Flemyngs laitlie cum hame fra Flanders . . quhairof sex to mak seyis and the sevint for brayd clayth; 1601 *Edinb. B. Rec.* V 287.

2. *attrib.* and *comb.*

Also SAY-BUMBACIE *n.*

(1) A pare of se peis; 1493 *Prot. Bk. J. Young* 132. Ane greine sey approun; 1658 *Lanark B. Rec.* 169. She never took a sey apron off the dyke; 1673 Paterson *Ayr & Wigton* II 546. — Ane blew say bed; 1653 *Soc. Ant.* XXIII 300. A rid pock for a sea bed, a pair blankets [etc.]; 1659 Craven *Ch. in Orkney* II 197. — v pair of sey cowrtingis; *Will A. Betoun* 224. Ane stand of blew sey courtenes; 1635 *Haddington Corr.* 302.

(2) Ten ellis of quhyt sey clayth to be littit blak; 1555 *Dundee B. Ct.* III fol. 19 (17 Dec.). My rid seaclaith [*blank in ed.*] petticot; 1601 *Edinb. Test.* in *Hist. Fam. Seton* 585. [3½ ells] of sey cloath for ane cott; 1684 Reid *Forfar* 193.

(3) Of spune sey brechane ȝarne ane stane wecht; 1576 *Edinb. Test.* V 56b.

(4) Ane uther sey lume with the furniture; 1624 *Edinb. B. Rec.* VI 259.

Say, Sey(e, *n.*[2] Also: **saye, seay, seie, se, sye.** [ME and e.m.E. *say* (Gower), *saie* (Manning), *sey* (1600); aphetic f. Assay *n.*]

1. The action of testing the quality or fitness of a person or thing; a test or trial of quality or fitness by use, a try, a try-out. Cf. Assay *n.* 3, Essay *n.* 1 a.

(*a*) The best and the gay of it, myself tuk a say of it, A wylie-coit, I will nocht lee; 1568 Sempill *Sat. P.* xlviii 93. They .. come wastwart fornent Sanctminians and thair persaueit it ane small village, gaif it ane say and landit thair; Pitsc. II 108/20. The Lord .. sufferit mekill woe and paine With Jews that wer hard of say; *Maitl. Q.* 42/4.

(*b*) All the personis that ar clengit [*sc.* of infection] .. till tak the sey of thar geir .. that the ballie .. tak thai persons and thar geir furth .. and clenge all thar geir agane; 1530 (*c*1580) *Edinb. B. Rec.* II 39. Goe, tak the first seye of hir yoursell; 1584 Sempill *Sat. P.* xlv 406. He .. contented to teatche ther and tak a sey what God wald work with him; Melvill 255. I cannot grant this your humble and earnest supplication in Mr. Andro Melvill's favours, for I have alreadie given him to the Duke of Bulloigne .. ; but if that had not been, it might have been that yee should had a sey of him; Row 299. We must take a sey of the things alone that we have ouned with companie; 1654 *Laing MSS* 295.

b. *specif.* The trial or test of combat; the combat itself. Only in Hay *Alex.* Cf. Assay *n.* 1, 2.

Ȝoung .. men wenis it is bot play Quhill thai haue prefit and anis bene at the say; Hay *Alex.* 1274. Antigone .. Gart his futmen ga mak the first assailȝe .. Than Nicolas quhilk saw thame at the sey To confort thame brak of his array; *Ib.* 1499. Wely men that oft was at the say Forsaw that cast and mett thame in the front And wele tua hundreth speris .. Straik of the formest doun; *Ib.* 3607.

c. *specif.* The testing of the fitness of a tradesman prior to his being accepted into the craft. Cf. 7 below.

That all skynneris .. sall come in orderlie be form of say and trying of their gratification as becometh and to pay thair composition .. and to mak ane essay as follows; 1633 *Aberd. Trades* 204.

d. A testing or trying situation.

Alwayis they would not wis it to cum to that seay, gif eather your gud circumspectione or thair aduertisement myght preserve it; 1571 Bann. *Trans.* 142. The Lord God opin my eies by a second sight to seie His favor, aime, and end, with my use of this od seie of transporting motions quherby .. He comunicateth Himselth; 1637 Johnston *Diary* I 253.

e. *attrib.* in *fig.* context. A testing shot, something intended to test opinion.

This letter alarumed some peeple, as an say-shot that the king intended .. to put papists in the government; 1685 Lauder *Notices Affairs* II 677.

2. Trial; trouble; tribulation.

For than ȝe knew thay wer ȝour fais, Bot now thay cum in freindis clais, Quhilk is ane sairer sey; 1572 *Sat. P.* xxxi 78.

3. The testing of metals, in order to ascertain their standard of purity.

That it [*sc.* pewter] sall keip this sey .. howsone it beis twechit with ane het yrne it salbe cleir and gif it beis cleir tobe haldin sufficient; 1567 *Acts* III 43/1. Men mettall tryis by sey, and not by slight; Montg. *Misc. P.* xxxii 23. And ane perfyte saye takin be thame of the samin [*sc.* ore] and reportit be thame to the saidis Lordis of Secreit Counsell; 1608 *Mining Rec.* 121.

4. The act of tasting food or drink (for poison) before presenting it to a person of rank.

A say scho tuk off all thyng at thai brocht; *Wall.* viii 1274.

5. *fig.* A taste, a foretaste; a sample.

Bot I, having intended to sett downe ane essay, have scarce given you the say of a perfyt letter; *c*1600 M. Napier *Mem. J. Napier* 297. I wishe any man who thinks more might have bein got done .. had got a sey of the worke wee had; 1680 *Aberd. Council Lett.* VI 219.

6. a. A sample of a precious metal, kept as a record of its fineness. Cf. Assay *n.* 3 b, Essay *n.* 1 b. Also, *attrib.* with *balance, box* (= the box in which the samples in a above were kept), *hous* (= the assay house), *piece*. **b.** Other material for testing, prob. in the making of gunpowder. **c.** *attrib. Pece* (Pece *n.* 14) *of sey copper*, prob. = a piece of artillery (made of copper) intended for testing. **d.** *Ane sey pece* (Pece *n.* 18) or *tas*, ? a cup or goblet of some sort made of precious metal of a particular fineness.

a. The sayis to be tane of the said gold and silver at every melting and putt in the said box sa that the fynes tharof may be perfitelie knawin; 1525-6 *Acta Conc. Publ. Aff.* 240. The said Harbert generale of hir hienes cunye and cunyehous .. sall .. hald gude and faithfull register .. of all gold and silver that sal be cunyeit alswele of the quantitie and wecht as of the fynes and to assist to set the seyis and deneris callit the seyis of the box to be put in the said box; 1560-1 *Reg. Privy S.* V 1 190/2.

attrib. (1) I went to the assay maisters of the banke of Amsterdame to buy a fornace and ane say ballance with its piller and weights; 1646 J. Hope *Diary* (1958) 163.

(2) To ane smytht for the of taking of the lok of the say box; 1532 *Treas. Acc.* VI 103. The generall and maister of his hienes cunyiehous to transport thameselffis, the prenting irnes, say box, and all officiaris furniture, warklumis and necessaris belanging to the said house .. to .. Perth; 1585 *Reg. Privy C.* IV 26. That the say box belong to him when it is broken vp quhich will not be much, because that it conteins bot ane quartre of everie say piece; 1641 Cochran-Patrick *Coinage* I xxx. That it is 27 or 28 yeirs since the say box was closed; *Ib.* xxxi.

(3) Glas .. for the new windok in the chalmer abone the say hous; 1608 *Hilderstoun Silver Mines* I 59b.

(4) Say piece; 1641 Cochran-Patrick *Coinage* I xxx (see (2) above).

b. [Inventory of the goods, etc. of .. Michaell Gardener master cannoner (gunner and powder-maker) to the king.] Ane kist for keping of poulder with ballands and wechtis seis and rangis; 1584 Whitelaw *Sc. Arms Makers* 299.

c. Munitionis .. Thre pece of sey copper .. wes brocht hame .. for ane sey; 1541 *Treas. Acc.* VIII 119.

d. Gevin to eyk ane sey pece of the kingis grace, iij unce Inglis grotis; 1535 *Treas. Acc.* VI 248. For ane kais to ane sey pece, ane kais to ane toysting pan, and for ane kais to four ladillis; 1541-2 *Ib.* VIII 51. To Jhonne Mosman to gylt ane cowp and ane say pece; 1547-8 *Ib.* IX 166. To Jhonne Mosman xiiij unce wecht grottis and testanis to be ane sey pece to my lord governoures grace and to mend certane coupis of his graces; 1550 *Ib.* 407. For the making and furnissing of ane laver of silver dowbill ouregilt with gold with ane say tas to the kingis majestie; 1575 *Ib.* XIII 63. Ane sey tas of siluer weyand four vnces; 1609 *Edinb. Test.* XLV 240b. The littill siluer peice the silwer goblet and littill sey tas; *Ib.* XLVI 36b.

7. A demonstration of a craftsman's skill; the test piece made by a craftsman and submitted to the craft as proof of his capacity. = Assay *n.* 3 c, Essay *n.* 1 b. **b.** *attrib.* With *schounes*, etc.; *peice* (cf. Pece *n.* 11); *mony*.

(*a*) His say to be ane utred pistolat; 1623 Whitelaw *Sc. Arms Makers* 207. That no prenteis tak upon hand to set up and work at his awin hand .. without he give his say to the saides two men and craft; 1631 *Kelso Glovers* 9. All those who sall enter frieman of the said [*sc.* baxters'] traid hereafter sall have for his say, first to peit the oven, and his say to be an ait bakin, and a wheat baikin, with ane pye with six corners [etc.]; 1669 *Aberd. Trades* 230.

(*b*) Quhat tyme the said Waltir desyris to gif in his sey; 1560 *Perth Hammermen* 90. [To] produce ane sufficient sey of ane maister stik of wark to be ane sufficient craftisman to serve the queenis grace legis; *Ib.* 91. 1620 *Glasgow Hammermen* 9. 1623 Whitelaw *Sc. Arms Makers* 190. Except he have servit out his prenteisschip and theireftir give in his sey or tryell of the said craft; 1639 *Lanark B. Rec.* 133. Johne Ker hes giwen in his sey of craft and is accepit be the craft; 1646 *Dunferm. Hammermen* MS p. 53. No printis nor frimane sall .. be .. admittit .. vntill such tyme that he give in ane sufficent seye of quhat traid so ever they be .. and quhen they goe to mak ther sey it is statut and ordainit that ther be tua of the traid he is of and to put him in ane close hous and seightit twa tymes evrie day; 1654 *Rec. Old Aberd.* I 328. 1656 *Dumbarton B. Rec.* 73.

(*c*) But [he] also sall asstrick himselfe to any ane pairt of the said wright craft .. of which he can give the best sye, and is most propitient in; 1670 *Dundee B. Laws* 588.

b. (1) Vellie Sadller is admitted master to the cordenar craft, and hes gifyn hes sey schounes, buttis, [and] mullis; 1590 *Dundee B. Laws* 390.

(2) [Sey masters] sall not apoynt him [an apprentice] any sey peice quhill they sie timber in readines for such work as they mynd to be his sey; 1650 *Glasgow Wrights Acts* 10.

(3) And also it is ordained that the bill money and sey mony be allowed in pairt of payment; 1698 *Aberd. Trades* 279.

Say, *n.*[3] Also: **saye.** (Of doubtful etymology. Also in the mod. dial. 'The part of a carcase of beef extending from the shoulder to the loin' (SND, s.v. *Sey* n.[3]). Cf. ON *segi, sigi* a strip or slice of flesh.) — Denner and supper .. for toistis trinsch bread and sayes, iijb.; 1593-4 *Mar & Kellie MSS* 41. For ane say of beife; 1689 Lorimer *St. Cuthbert's* facing 112.

Say, Sa, *v.*[1] Also: **saye, sai(e)-, saii-, sae; sey, se; saine, sayn(e, sane, say(i)th, sayeth, sene.** P.t. and p.p. **said(e, sayd(e, sayit, -id, -ed; sad(e, saed, seid, seyd, sed(e, sedde, sead, seyden.** [ME and e.m.E. *seggen* (*c*1175), *sai, sayne* (both Cursor M.), *seyen* (*c*1300), *sayen* (*a*1547), etc. Pres. t. 2nd pers. *seist* (*c*1175), *seggesst* (Orm), *sais* (Cursor M.), 3rd pers. *seied* (*a*1175), *seȝȝþ* (Orm), *seyþ* (*a*1250). P.t. *seide* (*c*1175), *seȝȝde* (Orm), *sade* (*c*1200), *said* (*a*1352). P.p. *iseit* (*c*1175), *ysed* (1297), *said* (Cursor M.), etc. OE *secgan* (pres. t 2nd pers. *segest, sægst*, 3rd pers. *segeð, sæg*, p.t. *sægde*, p.p. *(ge)-sægd, (ge)-sǽd.*

See also SAID *ppl. adj.* and SAID *verb phr.*]

1. *tr.* To utter, pronounce or articulate (an utterance); to utter (a word or words, a saying, etc.). Chiefly followed by direct quotation of the words uttered.

Also *to say nocht; to sa sorow* = to mourn.

(1) *pres.* This cane thai sa sorofully, 'Gais [etc.]'; *Leg. S.* xix 516. Sanct Machor .. Sayand, 'A wis man sad wmquhill, Sek na heyere thing than [etc.]'; *Ib.* xxvii 1127. Thane sais the wich, 'Gyf [etc.]'; *Ib.* xxxiii 501. Quhen ony thing he chargis thé Say nocht, 'I wat it wyll nocht bee'; *Consail Vys Man* 108. Sanct Dauid sais in haly prophecy, 'The man is blissit that dois bayth dome et law'; *Bk. Gude Counsale* 197. Thus gait cowth he sane [: agane; *v.rr.* sayne], 'Ga doun Schir Kid'; HENR. *Fab.* 750 (Bann.). He sayis, 'Men callis me Rauf Coilȝear'; *Rauf C.* 46. Nane suld say 'My fadere,' bot he that is his natural sone; IRLAND *Mir.* I 26/34. Says all, 'Amen for cherite'; *Seven S.* 2782. Saye; *Ib.* 361. I dremed ane angell came fra hevin With plesand stevin sayand on hie, 'Telȝouris and sowtaris, blist be ȝe'; DUNB. (OUP) 158/3. Now baldly says he, 'Citezanys haue doyn'; DOUG. XI ix 34. He begynis with a stuir .. voice, and sayis, 'I praise my God grittumlie [etc.]'; 1569 Bann. *Memor.* 7. As quhene Dawid sayith, 'I thocht [etc.]'; PITSC. II 65/9.

p.t. 'Schir,' said the child [etc.]; *Alex.* II 2885. And courtasly, foroutin hone, He said, 'For ane sicht [etc.]'; *Ib.* 7092. 'Bot God,' seyd he in-to that stryve, 'Myght (he) be mayd sekyr of hys lyve!'; *Troy-bk.* II 27 (C). A Roman .. Till Silla sayde apon this wyis [etc.]; WYNT. IV 2139. 'Be Goddis blud,' he sayd, 'yhe sang'; *Ib.* VIII 915. In secrete said he to the quene, 'Madame gif ȝe desyre [etc.]'; *Seven S.* 654. And forthir eik onto thame thus he said [: onglaid], 'O Latyn folkis [etc.]'; DOUG. XI iii 17. He brak the breid and said, 'Tak ȝe and eit'; Q. KENNEDY *Breif Tract.* (ed.) 113/10. He [*sc.* Timothy] sayd nocht, the antiquiteis, he sayd nocht, the thingis haldin of hald; WINȜET II 55/6, 7. 'This helper .. Hes cawsit me transgres,' Sayd scho; MAITLAND *Bann. MS* 13a/107. I [*sc.* Atholl] said in Yrishe to him [*sc.* Argyle], putting my hand to my breist, 'The wrang I haue gottin is in this pairt'; 1579 *Acts* III 176/2.

(*b*) He sad on hicht, that all mycht heyre, 'Pece be till ȝow'; *Leg. S.* i 249. Than the pupill sad he till, 'Gyf he liffis [etc.]'; *Ib.* 482, etc. Thairwith come Eis, sad, 'I sit warme and hait [etc.]'; *K. Hart* 831.

(2) *pres.* In harte thai had sike wgrines, That thai had no word for to say; *Leg. S.* vii 717. Sa to hym thus my wordis on syk wys; DOUG. IV v 98. *Ib.* XI xiii 80. And with them passit brocht againe Rose water cleir, doing thir wordis saine, 'I am callit [etc.]'; *Clar.* II 150.

p.t. Criste .. sad til hyme this wourd; *Leg. S.* i 11. To quhome Eccuba .. Said [*C.* Seyd] to him with full yrous chere Suche lyke wordis .. 'O felloun tratour [etc.]'; *Troy-bk.* II 835 (D).

p.p. Quhen Bantillas had said that sawe; *Seven S.* 493. Quhen that the semely had said hir sentence to end; DUNB. *Tua Mar. W.* 146. Quhen thir wourdis war sade; *Boece* 160. Quhat planar vordis culd the apostil hef said; Q. KENNEDY *Breif Tract.* (ed.) 115/1.

(3) Gyf thai twa sais thé nocht bot nay, I pray thé vryth thi wyll away; *Ratis R.* 1048. Of meit and drink sayid neuer nay; MAITLAND *Maitl. Q.* 17/46.

(4) He that slew his wife yesterday may sit and sa sorow on hir this day; CARMICHAEL *Prov.* No. 780.

b. Of the law, writing, etc.: To pronounce, state (something specified). Cf. 2 b.

And the lawis sais, 'Quod morte moriatur [etc.]'; HAY I 109/6. There was a writ fund .. sayand: In hac boxa [etc.]; 1488 *Treas. Acc.* I 87. For the writ sayis In eternum [etc.]; IRLAND *Asl. MS* 57/2. Conforme to his vord saynge, 'He that eatis my flesche [etc.]'; Q. KENNEDY *Breif Tract.* (ed.) 119/29.

2. To declare or state (a fact or opinion) in speech or writing (*to* another or with indirect object). **a.** With a clause (introduced by *that* or *(bot) gif* (= whether, unless), or with ellipsis of *that*) as object.

Also with noun phrase and infin. as object, rendering L. accus. and infin.

(1) *pres.* 'Schyr perfay,' Quod the gud-wyff, 'I sall ȝow say The king Robert the Bruys is he [etc.]'; BARB. VII 250. And say this bischope als that schow .. Is the fende in wemanis schape; *Leg. S.* iii 1093. Thar-for sais to me now fyrste, Gyfe ȝe forout corrupcione Of flesche kepit myn benysone; *Ib.* vi 152. Of his halynes als sais vs The doctoure Crisostomus That [etc.]; *Ib.* xxxvi 661. Sais .. to the king Bot gif he succour ws in hy All dede in Gaderis mon we ly; *Alex.* I 254. Gif a man cummys till a merchand .., and sais him that his fader aucht him a thousand nobleis [etc.]; HAY I 262/19. I say yow in certane He salbe set agane; *Gol. & Gaw.* 167. First say I thé, that twichand Italy [etc.]; DOUG. III vi 53. That William .. accusit Edwart .. opinlie in fensit curt, saing to him that [etc.]; 1583-4 *Inverness Rec.* I 294.

p.t. Fore myn maystere, the feynd, sade me That prince oure al the warld is he; *Leg. S.* xix 163. The Gregeois .. Seyd to Pryame that pas wald thai To Thenadone; *Troy-bk.* II 700 (C).

(2) *pres.* Sayand, that nother dede na dreid Till sic discomfort suld thaim leid, That [etc.]; BARB. XI 419. Catone sais .. that idilnes giffis novrysingis to vicis; *Leg. S.* Prol. 1. Sume men sais he crabyt is; *Ib.* xxii 786. Saiand that he had [etc.]; 1391 *Red Bk. Menteith* I 160. At the quilk acord he sayis he wald be gladly; 1401 MacRae

Early Sc. Texts No. 8. 1419 *Wemyss Chart.* 45. *Burgh Laws* c. 41 (B). I say, treuly he awe to save his lyf and defend him; HAY I 200/28. For wyfis sayis that lukand werk is licht; HENR. *Fab.* 102. Gif that ony speiris [etc.] .. Say thow, I left the laif vnto the freiris; *Ib.* 2971. Say ȝe nocht this clerk ill him bure That [etc.]; *Seven S.* 1639. The quhilk schiref execut and indorsat the sade letter, sayand on the bak of it that [etc.]; 1497 *Acta Conc.* II 75. Thai say, allon me challancis Ene; DOUG. XI viii 159. Men wald say I sa this for invy; STEWART 49487. Saye; LAUDER *Off. Kings* 330. ROLLAND *Seven S.* 683 (see SAID *verb phr.*). Men sayis als gud luve cumis as gois; *Bann. MS* 257b/63. Sum says it is a mater that wirkes out of the stanes; DALR. I 47/33. Thomas Symond, cutlar, .. sayis that he .. sayins the barne with the saying 'In nomine patris et filii et spiritus sancti, etc.'; 1601 *Elgin Rec.* II 96.

(*b*) And thow may nocht asonȝe thé And sa for caus it suld swa be; *Ratis R.* 152. And thai sae that daye wyll be in Scotland or Sateruris day; 1545 *Corr. M. Lorraine* 141.

(*c*) The Grekis he sayis .. Entered nocht at the brokin wall .. For of the hors no mencioun He maid .. Bot Vergill of the hors of bras, As Dytes seis, he sais it was. Bot Dares sayis that [etc.]; *Troy-bk.* II 3098. Thow quho knawis his birth his lyfe his end Mey sey his saull to lesting lyfe is send; HUDSON *Maitl. F.* 448/10. Nau seeth that he vnderstood not be yowr sister that [etc.]; c1580 Q. MARY in *12th Rep. Hist. MSS* App. VIII 9.

(*d*) At thar bene mony goddis I wil not sane [: remane]; DOUG. VI Prol. 133.

p.t. and *p.p.* He sad, he subuertit nocht; *Leg. S.* i 83. He seyd, the Gregeois thynges two Asked; *Troy-bk.* II 95 (C). Inne hye thame answered than Calchas And seyden that [etc.]; *Ib.* 1020. For-thi thai sayde that scho thare wes Amang thare goddys as a goddes; WYNT. II 255. Arrius the gret herretik said and teichit that [etc.]; IRLAND *Mir.* I 23/16. He stinckett lyk a tyk sum saed; DUNB. (OUP) 101/48. And thair in presens foirsaid the said Flemyngis [*pr.* Flemyng is] alledgit and sayit that the bill gevin vpoun thame was nocht of veritie; 1515 (c1580) *Edinb. B. Rec.* I 156. For dyverse tymes ye have seid you wolde be glad to speke with your sonne my lord; 1548 *Cal. Sc. P.* I 172. And without doubte thay sayd walde be, gif he did; DALR. I 136/3. Sayed; 1607 *Soc. Ant.* XI 414.

(3) Sayand .. Troyes oost to haue venquest bene; *Troy-bk.* II 1276.

b. Of a law, writing, proverb and the like: To state.

This endenture made .. betuixt Alexander of Irwyne .. on a part and John Moigne .. on the tothir part saies [etc.]; 1388 *Antiq. Aberd. & B.* III 294. The law writtin sais that quhen were is decernit [etc.]; HAY I 158/21. Ther is ane ald prouerb that says that ane herand damysele and ane spekand castel sal neuyr end vith honour; *Compl.* 108/12. Saiis; 1559-60 *St. A. Kirk S.* 14. Thair is na law that sayis that Frenchmennis gudis .. shall [etc.]; 1561 *Reg. Privy C.* I 181. *Acts* (1566) To the Reader iij.

c. With an indirect question as object.

Thy suith-fastnes sais [F. *dites vérité*] to ȝour king Quhare ȝe think maist to lufe lelely; *Alex.* II 2230. Schir sayis me .. Quhat tua thingis dois ȝow to dre Sorrow and pane; *Ib.* 2476. As was said long before gone by .. How Achilles [etc.]; *Troy-bk.* II 2786. For quhow grysly and how gret I ȝou sayn Lurkis Polyphemus; DOUG. III ix 96.

d. With noun, pronoun, phrase or infinitive as object.

See also SAID *verb phr.*

(1) *pres.* Thow may say quhat thow will on fer; BARB. V 620. For thu leis all that thu sais; *Leg. S.* i 512. My sueit dam Ydory Sais sum comfort now in hy! *Alex.* II 2252. And quha forsuith suld the richt sane [: mane]; *Ib.* 9415. The burgesis of the burgh of Dunde sayand the contrare; 1402 *Dundee Chart.* No. 22. To say thé thar awys; *Ratis R.* 783. Say me ane chois; *Gol. & Gaw.* 1181. Scho said, 'Say ȝe that in effek Thy hed sall gang than of thi nek'; *Seven S.* 673. This certanly scho syngis & sayis That ȝe sal me the basyng hald; *Ib.* 2584. Sittand at eys ilkane say his entent; DOUG. XI ix 36. He slew his mother creuellye .. Sum sayis to be at lybertie Sum sayis for hir adultrie; LYND. *Mon.* 3219. Quhat wald ȝe say gif [etc.]; ROLLAND *Seven S.* 4450. Weil war he all his dayis That vnderstude quhat ȝone bird singis or sayis; *Ib.* 8393. Thir ar the wordis of ws al; we al says the samyn; WINȜET II 76/26. Pope John .. left behind him .. more aboundance of gold as sayeth Baptista Platyna nor any .. did before him, some sayth 25 mylȝeon; FERG. *Answer* in *Tracts* 11. Of the fishes, how copious thair thay ar, I neid to say naything; DALR. I 63/27.

(*b*) Venus na mair [L. *plura*] sufferit him plene nor sa [: Asya]; DOUG. I vi 138 (Sm.). 'That thing,' quod thai, 'quhilk Apollo wald sa [: Ortygia]'; *Ib.* III iii 13 (C). Thow sall se at it is trew at we sa; ABELL 18a.

(*c*) Enewch tharof, now will I na mor sayn [: agane]; DOUG. I Prol. 219.

p.t. and *p.p.* Quhen this wes said that er said I; BARB. XII 477. & of his modire syne sad I Sume thing; *Leg. S.* Prol. 79. That I haue said I sall hald; *Rauf C.* 315. And in it that the kingis hienes sal mak be sade to thame in that mater; 1496-7 *Acta Conc.* II 69. He to mak ansur till ony thing that ma be sedde to hym off resson; 1550 *Corr. M. Lorraine* 339. Wat euer bis sayed bi sur off my gud mynd; 1566 Q. MARY in *Inv. Q. Mary* xxxii n. Christ hes it sed; 1570 *Sat. P.* xix 99. I .. sayed no more; c1695 *Anal. Scot.* I 154.

(2) For ocht that he cuth sa ore do; *Leg. S.* xl 276. Betys .. thinkis to defend him sa That he nouther skaith sall do nor sa; *Alex.* I 2314. Unquyt I do no thing nor sane [: vane]; DUNB. (OUP) 66/67.

e. *To here say.* See HER(E *v.* 2 c for further examples.

Man is in-to dreding ay Of thingis that he has herd say; BARB. IV 763. And quhen the paianis hard this sa [: ay] Thai sad 'sik merwal neuir befel As [etc.]'; *Leg. S.* xv 84. Robene, thow hes hard soung and say, .. 'The man that [etc.]'; HENR. *Robene & M.* 89. I here say that Ciuilius .. was ane grete occasioun of this batall; BELL. *Livy* I 53/7. Quhowe beid I heir sayeand large report cumin in this partis be dyvers marchandis; 1543-4 *Corr. M. Lorraine* 68. Arristotyll nor Plato, I heir sane [: Italiane], Wrait nocht .. In Duche nor Dence [etc.]; LYND. *Mon.* 566.

f. Quasi-*impers.*, in the passive, with following clause, also noun, etc., complement specifying what is stated.

(1) For-quhy it sall be seyd but lese That [etc.]; *Troy-bk.* II 359 (C). The which is sooth syth as it is Inne-to proverbe seyd oft I-wys; *Ib.* 590. Ther is ane Scottis man callit Sanders Prinille, and is [*sic*] it is sede betrasit all the men off the Mers; 1544 *Corr. M. Lorraine* 94. It is sed Seterday my unfrinds wil bi v[i]th ȝou; 1568 Q. MARY in Ellis *Orig. Lett.* 1 Ser. II (1824) 253. Ther is that to be sead which I should heve bein gled to heve hard from you; 1666 *Laing MSS* 351.

(2) Tharfor the towne Was in-to pece—gif it mycht be Said pece in suche aduersite; *Troy-bk.* II 2242.

3. To give a recommendation, instruction or command (*to (till)* a person or with indirect object) (to do something). Const. noun clause with subjunctive verb or with *suld.*

And quhen myn moder sad me till .. I suld ga To schere hire corne .. I did as myn moder saide [: lade *p.t.* of lay *v.*]; *Leg. S.* i 220. Quharfor he bad me to ȝow say That ȝe in hast hald hyne ȝour way; *Troy-bk.* I 49. Syk cummyrryt wightis suffir, I thé say, To haue ane hald and dwel heir; DOUG. V xii 94.

4. *ellipt.* or *absol.* In uses of prec. senses: To declare or state; to recommend or instruct. **a.** With the adverbs *sa (so), sic, thus(gat).*

He .. thusgat sade, 'Nero [etc.]'; *Leg. S.* i 430. And vther prynces .. Apponed ageynes Vlixes, Seyand thus that [etc.]; *Troy-bk.* II 1192 (C). He left behynd him in his steid This Egra—the buke sayis thus—And a sone [etc.]; *Ib.* 2095. Rycht thus say we of testament, and thingis left to men be testament; HAY I 144/11. HENR. *Fab.* 750 (Bann.) (see 1 (1) *pres.* above). For thus myn auctor sayis; *Wall.* IX 25. Quhen Amipullus had said thus Than [etc.]; *Seven S.* 43. Scho said 'My deir quhy say ȝe sa?' *Ib.* 1093. With petous voce thus cuth I sene [: pleyne]; DUNB. (OUP) 26/31. Sa sais the curat of our kirk; Id. *Tua Mar. W.* 306. Says nocht ȝour sentens thus, .. 'Quhat honeste .. is tobe dram?' DOUG. IV Prol. 156. So my author sais; STEWART 1422. Sanct Paul sais thus [etc.]; HAMILTON *Cat.* 101. For the injurius vordis saing to the said Anne .. and gyf he beys fundyn saing sic to onne nychtbor wyf in tyme cuming [etc.]; 1559 *Inverness Rec.* I 35.

b. In clauses introduced by *as (efter as, that)*. Also *passive*. Also const. *to* a person or with indirect object.

active (1) As the buk sais [: betraise]; *Leg. S.* i 23. *Ib.* 127 (see c below). The wattyre off Jordane in his dayis Was dry, as the story sayis; WYNT. II 1070. As I said before a man wourthy Is worth ane hundreth in sic ane ieopardy; HAY *Alex.* 3369. This beand done, as ȝe haif hard me sa [: wa (= way)]; STEWART 29931. The land[is] liand all In Cuninghame, that my author did sa; *Ib.* 41159. The lard of Howpaslot left ay ane to keip the feild as thai say; 1541 *Ex. Processes* (Reg. H.) No. ix (Scott v. Scott & McDowall). Than sall one fyre, as clerkis sane, Mak all the hyllis and valais plane; LYND. *Mon.* 6032. FERG. *Answer* in *Tracts* 11 (see 2 d (1) *pres.* above). As we say, ane man is at hame, or not at hame; SKENE *Verb. S.* s.v. *Haimsuken.*

proverb. As the gude man sayis, sa say we, bot as the gude wife will al mon be; CARMICHAEL *Prov.* No. 176.

(2) On this maner assentyt war The barownis as I said ȝow ar; BARB. I 136. At gret myscheiff as I ȝow say [: thai]; *Ib.* XI 611. Gyf it sa be Of the pure man as thu sais me; *Leg. S.* i 130. Than Petir his staff .. bad hym .. Lay it on his falow sowne, Than as he sad hym, he has downe; *Ib.* 202. *Ib.* 223 (see 3 above). Thai grantyt all, as Wodstok can thaim say; *Wall.* VIII 703. Out off the toun as Scottis men till hym said, That serwandys oysyt with cartis hay to leid; *Ib.* IX 700. Sua we may sa in al our aduersite as the propheit Heliseus said to his seruand, .. 'Be notht red [etc.]'; GAU 59/18. Gif ȝe will do efter as I ȝow say; ROLLAND *Seven S.* 3686.

passive (1) Vnder silens of nycht as said is; 1471 *Acta Aud.* 20/2. Togidder with the laf .. gevine in varandice thairof as said is; 1501 *Stirlings of Keir* 274. DOUG. IX vi 139. The gallows .. quhairof Alester M'Gregour .. himself wes mountit abone the laiff as said is; *c*1605 *Black Bk. Taymouth* 40.

(2) All this grit brawling .. Wes for a pig as ȝe hard sayn; *Colk. Sow* I 431. As I hard sane; LYND. *Meldrum* 999.

c. In parenthetic clauses.

(*a*) And sa this Petir here, I say, Hym-selfe will nocht helpe .., Bot, as men sais, has no kin dowt To thole for Criste as vnderlowte; *Leg. S.* i 125. A barge of werre garnyst with Inglischmen in certane & sum men sais of the Newcastell; 1405 Slater *Early Sc. Texts* No. 58. For fra that to the hart he sayde Ane ewyn strekande wayne wes layde; WYNT. II 431. Thai ware Schyre Adam Mwre, thai say, And Schyre Alexawndyre the Ramsay; *Ib.* VIII 3565. Ladie, I say ȝow in certane, ȝe sall haue lufe for lufe agane; LYND. *Meldrum* 207. Was neuer man .. Nocht Hercules, I dar weill say, That euer faucht better; *Ib.* 1403. The knicht, I say ȝow sure, .. is in my chalmer flure; ROLLAND *Seven S.* 6633. Raigeing I say, nocht only aganis our mother the haly, catholik kirk, bot [etc.]; WINȝET II 7/30. 'Departit, oh!' quod he, 'and deid, I say ȝow'; 1570 *Sat. P.* x 388. [They] now haueand the oppertunitie, sais scho, of James absence; PITSC. I 125/22. Thow wes begottin, sum sayis to me, Betuix the devill and ane duin kow; MONTG. *Flyt.* 169 (T). Another thing nocht vnlyke, he sayis, was seine, twa ȝeirs eftir this; DALR. I 61/23.

(*b*) Evin as Meffan, his scuill maistre, thai se, schew him the way; 1571 *Sat. P.* xxix 28.

5. *intr.* To impart (chiefly, spoken) information to or to address another or others in response to an invitation, etc. to do so or in one's turn; to speak, answer, tell.

Sen ȝour willis is That I say, wyt ȝe sekyrly [etc.]; BARB. XIX 159. Thane sad the kynge, 'Me think that thu As pere & peyre sais to me now'; *Leg. S.* vi 634. I intromettit me neuer are In message for to do nor say; *Alex.* I 791. And I to say is richt ioyous [F. *Et moy .. du dire tous joious*]! *Ib.* II 2076. Quharfor bettir to be styll Than say vnknawandly thar-tyll; *Ratis R.* 904. For-thi chargyd than war thai, And fullyly instruct to say, And the grevys for to schawe To the Pape; WYNT. VIII 2364. That thai sal nocht say for haterent of the ta part na lufrent of the tother part bot for the suthfastnes for to say; *Burgh Laws* c. 22 (A). I can nocht say sudanelye .. Bot I sall call my cardinallis and my counsall; *Howlat* 120 (A). *Gol. & Gaw.* 409 (see SAW *n.*[2] 1 (2)). My hony hart how sayis the sang; *Peblis to Play* 247. And all was one to saye be myne awys; *Bk. Chess* 97. *a*1510 *Aberd. Univ. Rev.* XXXVI I 51 (see b (1) below). The fader than omnipotent maist hie .. Begouth to say and quhen he spak, all cessyt; DOUG. X ii 109.

b. With the adverbs *hardely, il, schortlie, suthly, wele*, describing the manner of speaking or the content or nature of what is said.

Also *Sayweill* as an allegorical character, orig. after *Piers Plowman.*

(1) I say ye soithly; BARB. VII 260. Certis, thu seis [L. *dicis*] ful il Quhene thu thaim callis dissawouris; *Leg. S.* xxxviii 370. *Ib.* xix 516 (see 1 (1) *pres.* above). Quod the Bauderane, 'Sais hardely!' *Alex.* II 2289. To say schortlie thair was bot few left haill; HAY *Alex.* 2169. Suthly to sane; *Gol. & Gaw.* 4. Woman cane say ande gane say als [Ande] worde nocht reide thocht thai sa fals; *a*1510 *Aberd. Univ. Rev.* XXXVI I 51. Treit wele Maxwell and sa wele to hym; 1543 *Corr. M. Lorraine* 18. He told him .. Of all the weiris schortlie for to saine; *Clar.* III 1355.

(2) Quhen say weill at sumtymes salbe brocht base, Do weill sall triumph in euerie place; *G. Ball.* 208. Sayweill but doweill is nocht worth a straw; *Bann. MS* 52a/95.

c. ? Nonce, const. as *reflex.*

For all to hym thar couth na man thaim say; *Wall.* I 200.

6. To say (something) *agan(is (in contrar), of* (= concerning), *on* (= about), *to* (= in reference to), *with* (= in favour of, in accusation of) a person or thing. **a.** *tr.* **b.** *intr.*

a. (1) As for ony civile accione that ony party has to say again thir persons, that [etc.]; 1496 *Acta Conc.* II 29. Any clayme .. Jhone Stevynsone hed to persow and say aganys hym; 1532-3 *Dunferm. Reg. Ct.* 90.

(2) God off the Dewyll sayd .. He is .. a leare fals; WYNT. VI 2169. Off this sentence, allace, quhat sall I say; HENR. *Fab.* 1248. Gret is the dignite and lovage that Sanct Johne sais of this haly spous of Jhesu; IRLAND *Mir.* II 66/16. Quhat sal I of this wondir warkis sayn? DOUG. VI Prol. 99.

(3) Gif ony man wald say on ane othir that he had [etc.]; HAY I 262/28. I am saikles of ȝone he sayis on me; ROLLAND *Seven S.* 9177.

(4) Gif the said George has oucht to say to the said oxin to persew thaim as law will; 1478 *Binns P.* (SRS) 7.

b. (1) The saide Wilȝeam had na place til impung the sade assignacione nor say agane the sammyn; 1496 *Acta Conc.* II 19. To obiect or say aganis the said assis; 1532 *Ex. Processes* (Reg. H.) p. 5 (Betoun v. Melville). 1553 *Lanark B. Rec.* 27. Na parte comperit to obiect nor say in contrar the personis of inqueist; 1558 *Inverness Sheriff Ct.* 38. BALFOUR *Pract.* 383. The king will that it sall not be lesome to the inferiour court to say against that record; SKENE *Reg. Maj.* I 55b.

(2) Of al payne, That men ma of think or sayne; *Leg. S.* xxix 266. Of which ȝow till I shall ȝow sey, if that ȝe will; *Troy-bk.* II 188 (C).

(3) Except the pannell wold pas from the repledgiatioun or say to it presentlie quhilk he hes done; 1641 *Justiciary Cases* II 447.

(4) The judge sall gar raise ane unlaw .. of ilk soyttoure that sayd with the dome that is falsed; BISSET I 286.

7. (To be) *alsmekle to say as*, to be equivalent to, to mean the same as. **b.** *That* (also *quhilk*) *is (als mekill) (for) to say (as), thay ar to say*, to wit, namely.

And that cat in Grece is als mekle to say as ingenious in Fraunche; *Loutfut MS* 18a. Ysaac is alsmekle to say as lauching; *Asl. MS* I 313/24. Innatyve is alsmekil to say as inborn; DOUG. *Comm.* I Prol. 97. For that war als mekle to say as [etc.]; DAVIDSON *Answer* 192.

b. (1) Makyne Drosser, That is al-so mekill to say her As the Durwarth sonnys; BARB. III 100. Julis hoff als mekill to say as the hall or clois of Julius; *Boece* 98b. Tanquam tabula rasa Quhilk is als mekill for till sae Rady for gud and ill; LYND. *Sat.* 225 (B).

(2) The forsayd lordes are acordit in specialte as efter folowes that is for to say, The erle of Northumbr' [etc.]; 1385 *Rot. Sc.* II 73/1. Beistis .. of syndri eildis and syndri hewis that is to say blak broun rede and braundit; c1390 *Reg. Morton* I xl. That is to say that the forsayd Robert sal [etc.]; 1395 MacRae *Early Sc. Texts* No. 4. Thir witnes that is to say Sir Wilȝame of Dissynton .. Sir Water Olyfand [etc.]; 1400 *Ib.* No. 7. WYNT. VIII 825. *Acts* I 711/2. In lissis that is to say in barreris; HAY I 255/35. Saye; *Bk. Chess* 2074. Cantere, quhilk is to say the hede and hiechest part of the erde; *Boece* 53. I leyff all my half of my hous to byde about my hous that is to say the insycht .. to remane; 1549 *Glasgow Test.* I 49b. The heicht hows, that is to say with hall, chalmer of deis, and the heycht loftis; 1564-5 *Glasgow Prot.* III 66.

(3) The annuelrentis following .. Thay ar to say ane annuelrent of twentie pundis [etc.]; 1569-70 *Canongate Ct. Bk.* 175. The remanent subjectis of this realme, thay ar to say Capitane James Bruce, Johnne Hammiltoun [etc.]; 1572-3 *Reg. Privy C.* II 199. Landis .. layand withtin the burght of Aberdene they ar to say [etc.]; 1587 *Antiq. Aberd. & B.* II 438.

8. *tr.* To tell or speak (the truth). Also in the parenthetic phrase *(the) suth to say*. Also const. indirect object.

(1) Giff I the suth sall say; BARB. X 298. Thai sad hyme the suth alswith; *Leg. S.* xxvii 747. Or thu gange away, Alhale the suth I sal thé say; *Ib.* xix 136. Say me the suth without lesing; *Alex.* II 2442. Do tary nocht it is suth I thé say; *Wall.* VI 451. Suth said he; *Seven S.* 1657. I sall say furth the south; DUNB. *Tua Mar. W.* 157. The suith if I sould saine; *Clar.* II 840.

(2) Thai war the sagis suth to saye That [etc.]; *Seven S.* 1837. The salt was all the sarar suth to sayn [: rayn, slayn]; DUNB. *G. Targe* 198. Sayn; DOUG. II ii 103. Thay him prayit .. to remaine With tham, quhilke he grantit, the suth to saine; *Clar.* II 322. Sen scho hes nowthir rewth Nor mercy suth to sane [: agane]; SCOTT xxiii 34.

9. To speak of, mention.

To se the maner certanely Of the gold at the letter said; *Troy-bk.* II 1883.

10. a. To relate, esp. out loud (a story, message, etc.), to deliver in speech.

And thai [*sc.* stories] war said on gud maner; BARB. I 4. Goddis word thar to say; *Leg. S.* iii 292. [The wife] quhene scho vaknyt can it [*sc.* a dream] say Til hyre howsband that by hir lay; *Ib.* xii 17. I am send .. ane charge for to say; *Gol. & Gaw.* 137. Master Barbour .. said the Bruce amang his othir werk; *Wall.* XI 1214. He suld .. say his legatioun or gif his lettres; *Lindsay MS* 17b.

b. To repeat or recite (a mass, prayers, etc.). Also *absol.* **c.** Of schoolchildren: To recite (a lesson) from memory.

b. The fyrste he was That euire be-guth to sa messe; *Leg. S.* vii 68. He suld ilke day say the bed; *Ib.* xxii 674. 1481 *Misc. Spald. C.* II 324. Sayand, 'Fiat voluntas tua [etc.]'; IRLAND *Mir.* I 42/3. To ane preist said mes to the king; 1501 *Treas. Acc.* II 71. Thou sais for thame few psaltris, psalmis or credis; KENNEDY *Flyt.* 318. Dauid Duly tailyour .. past to Sanct Gelis kirk quhilk was Sonday and thair said mes; 1530 (c1580) *Edinb. B. Rec.* II 36. The Houris of the glorios Virgine .. to be daly sade; *Boece* 479b. I gat no tyme .. To saye In Manus Tuas Domine; LYND. *Trag. Card.* 252. With beidis in hand, crosand and sayand creid; ROLLAND *Ct. Venus* II 565. That he that aught the house .. sould say the grace and blise the meate; PITSC. I 249/5. Thair durst nocht ane mese be said in Scotland; *Ib.* II 143/27.

absol. Quhen he had said, ane sleip he tais; *Alex.* II 433. To sing or say for thy saluatioun; HENR. *Fab.* 836. We haue said, prait, askit and desirit; IRLAND *Mir.* I 56/31. To the honest sustentatioun of ane chaplane daylie to sing and say at the said altar; 1525 (1527) *Reg. Great S.* 96/1. LYND. *Trag. Card.* 382.

c. They who learns Latein most have a pense of that quhilk they have learned before to saye everie morning; 1640 *Dundonald Par. Rec.* 465.

11. *tr.* and *intr.* *(To) say furth, say (sais) on*, (to) speak out, say (what one has to say). Chiefly imperative.

(1) 'Sa furth,' quoth he; *Wall.* XI 471. DUNB. *Tua Mar. W.* 157 (see 8 (1) above). Quha can do bettir sa furth in Goddis name; DOUG. I Prol. 478.

(2) Tharfor sais on ȝour will planly; BARB. XII 199 (C). Dame Fesonas say on ȝour thocht! *Alex.* II 2435. 'Now say on smertly!' said the king; *Ib.* 5979. Schir, say on; 1573 DAVIDSON *Sat. P.* xlii 136.

Say, Sey, *v.*[2] Also: **saye, seay; se, seie.** [ME and e.m.E. *saye* (c1380), *saie* (Lydgate), *say* (c1475); aphetic f. ASSAY *v.*]

1. *tr.* To put to trial or proof; to test, try out. = ASSAY *v.* 1.

(1) The luits beine sayit and the strings; *Clar.* II 1645. Eftir that he had sayit hir [*sc.* the ship] lang for hyir, .. he coft hir; 1567 *Reg. Privy C.* I 517. To preif my spreit and say my scharp ingyne; 1571 SEMPILL *Sat. P.* xxviii 41. His thovmbis was on the soillis to say them [*sc.* shoes]; 1584 *Ib.* xlv 796. For viii pound poulder to saye tham [*sc.* guns] with all; 1599-1600 *Skipper's Acc. (Morton)* 52b.

(*b*) For monting and drawing of the samin [gun] to the Cragheid to be seyit and schott; 1542 *Treas. Acc.* VIII 127. To seay a bow that was sumthing wicht; 1584 SEMPILL *Sat. P.* xlv 930. Giffen Robert Stewart to summond theis that seyit the wattir laitlie 13 s. 4 d.; 1613 *Inverness Rec.* II 117.

(2) The king gart schot ane cannon at hir [*sc.* the ship] to say hir gif scho was wicht; PITSC. I 252/4.

b. *intr.* *To say to*, appar. to make trial of, have recourse to.

Bot gyf thyne amouris .. be vedit .. Say to gud hop and resoune than. Gyf thai twa [*sc.* hope and reason] .. can nocht find That scho may syk a band wnbynd I pray thé reul thé as thai red; *Ratis R.* 1065.

c. *tr.* To try by tasting; to consume. = ASSAY *v.* 1 c.

That nane of the tavernaris wyffis, nor wemen seruandis .. waill, sey, or by ony wynis; 1520 (c1580) *Edinb. B. Rec.* I 202. Butter, new cheis [etc.], .. Grene leikis and all sic, men may say; SCOTT v 34.

d. To test or assay (metal). = ASSAY *v.* 1 d.

Every pudermaker .. sall say the tyn weschell maid be utheris .. be tuiching thairof with ane het irne; 1562 *Edinb. B. Rec.* V 371.

e. To try (*on*) (a garment); to try (a garment) *on* (a person or animal).

That my lord wes seyand on a py dowblett with the taill pecis for the wamb; 1600 *Crim. Trials* II 158. My lord .. seyed on your worship's horse both the fute mantells; 1633 *Maxwell Mem.* II 223.

2. a. To put to the test, to test out (an opponent). Cf. ASSAY *v.* 3.

This realme, quhilk enemeis oft sayit With cruell weir; 1570 *Sat. P.* xvii 137. The blak knicht sayit thame all bot thair was nane that mycht war him; PITSC. I 243/7. I will haue them all befoir me .. and say thame quhat they can do all befor me; *Ib.* 270/21. Bot they war weill sayit or they past out of Scottland; *Ib.* 340/6. [It may chance hereafter when they have] seyit [and touched others well as they will if matters be well guided .. then your men may the rather enter to the place and dispatch them]; 1602 *Cal. Sc. P.* XIII 951.

b. To test, try out, try (a person or thing) as a source of support in the performance of an undertaking.

Obey God, go say God, with prayer and fasting; 1572 SEMPILL *Sat. P.* xxxviii 104. And chance and fortoun quhilk we think to sey will supplie the rest; PITSC. I 19/29. 'Now it may be thought I am liberall .. because na minister will venture to com amangs us, therfor get me men and sey me'; MELVILL 434. Men sould sey they freinds or they mister; CARMICHAEL *Prov.* No. 1108. When the duke hath seyed the nobilitie of Scotland for his returning with small profite and no resolute answere now he turns him to insinuat himself in the favours of England; CALDERWOOD VIII 240.

c. *reflex.* To make trial of oneself, to test one's mettle, in combat.

This turnament and iusting beand indureit the space of xl dayis .. befoir that evirie man had tyme and lasour to sey him sellff gif he pleisit; PITSC. I 244/3.

3. To try out in performance, to attempt; also const. *on* another.

(1) He .. come agane, and seyit all his cure; *Freiris Berw.* 347 (B). He satt doun to say the spynning; *Wyf Awcht.* 62. Nor ȝit haue sayit the archere craft, nor schot my selfe with sic ane schaft; MONTG. *Ch. & Slae* 186 (W). I freyit to sey it, I luikit vpon loft; *Ib.* 347. Go, sey thy science; Id. *Flyt.* 64 (T).

(2) They could not choose a more proper subject .. to sey their craft on nor us; CALDERWOOD V 557.

b. *ellipt.* To make trial of as an experience, try.

If any of my dolour dout, Let ilkane sey thair tym about; MONTG. *Misc. P.* v 38.

4. To attempt or endeavour (*to* do something); to attempt to bring (something) about. Also *absol.* = ASSAY *v.* 2.

(1) War not thow seis full ill the band to leid; 1581 *Sc. Ch. Hist. Soc. Rec.* IV 125. Quhairwith befoir he seyit to schuit, He [etc.]; MONTG. *Ch. & Slae* 735 (W). Thoght .. death with deidly dairt Suld sey to caus me smart; Id. *Misc. P.* xx 41. Id. *Devot. P.* iv 38, etc.

(2) There ar of minister [*sic*] that say a confedaracy with them; 1680 *Soc. Ant.* XLV 233.

absol. Yee should not say, sir, till yee see; But, when yee see it, say; MONTG. *Ch. & Slae* 1155, 1156 (Wr.).

5. To find out or discover by trial or examination; to ascertain.

(1) The lerit sone off erll or lord .. Saying his odius ignorance Panting ane prelottis countenance; DUNB. (OUP) 132/49. To sey and feill thair minds to thaim and se quhome in thai ma lippin in; 1525 *Aberd. B. Rec.* I 113.

(2) He suld seay Gif he culd lerne him to obey; 1584 SEMPILL *Sat. P.* xlv 259. Seie; 1637 JOHNSTON *Diary* I 253.

b. *specif.* To search or examine (a market) for badly made goods.

1694 *Glasgow Hammermen* 49 (see SAY-MASTER *n.* b).

Say, var. of SA *n., adv.* and *conj.*, SAW *n.*[2], SE *v.*[1] **Sayance,** var. of S(C)IENCE *n.*

Sayar, -er, *n.*[1] [ME and e.m.E. *segger* (Manning), *seiere* (Wyclif), *seyere* (*a*1400), *sayer* (Caxton), *saier* (*a*1533); SAY *v.*[1]]

See also *mes-sayer* MES *n.*[1] 2 b and NA-SAYER *n.*

1. A narrator or author.

Be the documentis of wys clerkis that knawis the subtill ymaginacione of the sayar; *Wisd. Sol.* (STS) 522. The sayar eik suld weil consider thys, Hys mater, and quhamto it entitilit is; DOUG. IX Prol. 27. Far eithar is quha lyst syt doun to moyt Ane otheris sayaris faltis to spy and noyt; Id. *Exclam.* 29.

2. One who says or celebrates mass, the celebrant.

The sayar of it [*sc.* Mass] to committ horrible blasphemye in usurping upoun him the office of Christ; KNOX II 139. The lawis of the realme had appoynted punishment for sayaris and heararis of masse; *Ib.* 274. 1587 *Acts* III 430/1. Aganis the sayaris of messe and ressattaris or interteneyaris of excommunicat papists; 1593 *Ib.* IV 17/2. 1600 *Bk. Univ. Kirk* III 961.

3. *Evil-sayer*, one who speaks evil.

Quhisperars, bakbytters and euil sayers; KING *Cat.* 40.

Sayar, -er, *n.*[2] Also: **syer.** [ME *sayer* (?1370), *seyer* (*c*1460), 19th c. Eng. *sayer* (1835); SAY *v.*[2]] One who assays or tests; an assayer of metal. = ASSAYOUR *n.* — The wardanis fe the sayaris fe and the sy[n]karis of the irnis fee; 1526 *Acts* II 317/1. The generall of his cunyehous maister cunyear wardanis sinkar syer prentaris forgearis and utheris; 1579 *Reg. Privy S.* VII 321/1. Sayer; *Ib.* in Cochran-Patrick *Coinage* I xxxiv. **Saybo,** var. of SYBO(W *n.*

Say-bumbacie, Sey-bombasie, *n.* Also: **sea-, sy(e)-** and **-bombacie, -sis, -bumbas(s)ie, -bumbeshe, (-bamsey).** [SAY *n.*[1] and BOMBASIE *n.*] A type of bombasine, ? finely woven like serge. Also *attrib.*

For further examples see BOMBASIE *n.*

viij elnis of say bumbacie .. at xx s. the eln; 1584 *Edinb. Test.* XIV 132. Aucht elnes of blak seybamsey to be ȝour sone ane stand of claithis; 1610 *Tailor's Acc. Bk.* A 84. Ane lowise goun of sey bumbasie; 1612 *Edinb. Test.* XLVII 87. To acquent him and instruct him with sey bumbasie that is brocht to him to work; 1615 *Glasgow Weavers* 26. Of gingillane saittin seybumbassie; 1632 *Inv. Newark* 2a. Ane wob of sey bombasie; 1639 *Justiciary Cases* II 371. Seabumbacie; 1643 *Dunferm. B. Rec.* II 194.

attrib. For balling to hir sey bombasis goune xiij s. iiij d.; 1610 *Tailor's Acc. Bk.* A 84. To ȝour seabumbeshe clais 9 ellis of pleding; 1620 *Ib.* 103b. Ane pair of grein sey bumbasie courtenis; 1621 *Edinb. Test.* LI 66. To .. Marioune Littill ane blak sybumbasie gowne; *Ib.* 86. Ane stand of gingiling sybombasie courtenes; 1635 *Haddington Mem.* II 303. Ane reid sye bombacie coatt; 1648 *Edinb. Test.* LXIII 349. — To teache .. James Calder in thir four poyntis of his weifer craft, viz. the wound loome, sea bombacie loome, the playd loome, and playding; 1621 *Glasgow Weavers* 25. Of grein seybumbassie skreines; 1632 *Inv. Newark* 3b.

Saycht, var. of SICHT *n.*[1] **Saycretar,** var. of SECRETAR(E *n.* **Sayddis,** pl. of SAID *ppl. adj.* **Sayd(e,** varr. of SAID *verb phr., ppl. adj.* and p.t. and p.p. of SAY *v.*[1] **Saydlar,** var. of SADILLAR(E *n.* **Saydly,** var. of SADLY(E *adv.*

Say-drink, Sey-drink, *n.* [SAY *n.*[2] and DRINK *n.*] In Glasgow and in Fife: A charge levied from a newly admitted entrant to the freedom of a craft, presumably to defray the cost of drink provided to those present at his assay (ASSAY *n.* 3 c, SAY *n.*[2] 1); later commuted to a fee. — [He] sall pay .. the sey drink at the will of the craft; 1585 *Dunferm. Hammermen* MS p. 1a. [The entry money was fixed at 16 punds with] bankett and say drink; 1608 *Glasgow Bonnetmakers* 35. In all tymes cuming thair banquettis and say drink .. sall be convertit in money to the commoune weill of the craft; 1612 *Glasgow Weavers* 26. Quha hes payit speaking drink, sey drink, and tuentie li., conforme to ane act maid of befoir insert in this buik; 1622 *St. A. Baxter Bks.* 85. Johnne Jaiphray is admittit frieman .. and hes satisfeyit the sey drink; 1629 *Glasgow Hammermen* 15. Everie stranger entering ane frieman heireftir, sall pay of wpsett at his entrie four scoir markis, by and besyd the dewis to be payit for the say drink, mortclothe [etc.]; 1650 *Glasgow Weavers* 28. Each person intending to mak himself friman being friemens sons and goodsones sall pay at thair admission eight pundis Scotis quhairin ther say drink is to be includit; 1683 *Ib.* 34.

Saye, var. of SA *n.* and SAY *n.*[1], *n.*[3], *n.*[4], *v.*[1] and *v.*[2] **Sayed,** p.t. and p.p. of SAY *v.*[1] and var. of SID(E *n.* **Saye(o)r,** varr. of SYOUR *n.* **Sayer,** var. of SAYAR *n.*[1] and *n.*[2] **Sayeth,** var. of SAY *v.*[1] **Sayffande,** var. of SAV(E)AND *pres. p., prep.* and *conj.* **Sayf(f)(e,** varr. of SAFE *adj.* and SAV(E *v.* **Sayhcment,** var. of SITHEMENT *n.* **Sayid,** var. of SAID *ppl. adj.* and p.t. and p.p. of SAY *v.*[1] **Sayin,** var. of SAIN(E *v.*

(Saying,) Sayng, Seying, *vbl. n.*[1] Also: **seing.** [SAY *v.*[2]; cf. 19th c. Eng. *saying-knife* (*a*1858).] The action of putting to trial or proof; testing, assessing, examining. — Compt maid with William Striveling for expensis maid be him apone the sayng of the led mynd of Ilay; 1511-12 *Treas. Acc.* IV 274. To Robert Fynder [etc.] .. at the schutting and seying of the chalmeris quhilk wes maid be Johnne Kyll in Leith; 1557-8 *Edinb. Old Acc.* I 255. xiii seing of vrenis, the pece xvi d.; 1579 *Misc. Bann. C.* II 197.

Saying, *vbl. n.*[2] Also: **-inge, sayng(e, saing, sying.** [ME and e.m.E. *saying* (Cursor M.), *sayng, seyyng, sedgeyng* (all in Manning), *sayinge* (1387-8), *saynge* (Caxton).]

1. The action of uttering words, also expressing (something), counting, etc. in words.

To .. fle ewill deidis or reprevable sayng or sweryng; *Porteous Noblenes* 176/7 (Asl.). The sade Peter allegit that the sade Jhone had varyit in the saying of the sade vij Franch crounis .. in the sade purse, .. sayand sumtyme that thair was .. bot thre .. and uther tymes v; 1496 *Acta Conc.* II 39. Or than to mak my sayng schort sum tyme; DOUG. I Prol. 123. [She] refusit the saying off the wordis of sclander conteinit in the bill; 1628 *Fraserburgh Kirk S.* II 18 June.

b. The action of reciting (the mass).

For the saynge of the said mes; 1473 *Lanark & R.* 194. For sying and ministratioune of the mes; 1572 *Crim. Trials* I II 31. Of the saying of messis within the burgh of Dumfries; 1602 *Reg. Privy C.* VI 716.

2. Something said; a statement, pronouncement, dictum.

Ȝour saying I haif sene and on syd set it; HENR. *Practysis* 5. Thar sayng and promys [should be] fermely haldin and kepit; *Porteous Noblenes* 174/13 (Asl.). Thocht in our leid hys [*sc.* Virgil's] sayngis to declar I haue .. tane the cur; DOUG. IX Prol. 64. Thome Bell .., restit and sworn, concordis with Jhone of Murraye word be word of his sayingis; 1525 *Selkirk B. Ct.* (ed.) 75. At your maiestie pleysse, rattefy and afferme the saying ye promest ws the tyme of our chesing in office, that is to say .. that we hef securite in your bukis thairof; 1539 *Aberd. B. Rec.* I 160. The said .. Erle of Cathenes schamis nocht to allege vntrew saingis to glois his faling in the article; 1549 *Soc. Ant.* XI 95. LYND. *Mon.* 1339. To absteyne .. fra sic blasphemous sayingis; 1560 *St. A. Kirk S.* 36. Tyme beiris witnes my saying is trew; *Bann. MS* 82a/10. The writtinges .. vsis to be stufit .. with schoring and hedinfull saingis; 1573 *Cath. Tr.* 3/3. At this saying the erle was verray wraith and cryit for hors; PITSC. I 92/13. Lipning .. that thair awin sayinges .. wald discower thair malicious myndes; 1606 *Lett. Eccl. Affairs* I 52.

3. A wise or authoritative dictum or precept (either spoken or written).

Salomonis saying thow may persaif heirin; HENR. *Fab.* 1130. The dictis or sayingis of the haly faderis & wys philosophouris; MYLL *Spect.* 273/20. Quhy suld nocht thare sayng be tyll ws schawin In our language? LYND. *Mon.* 669. Now wyll we rehers the commoun sayingis .. for trewlie, albeit the sayingis be barbour and commoun, the rycht understanding of the samyn servis mekle for men unlearnit; Q. KENNEDY *Tractive* 131. Thay pervert and thraw the sayingis of the doctores; *Ib.* 166. The sayingis of the samin wtherways this man, and that man interpretis; WINȜET II 18/24. Gret authoritie was gyuen to his [*sc.* Thomas the Rhymer's] sayngs; DALR. I 341/5.

b. A proverb.

It is ane auld saying in Scotland, 'I may weill sie my freind neid bot I will not sie him bleid'; PITSC. II 157/14. The wise saying of the borowers being a servant to the lainer; 1677 *Red Bk. Grandtully* II 233.

Sayis, pl. of SITHE *n.*[2] **Sayit,** p.t. of SAY *v.*[1] **Sayith,** var. of SAY *v.*[1] **Sayk,** var. of SAK(E *n.* and SEK *n.*[1] **Saykles,** var. of SAK(E)LES *adj.* **Saylage,** var. of SAILAGE *n.* **Saylar, -er,** varr. of SAIL(L)AR *n.* **Saylche,** var. of SELCH *n.* **Sayl(e,** varr. of SAIL(L *n.*[1] and *v.* **Say-lik,** var. of SALYKE *adv.* **Sayline,** var. of SILING *vbl. n.* **Sayling,** var. of SAIL(L)ING *vbl. n.* **Sayll(e,** varr. of SAIL(L *n.*[1] and *v.* and SAL(E *n.*[1] **Sayloured,** var. of SILO(U)RIT *ppl. adj.* **Saylȝe,** var. of SAILȜE *n.*

Say-master, *n.* Also: **sey-** and **-maister.** [SAY *n.*[2] and MAISTER *n.* 12 d; cf. e.m.E. *assay master* (1647).] **a.** An officer appointed by an authority to assay (coin) or test (metal) for quality. **b.** A master of craft appointed by the craft to examine the essays of applicants for admission to the craft, also to examine the work of fellow craftsmen for quality. — **a.** Andro Balvaird, sey-master to his majesty's cunȝea-house; 1646 *Edinb. Test. Index* II 25. That ther be a saymaster appointed by the magistrats of each burgh for tryeing of the same [metals]; 1663 *Acts* VII 458/1. — **b.** That quhasumever presumes to receave the benifyte of ane frieman sall have twa sey maisteris to awaitt on hym quhill his maisterstick be maid, and to mak ane pair buitts [etc.]; 1634 *Aberd. Trades* 281. [He] sall .. produce both his discharge from his maister and burgese tickett .., and if they be valeid, then his sey maisteris to be choysine in face of court; 1650 *Glasgow Wrights Acts* 10. That ane timber chest be made for .. keeping of the insufficient work that shall be found amongst hammermene .. in the mercat and ordains the sey masters to search and sey the mercat ilk mercat day; 1694 *Glasgow Hammermen* 49.

Sayme, var. of SAIM *n.* and SAME *adj.* **Saymen, Saym(y)ne,** varr. of SAMIN *adj.*

Sayn, var. of SAIN(E *v.*, SANCT *adj.* and *n.* and SAY *v.*[1] **Saynct(e,** varr. of SANCT *adj.* and *n.* **Saynd,** var. of SAND *n.*

Saynd(e, Send, *n.* Also: **saind.** [ME and e.m.E. *sande* (Layamon), *sonde* (Ancr. R.), *saand(e* (Cursor M.), *sende* (York Myst.), OE *sand, sond.* Forms in *-e-* may be derived from SEND *v.*] **a.** A person or persons sent with a message; a messenger or embassy. **b.** A message sent by this means.

a., b. For his saynd till thaim send he; BARB. V 196. Quhill the saynde of God, the quhilk was to be send fra the fader of hevyn, war cummyn; HAY I 189/31. I rede we send furth ane saynd to yone ciete; *Gol. & Gaw.* 47. Thai wyis .. Sent ane saynd to the souerane sone; *Ib.* 367.

(*b*) I traist thow be ane send of God almycht; HAY *Alex.* 9415. The king was nocht in habit ryall thare Bot as a send of Alexander he ware; *Ib.* 16233. In Scotland he hes send, To king Robert his mynd for to mak kend .., Desyrand hir as quene in mariage. And or this send come fra the king of France, .. The ȝoung Douglas had weddit that lady. This messinger so frustrat .. Ȝeid hame agane; STEWART 56423. Not for ane ambassadour to the baptisme bot as ane send to execute the neglecting of doing that kyndnes; BUCH. *Detect.* (1727) 59. And handis hes brocht This bill vnto ȝour guidlie heidis; Ȝour guidlie heidis this send hes socht; *Montg. Suppl.* viii 7. I think he might hau answered the first send; 1660 *Red Bk. Grandtully* II 151.

c. Something sent or given, a token or gift.

Sir Eger quot. may belong in b above.

I know the token .. And if he .. likes better then come to me, Let him alone .. For he shal have no send of mine; *Sir Eger* 1810. For the gude saind that ȝe haue send to me, Ȝe serue all seuin on gallous for to die; ROLLAND *Seven S.* 3942.

d. The, or a, deliverance or dispensation of God. Also *pl.*

(*a*) Walter throw the saynd of God abbot .. of Abirbrothoc; 1428 *Liber Aberbr.* II 58. To luf God abwne al thinge & to thank hyme of al his sayndis and gyftis; *Cr. Deyng* (STS) 110. The saynd of God [thai] ay reput myschance; KENNEDY *Pass. Christ* 914.

(*b*) Geve it [*sc.* pain] cummys be the send of God it is souuerane medycyn for the saule; IRLAND *Mir.* III 60/10. Thair is na evil of payne or trubil in the pepil, bot it cummis be the send of God; HAMILTON *Cat.* 137.

Saynde, p.t. of SAIN(E *v.*

Sayndis-man, *n.* [ME *sondes-mon* (Layamon), *sondez-mon* (14th c.), *sandisman* (*a*1400-50), *soundisman* (*c*1400), only in allit. verse; f. possess. of SAYND(E *n.* and MAN *n.*] A messenger. — I rede ane sayndis-man ye send to yone senyeour; *Gol. & Gaw.* 326.

Sayne, var. of SAIN(E *v.*, SAY *v.*[1] and SYN(E *adv.* **Sayng(e,** varr. of SAYING *vbl. n.*[1] and *vbl. n.*[2] **Sayning,** var. of SAINING *vbl. n.* **Saynit,** p.t. and p.p. of SAIN(E *v.* **Saynt(e,** varr. of SANCT *adj.* and *n.*

Sayp, var. of SAPE *n.* and SCHAPE *n.*[2] **Sayr(e,** varr. of SAR(E *n.*[2], *adj.* and *adv.* **Says,** var. of SISE *n.*[1] **Sayse,** var. of SES(E *v.*[1] **Saysine, -ing, -yng,** varr. of SESIN(E *n.* **Saysowner,** var. of SESSIONER *n.*[1] **Say-sting, -styng,** varr. of SASTEING *n.* **Sayt,** p.p. of SET *v.* and var. of SETE *n.* **Sayth,** var. of SAITH *n.* and SAY *v.*[1] **Saythles,** var. of SCATHELES *adj.* **Sayx,** var. of SEX *num.* and *n.*[3] **Sayzt,** var. of SITH(T *n.*

Sazarene, var. of SARACEN(E *n.* **Sazion,** var. of SESIN(E *n.*

(Sca,) Ska. Altered var. of SCATH(E *v.* to injure. — In sanctuar sicklyke that nane war sene, Fleand folkes thair into hurt or ska [: pra (= prey)]; STEWART 20749.

Sca, var. of SCALL *n.* **Scaap,** var. of SCALP *n.*

Scab, Skab, *n.* Also: **skabb(e, scabe, skap; schabe.** [ME and e.m.E. *scab* (*c*1250), *schabbe* (*c*1290), *scabbe* (Chaucer), *shab* (Wyclif), *skabe* (1538), OE *sceabb*, ON **skabbr.*]

1. (The) scab, one or other of a variety of skin diseases affecting animals, esp. horses or sheep. Also *pl.* and *fig.*

Quots. *Exch. R.* and *Carnwath Baron Ct.* may properly belong to 3 below.

(*a*) The punctis to be inquirit at the inqueist. Scab, waitht, lippir, pikry. Thir ar the iiij hed punctis; 1499 *Exch. R.* XI 393. To the inqueist scab vayth & leipare for pykre and the presentis of heid court; 1528 *Carnwath Baron Ct.* (SHS) 73. My lord gart put in to his heid court skap lypyr and waith, the inquest fyndis nane; 1536 *Ib.* 186. That the said fille deit in the scab; 1552 *Elgin Rec.* I 114. The .. horse was foull of the scab; 1568 *Inverness Rec.* I 163. Conforme to the auld statutis maid anent scab and fairsy; 1573-4 *Glasgow B. Rec.* I 1. The sheepe doth feare the skab; JAMES VI *Poems* I 172/1121. All hors and meares having the mortiecheane or scab to be removit out of the town; 1635 Cramond *Cullen Ann.* 39. 1664 *Rec. Old Aberd.* I 105. The scab was rageing in that countrey .. which made him .. greatlie afraid he might .. endanger his masters horses; 1701 *Rothesay Par. Rec.* 140.

(*b*) The inqueist fyndis that the hors that David Brydin sauld to Thome Ker vas infekit with schabe and unabyll; 1531 *Selkirk B. Ct.* (ed.) 117.

pl. The falt of .. watteris did sic dammage to [the] beistis that infinite nowmer thareof grew first full of scabbis [L. *scabie*], and sone eftir deit of thrist; BELL. *Livy* II 93/27.

fig. Lest the contagious scab of ane scheip .. mot smit the hail innocent flok of Christe; WINȜET II 30/11.

2. a. A pock or pustule. **b.** *sing.* and *pl.* A skin disease in which the skin breaks out in pustules or becomes scaly; scabies; ? syphilis. **c.** One of the plagues of Egypt, ? = b above.

Quot. HAY *Alex.* may be an instance of a rather than b.

a. Ewin as anne ewill skabbe or anne poke cann not alwayis be keipit in with the violence of medicynne; NISBET III 318/14. Thois men are maist subiecte to the pest .. without thay be euacuat be opining of ane vaine .. or sic corruptioun expellit be scabe or hulcer, quhilk is leist sure preseruative; SKEYNE *Descr. Pest* 12.

b. He brekis out of scabis all his bodie; HAY *Alex.* 12289. & .. it perfytlie cuiris (being drunkin) the exteriour scabbis, wyldefyre, darteris & vther filthines of the skyn; SKEYNE *Descr. Well* Sig. A4. That scab that ȝe beir will scall thé to the schoulderis; MONTG. *Flyt.* 94 (T). The mischief on thy melt & maw The scabbis and the canker; *Ib.* 305. God nor he die in the skab and glengon as his brother .. did; 1643 *Rec. Old Aberd.* I 72. I stayd a month in that toune, partlie to .. cure myself of .. the ich or scab; TURNER *Mem.* 116.

c. Fyrst thare wattyre turnyt in blude .. Syne in thare bleddyris boldnyt bylys, And alkyn bruke and scab that vyle is; WYNT. II 578.

3. The hard crust, or scab, that forms over a wound or sore. Also *fig.*

The flatterer ay The owne skabb clawes [etc.]; JAMES VI *Poems* I 188/1411. Like a flee, happing from scab to scab; BOYD *Last B.* 102.

fig. The Moderator .. said, man, take off the scab of the wamb of it, .. for so .. he called the king's seal upon it; 1691 *Continuation Historical Relation Late General Assembly* 45.

Scab, *v.* [17th c. Eng. *scabbe* to form scabs upon (1632), to become encrusted with scabs (1683).] *intr.* To develop symptoms of 'the scab', see prec. 1. — To put sheep in houses makes them scab; DONALDSON *Postscript Husbandry* 38.

Scabard, var. of SCAUBERT *n.*

(Scabarting,) Skabarting, *vbl. n.* [e.m.E. *scabarde* v. to sheathe (1579-80); SCAUBERT *n.*] The act of furnishing a scabbard for a sword. — For ȝour sword dressing and skabarting, ii s. vi d.; 1621 *Maxwell Mem.* I 330.

Scabbed, var. of SCABBIT *ppl. adj.* **Scabbert,** var. of SCAUBERT *n.*

Scabbit, Skabbit, *ppl. adj.* Also: **-ed; scabit, -yt, skabit, -ett; skebbat, skapped; s(c)hab(b)it.** [ME and e.m.E. *scabbed* (Manning), *schabbede* (Piers Plowman); SCAB *n.*]

1. Of an animal, esp. a horse: Suffering from scab (SCAB *n.* 1). Also in *fig.* context and *proverb.*

(*a*) To .. vpricht the skaitht done .. be his skabbit hors to .. al vytheris gudis becaus the said hors ar forbodyne gudis; 1492 *Aberd. B. Rec.* I 420. Colleveris .. That scabbit ar, hes cruik and cald; DUNB. (OUP) 128/60. The inqueist ordanis Adam Hag to bring his scabbit hors to the kyrkstyll; 1526 *Carnwath Baron Ct.* (SHS) 55. Quho wyll go sers amang sic heirdis scheip May habyll fynd mony pure scabbit crok; LYND. *Dreme* 893. Ilk ny[ch]tbor that hes .. skabbit hors .. and puts the samyn .. in ony pasteraig, the gudis to be eschett or cassing our the craig; 1549 *Banff Ann.* I 27. Inhibetit .. for to lat his scabbit horses amangis .. his cottars horses; 1565-6 *Inverness Rec.* I 133. That na skabbit hors be fund on grenis or ony vthir place bot speciale to be haldin within howsis quhill thay be sufficientle laid; 1585 *Forres B. Ct.* 21 Nov. xix gaittis .. quhilk gait wer scabbit; 1594 *Breadalbane Ct. Bk.* 75b. 1598 *Kirkcaldy B. Rec.* 150. Complenit vpoune for having .. of scabbit hors on the comoun grass; 1629 *Dumbarton B. Rec.* 31. That nane keippe any caldit, scabbit, or other seik bestis within this bruch and freedom thereof; 1640 *Hawick Ann.* 50. Thrie meiris and thrie naigs all scabbed; 1678 *Kirkcudbr. Test.* (Reg. H.) 26 May.

(*b*) For the skath sustenit be hir of a mer and a stag skabit; 1493 *Acta Conc.* I 300/1. Ony scabyt horse, or ony wdir wnlauchful geyr; 1541 *Prestwick B. Rec.* 57. Tua skabit staigis price of tham baytht vj li.; 1592 *Edinb. Test.* XXV 15. Skabett ky; 1630 *Douglas of Morton* 739.

(*c*) Ane falty skebbat mayr; 1533-4 *Dunferm. Reg. Ct.* 106.

(*d*) That all those who hes skapped horse keip them upon ther owne grasse; 1674 *Rothesay B. Rec.* 329.

fig. He will nocht feir .. the censure of the churche, rejecting him as the scabbit scheip from the rest of the floke, be excommunicatioune; 1570 *Misc. Bann. C.* I 43.

proverb. A skabbit scheip infeckis All the haill flok; *Bann. MS* 135a/35. The proverb sayis of auld: ane scabbit hors will feill quhair he is sair; 1570 *Sat. P.* xxii 65. A scabbit scheip wald fane infect the lave Causing seditione into the kirk to ryse; 1584 SEMPILL *Sat. P.* xlv Prol. 21. A skabbed horse is good enough for a skald squir; FERG. *Prov.* No. 69. CARMICHAEL *Prov.* No. 175.

b. Of a thing: Contaminated with the scab.

For wranguslie taking of ane meir .. for ane daye and pat ane scabit summok on hir; 1593 *Breadalbane Ct. Bk.* 44.

2. Of a person, or a part of a human body: Suffering from scabies or syphilis; covered with pustules or scabs. Also *proverb.* and *transf.* **b.** A term of opprobrium, expressing contempt, disgust or revulsion.

a., b. A crabbit, scabbit, evill facit messan tyke; KENNEDY *Flyt.* 495. Ane skabbit skarth, ane scorpioun, ane scutarde behind; DUNB. *Tua Mar. W.* 92. Calling of the said James scabbit lyper carlle; 1541-2 *Elgin Rec.* I 67. I sau cammauyne, quhilk is gude for ane scabbit moutht; *Compl.* 67/15. Blynd, crukit, and seik, .. deformit, scabbit Moabites; WINȜET I 43/31. Flour the pin, scabbit skin, eit it in that thow spak; POLWART *Flyt.* 766 (T). Alledging scho callit hir commoun theiff skabbit glengorie quyne; 1630 *Fraserburgh Kirk S.* II 25 Aug.

(*b*) Calland hym dovr drunkyn beist and schabbit carlis get; 1541-2 *Elgin Rec.* I 68. Calland .. Ellen and Margarat shabbit, clangorit carlis birdis; 1542 *Ib.* 72. Calling Megot Stuart schabit, blerit, clangorit carling; *Ib.*

proverb. He cals me skabbed because I will not call him skade; FERG. *Prov.* No. 375.

transf. This burges .. promeist hir .. ane new garmoun For hir huredom and harlatrie And hir scabbit scurilitie; ROLLAND *Seven S.* 3541. Our neighbour nation will say of us, Poor Scotland, beggarly Scotland, scabbed Scotland, lousie Scotland; *Presb. Eloq.* (1693) 80.

Scabboird, -o(o)rd, varr. of SCAUBERT *n.* **Scabe,** var. of SCAB *n.*

(Scabell,) Skabell, *n.* [? Altered form of SCAPELRY *n.* Cf. e.m.E. *scapple* (1578), anglicised version of L. *scapula* the shoulder.] ? A (short) cloak or scapular. — Ane cepe almerye with ane skabell; 1562 *6th Rep. Hist. MSS* App. 648/1.

Scaberd, -ert, varr. of SCAUBERT *n.*

(Scabin,) Scabyn, -un, *n.* [Med. L. *scabinus* (1213 in Latham). Cf. e.m.E. *scabin* (1617).] A municipal official in a Flemish town. — In the behalf of the burrow masteris scabynis and consale of the toune of Mydleburgh in Zeland; 1526 *Acts* II 305/1. Comperand befoir the burromaisteris, scabuns, and counsalouris [of Antwerp]; 1564 *Reg. Privy C.* I 309.

(Scabiose,) Scabius, *n.* Also: **schabyose.** [ME and e.m.E. *scabiose* (*c*1400), *scabyous* (1526), etc., med. L. *scabiosa* (*a*1250 in Latham); f. as next.] Scabious, a plant of the genus *scabiosa*, believed to be efficacious in certain skin diseases; — Thir iiij herbys that is with dytane [etc.] .. and scabius; *c*1420 *Liber Calchou* II 451. Schabyose; *Ib.*

Scabiose, *adj.* [e.m.E. *scabious* (1603) pertaining to scabies, F. *scabieux* (late 14th c. in Larousse), L. *scabiōsus.*] ? Suffering from scabies or some other skin disorder, or ? used as a term of general opprobrium or contempt. — Our scabiose knight sayed God made all well; 1661 *Lauderdale P.* I 81.

Scabit, var. of SCABBIT *ppl. adj.* **Scabius,** var. of SCABIOSE *n.*

Scab pikar. (?) — That na forestar hald undir him in his steid haggar flegeour turnour barcar colebyrnar pelar of bark scab pikar wrychtis swyne nor gait [etc.]; 1499 *Exch. R.* XI 394.

Scabrous, Skabrous, *adj.* [17th c. Eng. *scabrous* (*a*1637), F. *scabreux* (*c*1500 in Larousse), late L. *scabrōsus*, L. *scabr-*, shortened stem of *scaber* roughly uneven, etc.] **a.** Rough and uneven in surface. **b.** Of style: Crude; unpolished. **c.** Of questions, etc.: Difficult to resolve. — **a.** [Ane rockie scabrous iyle callit] .. the Gaytis Ile; MONRO *W. Isles* (1961) 53. — **b.** Thy raggit roundaillis .. Sum .. [out] of lyne With skabrous collouris; POLWART *Flyt.* 25 (T). His scabrous compend .. did in weary affternoins refresch vs; FOWLER II 28/30. — **c.** We hope to give light to these scabrous questions; 1643 *Rec. Kirk Scotl.* 381. Finding the letter .. to be so hask, and the instructions so scabrous, that [etc.]; 1649 BAILLIE III 99.

(Scabrously,) Scabruslie, Scabrushly, *adv.* [f. prec.] ? Crudely, badly; in a manner not easily understood, incomprehensibly. — He appeareth to us rather scabrushly to have translatit that which he wrytis furth of Latine; 1565 KNOX VI 485. Albeit that some thingis be obscurly, and some thingis scabruslie [*v.r.* doubtfully] spokin; KNOX I 10.

Scabun, var. of SCABIN *n.*

Scabwort, *n.* (Late ME and e.m.E. *scabwort* the plant Elecampane (*c*1450).) — *Scabiosa,* scabwort, curledodie, mors diaboli the devillis bitt; *Herbarius Latinus Annot.* cxxxix (Adv.).

Scabyn, var. of SCABIN *n.* **Scabyt,** var. of SCABBIT *ppl. adj.*

Scaccar, Scacker. [Varr. of CHEKKER *n.*, after med. L. *scaccaria* (*c*1250 in Latham).] = CHEKKER *n.* 2, the annual audit of the royal revenues. — To the saidis foure violaris for thair service maid at the last scacker; 1550 *Treas. Acc.* IX 461. For the satisfying his [*sc.* the Sheriff of Moray's] comptis aucht to oure soverane ladeis scaccar; *Ib.* 465.

Scad, var. of SCALD *n.*[2], *adj.* and *v.*[1] **Scadding,** var. of SCALDING *n.* **Scade,** var. of SCALD *v.*[1] **Scadlips,** var. of SCALDLIPS *n.* **Scadyng,** var. of SCATHING *vbl. n.* **Scafalding, -alling,** varr. of SCAFFALTING *vbl. n.* **Scafanger,** var. of SCAVANGER *n.*

Scaff, Skaf(f, *n.* Also: **scaffe, skaffe; skaif.** [Late ME *scaphe* (*a*1450), OF *scaphe, scaufe, escaf(f)e* (all in Godefroy), L. *scapha.*] A light boat; a skiff.

(*a*) Ad portum de Coldyngham .. j scaffe; 1374 *Coldingham Priory* lxxvii. In a skaf a-pone the se Sayland [etc.]; *Leg. S.* xxvi 274. Skaff; 1512 *Treas. Acc.* IV 373. In botis and skaffis all his armye eselie tuke land; *Boece* 445b. And tuke the se thair in ane litill skaffe; STEWART 51975. For hardis to the skaf .. for treis to the skaff cutting .. burdis irne work talloun & haddir to the skaff & for making of the samin; 1539 *Ayr Common Good Acc.* Quhen thai dar not travell to far seis in fisching, for storms of weder upon thair small scaffis; 1576 *Orkney Oppress.* 66. Ane gift of licence .. to inhibite .. skaffis, skeldrykeis and ȝowis to transporte ony passingeris .. in sic weschellis [etc.]; 1600 *Conv. Burghs* II 81. *Ib.* 82.

(*b*) To .. marynallis in the Inglis skaif for wittaling and cariag to the greit schip; 1512 *Treas. Acc.* IV 453. For ane litill boit to the greit schip, callit ane skaif, for erandis; 1512-13 *Ib.* 472. Ane dra bait and ane skaif price of thame bayth fyfty li.; 1568 *Edinb. Test.* I 107.

Scaff, Skaff, *v.* Also: **scaffe; scoff.** [Cf. MFlem. *schaeuen* to beg importunately, to sponge (Bense).] *tr.* and *intr.* To obtain (something) by begging or scrounging; to beg or ask for in a cringing, importunate or bullying manner. — *tr.* He sayis, thow skaffis and beggis mair beir and aitis Nor any cripill in Karrik land abowt; DUNB. *Flyt.* 133. I purpose not to scaffe my meete at my cheese, nor am I now to learn at Doctor Lindsay; 1620 Calderwood VII 424. The saids provest [etc.] .. oblegis them .. to caus restraine all extrainier or former touns posts to scaff or carie leters from this brughe; 1667 *Aberd. Council Lett.* IV 298. — (*b*) The petitioner haveing hade nothing .. to live upon but what he scoffed and begged by way of charitie; 1689 *Reg. Privy C.* 3 Ser. XIV 98. — *intr.* They scaffed throche all Scotland, oppressand the leall men als weill as the theiff; PITSC. (1814) 512.

Scaffald, Scaffall, Schaffold, *n.* Also: **scaf-, skaff-** and **-alde, -auld, -old(e, -uld, -eld, -alt, -olt, -illt, -hold, -huld; -ell, -uill, -oll; scaffehold; schaffel(d; scaiffald, skaifell, -ill; scalfold, scalffauld, skalfolde, scalffet, skalfet; scaffat, -ett, skaffat, -et(t, -it; (sklaffauld); scoffold.** [ME and e.m.E. *schaffald* (north., 1341), *scaffot* (1349), *skaffald* (1354), *scaffold* (1435), *schafhold* (Malory), *chaffold* (1514), *schaffold* (*a*1552), OF *cha(a)faut, eschafalt, eschafaut* (all 1160 in Larousse), vulgar L. *catafalcum, med. L. *scafaldus, scalfaudus* (15th c. in Latham).]

1. A military engine for use in assaulting a walled fortification.

The Inglis ost .. assembill .. With instrumentis of ser maneris As scaffaldis [*C.* scaffatis] leddris [etc.]; BARB. XVII 343. Syndry scaffaldis [*C.* scaffatis] thai maid with-all That war weill heyar than the wall; *Ib.* 601.

2. A temporary platform, set upon supports or (later) suspended, to hold workmen and their materials, etc., during work of construction, repair, alteration, etc., of a building.

Sometimes, ? esp. in *pl.*, ? referring to the whole structure, supports as well as platform. = SCAFFALTING *vbl. n.*

[Pro meremio empto pro scaffaldis faciendis pro opere noue camere; 1329 *Exch. R.* I 125.]

sing. For making of ane scaffald for dountakin of the samin [*sc.* a bell] furtht of the stepill; 1541 *Treas. Acc.* VII 499. Payit for the making of .. the playaris hous, the jebbettis and skaffauld about the samyn; 1553-4 *Edinb. Old Acc.* I 110. Men to beir up tymmer to mak the skaffald .. xvj punsionis to the skaffald; 1554-5 *Ib.* 148. To sex men .. [for] strikin of ane skaffald; *Ib.* 149. For bigging ane scafhvld wpon the head of the scafhold; 1595-6 *Dundee Treas. Acc.* Aucht dalis to the clarkis chalmer .. quhilk was scaffuld at the painting of it; 1598-9 *Misc. Spald. C.* V 71. For towis for hinging over the skaffold for making in of the holl for the lead pype; 1625 *M. Works Acc.* (ed.) II 177. For the skafillt of the new sink; 1645 *Tulliallan Coal Wks.* 110b. To mak thrie holls for ane scaffold that Robert Mein or his tennents to [*sic*] turffe on; 1663 *Melrose Reg. Rec.* II 69. Scafold; 1680 *Sheriffhall Coal Accompt* 19 June.

(*b*) xij dubill garrounis for the scaffat .. iiij s.; 1560-1 *Edinb. Old Acc.* II 139. To the varkmen and cartairis for carting the dailis to the scaffett off the hospitall and to the stepill; 1601-2 *Montrose Treas. Acc.* 7.

(*c*) The setting of ane scaiffald in the new kirk; 1582 *Edinb. D. Guild Acc.* 126.

(*d*) To the measons at the vppseting and down taking of the skalfet; 1587 *Dundee B. Laws* 143.

(*e*) Skaifell; 1628 *M. Works Acc.* (ed.) II 260. For doun taking the skaifill of his maiesties bed chalmer; 1629 *Ib.* 268. To mak ane skaffell .. abone the yett on the west gavill of the Metrapolitane Kirk, that the said John may build and repair the brek of the wall; 1631 *Glasgow B. Rec.* II 2.

pl. For making skaffaltis gyraippis traissis and grayth to cartis; 1538-9 *M. Works Acc.* (ed.) I 258. For the binding of the cuppils of the queir, making skaffats thereto [etc.]; 1552-3 *Old Dundee* II 127. Ane dusin garronis to be skaffetis; 1559 *E. Loth. Antiq. Soc.* VII 58. For making of skaffattis to the eist galf wyndoke of the kyrk; 1574-5 *Haddington Treas. Acc.* 8.

(*b*) For the scalffettis away bering; 1505-6 *Edinb. Hammermen* 54.

(*c*) For making of skaffellis to put upe the said glas xviij d.; 1563-4 *Edinb. Old Acc.* I 448. That all maisteris .. of warkis .. sie thair skaffellis and futegangis surelie sett and placit; 1598 *Edinb. Masons* 2b. Naillis to mak the skaffuillis; 1618 *M. Works Acc.* (ed.) II 112.

(*d*) Ane wricht to help to mak the schaffels and futegangis; 1611 Macgibbon & Ross V 7.

3. A platform, stand or seat sufficiently elevated to give prominence to persons of rank, officials, judges, etc. on public occasions, also to persons undergoing public punishment.

(1) About the listes salbe .. maid .. scalffauldis, part maid to princes or vtheris lordis or jugis ordanit thair to; *Loutfut MS* 2a.

(*b*) Giff the king be present in propir persoun thair salbe maid att the fut of his scaffald ane place quhar the .. marschall salbe; *Loutfut MS* 2a. This jurnay chosin was in the North Inch of Perth, and jugis war sett on ane skaffald tharfor dewisit; *Boece* 617b. That scho be putt in the irnis .. aboue the croce on the skaffald, that all the pepill may se; 1538 (*c*1580) *Edinb. B. Rec.* II 88. Debursit in preparing of ane scaffald the tyme of the accusatioun of Wallace; 1550 *Treas. Acc.* IX 435. He ordanit eike two skaffaldis .. The ane .. for the kings majestie, Ane uther for the queinis royaltie And for hir dochter; *Clar.* II 1343. The king gart in ane scaffold by him neir .. sit .. lordis seir, For to be judge quha provit knightliest; *Ib.* IV 981. The said Sir Williame .. apoun ane publict skaffald .. wes degradit; LESLIE 143. A king is as ane sett on a skaffolde quhase smallest actions & gestures all the people gazinglie do behoulde; JAMES VI *Basil. Doron* 162/2. Scaffeholds wer erected in 3 places of the toune; BALFOUR *Ann.* II 184. Scaffhold; *Ib.* 185.

(*c*) The quenis grace come doun to the tolbuith, at the quhilk was .. twa skaffattis, ane abone and ane vnder that; *Diurn. Occurr.* 60. She sall be had to the suck stule, and upon the skaffet thereof her hair sall be cut off; 1580 *Old Dundee* I 84.

(*d*) For dressing wpe the scaffallis about the corss .. 10 s.; 1623-4 *Misc. Spald. C.* V 98.

(2) Ane prince .. sal sit juge in scaffald above the entre of the listes; *Maner of Battale* 226. The juge suld gang in the listis .. and suld entir in scaffald [*Lindsay MS* skaffauld] his counsalouris eftir him; *Loutfut MS* 2b.

b. *specif.* An elevated platform on which a public execution takes place.

(*a*) For making of the skaffald to justifie the Franche men; 1558-9 *Edinb. Old Acc.* I 280. Thre gaittis .. to mak the skaffald and set up the jebait; *Ib.* 281. To the pynouris for the bering of dailles and pouncheonis .. with the gibbet and madin to mak ane scaffald .. the day quhen Thomas Scot was justefeitt; 1565-6 *Ib.* 504. For making of the skaffald quharvpoun Thomas Hoppringill wes heidit; 1566 *Edinb. B. Rec.* III 217. James Hamiltoun .. was .. heidit at the skaffald at the trone of Edinburgh; *Diurn. Occurr.* 23. Scott wes hangit [etc.] .. vpoun ane skaffalt at the mercat croce; *Ib.* 98. Alexander quhene he was wpoun the skaffold .. maid ane singular goode wriesone; PITSC. I 59/27. The lordis .. had maid ane scaffauld to have heiddit the provest of Dundie; *Ib.* II 265/13. The scaffeld .. was covered with linin cloth, then sand, next cloth, then scarleth; 1584 *Gowrie P.* 36. Quhill we sie him die on the scaffald for it [etc.]; 1600 *Crim. Trials* II 254. He was condamned to be headit on a sklaffauld; MELVILL 117. Skaffalde; *Ib.* Scaffolt; BIRREL *Diary* 49. Sprottis depositioun vpoun the skalfolde: .. Sprott .. at the foure nvkeis of the scaffold ratified his formar depositiouns; 1608 *Crim. Trials* II 260.

(*b*) And thair, vpoune ane scaffall, his heid to be strikin fra his bodie; 1565-6 *Crim. Trials* I I 475. The said aduocat .. produceit the depositiones .. wpoun the skaffell the tyme of his executioun; 1581 *Acts* III 197/1. Scaffoll; 1594-5 *Misc. Spald. C.* V 59.

(*c*) [He was] conveyed to the schaffeld by his kepar; 1584 *Gowrie P.* 36. Our king was brought to a schaffold, the other slain be a assasin; 1665-7 LAUDER *Jrnl.* 91.

c. (To go) *till .. skaffauld,* to be executed.

They ar seditius, .. and thairfor man till exyll, banissment, preasone, and skaffauld; MELVILL 202.

4. A platform or (temporary) stage on which a performance or public entertainment takes place.

(*a*) In diuers rewis, playis and personagis, Sum contenance on scaffaldis and on stagis; HAY *Alex.* 17458. Thocht sic thingis semys mare ganand to be rehersit on scaffaldis for .. delite of pepill; BELL. *Livy* II 176/17. Thow saw .. Disagysit folkis .. On ilk scaffold, to play ane syndrie storie; LYND. *Depl. Magd.* 111. Heir sall the karle lowp of the [s]caffald; Id. *Sat.* 1953 stage direction. For making of ane skaffalt on the hill to the clerk play; 1552-3 *Edinb. Old Acc.* I 76. *Forus,* the overloft of a ship, or scaffald for players; CARMICHAEL *Etym.* 33. For bigging the scafhold the tyme the barnis of the school playd; 1595-6 *Dundee B. Rec.* in Mill *Mediæv. Plays* 175. As sum dayes to the hunteng, playes, scaffoldis [etc.] .. tha gaue thame selfes to; DALR. II 456/14. Make scaffalds clare for cumlie comedies; HUME 53/45. The montebancke .. had his peopell that played on the scaffold, ane ay playing the foole [etc.]; 1662 LAMONT *Diary* 158. He [*sc.* the mountebank] sunk all most dead upon the scalfold or stage; 1665-7 LAUDER *Jrnl.* 69.

(*b*) For .. scheiratts till the skaffetts for latting of dyn to be maid be the playars feit; 1557-8 *Edinb. Old Acc.* I 272. The counsell ordanit the thesaurar .. to sett skaffitis for to the play; 1575 *Haddington B. Rec.* (Robb) 21 July.

(*c*) In the Poultrey a scoffold was erected, where .. to delight the queen .. nine trumpets .. did actually sound the Danish March; ? *c*1605 *Pitsc. Add.* (1728) 233.

5. A raised stand for spectators.

Thai war constrenit to mak public setis and scaffaldis in commoun placis quhare playis war devisit; BELL. *Livy* I 119/21. To caus be maid, For sight of lordis, skaffaldis heich and braid, On ather half, .. Hecher and lower efter thair degree Of nobillis and barrouns on tham sould stand; *Clar.* II 1334.

(Scaffald,) Scaffold, *v.* [e.m.E. *scaffold* to furnish with a platform or the like (*a*1548), in this sense (*a*1662).] *tr.* To furnish (a building) with scaffolding. — Appoyntis Johne Scott master wright to scaffold the kirk at the Trone; 1649 *Edinb. B. Rec.* VIII 209.

Scaffall, var. of SCAFFALD *n.*

Scaffalting, Scaffal(l)ing, Schaffaldin(g, *vbl. n.* Also: **scaf(f)ald-, -old-, skaf(f)alt-, -elt-, -olt-, -ald-, -eld-, -old-; skauffauld-; scalfatt-, skalfat-, -falt-, -ffauld-, -fett-; scaffat-, -et(t)-, -yt-, skaffat-, -et(t)-, skeffett-, skavet-, skevat-, -et-; scaffol(l)-, -el(l)-, -ull-, scaffl-, scafall-, skaffall-, -el(l)-; skaifill-; schaffalt-, -elt-, -att-; scofold-** and **-yng, -yn(e, -ene, -ein; (skaftolting).** [ME *skaffalding* (once, north., 1347-8), e.m.E. *scaffoldyng* (1512), *skaffollyng* (1531), *scaffling* (1663); SCAFFALD *n.* and *v.*]

1. A temporary framework consisting of a platform held in position by props, etc. (or, later, suspended from above) to accommodate workmen engaged on construction, decoration, etc. and their materials. Also, once, *pl.*, in the same sense.

Quot. 1687 in (*d*) below may rather be an example of SCAFFALD *n.* 2 with erron. word-spacing.

(*a*) As to the making of scaffalting . . the . . cheptour findand the stuffe . . thar to; 1506 *Glasgow Dioc. Reg.* II 153. Rauchteris for skaffeltein; 1512 *Treas. Acc.* IV 279. Skaffaltyng; 1529-30 *M. Works Acc.* (ed.) I 25. For . . aikkin sparris for trestis and nedlis for scaffalting; 1530 *Ib.* 29. For ii^m dure nalis . . for syle and skaffolting; 1532 *Ib.* 80. Corvaill naill to the vppsetting of the grete skaffalting; *Ib.* 85. Skaffelting; 1537-8 *Ib.* 220. Skaffalting; 1555-6 *Edinb. Old Acc.* I 181.

(*b*) For xii faldomis of cord to be skalfalting to the sklateris; 1532 *M. Works Acc.* (ed.) I 109. For sawing of uther small geistis to be lycht barkatts treistis scalfatting to the vindes; 1555-6 *Edinb. Old Acc.* I 193. For pyners of timmer to be skalfetting; 1587 *Dundee B. Laws* 143. For aucht menis disiwnis that maid the skalfating in the stepill; 1587 *Edinb. D. Guild Acc.* 284.

(*c*) Skauffauldin; 1552 *Edinb. Hammermen* 191.

(*d*) Scaffalding; 1513 *Aberd. B. Rec.* I 85. Twa pypis to be skaffaldene; 1558-9 *Edinb. Old Acc.* I 298. Skaffaldin; 1559-60 *Ib.* 334. 1578 *Edinb. B. Rec.* IV 75. Ane auld schipps balk coft to the skaffelding of the kirk; 1591 *Edinb. D. Guild Acc.* 428. iij 100 grit nailles . . for skafalding and foote gangis; 1629 *M. Works Acc.* (ed.) II 300. Scafalding; 1672 *Soc. Ant.* XIV 333. For putting in a hinging scafoldin the milnholl; 1687 *Sheriffhall Coal Accompt* 9 April. For scofolding to the dressing of the laigh kirk; 1722 *Glasgow B. Rec.* IV 154.

(*e*) To x werkmen to help vp with the ruf of Hannis toure and the skaffating of it; 1497 *Treas. Acc.* I 355. As to the making of scaffating and wpputting of the said silouris the said Mychell sall mak the samyn; 1506-7 *Reg. Episc. Glasg.* 613. Scaffytyng; 1529-30 *M. Works Acc.* (ed.) I 20. Skaffatyn; 1530 *Ib.* 40. Scaffatyne; *Ib.* 50. [That they] awaitt vpoun the grathing of thair rowmes, in skaffetting, personages and ordour thairof; 1538 (*c*1580) *Edinb. B. Rec.* II 90. Skaffeting; 1557-8 *Edinb. Old Acc.* I 261. For setting vp the scaffetting; 1571-2 *Haddington Treas. Acc.* For . . strykin of the cintreis & scaffeting; 1578-9 *Ib.* Skaveting; 1586-7 *Dundee Treas. Acc.* The said Mr. Dauid sall furneis . . skaffatene and sentreis for the pend of the said port; 1589 *Trans. Edinb. Archit.* IX 131.

(*f*) Skeffetting; 1557-8 *Cromarty Corr.* I 249. Skevating; 1586-7 *Dundee Treas. Acc.* Skeveting; *Ib.*

(*g*) xiiij mastis to be scaffaling for up[p]utting of the ruiff; 1539 *Treas. Acc.* VII 221. To vj pynouris . . in servyng of the masonis dounetaking of the scaffaling; *Ib.* 229. The saidis . . sall fynd to the said werkmen ledderis laithe & skaffellyng; 1561 *Edinb. B. Deeds* 12. For berin the tymmer to be Jon Melvill skaffelin quhen he pantit the beleue in the kirk; 1585 *Cart. S. Nich. Aberd.* II 387. Skaffalling; 1593-4 *Ayr Common Good Acc.* Ten daillis to be lath and scaffalling to the belhous; 1609 *Glasg. Univ. Mun.* III 558. For sax common deills to mak the first scaffeling; 1624-5 *Aberd. Shore Wk. Acc.* 135. For scaffolling to the upputting the windowis; 1624-5 *Misc. Spald. C.* V 101. Scafalling; 1629 *M. Works Acc.* (ed.) II 293. Scaffalin; 1629-30 *Misc. Spald. C.* V 103. Scaffullin; 1635 *Dumfr. & Galloway Soc.* 3 Ser. III 315. John Mylne, maister measoun, . . to furnish all necessaries, scaffoling & all, . . for the ground & soll of those roundis; 1644 in *Dundee Chart.* 173. The said scaffalling is rowinous; 1675 *Edinb. B. Rec.* X 214.

(*h*) For . . doun taking of the windowis for the wrichtis skaifilling and putting of thame up againe; 1628-9 *M. Works Acc.* (ed.) II 273.

(*i*) Scaffling; 1675 *Edinb. B. Rec.* X 430.

(*j*) To twa pynouris for berin of the pipis to the schaffeltin; 1509-10 *Edinb. Hammermen* 68. Schaffaltin; 1515 *Treas. Acc.* V 24. Sax dosone of gret birkin treis . . to be schaffaldin and trestis to the chapell syloring; 1534-5 *M. Works Acc.* (ed.) I 124. Ledderyng, schaffatting, lachter, fyr and fewell neidfull to the owtreddyng of the said werk; 1532 *Reg. Cupar A.* I 314. To the wrychtis for sawing and vpputting of schaffalding; 1581 *Red Bk. Menteith* II 420.

(*k*) Skaftolting [? *misreading of* skaffolding]; 1573 *Selkirk B. Ct. MS* 12 July.

pl. To the furnessing of the wrychtis fluringis sylingis . . skaffaltingis; 1537-8 *M. Works Acc.* (ed.) I 217.

2. A platform or stand.

Thow saw makand rycht costlie scaffalding [*Pitsc.* I 373/22, scalffaulding], Depayntit weill with gold and asure fyne, Reddie preparit for the vpsetting; LYND. *Depl. Magd.* 106. For scaffalding & vther neidfull provisioun at hir maiesteis entres; 1590 *Edinb. B. Rec.* in Mill *Mediæv. Plays* 197.

3. The action of providing (something) with scaffolding.

For scaffelling the cross on the coronatioune daye; 1632-3 *Misc. Spald. C.* V 104. To goe speidilie about the provyding of trees . . for skaffolding the wallis of this toun; 1650 *Edinb. B. Rec.* VIII 249.

Scaffar, Skaffar, -er, *n.* [SCAFF *v.*] One who lives at the expense of others by extortion, sponging, etc.; a parasite.

Scaffaris and scamleris in the nuke, And hall huntaris of draik and duik; DUNB. (OUP) 130/45. Eugenius . . commandit all idill pepil, as juglaris, menstralis, bardis, and scaffaris [*M.* skafferis] othir to pas out of the realme, or ellis to find sum craft to win thair leiffing; BELL. *Boece* II 99. Lugthak . . vsit stile sic skaffaris and rubbaris his tendir freyndis and bredir; *Boece* 162b. Sayand common beggaris and skafferis, thair meltyd wes bot small for all thair cuttit out hoys; 1542 *Aberd. B. Rec.* I 180. He had mispersonit the bailye, calland him skaffar; 1543 *Ib.* XVIII (Jam. s.v. *Misperson*). Thair is sic a multitude of regrateris and skafferis of the wymbes of the puir rissin vp within this burght; 1598 *Ib.* II 167. He banisched all tavernaris, druncardis, skaffaris . . out of his house; BISSET I 49/14.

Scaffat, -auld, Scaffehold, Scaffeld, varr. of SCAFFALD *n.* **Scaffating, -yne, Scaffel(l)ing,** varr. of SCAFFALTING *vbl. n.* **Scaffe,** var. of SCAFF *n.* and *v.*

(Scaffery,) Scaffry, Scaff(e)rie, *n.* Also: **scaffiré; skaf(f)rie; skaiffry.** [SCAFF *v.*] The action of obtaining improperly money, goods or benefits, as by wheedling, importuning, threatening or cheating. **b.** An instance of this. **c.** Something so obtained.

(1) Certane nychtbouris . . vsit sic ane scaffire vpon estrangieris & nychtbouris that thai micht nocht gudlie resort to this merkat without . . exhorbitant expensis; 1551 *Dundee B. Ct.* II fol. 85 (11 Aug.). Lyke as the poorer sort in the burrowes sould not have escaped the importable scafferie intended; 1582 Calderwood III 662. Being accusit . . for ganging to vther townes and brydellis be waye of scaffry [*ed.* staffry] and begging; 1600 *Glasgow B. Rec.* I 203. Against the libertie . . grantit to sic schamefull scafferie and extortion [*supra* the great extorsion used be the . . clerks of all judicatores . . in extorting . . exorbitant pryces for the wreatis]; 1606 *Reg. Privy C.* VII 165. The former abuse, scaffrie, and extortioun not onlie renewit bot rissin to a fer griter heicht; 1612 *Ib.* IX 323. Thay daylie committ schamefull extortionis, oppressionis, and scafferie upoun his majesteis good people; *Ib.* 365. The native poor . . ar prejudged of thair almous by the scafferie and oppressioun of thir . . beggars; 1629 *Ib.* 2 Ser. III 354.

(*b*) Of men a weir the bandis Quhilk fra all kynd of skafrie can not ceis; MAITLAND *Maitl. F.* 327/10. And als be maner of skafrie and mere extortioun . . takis furth of euerie laid of cornis brocht in the said cietie . . ane ladill full of cornis; 1575 *Glasgow Chart.* I II 165. The said Laurence . . with . . armit men cummis to the housses of us . . and remanis thairintill quhill our . . provisiounis be speindit compelling us to furneis him . . for feir of grittar oppressiouns; and swa depairtis but paiment . . and that in maner of sorning and skafrie; 1576 *Orkney Oppress.* 87. The man is . . a deboschit drunkard, and . . goes about to get sustenance by unlawfull skafrie and malitious informationis; 1616 *Melrose P.* 260.

(2) The wemen perturbatouris for skafrie of money or vtherwyse salbe . . put vpone the cukstulis; 1555 *Acts* II 500/1.

b. Sundrie vther incommoditeis and skaffries fallis out be this abuse; 1622 *Aberd. Eccl. Rec.* 110.

c. That the malt and meill . . sall be . . mesourit in the gray freiris place, and nocht put in na wther houssis, nor na skaiffry, sic as sampill and scheit schakin, to be tane thairof; 1561 *Aberd. B. Rec.* I 335.

Scaffett, Scaffhold, varr. of SCAFFALD *n.* **Scaffet(t)ing,** varr. of SCAFFALTING *vbl. n.*

Scaffing, Skaf(f)ing, *vbl. n.* [SCAFF *v.*] **a.** The action of SCAFF *v.* = SCAFFERY *n.* **b.** That which is obtained by scaffing. = SCAFFERY *n.* c. — **a.** Think that this lyfe is nocht thé lent For skafing heir of skruf and skum; *Bann. MS* I p. 17/30. — **b.** Quhat sall we commone of composicionns Quhare nother is kepit iustice nor mercy? .. Sic scaffing [*Arund.* skaffing] may be callit mydding corne; *Contempl. Sinn.* 174 (Harl.).

Scaffing, Skaffing, *ppl. adj.* [SCAFF *v.*] Sponging; open to bribery, corrupt. — Ane scaffing warlot, wanting schame; 1584 SEMPILL *Sat. P.* xlv 904. Skaffing clarks, with couetice inspired, Till execute their [*sc.* lawyers'] office man be hyred; HUME 73/163. Sum senators als weill as skaffing scribes, Are blinded oft with .. bribes; *Ib.* 74/209.

Scaffinger, var. of SCAVANGER *n.* **Scaffiré,** var. of SCAFFERY *n.* **Scaffling, Scaffol(l)ing, -ullin, -ytyng,** varr. of SCAFFALTING *vbl. n.* **Scaffold, -oll, -olt, -uld, Scafhold, -huld,** varr. of SCAFFALD *n.* **Scaffold,** var. of SCAFFALD *v.* **Scaffrie, -ry,** varr. of SCAFFERY *n.*

Scafin-feast, *n.* [? SCAFFING *vbl. n.* and FEST(E *n.* Cf. the later Sc. dial. *scaff* v. to eat or drink greedily, *scaff* n. food or provisions, also *scaff and raff* plenty, esp. of food. Cf. also 18th c. Eng. *scaff* to eat voraciously (1797).] ? A banquet. — In his time, the Scafin-Feasts were converted into money for the use of the magistrates that .. they may call for any refreshment upon that account [etc.]; 1685 *Macfarlane's Geog. Coll.* III 90.

Scafold, var. of SCAFFALD *n.* **Scafoldin,** var. of SCAFFALTING *vbl. n.*

Scahkill-bane, see SCHAKIL(L)-BANE *n.*

Scaid, var. of SAID *ppl. adj.*, SCALD *adj.* and SCATH(E *n.* **Scaiffald,** var. of SCAFFALD *n.*

Scaikit, Skaikit, Skeikit, *ppl. adj.* [Of unknown origin. Cf. mod. Sc. dial. *skaikit*.] ? Striped, streaked. — xij nycht bonnettis of taffeteis skeikit with silk; 1570 *Edinb. Test.* II 80b. Ane blak humill kow and ane skeikit stirk; 1581 *Brechin Test.* I 46. Ane broune skaikit [*ed.* skairit] ox; 1637 *Aberd. Sheriff Ct.* II 462. [Oxen described as] tua black ane scaikit ane uther tagat ane humbill and the rest hornit; 1640 *Ib.* 478.

(Scail,) Scaill, *n.*[1] [SCAIL *v.*, or ME *skaile* (once in Cursor M., 14th c.).] A dispersal or scattering; a separation. — There would be a scaill among our counsellors and the sessioners, 1625 in Laing *Hist. Scot.* (1904) III i 8. That they sall mak a scaill of thair counsell upone ther ellectioun day; 1652 *Aberd. Council Lett.* III 212.

(Scail,) Scaill, *n.*[2] Also: **skaill.** (? ME (north.) and e.m.E. *scale* a maniple, squadron, battalion (*c*1400, rendering OF *eschele*), med. L. *scala* (whence OF *eschiel(l)e, eskiele*), or ? an early instance of prec., here = 'a dispersed company, a scattered party' (OED). But perh. merely due to misreading of STALE *n.*[1]) — Schir Adam .. Saw bot the fleand scaill [*or* staill; *1571* scaill, *H.* skaill] perfay And thame that sesyt in the pray; BARB. xv 337* (C). Bot than baith forray, and the scaill, War knit in till ane sop all haill; *Ib.* (1571) xv 353*.

Scail, Scale, *v.* Also: **scaile, scayle, scaill, scayll-, skail(e, skaill, skayl(e, skayll, skale, scall, skall, skell, skeil-, skeill-, skeal-, skael; schaill.** P.t. also **schalit.** P.p. also **skelt; skail(l; schalit, schalyte.** [ME and e.m.E. *scail, skail* (v.r. *a*1375 *shail*) (north., all Cursor M.), *skayle* (Caxton's Trevisa), *scayl* (also *scail*) (*St. P. Henry VIII* V 215 n. and IV 510 n. referring to or quoting Scots), *skaile* (*a*1600, Floddan F.), *scale* (north., 1641), EIr. *scáilim*, Ir. *sgaoilim*, Gael. *sgaoil*. Cf. ON *skilja* to separate, divide.

The Eng. evidence strongly suggests that the word spread from north to south, perhaps a borrowing from EIr. into Sc. thence into northern Eng. Cf. SLANIS *n.*]

1. *tr.* To scatter (things, or the parts or constituents of a person or thing) abroad or in different places; to disperse by separating and scattering. Also *fig.*

(1) Here slane has bene mony men, Of quham the hedis in placis sere Ar scalyt; *Leg. S.* ii 378. Thar gert thai bryng in gret quantyte Gold & siluire & skalyt be; *Ib.* xxiv 534. Thai blissit banis .. Thai .. brynt in-to powdre smal, & skalit thaim in-[to] the wynd; *Ib.* xxxvi 573. In Goddis ȝard .. he saw sawyne il seide .. & gyf it gadderit wes il He skalit it ore it suld spil; *Ib.* xl 206. Gif the vassallis be nocht with thair lord, thai ar agaynis him and scailis his gudis; HAY I 123/9. Thy hurde to hid, to skaill I tuke na cure; KENNEDY *Pass. Christ* 1404. Sic dust that scalis sic fillokis with fuk salis Within this land wes never harde nor sene; DUNB. *Maitl. F.* 213/74. [The hire of horses to carry wrack, .. and to the men] that skalit it one the hillis; 1550-1 *Ayr B. Acc.* 113. And skaillit and sparpallit the meill .. in the said burne; 1578 *Reg. Privy C.* II 695. The said John haifing caused put some fulyie furth of his barne, and had scailled the same togidder besyd his said barne; 1595 R. Brown *Paisley* I 170. To Laurence Porter to skaill the townis red vpoun the hillis; 1610-11 *Ayr Common Good Acc.*

(2) Haill, beyme to skaill of ded the dirk vmbrakill; KENNEDY *Pass. Christ* 14. Lyk a certane sone, new risen to skail and skattir the cloudis of al tumulte; DALR. II 453/7. Bright Apollo .. Quhais glorious glance ȝit stoutly skaillis the sk[yis]; MONTG. *Sonn.* xiii 3.

fig. And fellon deid and word all way He preissit to scaill and put away; *Alex.* I 1422. Quha skaylis his thoucht in syndrynes, In ilk thyng it is the les; WYNT. VIII 2443. Terroure and fere scalit the curage of mony; *Boece* 181. That all my ioyes ar skaild; FOWLER I 275/2. Although .. the brightnesse of the gospell at His cumming scaled the cloudes of all these grosse errors in the Gentilisme, yet [etc.]; JAMES *Dæmonol.* (STS) 37/6. Happy are they .. who .. gives not their sins leave to fill the cup, but are daily skailing some of them by repentance; HENDERSON *Serm.* 159.

b. *specif.* *To scail* (one's) *hous (houshald)*, to dispose of one's dwelling, furnishings and belongings, sometimes by selling them (see also SCAILING *vbl. n.* 1 a (2). Also *to scail hous* and, *fig.*, *to scail* (one's) *nest*.

(1) All licht weman .. to decist fra .. venerie, and al thair buthis and houssis skalit; 1497 *Aberd. B. Rec.* I 425. Men & wemen hes skalit thair hale houshald; ROLLAND *Ct. Venus* I 551. Diverse .. proffitabill landis are laid waist, and mony honest houshalderis constrenit to skaill thair houssis; 1567 *Reg. Privy C.* I 516. Twa women for povertie scalit thair housses; 1576 *Orkney Oppress.* 61. Mony honnest houshalderis hes bene compellit to skaill thair houssis and beg thair meit in the cuntrie; 1580-1 *Reg. Privy C.* III 361. He .. schew the dischairge quhilk he had obtenit .. or utherwayis the hous wald have bene skaillit; 1588 *Comm. Univ.* III (St. A.) App. 195. Tennentis .. have bene forceit to skaill thair housis and ar put to beg thair meate; 1612 *Reg. Privy C.* IX 501.

(2) That we scalit house thair; 1559 *Rentale Dunkeld.* (SHS) 358. My harlot-mother and my sweet Father .. are going to skail and give up house; 1637 RUTHERFORD *Lett.* (1894) 297. *Ib.* (1664) I clxxxii 352.

(3) Judas sall juge ȝou, and God sall scale ȝour nest; 1581 *Sat. P.* xliv 328. Some of the counsell said they sould skaile their nest; CALDERWOOD III 647.

2. To spread (rumour, news, fame, etc.) far and wide.

Bot the tithandis var scalit sone, Of the deid Douglas had done; BARB. V 447 (C) (see also 13 below). His renoune sa scalit was Throw-out the marchis of Ingland; *Ib.* xv 554 (E). Throuch braid Bretane the woice tharoff was scalyt [: alyt (= ailed)]; *Wall.* XI 1300. Thys ilke wensch .. this rumour gan throu the pepill skaill [: taill]; DOUG. IV V 32. Thys menskles goddes .. Skalys thir newis est, west, north and sowth; *Ib.* 42. *Ib.* XII V 15.

3. To cause or allow (heat or poison) to spread throughout the human body; to cause (air) to disperse.

Plesancis and displesancis .. that drawis the hete fra the stomak and scailis it our all the body; HAY II 125/13. Quhen Ambrose be unhappy sleip had skaillit the vennom throw the vanis of his body, he .. deceissit; BELL. *Boece* II 62. A smal prick of a preine may scail the wind in blawin bleddars; FOWLER II 33/31. The saifrounid gulsoch .. doth his kendlid uennum skaile throuchout the corps; JAMES VI *Poems* I 159/417.

4. To pour out, spill (liquid); to shed tears. Cf. OURSCAILIT *p.p.* **b.** *To scale .. blude*, to shed blood; to kill.

I sal apon thame a myrk schour down skaill [*Sm.* schaill; L. *infundam*]; DOUG. IV iii 59. Quhen donc Awrora .. Hir siluer teiris skalit on the schowris; BELLENDEN *Bann. MS* 1a/8. Scho anserit leist that it [*sc.* water] being skaillit, aither beist or bodye suld go ower it and so gett ill by it; 1623 *Perth Kirk S.* MS 14 May. In the going scho skaillit swm [water]; 1623 *Crim. Trials* II 537.

b. Thai suld be of grete drede to scayle and sched mannis blude oure reklesly; HAY II 102/2. Besyde the altar blude sched and scalit new, Beand lew warm; DOUG. VIII iii 24. Them that settis ther felicite to skattir & to skail blude; *Compl.* 27/16. Bothe the nations noblest blood For others quarrels skeald [: seal'd]; GARDEN *Elphinstoun* 1036. Great blude skeild on both sydes; 1644 *Chron. Perth* 40.

c. To allow (water) to flow away.

In case it [*sc.* the mill dam] be insufficient and skell water; 1591 *Old Dundee* I 169. Thair names who refussit to concur with me and skaillet my watter; 1615 *Denmylne MSS* in *Highland P.* III 260.

d. *fig.* To confer; to use wastefully.

Albeit thou scailes them [*supra* the gifts of heaven and earth] upon us who are not able to receave them; 1619 *Sel. Biog.* I 104. I have skailed more of Christ's grace .. than I have brought with me; 1637 RUTHERFORD *Lett.* (1894) 243.

e. To use (ink) unprofitably, to waste (one's efforts as a writer).

Sinc .. the prophets, and all quho wrote since hau scal't ther ink .. I leiv the dispensing off thes precepts to tyme; 1661 *Red Bk. Grandtully* II 167.

5. To spread out, distend, slacken.

Cleopatra .. mycht thou se Wynd saill about .. Scaland [*C.* Sclakand] schetis and haldand rowme at large; DOUG. VIII xii 83 (Sm.).

b. To let (one's hair) hang loose, in a dishevelled manner.

The Troian madynnys quhite, With hair doun skalit, all sorofull gan pas; DOUG. I vii 109. The religyus nun with hair down skaill [*Ruddim.* skail]; *Ib.* IV ix 74. Religious wemen quhilk in the tempill made seruice .. with haris skalit out of plett; *Boece* 130.

6. To send off, or draw off, (soldiers) in different directions, also *to* a place; to disperse for tactical reasons.

He .. skalyt his men here & thare, .. Til he belewit al hyme ane; *Leg. S.* xxix 207. Hyr ost scho scalyd here and thaire, Lyk as scho affrayid ware; WYNT. IV 525. Ryd sarralie, skaill nocht thi men oure wyde; HAY *Alex.* 452. To skail thame in sindry partis he began to fle, traisting ilkane of thame be this way ay to follow on him; BELL. *Livy* I 57/23. xxij carage hors send to the campe eftir the Scottis men war scalit to the Franche campe witht victuellis iiij li. viij s.; 1549-50 *Treas. Acc.* IX 376.

b. *reflex.* Only in BARB.: To break ranks; to go off in different directions.

Ȝe sall se alsone that thai .. sall brek aray And scaile [*C.* scale] thaim our harnays to ta; BARB. XII 467. That thai thaim scalyt euerilkane; *Ib.* XIII 73. That thai scalyt thaim in troplys ser And till discomfitur war ner; *Ib.* 275.

7. a. To disband (an army or fighting force).

Reddy to obey the massouris .. to fulfill the kingis command .. and to remove hame and skale his folkis or byd with his grace at his plesure; 1528 *Buccleuch Mun.* II 152. Gif Malcolme wald skaile his army .. he wald heir peace; BELL. *Boece* II 233. [They] Skaillit the oist and hamewart all is gone; STEWART 36202. We ar contentit that baith the armyis in all partis be skalit; 1541 *St. P. Henry VIII* V 199. Anent the deliwing of the presoneris .. the armyis being skalit, your grace suld send thame .. ransoun free; 1542 Craven *Ch. in Orkney* III 166. Ye haif begylit thaim and causit thaim to skayll thair fokkis; 1543 *Corr. M. Lorraine* 15. The armie and navie of Englishmen and Scotishmen sall be skailled both by sea and land; 1560 Calderwood II 2. The saidis lordis .. skaillit thair folkis and depairtit; *Diurn. Occurr.* 137. King James .. obeyit the charg and skaillit his airme; PITSC. I 139/16. Chargeing ayther party to skaill thair forces and freindis; 1583 *Reg. Privy C.* III 571. MELVILLE *Mem.* 25.

b. To break up (a meeting, assemblage, or gathering of any sort) freq. by dismissal.

Bot ay tha qwyte yow lill for lall Or that thai skale thare [*Maitl. F. 126/52* ȝe skaill the] markat all; WYNT. IX 1508. We sall gar scale our sculis all thé to scorne; DUNB. *Flyt.* 215. To Archibald Lavingston to pas to the Castell of Levingstoun to skaill ane gadering; 1512 *Treas. Acc.* IV 379. Als sone as this conventioun was skalit; BELL. *Boece* I 36. Id. *Livy* II 118/22. Thai .. decreittit .. To skaill the court, and ilk man hame is past; STEWART 18520. In the henderend of the convenshyon, as tha war skalit, ther cam ane ship from France; 1547 *Cal. Sc. P.* I 5. Minassing ane officiar of armes with maist contumelius wordis and bodely harme. .. At the occasioun of the quhilk tumult the court was scalit and the electioun of officers left of; 1569 *Reg. Privy C.* II 59. Vnderstanding the lordis to remove and skaill the sessioun; 1569-70 *Edinb. B. Rec.* III 268. Quhill the said convocatioun be scaillit; 1579 *Reg. Privy C.* III 173. The king schortlie skaylet this assemblie; DALR. I 292/3. Gif they refuse to skaill their gadderinges and convocation he may stay or continue the court; SKENE *Verb. S.* s.v. *Schireff.* Gif the said schooll be dissolvit and skaillit, it will tend to the preiudice of the youth of this town; 1598 *Aberd. B. Rec.* II 174. The first preiching day at his awin parochin kirk so to stand fra aucht houris in the morneing till the people be sk[e]lt; 1615 *Orkney & Shetl. Ct. Bks.* 55. The suoning uann & skunnering euill that skaillis conuentionis thikke; JAMES VI *Poems* I 169/524. After the kirk was scailed; 1665-7 LAUDER *Jrnl.* 120. They .. break up and skailled the fair; 1671 *Reg. Privy C.* 3 Ser. III 395. When kirk was skael'd and preaching done; PENNECUIK *Poems* 385.

c. To break up or disperse (one's family). Cf. 1 b above.

Quhairthrow honest ho[*gap in MS*] .. and famileis ar skalit and put to begartie; 1567 *Acts* III 45/1. And thairthrow causit the said Patrik to skaill his house and familie thairof; 1570 *Crim. Trials* I II 13. Would he .. put away his wife .. and skail his family again? 1661 W. THOMSON *The Churches Comfort: Sermon Preached in the Old Church of Edinburgh* (1706) 18.

d. ? To cast off (one's friends).

Ȝe desyre me to stay or scale my freindis .. Gif ȝe will assure me that my lord .. will stand my gud freind .. I will stay my freindis .. in all tymes cumyng; 1568 *Breadalbane Lett.* 1 Nov.

e. To dismiss (a person or persons) from a place or employment.

I sall degrade thé, .. Scaile thé for scorne and shere thé of the scule; KENNEDY *Flyt.* 398. Sabbato ii° Januarii certane masounis skalit; 1535-6 *M. Works Acc.* (ed.) I 162.

8. To cause (a body of armed men) to scatter or withdraw in disorder by (chiefly) force or other means; to put to flight; to rout. Also, once, with the place attacked as object. Also *fig.*

(1) Thar small folk .. fled all skalyt her and thar; BARB. II 397. Dowglas .. full egrely Assaylyt & scalyt thaim hastyly; *Ib.* VI 430. Fortoun ws trawaillyt fast today That scalyt ws sa sodanly; *Ib.* VII 301. Fra thar fayis archeris war Scalyt .. Swa that thai dred nocht thar schoting; *Ib.* XIII 84. Thair best battellis are Skaillit and broken heir and thair. The laif ar nocht to dreid greatly; *Alex.* I 1926. The pepill he scalit and all to-frushit, For thay war pure, small mardale; *Ib.* II 9153. Swa was skallyd all that rowt That off the lawe wes had na dowt; WYNT. II 1439. The kyng .. Swne skalyd all that cumpany; *Ib.* VII 1396. Quhare thai with stanis and gynis .. Micht skaill his oist and tak na skaith agane; HAY *Alex.* 6128. He .. on them ruschit .. And skaillit them full wyde before his face; *Clar.* V 1195. The

morne sall I with my wark and cunning Skaill all ȝone hoist; ROLLAND *Seven S.* 6178. [He came furiously upon them, he] brak [their drums and] scailit [their footmen]; 1597 *Cal. Sc. P.* XIII 134.

(*b*) As tempest that garris woddis fall, He schalit and thirllit the mekill stour; *Alex.* II 1277.

(2) The lords marched towards the towne, with purpose to scale the same; 1585 *Misc. Bann. C.* I 136.

fig. One sight of Jesus Christ would even skail the school of them; GRAY *Gospel-Call* 4.

b. To disperse (a fleet).

A gret wynd rase, & to the trybune contrare was, & scalit his nawine to & fra; *Leg. S.* xxxviii 603.

c. To drive away (insects).

Thir fleis .. war all full of hir blud .. and gif thay war skalit utheris .. suld .. souk out the residew of hir blude; BELL. *Boece* II 272.

9. *To scail the* (one's, etc.) *sege* (*assege*). **a.** To force a siege to be raised by scattering the besiegers. **b.** To raise a siege by dispersing or withdrawing the investing force.

a. Duke Betys .. Richt towart Tyre the way hes caucht To skaill the sege of that citie; *Alex.* I 185. Wyth that herschype swa he thowcht .. That that assege suld skalyd be; WYNT. VIII 3854. Quhidder gif that thou list suffir neuermar Thar sege scalit nor thame fre of dangar? DOUG. X i 58. And my said lord regent .. purpos .. to scaile the said sege; 1517-18 *Acta Conc. Publ. Aff.* 118. How Archibald Dowglas .. devissit skall the asseig of Berwik; *Boece* 26. To the end they mycht skaile the seige about Edinburgh; BANN. *Memor.* 194. To Glasgow past .. , Thair skalit the sege, releuit the castell sone; 1573 SEMPILL *Sat. P.* xxxix 60.

b. How Caius Mucius .. causit Porsena to scale his sege; BELL. *Livy* I 147/5. *Ib.* II 66/11. The .. saige was skalit; *Diurn. Occurr.* 12. The Inglischmen .. skaillit thair seige; PITSC. I 155/21. The governour behovit to skaill the seige and reteir and skaill his airmie; *Ib.* II 86/21. [The queen] persuades him [*sc.* James I] to skail and rais the seige and hame to returne; DALR. II 42/23.

10. To remove or take away (a responsibility); to annul (a proclamation); to change (a decision).

All hie chargis suld be scalit & repute of na effect; BELL. *Livy* I 85/9. To proceid aganis ony persoun that aucht to keip the proclamatioun, the samin standand undischargit or skaillit; BALFOUR *Pract.* 345. He skaillit [*I.* stableid] his mynd and held out his porpois in France to haue his pensieoun; PITSC. II 44/12.

b. To frustrate (an enterprise).

For the men of Leyth came furth, and scaillit that interpryse; *Hist. Jas. VI* (1825) 104.

11. *passive.* (*To be*) *skail(l)it*, (to be) in a dispersed or scattered condition. With and without adv. compl.

There is some ambiguity with passive uses of 7 and 8 above.

a. Of people or animals. **b.** Of things.

a. (1) Thai ly scalyt as thaim lest; BARB. VII 321. Thair fayis, that in beddis war, Or scalit, fleand heir and thair; *Ib.* IX 429 (C). Bot thai war scalyt that thai mocht On na maner assemblyt be; *Ib.* XVII 118 (E). Thai [*sc.* the deer] war red & skalyt wyd; *Leg. S.* xxix 91. Quhan Emynedus .. Saw how [his] fers .. War skalit with thair shot of far [etc.]; *Alex.* I 1629. Bot skalyd swa that tyme war thai, That thai togyddyr ware noucht then Owt owre aucht hwndyr; WYNT. VIII 3488. The Franche men that thaim scalyt saw Ruschit on thaim; *Ib.* 6785. Thair innemeis ar dispersit & skalit; *Loutfut MS* 49b. And the part of the front of the batell that is scalit and nocht evinly wil be soner sloppit be thair inemys & deuidit & vencust; *Ib.* 129b. Sanct Colm .. reduceit monkis and religious that war skalit and vagabund; *Boece* 311b. Quhen Pichtis war skalit he dissoluit his armye; *Ib.* 335b. Item, we ar contentit that baith the armyis in all partis be skalit; 1541 *St. P. Henry VIII* V 199. That nayne hukstar by nayne [*sc.* meal] in the mercat quhill the toun be scali [?*for* scalit]; 1566 *Kirkcaldy B. Rec.* 69. Till Adam of Gordonis cumpany were scalet [*Trans. 303* -eth] furth of Abirdeine (wha wer thair for the tyme); BANN. *Memor.* 212. The great numbers convenit at Edinburgh ar skaled; 1595 COLVILLE *Lett.* 173.

(*b*) And thus had .. all thy company [been] destroyit and scailit; HAY II 110/22. Seeing that our forces ar skaillit, thai intend to put to execution thair pretendit processes; 1568 Anderson *Collect. Mary* IV I 126. Sche had gottin assured knowlege that our cumpany was skailled; KNOX I 370. The scheipe to abyde equallie betuix the haill bairnes vndiuydit or skaillit to the hous; 1593 *Edinb. Test.* XXV 105b. Than skailit ar the Scotis host; DALR. II 288/28. The infrequencie of the session, who for the most part wer skailed; 1628 *Dundonald Par. Rec.* 251.

(*c*) Al the Jowis .. In-to the kynrike of Juda Vare wyd schalit to & fra; *Leg. S.* vii 238. The Jowys that ware schalyte wyde, Assemblyt; *Ib.* 807.

(2) His men fra him sa scalit war To purches thar necessite; BARB. VI 28. Thai that the formast war War skalyt in soppys her & thar; *Ib.* VIII 326. *Ib.* XV 480. Qwhene the appostolis scalyt ware In the warld oure all wyd-quhare; *Leg. S.* i 67. He saw thame fle but areste, Vnknyt scalit in the foreste; *Ib.* xl 914. Emynedoun .. saw how his feris war Scalit and soupit heir and thair; *Alex.* I 1974. Now will thy scheip be skaillit in the feild; HAY *Alex.* 18604. The bissy beys .. On diuers colorit flouris skalit wide; DOUG. VI xi 58. [They] set on the Romanis, quhilkis wer that tyme skalit throw the countre; BELL. *Boece* (M) I 134. Many of the earl's friends being scaled off the town; PITSC. (1728) 17.

b. Ane angell .. thai quhelis in pecis small Brak, & scaillit was oure-all; *Leg. S.* l 891. The schippes that scaled wor Draif in one heipe; *Troy-bk.* II 1732. Be that Noe .. vnderstude That oure all scalit wes the flude; WYNT. I 416 (W). The schippis war scaillit throw tempest; *Florimond* 329. Claret wyne .. is .. clere, but the moder scailde [F. *toute la lye soit au fons du vaissel*]; HAY II 140/33. The chas was gret, scalyt our breid and lenth; *Wall.* V 282. Lyke as the huge forest can bewaill Hys granys doun bet and his branchis skaill; DOUG. XIII i 14. To the castel of Calidone wod .. was the slauchter and chais skalit; *Boece* 69.

12. *intr.* To separate, one from another; to disperse or scatter; to break up, disband. With *pl.* or *coll.* subject. With and without adv. compl. Also *proverb.*

(1) He bad his men .. Scaile and ilkan hald his way All him selff; BARB. VI 577. And the remanand suld all hale Skaill throw the toun; *Ib.* XVII 99. The folk of Ind .. scallit in troppellis heir and thair; *Alex.* II 4338. Bot a gret storme sa gert thaim [*sc.* ships] scaill That thai were drevin haill away; WYNT. VIII 6322 (W). Ordanis all .. liegis quhilkis war chargit of befor for to keip the said diett .. scaile and pas hame; 1523 *Acta Conc. Publ. Aff.* 175. [They] sall .. skaill als fast sindry .. as thay assemblit togidder; BELL. *Boece* I 150. Thame quhilk war bot gaderingis and daly disseuerit and skalit; *Boece* 467b. To caus his men to sched and skaill in twyn; STEWART 56827. And quhen we skall .. thay will tak our freindis geir or thay want; 1568 *Reg. Morton* I 34. As ship but ruther sa ga we now but gyde: We skail, we scatter, we wait not quhair we go; 1570 *Sat. P.* xvii 123. The erle of Schrewisbarie [etc.] .. come with xv^{m} men .. quhais cuming causit the Scottis and Frenche skaill; *Diurn. Occurr.* 47. They skayled ane and ane and past away; MOYSIE 75. He .. allegit that the ennemys wer skailen and fleing away behind the litle vilage; MELVILLE *Mem.* 202. For which effect your lesson must be hastened before the schollars begin to skaill; 1642 BAILLIE II 20.

(2) The assege than scalyd swne [*W.* Thai .. scalit all that segis sone]; WYNT. VII 2571. In passing of wattiris .. the ost mon synd & scaill out of aray; *Loutfut MS* 132a. That all scuillis scail and nane to be haldin; 1498 (*c*1580) *Edinb. B. Rec.* I 75. The Perse .. sufferit his army to skaill to thair loginnis for the nicht; BELL. *Boece* II 478. I traist the army sall skaill for this wynter; 1541 *Sc. Hist. Rev.* VI 156. That the futt band suld skaill incontinent; 1543 *Corr. M. Lorraine* 23. Quhilkis personis .. throw their inobedience constreinȝeit the prowest and cunsall to skaill and depart; 1561 *Inverness Rec.* I 65. The same day, the conventioun skalit; *Diurn. Occurr.* 346. Hou the congregatioun scailled; PITSC. II 150 h. of ch. The colleges skaillis ane monethe for necessitie; 1588 *Comm. Univ.* III (St. A.) App. 195. Ane fellon tryne come at his taill Fast

flichtren through the skise Bot suddenly that skull did skaill; BUREL *Pilgr.* I 247. Skalled; MELVILL 17. As the people war dissolueing and skailling fra the sermone; 1618 *Crim. Trials* III 451. Befoir the counsell skeillit frome chusing of the magistratis; 1624 *Perth Kirk S.* MS 4 Oct. Let the north [communion] table and north syde of the middle table skaill that thes tables may be furnisching againe .. quhill [etc.]; *c*1640 *Dundonald Par. Rec.* 522. Fund drunk when the efternoone sermone skaillit; 1655 *Dumfries Kirk S.* 22 Feb.

proverb. Breads house skaild never; FERG. *Prov.* No. 205.

Passing into: **b.** To depart; to go away from one's fellow(s); to leave the main body of an army or herd.

Hame wele cummyn ar thai & scalyt syne ilk man thar way; BARB. XIX 814. That na persoun of the oist in Ingland sall scail or pas othir to forray or speking withoutin .. bidding of the chiftane; *Harl. MS 4700* 259b. Wolle Craw [saw] .. the kow scall forther; 1522-3 *Selkirk B. Ct.* (ed.) 67. Syne scaillit hes and passit hame ilk man; STEWART 31611. Tyme is vntrusty and dois skaill from thé away; *Bann. MS* 82b/64. [To forbid any] to skaill or depairt fra the attendence of his majesteis persoun; 1584 *Reg. Privy C.* III 659.

c. *To scail abak*, to withdraw, retire.

Rycht softlie than thai skalit bayth abak; STEWART 2436.

13. To spread (far and wide).

Bot tithandis that scalis sone Off this deid .. Come to the Cliffurd his ere; BARB. V 447. A ful gret fyre he gert mak thare .. That scalyt fare the low, & wyd; *Leg. S.* xliv 284. Thane the gret lo[w] scalit on heid; *Ib.* xlv 167. *Ib.* xlix 87. The stynk scalyt off ded bodyis sa wyde [etc.]; *Wall.* VII 467. That the vennom micht skail throw all the vanis of his body; BELL. *Boece* II 62.

14. To scatter, throw away; to disperse.

And he that gaderis nocht with him scaylis; HAY I 123/8. The myst scalyt, the son schawyt fayr and brycht; *Wall.* XI 528. Quhill all the skyis skaillit violat; HUME *Promine* 67. Nou skaillis the skyis, The nicht is neir gone; MONTG. *Misc. P.* xli 7.

15. To fall in drops; to pour (down); to spill.

The pellican .. strikis hir self in the brest with hir neb & garris the blud scaill apon hir birdis; *Loutfut MS* 24a. Amyd the hawchis .. The recent dew begynnys doun to scaill; DOUG. XIII Prol. 22. The clouds of raine From tops of mountaines skails; HUME 27/62.

b. To spill contents; to leak. Also *proverb.*

Two sufficient sives for each multur .. whoso holds these sives and wilfullie skaills by the sive or defrauds the multure in any way is to pay 10 groats; 1669 *Holmains Baron Ct.* 11 Oct.

proverb. An old seck is ay skailing [*MS* skelling]; FERG. *Prov.* No. 82.

16. ? To come apart.

Gar heis hir [*sc.* the Fleming bark] quhill hir howbandis skaill; *a*1568 SEMPILL *Sat. P.* xlvi 12.

Scail, var. of SCALE *n.*[1] and SKELE *n.*

(Scaildrake,) Skaildrake, -draik(e, *n.* Also: **skeldraik, -dryk(e.** [The first element is of doubtful origin, but is perh., ? partly, f. north. ME *skell* (Manning), ON *skel* a sea-shell. For further conjecture, see the etymology to *Skell-* in SND. Cf. mod. Eng. *sheldrake*, also SKILLING-GUIS *n.*] **a.** The sheldrake. **b.** Some sort of boat. See also SCAFF *n.* for a further example. — **a.** Empt. .. v skaildrakis x s. .. ij hammis xij s.; 1533 *Household Bks. Jas. V* 232. They discharge ony persone .. to sell .. skeldraikis herroun buttir or ony sic kynd of foullis commounlie vsit to be chaisit with halkis; 1599 *Acts* IV 180/2. Skeldraikis herroun butter or ony sic kynd of foullis; 1600 *Ib.* 236/2. Skail-draikes; SKENE *Reg. Maj.* II 139b. — **b.** That the brugh of Kinghorne and ferre thairof .. is now laitle hewele trublit and hurt be the skaffis, skeldrykis and ʒowis of vnfre touns of Leith .. and of Newhewin; 1600 *Conv. Burghs* II 81. Skeldrykeis; *Ib.*

Scaile, var. of SCAIL *v.*

(Scailer,) Skailler, *n.* [SCAIL *v.*] One who scatters or puts to flight. — Breckeris of dowcatis and cranning hares, skaillers of bees and bee skepes; 1612 *Inverness Rec.* II 89.

Scaile-stair, var. of SCALE-STAIR(E *n.* **Scailie,** var. of SCAILʒE *n.*

Scailing, Scaling, -ynge, *vbl. n.* Also: **skail(l)-, skayl-, skaling; scalen; skeilling.** [SCAIL *v.*]

1. a. The scattering or dispersing of (a thing) in sense 1 of the verb. **b.** The action of allowing water to flow away in sense 4 c of the verb. **c.** The action of dispersing, disbanding, etc. (persons) in senses 6, 7, 8 and 12 of the verb. **d.** The action of raising (a siege) in sense 9 of the verb.

a. (1) Adiugit to have done wrang to Jhone Carmichel in skaling of his girs ourethwort the cawsay; 1554 *Dundee B. Ct.* II fol. 327b (8 Aug.). Fund in the wrang .. in castyng of hir downe .. and skailing of hir sybois; 1575 *Glasgow B. Rec.* I 38. For skaling of the middingis; 1604-5 *Ayr B. Acc.* 224. For skailling of fuilʒe upoun the hillis; 1609-10 *Ib.* 244. Fined for .. spoilling of the new sawn beddis and skailling of the seids; 1615 *Inverurie* 197.

(2) He ma nocht gudlie pay the saidis dettis .. without skaling of his hous; 1544 *Reg. Privy S.* III 115/2. For the wrangus skaling and sellin of thair geyr in opyn mercat; 1556 *Inverness Rec.* I 3.

b. To visite .. the Bridge Milne .. where the spout is and to ordor effectual course .. for preventing the scailing of the water; 1708 *Stirling B. Rec.* II 114.

c. (1) Tyme of scalynge, tyme of gaderinge; *Wisd. Sol.* (STS) 172. Lettres .. for prolongatioun of the pece and scailing on the bordouris; 1519 *Acta Conc. Publ. Aff.* 149. The scolars .. compellit be .. thair maister to conveine at sick tymes as thai war in wse and preuelege of libertie and skaling obefoir; 1569 *Aberd. B. Rec.* I 366. Under .. the skaling of the power foresaid, thair was incontinent greit .. distructioun done; LESLIE 46. [At the] scaling [from the Parliament]; 1572 *Cal. Sc. P.* IV 483. Willing them befor thair skailing to direct two or thrie of thair number vnto him; 1593 *Bk. Univ. Kirk* III 805. Let the hour of gathering [*sc.* at the school] in the morning be seaven of clock and the hour of skailing six; 1640 *Dundonald Par. Rec.* 465. At the kirk scailing; 1645 *S. Leith Rec.* 54/1.

(2) Better is scalynge [*pr.* stalynge] of thar barnagis In to the warlde [L. *per mundum dispersio*] .. Na [etc.]; *Bernardus* 386. Tithingis of the skaling of the Inglismen; 1497 *Treas. Acc.* I 354. Eftir the skaling of the said army; 1523 *Acta Conc. Publ. Aff.* 180. At the skaling of the nycht wasche, viz., be sex houris in the mornyng; 1569-70 *Edinb. B. Rec.* III 269. To vse the stryking of the swische nychtlie to the wauche and in the mornyng at skaling of the watsche; 1572 *Peebles B. Rec.* I 341. The scailling of the hie schole for feir of the pest; 1598 *Edinb. B. Rec.* V 232. In the skailing of the pepill from the execution, thair wes .. letters fund in proes; MOYSIE 24. Mr. .. Boyd to be spoken to about the soon scaling of the Barony Kirk on Sunday afternoon; 1651 in Boyd *Zion's Fl.* Introd. 53. Skayling; 1656 NICOLL *Diary* 175. At the skailing of the kirk; 1661 *Dumfries Council Min.* 14 Jan. After the skailling of ony buriall, to sight the calsay; 1664 *Glasgow B. Rec.* III 35. At skeilling of the preaching; 1683 R. Brown *Paisley* I 302.

d. The Ballioll eftir the skaling of the seige .. come .. to Skoyne; BELL. *Boece* (M) II 301.

2. *To tak scailing*, to cease hostilities (? by withdrawing one's forces).

The quhilk wes send .. With speciall power for to purches peice And scalen tak fra all battell and weir; STEWART 8938. Tha tuke na scaling nother nycht nor da; *Ib.* 10748. *Ib.* 8956.

3. ? Goods or articles to be disposed of (? by sale). Cf. sense 1 a (2) above.

The outreid & skailling of his quarter forsaid of the said schip; 1638 *Brechin Test.* V 257.

(Scailit,) Scailt, *ppl. adj.* Also: **skaild, skayld, scaillit, scalyt.** [SCAIL *v.*] **a.** That have, or has, been scattered or dispersed; ? in different places (rather than concentrated in one area). **b.** *fig.* Of legal lore: Uttered piecemeal rather than as a coherent statement. — **a.** Feill scalyt [*1570* scaillit] folk to thaim will son ranew; *Wall.* x 691. The prouest .. and counsall .. ordanes ane scailt watche to be kepit bayth day and nycht in the town, ane quarter euery xxiiij houris, and that ilk persoun that is haill and present watche in his awin persoun; 1582 *Edinb. B. Rec.* IV 240. — **b.** Sum castis summondis and sum exceptis; Sum standis besyd and skaild [*M.* skayld] law keppis; DUNB. (OUP) 200/30.

Scaill, var. of SCAIL *n.*[1], *n.*[2] and *v.* and SCALE *n.*[1], *n.*[2], *n.*[3] and *v.* **Scailling,** var. of SCALING *vbl. n.* **Scaill-stair,** var. of SCALE-STAIR(E *n.*

Scailʒe, -y(i)e, -ʒie, -ʒee, *n.* Also: **skail-** and **-ʒe(e, -ʒie, -yie, -ye; scail(l)ie, skail(l)y, -ie, scalie; skealye; scalʒe, -yie, skalʒie, -ʒe, -y(i)e; skeilyhe, skelʒe, skellie; skilly; skila.** [MDu. *schaelie* (mod. Du. *schalie*), OF *escaille* (13th c. in Larousse).] Slate, the material; blue roofing-slate. Chiefly *coll. sing.*, but sometimes with *pl.* concord. Also *attrib.* and *comb.*

(1) To the man that brocht vp the scailʒee; 1496 *Treas. Acc.* I 275. To the men that set vp the scailʒe; *Ib.* 276. The man that thekis the scailʒe; 1497 *Ib.* 359. To the man that thekis the gallory in Striuelin with scailʒee; *Ib.* 370. Payit .. for .. fraucht of ane bote full of scailʒe fra Strivelin to Leith; 1502 *Ib.* II 155. Ane dumb man in Strivelin that wrocht scailʒe; 1507 *Ib.* IV 84. Cum tegulis aut le scailʒe derect.; 1508-9 *Reg. Great S.* 702/2. Skailʒe; 1515 *Rentale Dunkeld.* (SHS) 296. Skailye; 1531-2 *M. Works Acc.* (ed.) I 56. j[m]lx of skailʒe coft in Peblis to mend the ruiff of the grete hall witht; 1539 *Treas. Acc.* VII 218. For vi ruidis new theking werk .. with skailye cariage and all costis .. xlii lib.; 1541 *M. Works Acc.* (ed.) I 285. Skailye and sclait; 1558 *Ib.* 293. Skailʒe .. for beating and making of the [castle] .. watterticht; 1574 *Red Bk. Menteith* II 408. For fourte menis mait that led the skailʒe out of the hilend; 1581 *Acc. Repair Doune* in *Moray Mun.* Box 1, No. 300. The thak of the same is off skailʒe; 1583 *Chapel Royal* lxxxvii. The sklaittis, skailyie, lyme, sand and tymmer; 1584 *Reg. Privy C.* III 678. For carrying hame of the lairdis skailʒe to Balloch; 1592 *Breadalbane Ct. Bk.* 38b. Ane .. Ile callit .. Ellan Slait, quhairin thair is abundance of skailʒie to be wyn; MONRO *W. Isles* (1961) 52. For thre hundreth nails to put on the skailʒe vpone the stabillis; 1603 *Argyll Acc.* 5 June. That .. no .. persones .. builde anye housis within the said burgh .. bot suche as salbe covered with sklaite or skailʒee leid tyild or thakstane; 1621 *Acts* IV 627/1. With skailʒe and fog theiking; 1635 *Chapel Royal* cxciv. The theiking of the rooffe with skailyie and the flatt rooffe with leid; 1663 *Edinb. B. Rec.* IX 327. I will have all the house .. covered with blew skailyie; 1673 Mylne *Master Masons* 184. The corneice betwixt the lead roofe and the skailʒie; 1679 *Ib.* 200. Quaedam [stones] minora; eaque vel vulgaria, nostratibus gray-sclate, vel nigri aut caerulei coloris, blew sclate, or scailʒie; SIBBALD *Scot. Illustr.* III 45.

(*b*) Riggyngtild [and] skalʒe; 1507-8 *Rentale Dunkeld.* (SHS) 170. Skalʒie; 1514 *Ib.* 290. [A dwelling .. covered with small slates or] scalʒe; *a*1517 *Ib.* 323. Skalye; 1539-40 *M. Works Acc.* (ed.) I 285.

(*c*) Sklaited with skaily; 1671 Mylne *Master Masons* 170. Scailie; 1693 *Logie Par. Hist.* I 333. They have freestone and hardstone, slate and scaillie, as good as is in Scotland; 16.. *Macfarlane's Geog. Coll.* III 142. Here is to be found marle .. slait and skailly; PENNECUIK *Tweeddale* 5. — Scalie; 1683 ERSKINE *Diary* 15.

(*d*) For the mending and reparacioun of the ouer kirk .. with skeilyhe; 1508 *Reg. Soltre* 162.

(*e*) So the same roof will serve again and the same skilly very near; 1673 Mylne *Master Masons* 185.

(*f*) Churches .. roofed with blue skellie; 1686 ERSKINE *Diary* 205. 1703 *Stirling B. Rec.* II 348.

(2) To the werkmen that had the skeilyhe abone the kirk and put thame in the wolt; 1511-12 *Reg. Soltre* 163. ii[m] skailye; 1531-2 *M. Works Acc.* (ed.) I 108. For ane thowsand skailye iii lib. vi s.; 1534-5 *Ib.* 129. For iij[c] skelʒe to mend the ruffe of the grete hall .. xviij s.; 1542 *Treas. Acc.* VIII 132. The wynning of ane thousand skailʒe; 1581 *Red Bk. Menteith* II 420. Certane skailʒie .. quhilkis wer transportit .. to the reparatioun of the palace; 1585 *Treas. Acc.* MS 86. To furneis als many skailyie as will theik sufficientlie with fog the weyhous of the burgh; 1612 *Edinb. B. Rec.* VI 86. For 200 skalyie £6; 1651 *Stirling B. Rec.* II 308. For 200 skailie .. £9 16 s.; 1673 *Stirling Common Good* 85b. The treasurer to send ane express to Deuchray .. anent 8000 scailyie to the tolbooth; 1703 *Stirling B. Rec.* II 100.

attrib. and *comb.* (1) Scailʒe man; 1505 *Treas. Acc.* III 144. To the scailye laider for the leding of xvj[c] scailye; 1513 *Ib.* IV 526. Johne Grahame skailyie man; 1628-9 *M. Works Acc.* (ed.) II 232.

(2) Ane skailyie stayne, 14 d.; 1589 *Edinb. B. Rec.* IV 540. For reparatioun of the sklait and skaillie work of the castell of Striviling; 1601 *Treas. Acc.* MS 106. Ane scalyie [*pr.* sealyie] table; 1616 *Edinb. B. Rec.* VI 379. The skailyie craig; 1617 *M. Works Acc.* (ed.) II 29.

(3) vii[c] skailyie nailles xlix s.; 1626-7 *M. Works Acc.* (ed.) II 201. ii[c] thin headit skealye naill for the wind schotes of the .. windowes; 1633 *Ib.* 366. 1000 skaillie naillis for mending the skaillie of the great haill at half a merk the hundreth; *c*1643 *Rothes P.* For 200 skailie naill £1 6 s.; 1673 *Stirling Common Good* 85b. Skailie pinnes; 1677 *Ib.* 114. 800 skailʒie nailes; 1681 *Highland P.* I 319. The counsell .. aproves of the dean of gild his buyeing of about 35,000 skila sclat for .. publict uses; 1702 *Aberd. B. Rec.* IV 332.

b. As a place-name element.

Skellietoun; 1662 Wodrow *Hist.* (1721) I App. 59. Skelliehil; *Ib.*

(Scailʒer,) Skailyer, *n.* [SCAILʒE *n.*] A slater. = *scailʒeman*, s.v. SCAILʒE *n. attrib.* and *comb.* (1). — Sclaiter and skailyer; 1618 *Edinb. B. Rec.* MS XII 265.

(Scaiographie,) Skaiographie, *n.* (e.m.E. *sciographie, sciagraphie* the branch of the science of Perspective which deals with the projection of shadows (1598), a sciagraphic delineation or picture (1611), F. *sciagraphie, sciographie*, L. *scia-, sciographia.*) — For the printeres and wryteres of this age, caring for noe more arte then may win the pennie, wil not paen them-selfes to knau whither it be orthographie or skaiographie that doeth the turne; HUME *Orthog.* 2.

Scaip(e, varr. of SCAP(E *v.* **Scair,** var. of SCAR *n.*[1], SCAR(E *n.*[1] and SCARE *v.* **Scairc(e)-, Scairs-,** varr. of SCARS-. **Scairch, Scairis, Scairs(e,** varr. of SCARS(E *adj.* and *adv.* **Scaire,** var. of SCAR(E *n.*[1] **Scairing,** var. of SQUARING *vbl. n.* **Scairré,** var. of SKERRY *n.* **Scairsbrough wairning,** var. of SCARRSBRUGH WARNING. **Scairth,** var. of SCARS(E *adj.* and *adv.*

(Scait,) Skait, -ytt. ? Var. of SCALD *adj.* b or of SCALDIT *ppl. adj.*, or a conflation of these. Cf., also, the first element of next. — Calland him skaytt karll; 1543 *Aberd. B. Rec.* XVIII (Jam., s.v. *Mispersoning*). Ar thow takin af thai bonnat to ane skait cretor? 1610 *Rec. Old Aberd.* I 49. Calling William McIlraith skait knawe and many vther injurious wordis; 1631 *Kirkcudbr. B. Rec.* II 434.

Scait, var. of SCAT(E *n.* and SCAT(T *v.*

(Scaitbird,) Skaitbird(e, *n.* [Cf. later Sc. dial. *skatie-goo* the Arctic skua, also known in later Sc. as *Dirten Allen, Dirty Allan*, from a former popular belief that it ate the excreta of other birds (cf. F. *chasse merde*). The first element may be connected with SKITE *v.*] ? The skua (*stercorarius parasiticus*). — Dirtin Dumbar .. Skaldit skaitbird [*M.* skaitbirde] and commoun skamelar; KENNEDY *Flyt.* 37. Quhen he come hame it wes ane howlat nest Full of skait birdis; *King Berdok* 27.

Scait-frie, var. of SCOT-FRIE *adj.* **Scaith,** var. and p.p. of SCATH(E *v.* **Scaith-,** var. of SCATH(E)-. **Scaithis,** var. of SCATH(E *v.* **Scaith(t,** varr. of SCATH(E *n.* **Scaitland,** var. of SCAT(T)LAND *n.*

Scakkes, see SCALP *n.*

Scal, var. of SAL(L *aux. v.*, SCAL(E *n.*[2], SCAL(L *n.* and SKELE *n.*

(Scaland,) Scayland, *ppl. adj.* [SCAL(E *n.*[2]] ? Covering with, or as with, scales. — Tha scayland schalis [of a dragon] set vp rath; *Leg. S.* xxviii 413.

Scalbart, -bert, -bourd, -burde, varr. of SCAUBERT *n.*

Scald, Skald, Scold, *n.*[1] Also: **scauld; skaul; scall; skold.** [Midl. and north. ME and e.m.E. *scald* (Orm), *skalde* (*c*1400), *skawde* (*c*1460), appar. obs. after 1483, south. and e. midl. ME and e.m.E. *scold* (*c*1275), *scolle* (1569), *skol* (1572), *scowld* (1611), ON *skáld* a poet, also ? a lampooner.] One given to the use of vituperation or abusive language; one who by the use of such language causes public disturbance; a foul-mouthed quarreller.

Applied to both men and women, but in the 17th century perh. increasingly to women.

(*a*) Scho callit to hir cheir .. A skeg a scornar a skald; *Colk. Sow* I 98. So suld thay [*sc.* women] nocht be nemit Skaldis baldis and thairthrow .. schamit; *Ib.* II 152. And knaw, kene scald, I hald of Alathya; KENNEDY *Flyt.* 322. I will not for none offence Flyt lyke ane skald; MAITLAND *Maitl. F.* 329/18. Scaid scald, our bald, soone fauld, or I melt thé; POLWART *Flyt.* 795 (T). As also of all skalds and flyettars within this burgh being tryit; 1604 *Paisley B. Rec.* 269. Margrat Tailyour to be .. banesit the toun as ane notorious skald; 1614 *Elgin Rec.* II 140. That all scaldis, bairdis .. and persons offensive to their neighbours be speiche, pay 40 s. and sit in the cockstuillis the space of four houris; 1618 *Orkney Rentals* App. 51. Beatrix Wichtane .. and Margaret Neisch .. are found bauld schameles skaldis; 1619 *Perth Kirk S.* MS 24 May. That scho was a gryt scald ane banner & sweirar & blasphemer of the name of God; 1629 *Fraserburgh Kirk S.* II 3 June. Childrein dissobedient to parents, lewd leivers, common scaldis and incorrigibill harlottes; 1632 *Edinb. B. Rec.* VII 108. Declarit to be a perjured woman and a wyld scald quho resett all sort of infamous persones in her hous; 1633 *Strathbogie Presb.* 6. The old act made against skalds, bards and slanderers, to be intimate the nixt Sunday out of the pulpit; 1643 *Orkney Rentals* App. 56. Marjorie Layng to be banisched scho being .. ane theif and a scald; 1647 *Elgin Rec.* II 258. Common flyters and skalds; 1656 *Lasswade Session Rec.* in Lee *Hist. Church* (1860) App. I 405.

(*b*) That na flitaris nor scauldis be hard on the gait .. vnder the pane of puttin of thame in the cuke stulis; 1559 *Peebles B. Rec.* I 256.

(*c*) [The town council ordained that] all skaulis and commoun flyteris be put in the clespis thairin to remane thrie houris; 1584 *Ayrsh. Coll.* 2 Ser. II 43.

(*d*) That all sik scallis and bardis quha heireftir hapins to abuse honest wemen with thair blasphemous langage .. be pwnist [*pr.* pwinst] in prissoun; 1611 *Glasgow B. Rec.* I 317.

(*e*) All common skoldis, flyttaris and bardis to be baneist the toun; 1562 *Aberd. Eccl. Rec.* 9. Isobel Gray ane professed scold and curser; 1655 *Ellon Par.* 117.

b. An act of scolding.

Delated Robert Milline and his wife for scold, flyting and striking one another most unchristianly; 1668 *Kingarth Par. Rec.* 52.

Scald, Scad, Skaid, *n.*[2] [17th c. Eng. *scald* (1601) an injury caused by hot fluid or steam; SCALD *v.*[1]] *fig.* ? A mark or blemish. Also in *hart-scald* or *hartskaid*, see HART *n.*[1] 4 and *HART *n.*[1] 4. — Your cross is of the colour of .. Christ .. and that .. colour will .. neither be stained nor cast the colour. Yet it reflects a scad like the cross of Christ; 1640 RUTHERFORD *Lett.* (1894) 580.

Scald, *adj.* Also: **skald, scauld; skawd; skade, scaid, skaid; scad.** [e.m.E. *scald* (*c*1500), later spelling of ME *scallede* affected with the 'scall' (1529), contemptible (*c*1500). See also SCALDIT *ppl. adj.*] Affected by a scabby disease of the skin or scalp; scabby, passing into an expression of general opprobrium or revulsion, foul, filthy. Also *proverb.*

DUNB. *Flyt.* 58 (M) (see SCALDIT *ppl. adj.*). Howbeid I se thy skalp skyr skawd [*Ch.* skyre skaid]; LYND. *Sat.* 2485 (B). Gum gait gallit and scald [*H.* Gumgad bald skade] foul fawit quhy flait thow? POLWART *Flyt.* 767 (T). Hellis spark skald [*H.* scalded] clark .. Scaid [*H.* Scad] scald, our bald; *Ib.* 794, 795. For sclandering the said William Sinclair, calling him ane skaid thaife, without ony liklines or probatioun; 1604 *Shetland Sheriff Ct.* (ed.) 111. Cristane Jack to intertenye a pure skade litillone; 1620 *Perth Kirk S.* MS 24 Jan. There were but three scauld, and one bald-pated legist in that place; URQUHART *Wks.* Rabelais II iii.

proverb. A skade mans head is soon broken; FERG. *Prov.* No. 24. A skabbed horse is good enough for a skald [*MS* (No. 66) scalled] squir; *Ib.* No. 69. He cals me skabbed because I will not call him skade; *Ib.* No. 375.

Scald, *v.*[1] Also: **skald; scawde; skaude; scall; scad, skad; scade; skaid; schald.** [ME and e.m.E. *schalde(n* (Ancr. R.), *skalde(n* (*a*1300), *scalde(n* (Rolle), *scolde(n* (1340), OF *eschalder, eschauder* (*c*1160 in Larousse), *escauder* (13th c. in Godefroy), to burn, scald, late L. *excaldāre* to wash in hot water.]

1. *tr.* Of fire: To scorch or burn.

Also of the fire of Purgatory.

The .. blesys of hait fyre Reddy to byrn thi schippys .. And al the cost belyve of flambys scald; DOUG. IV x 89. A byrnand schide .. That blesyt vp hys lang berd .. Quhilk scaldit thus a strang fleur dyd cast; *Ib.* XII v 199. — Bot all pure folk .. [in Purgatory] Was skaldit baine and lyre; *G. Ball.* 187.

b. With personal subject: To burn (a person) by way of punishment. Also *absol.*

Thow art .. Ane schameles schrew, the maister deuell mot scald thé; ROLLAND *Seven S.* 4020. ȝit war thay not cummit to that furie & rage as to bruyle and scald quha sa euer suld speik aganis thame, bot contentit thame selfis with presoun or banischement of sic persones as war contrarious to thame; *Lyndesay Pref.* (STS) 400.

absol. Christis trew disciplis all .. To byrne and skald thay neuer pretendit. The pope defendis his traditioun Be flammand fyre; LYND. *Mon.* 4642.

c. To hurt by, or as by, the action of hot steam or liquid. Also *absol.*

The sliddrie vreter, carier of Salt vrine skalding [*v.r.* skadding] sore; JAMES VI *Poems* I 158/864. The first handis that handles thé, Brunt and scalded sall they be; 1662 *Crim. Trials* III 613.

2. *transf.* To affect (something) in a way comparable to the effect produced by the action of hot water or steam; to damage severely. **a.** Of words, etc. **b.** Of a sore. **c.** Of heavy rain.

a. The quent and curyus castis poeticall [etc.] .. Caxtoun for dreid thai suld hys lippis scald [*Sm.* scawde, *1553* skaude] Durst nevir twich; DOUG. I Prol. 258.

b. With pottingeris poulderis except thow ovrsmeir That scab that ȝe beir will scall thé to the schoulderis; MONTG. *Flyt.* 94 (T).

c. For oft-times a sudden spait of rain scalds the tender grain; BELHAVEN *Rudiments* 12.

d. Of thoughts or cares: To inflame or irritate (a person, his or her mind).

Quhene sick thocht can me schald; *Leg. S.* xviii 961. The bissy curis of Turnus mariage Skaldyng hir breist and mynd half in a rage; DOUG. VII vi 10.

3. To cleanse, wash out or sterilise with boiling water.

[In time of plague persons taking money must put it] in an veschel and skaid the samyn thatt it may be purifeit fra all contagiousnes; 1568 *Canongate & Broughton Ct. Bk.* in *Misc. Maitl. C.* II 314. Syne sup sex soipis but sumthing thin Of the deill scad [*H.* scade] thy guttis with in To haill thé of the skitter; POLWART *Flyt.* 237 (T).

4. *proverb.* To cause (one's lips) to be scalded or burned by eating hot food. Also with reflexive object.

To forge a querrell .. and skad your lippes in other men's kaile; 1584 Calderwood IV 131. Scad not your lips in other folks caill; CARMICHAEL *Prov.* No. 1340. — And thow skip as thow skoupe thow wilt skad thee; CARMICHAEL *Prov.* No. 784.

5. *intr.* To suffer the effects of excessive heat; to faint or swoon in consequence of (*with*) this; to shrivel up.

Scho scaldis, and with mony wrikis and thrawys Presys forto eschew the feirfull heit; DOUG. XIII iv 86. Bot vthir sede .. quhen the sonn was risen, thai scaldit; and for thai had nocht rute, thai dryet vp; NISBET *Matth.* xiii 6. Men scaldit with gret heet; Id. *Rev.* xvi 9. The inchoatiue forme, as *feruesco calesco.* I begin to scald or to hete; VAUS (1531) 2 (de verbo).

b. *fig.* To burn with (*in*) strong emotion or desire.

Hys mynd scaldand in greif and egir will; DOUG. IX xiii 46. In our ardent desyre Of the bargan he scaldit hait as fyre; *Ib.* XIII vi 104.

6. To behave as if boiling, or about to boil; to froth; to bubble.

[The] flude of Acheron .. Drumly of mud and skaldand as it war wod Popland and bulrand furth on athir hand; DOUG. VI v 4. Catharin .. cam in to her hous when her milk was on the fyr skadding; 1650 *Brechin Presb.* 40.

7. To set fire to property, etc.; to burn.

They are burning and scalding, slaying and murdering, and using all kind of oppression; R. BRUCE *Serm.* 314.

Scald, Skald, Scold, *v.*[2] Also: **skauld; skall.** P.t. **scowld.** [ME and e.m.E. *scolde(n* (Piers Plowman); SCALD *n.*[1]] **a.** *intr.* To rail; to brawl; to quarrel noisily and in unseemly terms.

Freq. coupled with FLYT(E *v.*

In the quot. for 1613, perh. in *tr.* use. See also the quot. f. *Polemo-Mid.*

(1) Crystane Lyone and Elspet Edmond have acted ther selffis nevir to flyt, skald nor miscall wtheris heirefter; 1613 *Elgin Rec.* II 134. Cristane Andersone and Margrat Alves have acted them selffis nocht to skauld nor flyt with wtheris; 1621 *Ib.* 165. If scho skall and speek schamfull language [etc.]; 1629 *Glasgow Weavers* 116. The sessioune .. inhibit Margaret Sowter to scald or flyt; 1629 *S. Leith Rec.* 16/2. They schould not be fund scalding nor flyting; 1657 *Falkirk Par. Rec.* I 203. Helen Jameson and Grisall Whyt being accustomed to flyt and scald .. ar reconciled; 1671 Meikle *Old Session Bk.* 94.

(2) Misere scaldaverat omnes; *Polemo-Mid.* 120.

(*b*) The said Marione .. to act hirself not to scold or flyte hereafter; 1653 *Peebles B. Rec.* II 14. The said Andrew Hutson scolded and flett; 1662 *Soc. Ant.* XXII 232. He heard them flyting and scolding; 1669 *Corshill Baron Ct.* 86.

b. *tr.* To rebuke, chide.

[Tapestry of] the historie of Sussanna wherof there is fyve peissis .. 3. Daniell revyled the elders .. 4. schee scowld the people but schee was iustified; 1631 *Buccleuch Household Bk.* 29 Oct.

Scaldand, -ing, Schald-, Skaldand, *ppl. adj.* Also: **schaldane.** [SCALD *v.*[1]]

1. Ardent.

(*a*) In skaldand word luf God sal thay, Belyk, for thare word sal be ay .. To the folk as a brund of fyre; *Leg. S.* xxxvi 81. With eger willis and scaldand; *Troy-bk.* II 1674.

(*b*) Mene .. Of dedis gud & spyryt schaldand; *Leg. S.* xviii 98.

(*c*) Betuix sa manye scalding sighs and havie layes of woe; FOWLER I 82/147.

2. Scalding hot, scalding.

In the quhilkis [*supra* Grynland .. and Ysland] .. thar is .. scaldand watter that birnis baith stanis and irne; *Asl. MS* I 164/1.

3. Burning.

(*a*) Skaldand; DUNB. (OUP) 150/23 (see (*b*) below). Syrius .. Quhilk with the scaldand heyt at hys rysyng Byrnys the erth of drowth; DOUG. X v 145. Throw thir hait scaldand flambys brycht; *Ib.* XIII vi 226.

(*b*) Mony prowd trumpour with him trippit Throw schaldane [*B.* skaldand] fyr; DUNB. (OUP) 150/23 (M). Schaldand hellis flude, Flagiton, but lycht; DOUG. VI iv 59 (Sm.).

Scalded, var. of SCALDIT *ppl. adj.*

(Scalder,) Skalder, *n.* [SCALD *v.*[2]] = SCALD *n.*[1] — Janet Innes .. a commoun skalder, a blasphemar; 1602 *Elgin Rec.* II 104. Quhasaever salbe tryit to be the first skalder or speikar with any uther sall not onlie be put in the cockstuill bot [etc.]; 1625 *Peebles Gleanings* 62.

(Scalder,) Skowdr-, *v.* [North. Eng. dial. *scaldr-* (1600), *skawder* (1876); SCALD *v.*[1]] *tr.* To scorch; to cook by scorching in a fire. — He was fean till eat of a magre guse, skowdrit with bar stra; MELVILL 134.

(Scaldering,) Scowdring, Scouthering, *vbl. n.* [North. Eng. dial. *scalderings* (1804) 'the under-burnt cores of stone lime' (OED); SCALDER *v.*] **a.** The action of burning or scalding (as a means of cleaning). **b.** The drying up or shrivelling (of the skin, etc.). — **a.** For scouthering of the pan 16 s.; 1647 *Tulliallan Coal Wks.* 46b. — **b.** In sepulchrals .. some .. preferred to wither in the aire, .. preuenting the stinck of putrifaction by scowdring their skins in the sunne; BIRNIE *Kirk-b.* iv.

(Scalderit,) Skolderit, Skowdrit, Scouther'd, *ppl. adj.* [North. Eng. dial. *scaldered* (1796); SCALDER *v.*] Scorched; shrivelled. — (*a*) Fy skolderit skyn thow art bot skyre and skrumple; DUNB. *Flyt.* 122. Thy skolderit skin hewd lyk ane saffrone bag; *Ib.* 171. Lang he la in that infirmitie .. His skowdrit skyn wes blak as ony ruke; STEWART 36458. — (*b*) On shoulder clap made her Mess James embrace And lick the dreepings of his scouther'd face; *c*1690 *Bk. Pasquils* 183.

Scalding, Skalding, *n.* Also: **scadd-, skadd-, scawd-** and **-yng, -yn(e; schawdyn.** [ME *scaldyng* ? the carcass of a 'scald' sheep (1302-3), *skaldyng* (1338), *scaldyng* (once, 1429) a sheepskin of small value, ? one taken from a 'scald' sheep; ? SCALD *adj.*] ? The skin of a 'scald' sheep. Also (once) *coll.*, and *attrib.* with *skin.*

(*a*) Per vendicionem .. pellium, que dicuntur schorlyngis et scaldyngis; 1374 *Exch. R.* II 453. De custuma mille .. pellium que dicuntur skaldinggis; 1435 *Ib.* IV 604, etc. That naman by woll skynnys derar than x d. schorlynges vj d. scaldynes iij d. the dusane; 1442 *Aberd. B. Rec.* MS V II p. 654 (27 April). Pellium .. vulgariter dictarum schorlingis, skaldingis, futefellis, .. lameskynnys, todskynnis [etc.]; 1451 *Edinb. Chart.* 72. Skaldynis; 1455 *Exch. R.* VI 11. 1482 *Edinb. Chart.* 167. That na skynnaris .. by ony manner of skynnis bot schorlingis and scaldingis; 1529 *Conv. Burghs* I 511. A charter of Kyng James the Secound of the schowrlyngs and schawdyns; 1531 *E. Loth. Antiq. Soc.* VI 34. 20 s. custom of 160 woolfells; 6 s. 8 d. of 100 schorlings; 13 s. 4 d. of 200 scawdings and mortwells; 16 s. 8 d. of 1000 lambskins; 1534 *Exch. R.* XVI 360. 1549 *Stirling B. Rec.* I 57. Skynnis vndirwrittin callit in the vulgar toung scorlingis scaldingis futefaillis [etc.]; 1592 *Acts* III 580/2. [Also payments of feus in] truffes and scaldings; 1684 Jervise *Memor. Angus & Mearns* II 300.

(*b*) Skaddyngis; 1603 *Reg. Great S.* 515/1.

coll. To George Lovell for makyng of the expensis vpon the skadding & schorling; 1553 *Dundee B. Ct.* II fol. 246 (28 Aug.).

attrib. That na marchand .. by ony calfe skennis, kid skennis, na skalding skennis; 1527 *Stirling B. Rec.* I 31. Aucht scoir of beir bakis alias scadding skynnis at iij li. the scoir; 1594 *Edinb. Test.* XXVII 298b.

Scalding, Skalding, Scolding, *vbl. n.* Also: **scauld-, scall-; skaild-** and **-in.** [e.m.E. *scoldyng* (1486); SCALD *v.*[2]] The action of scolding in the sense of SCALD *v.*[2]; vituperation; violent abuse.

Chiefly, esp. in later use, of incidents involving women.

(*a*) And maid him be schakin up with wemenis scalding and that be ane of hir awin trane; BUCH. *Detect.* (1727) 42. That the said persouns .. gif na sclander be banning and sweyring, flyting, skalding, filthie speiking or vitious leving; 1591 *Edinb. B. Rec.* V 53. Faltis deferit: Johne Rodman .. to vse braulling, scalding and flyting; 1603 *Dundonald Par. Rec.* 40. Jonet Tod and Jonet Ayngill .. for their misbehaviour in flytting and scalding in the audience of their nybours .. to stand .. in the gougs; 1608 *Kinghorn Kirk S.* 14. Jonet Burne [is] .. the first .. movar of this trublance skaldin and flytting .. geving hir evill langage; 1609 *Dunferm. B. Rec.* II 56. Grisall .. convict in scalding & flytting & misorder; 1625 *Fraserburgh Kirk S.* II 31 Aug. Convict of mutuall skalding and slandering one of ane other; 1641 *Saltoun Kirk S.* in *Reliq. Antiq. Sc.* 154. Margrat Gray .. for scalding, railling and abuisseing of Issobell .. in calling her jad and queyn; 1662 *Banff Ann.* I 145. The said Katherin for hir scalding to remaine in prison during pleaser; 1672 *Kirkcudbr. B. Rec.* MS 11 May.

(*b*) Issobell and Margaret Tailyeour .. to do thair repentance .. for thair publict flytting and frequent skailding on the calsay; 1596 *Elgin Rec.* II 42.

(*c*) For the scalling and outrailling of Margrat Cuitlar; 1635 *Kirkcudbr. B. Rec.* II 511. Elspet Card delat for cursing, swearing and scalling; 1652 Cramond *Ch. Grange* 12.

(*d*) Scaulding; 1665 *S. Ronaldshay* 52.

(*e*) As also Isobel Gray ane professed scold and curser with Margaret Craigheid and Christian Duncan for scolding lykwys; 1655 *Ellon Par.* 117. In reguaird of ther inchristian cariage in flyting and scolding, banning and swearing; 1669 *Corshill Baron Ct.* 86.

Scalding, *ppl. adj.* [e.m.E. *scoldinge* (?1553); SCALD *v.*²] Vituperative; abusive. — That the said Christane callit the said Jonet witche & wther scalding speiches; 1626 *Fraserburgh Kirk S.* II 3 May.

Scalding, var. of SCALDAND *ppl. adj.*

Scaldit, Scallit, *ppl. adj.* Also: **scalded, skaldit; scalled, skallit, -et.** [ME and e.m.E. *scallede* (1340), *scaled* (Chaucer); SCALD *adj.*] Scabbed, scabby, chiefly as a term of abuse, despicable, contemptible. Also *transf.*

(*a*) Skaldit skaitbird and commoun skamelar; KENNEDY *Flyt.* 37. Revin, raggit ruke, and full of rebaldrie, Scarth fra scorpione scaldit [*M.* scauld] in scurrilitie; DUNB. *Flyt.* 58. Scalded; POLWART *Flyt.* 747 (H) (see SCALD *adj.*).

(*b*) Scalled; Ferg. *Prov.* MS No. 66 (see SCALD *adj.*). Archbald Jacksone depones she callit Helein Maxwell skallit mans get; 1655 *Dumfries Kirk S.* 26 April. William Boig his scalled head appointed to be cured; 1676 *Cramond Kirk S.* II 16 July.

transf. Dirtin Dumbar .. Pretendand thé to wryte sic skaldit skrowis; KENNEDY *Flyt.* 26. The Lord .. gave command to the pest that it should not come near to your ports when he scattered it over other congregations .. their skallet boards and skallet trenchers in Glasgow and Edinburgh, they may bear you witness; WELSH *Forty-eight Serm.* 480.

b. *specif.* Of a horse.

Appoints John Mwir .. to tak tryall of all scallit horses in this brughe; 1670 *Rothesay B. Rec.* 194.

(Scaldlips,) Scadlips, *n.* [SCALD *v.*¹ 4 and LIP *n.*] 'Broth containing a very small portion of barley, and on this account more apt to burn the mouth' (Jam.). — And there will be .. sing'd sheep-heads, and a haggize, And scadlips to sup till ye're fow; 16.. *Sempill P.* 69/65.

Scald market. (? Cf. SCALDING *n.*) — Be-west the castle burn quhair the myddings and the scald market stand; 1560 *Old Dundee* I 53.

(Scaldry,) Scaldrie, *n.*¹ [SCALD *v.*²] Vituperation; vulgar abuse. — Personis convict for flyting and scaldrie; 1502-3 (*c*1580) *Edinb. B. Rec.* I 97.

Scaldry, *n.*² Also: **scouldrie.** [SCALD *v.*¹] A scalding-house ('a room in which utensils or the carcasses of animals are scalded' (OED)). — To build the said kitchens baikhouss and pastrie rowmes scouldrie and others necessare office-houss; 1672 Mylne *Master Masons* 179. Kitchens paistrie roomes scaldries; 1676 *Ib.* 192.

Scald-stair, var. of SCALE-STAIR(E *n.* **Scaldyn(e, -yng,** varr. of SCALDING *n.*

(Scale,) Skale, Scail(l, Schale, *n.*¹ Also: **skail(l, schaill; shall.** Pl. also **skailse, skaillies.** [ME and e.m.E. *scale* (Layamon), *skale* (Ancr. R.), *skayll* (*c*1460), ON *skál*, also ME and e.m.E. *schale* (Ancr. R.), *schal* (*c*1325), OE *scealu*, of the same ulterior origin as S(C)HELL *n.*]

1. a. One of a pair of pans, or the like, in a balance or weighing-instrument. Also *fig.* **b.** Chiefly *pl.*: A weighing apparatus consisting of two such pans suspended from a finely balanced cross-beam. Also *a pair of scales* and (once) *a skailse.* = S(C)HELL *n.* 7.

Et pro costagiis factis super tronam, videlicet, skalys, per tempus compoti, xiij s. iiij d.; 1403 *Exch. R.* III 569. Allocate skalis factis pro trona de novo, xiij s.; *Ib.* 571. Pro reparacione trone, videlicet, in cordis novis et ferro ad easdem, et pro tabulis ad unum skale de novo factum; 1417 *Ib.* IV 274.

a. The mesur off the weyis .. The strengis off the balk sal be evyn elik lang with the balk and the skalis elik lang weyand hewy; *Acts* I 387/2.

fig. Quhene we wald in skale put don His ewil consawit suspicione .. & in-[to] the tothyre skale His gud dedis ware al hale; *Leg. S.* xxii 739.

b. (1) *sing.* Ane skaill for veyinge l s.; 1625 *Brechin Test.* IV 184.

pl. The thesaurer .. to tak and keip the braysin schales and irne wark to the behuif of the guid toun; 1583 *Bk. Old Edinb. C.* XV 14. The weights and skails; 1682 Cochran-Patrick *Coinage* II 185. John Aird caused make and provyde all weights and skaills necessarie for weighing all sorts of meall; 1696 *Glasgow B. Rec.* IV 224. For reparation of balks, skaills and weights in the meall mercatt; 1697 *Ib.* 239. Skaills and weights for weighing monie; 1701 *Foulis Acc. Bk.* 300.

(2) Ane balk & ane pair of skaillis price vj li.; 1609 *Edinb. Test.* XLV 347b. Small weihtis with ane pair of skaillies; 1640 *Ruthven Corr.* 68. Ane pair of skaills for weighting of money; 1699 *Edinb. Test.* LXXX 356.

(3) Ane skailse; 1658 *Edinb. Test.* LXIX 173.

c. ? As a representation of Libra, the sign of the Zodiac.

A half of the laich of a coit .. figurit with scaillis; 1561 *Inv. Q. Mary* 45.

2. A shallow bowl or dish; ? some sort of cup; ? the bowl of a chalice. Cf. SCOL(L *n.* and SCUL(L *n.* 2.

(1) For lxiij treyne platis, alsmony treyne stopis, samony cogis, xxx skalis, tua tubbis; 1512 *Treas. Acc.* IV 186. For ij^c scailis vj ladillis; 1513 *Ib.* 489. To hir [*sc.* the ship the 'Mary'] of scailis and platis, xxiiij pece; *Ib.* 501. *Ib.* 492, 495, 498, 500. Thretene tue pleatis .. ellevin skaillis pryce of the pice i s.; 1625 *Orkney & Shetl. Test.* I 138.

(*b*) 1 silver tasa or shall or cassa weighing 2M. 40 lod. f109⅓; 1680 *Scots in Poland* 126.

(2) To mak ane chaleis of xl vnce that is to say the futt and schaill contenene xxxviij and the patone of the same vj vnce; 1527 *Aberd. B. Rec.* in *Cart. S. Nich. Aberd.* II 359.

Scal(e, *n.*² Also: **skale, scaill, skail(e, skaill, skal; skell; schale.** [ME and e.m.E. *skale* (Chaucer), *scale* (14th c.), OF *escale* (1190 in Larousse), *escaille* (1260 in Godefroy), ultim. f. Gmc. **skalja.*]

Chiefly *pl.*

1. A scale on the skin of a fish or reptile, etc. Only *pl.*, the scaly surface of a fish or reptile's skin. Cf. S(C)HELL *n.* 12.

(*a*) The fische with thair scalis; *Colk. Sow* I 129. The serpent .. Hyr sprutlyt skalys vpset grysly to se; DOUG. XI xiv 69. Sum of thaim [*sc.* fish] armit with schellis; sum with hard skalis; BELL. *Boece* I xlv. I beheld the pretty fische .. vitht .. there skalis lyik the brycht siluyr; *Compl.* 37/18. Ane kind of fish the quantitie and shape of ane salmond without skaills; MONRO *W. Isles* (1961) 77.

(*b*) Tha scayland schalis [*sc.* of a dragon] set vp rath; *Leg. S.* xxviii 413.

b. *Scale-fish*, fish having scales [e.m.E. (1601)].

Scale [*pr.* seale] fish; 1612 *Bk. Rates* (Halyb.) 304. Salmonis piscatione de tugnet, tam lie reidfish quam skailfish intra fluxum maris; 1696 *Retours* I Inq. Spec. Elgin et Forres (178).

2. After *Acts* ix 18: Blindness represented as a covering of scales over the eyes.

And with that word fel fra his ene Mirknes as tha skalis had bene And he that sicht thane gat; *Leg. S.* ii 610. The skailes of darknesse which our eyes be-night He doth remove and maks vs clearly see; MURE *True Crucifixe* 972.

3. A thin plate of wood or metal used in making or ? ornamenting scabbards or the hilts of swords or knives. See also SCALIT *ppl. adj.* b.

For skaillis of scalbertis; 1573-4 *Elgin Rec.* I 144. The gros wairis [coming from France] .. sic as vynigar [etc.], .. schuellis, and suord skaillis; 1598 *Conv. Burghs* II 39. Ane 100 suord skaillis; 1605 *Edinb. Test.* in Whitelaw *Sc. Arms Makers* 296.

4. A particle of scaly or scurfy skin, scurf, dandruff.

Cf. SCAL(L *n.*, with which there may have been some confusion.

(*a*) *Furfures*, skails; CARMICHAEL *Etym.* 26. *Furfures*, pl., scaills of the heade; DUNCAN *App. Etym.*

(*b*) Ane Indian kome to taik skellis owt of Oncies heid; 1614 *Laing MSS* 139.

(*c*) *De Morbis. Porrigo*, scales in the head; WEDDERBURN *Voc.* (1709) 13.

b. Applied to the lichen *Lecanora tartarea*, so-called because of its scurfy appearance.

The muscus crustæ aut lichenis modo saxis adnascens stavus, which they call a skal, from the resemblance it hath to that; *Sibbald's Orkn. & Shetl.* 25.

Scale, *n.*[3] Also: **scalle, scaill, skaill.** [ME and e.m.E. *skale, scale* (both Lydgate), L. *scāla*, Ital. *scala.*] **a.** *fig.* A ladder. **b.** *specif.* A scaling-ladder. **c.** A flight of steps; a stairway. See also SCALE-STAIR(E *n.*

a. Sche is the scaill & leddir betuix the hevin and erd; IRLAND *Mir.* I 151/27. Mounting from degree to degree vpon the slipperie and vncertain scale of curiositie, they are at last entised .. to seeke to .. magie; JAMES VI *Dæmonol.* (STS) 7/23.

b. Preparatioun of scailles [*MS* scalles] and ledderis was maid for the assault .. The scailles war appointed to be maid in Sanct Gelis churche; KNOX I 452.

c. With ane convenient skaill to go to the dowcat and foirwall with ane yet throuche the samin of hewin wark to go to the yard; 1589-90 Mylne *Master Masons* 68. First ascending from the transe of the old gate by an strait scale of stone to the lower galery and from thence to the higher be an timber scale; CRAUFURD *Edinb. Univ.* fol. 71. The common loft .. with entrie by the scale and stair .. thereto; 1686 *Ellon Par. Rec.* 145.

(Scale,) Skaile, *n.*[4] [Of doubtful origin. OED takes as SCALE *n.*[1], in specif. use; but may be f. as SCALE *n.*[3] (e.m.E. *scale* = a graduated series (1605)).] ? A kind, or genus, of fish. — Heir of the seas (which dyuers skaile Of fish contenis) dois witnes beare; JAMES VI *Poems* I 87/68.

(Scale,) Scaill, Skail, Skayle, *v.* [ME and e.m.E. *scale(n* (Wyclif), *skayle* (north., *a*1400), *skale* (Lydgate); SCALE *n.*[3] Cf. MF *escaller* (15th c.) (OED).] *tr.* To climb (walls) by means of scaling-ladders; to attack, or enter (a city) by force, using scaling-ladders. Also *fig.* — Thay assailyeit the castell on the east syd whilk was easilie win; bot thais that scaillit on the wast syde thay war repulsit; *Hist. Jas. VI* (1825) 143. To see a bresh defended and assailde, A fort most bravelie forc'd, a city skailde; 1628 *Fugitive Poetry* II vii 4/34. — *fig.* Your beautye .. skaild the walls of my vndantond harte; FOWLER I 146/10. The rampiers of thy rigour for to skayle; *Ib.* 172/14.

Scale, var. of SCAIL *v.* **Scaleb,** var. of SCALOP *n.*

(Scaleful,) Skaillfull, *n.* [SCALE *n.*[1]] A bowl, cup or scoopful. — John Fogo .. being heavilie diseasit ȝe came to him and persuaded to cuire him .. And for affectuating the same he took ane skaillfull of water and set in a seif; 1633 *Orkney Bp. Ct.* 86b. *Ib.* 89b.

Scalen, var. of SCAILING *vbl. n.*

Scalerig, *n.* (?) — Subtill merchandes .., vnprofitable importers Of needlesse thinges, which men buy head-longes, rash, As scalerigs, wyne, tobacko and such trash; CRAIG IV 31.

Scale-stair(e, *n.* Also: **scaill-, scaile-, skaill-; scald-.** [SCALE *n.*[3] and *stair.*] ? A straight (as opposed to a spiral) flight of steps.

The south entrie thairof to be ane skaill stair of tymer cloislie cled beneth & on the sydes with aik; 1603 *Haddington B. Rec.* (Robb) 8 June. For the long window in the scald stair; 1633 *M. Works Acc.* (ed.) II 380. Ane skaill stair to the hich galrie abowe the great hall; 1643 *Thanes of Cawdor* 285. Twa housses ilk ane of them twa hous hight togidder with ane skaill stair betwix the saids twa housses; 1643 *Edinb. B. Rec.* VIII 28. [To erect] ane fyne handsome scale staire twentie-three foott squaire, all of aisler fyne hewen stone, to serve the secound storie, fynelie wrought with fyne railles and ballasters with handsome pilasters globes and pendents, each steppe being of one peece and lopped knitt and hung in the other, according to the draught and modell thereof done and wrought in timber; 1672 *Soc. Ant.* XIV 329. The scaill stair .. upon the east cheick of the enterie to Grayes Closs; 1674 *Edinb. B. Rec.* X 201. Wanscott to be mullerings for the new scaill stair; 1675 *M. Works Acc.* MS XXX 22. Wainscott ballasters, standing and hanging knups for the timber scaile stair; 1676 Mylne *Master Masons* 199. To make for his essy ane house .. with ane large scaill stair for ane entrie, with ane turnpyke in the back syd; 1686 Murray Lyon *Hist. Lodge Edinb.* 20. That .. entry to his said upper loft be by a scale staire .. stretching northward and landeing by ane platte above the lower doore; 1701 *Cramond Kirk S.* IV 23 July.

Scalfatting, var. of SCAFFALTING *vbl. n.* **Scalffauld, -et, Scalfold,** varr. of SCAFFALD *n.* **Scalie,** var. of SCAILȜE *n.*

Scalier, *n.* [F. *escalier.*] A staircase. — A private passage which led to a lanterne scalier; 1652 URQUHART *Tracts* (1774) IV 80. A wonderful scalier or winding staire; Id. *Rabelais* I liii.

(Scaling,) Skaling, *vbl. n.* Also: **scailling.** [e.m.E. *scaling* (*a*1513); SCALE *v.*] The action of scaling, in the sense of SCALE *v.* — Whair he had his preparatioune of ledderis for skaling the casle wall; 1596 *Warrender P.* (SHS) II 295. Thre quarters of the great towre .. fell doun .. and a great quantitie of the bak wall that lyis to the south and south west for scailling with ledders for thare easie entrie; *Hist. Jas. VI* (1825) 143.

Scaling, var. of SCAILING *vbl. n.*

Scalit, -yt, Skalit, *ppl. adj.* [SCAL(E *n.*[2]] **a.** Having scales. **b.** Plated (see SCAL(E *n.*[2] 3). — The syluer scalyt [*Ruddim.* scalit] fyschis; DOUG. XII Prol. 55. — **b.** Twa dosane of reid skalit knyfis price of thame xxiiij s.; 1575 *Edinb. Test.* III 387. Blak skalit knyfis price of the dosane x s.; 1578 *Ib.* VI 361b.

Scal(l, Sca(w, *n.* Also: **skall, skau, skaw.** [ME and e.m.E. *scall* (Cursor M.), *scalle* (*c*1374), prob. f. ON *skalle* a (naturally) bald head.] A scaly or scabby disease of the skin, esp. of the scalp, the scaly or scabby surface of the skin so caused.

Some examples may belong in SCAL(E *n.*[2] 4 with which there may be some confusion.

Thar-to sic a sca had he That of his body nocht wes fre & thar-with had he hard torment; *Leg. S.* xxxvi 1067. The lypir and the faland ill, Wild-fyre and scaw thai rakin till; *Ratis R.* 182. This oulie hes ane singulare virtew aganis all maner of cankir and skawis; BELL. *Boece* I xxxviii. The skall [*v.r.* skau] vpon their head doth of Their saltest fleumes proceed; JAMES VI *Poems* I 164/973. The combe is not commed to the hard sca yet; CARMICHAEL *Prov.* No. 1431. Their [*sc.* the inhabitants'] baldness proceeds from scals [*Descr. Zetland* scalls], for all of them being full of scals [*Descr. Zetland* scalls] while they are young, doeth necessarly occasion their loss of hair: Now their scalyness may be judged to proceed either from their nestiness, or .. from the dropping of the fishes, which hang above their heads [etc.]; *Sibbald's Orkn. & Shetl.* 53. William Boig .. haveing his head overgrowen with a scall came and entreated that the session would cause cure him; 1676 *Cramond Kirk S.* II 16 July.

Scall, var. of SCAIL *v.* and SCALD *n.*[1] and *v.*[1]

Scallag, -aig, Scolog, -ag, *n.* [Gael. *sgalag*, earlier *scoloc*, a landless farm-labourer, Ir. *scológ*, f. *scol* school, orig., the eldest son of a monastic tenant, handed over to the church to receive an ecclesiastical education, but still entitled to inherit his share of the family land; later, applied to all monastic tenants, and developing finally into farm-servant, generally. Also in the later (Highland) Scots dial.] A land-labourer or farm-servant in the Highlands. Also *attrib.* with *landis.*

[Vidit multitudinem scoloccorum in terra; 1206 *Misc. Spald. C.* V 210. Nativos et scoloccos a terra illa amovisse et istum terram coluisse; *Ib.* 212. Duncanus .. ex toto scoloccos amovit et terram illam preter episcopum coluit primus; *Ib.* 213. Dicta terra de Elon per rectas diuisas suas sicut Scoloci eam nunc tenent; 1265 *Coll. Aberd. & B.* 312.]

The very place named Bearn ni Scallag, that is, the Servants Gap, becaus the men who did the slaughter were servants and scallags; FRASER *Polichron.* 84. He and M'Kenzie mett quhair they fought hand to hand till M'Kenzie's scallaig more .. went betwixt M'Kenzie and him; 1667 *Highland P.* II 23.

attrib. Terre ecclesiastice de Elon que dicuntur le Scologlandis; 1387 *Coll. Aberd. & B.* 310. [Requiring the presence of more clerics than the Vicar of Ellon, and the] clerks [supplied by the] scolag lands; *a*1560 *Buchan Cl.* IV 102.

Scalle, var. of SCALE *n.*[3] **Scalled,** var. of SCALDIT *ppl. adj.*

Scallet, *n.* (e.m.E. *scalade* n. an assault by scaling-ladder (1591), also mod. Eng. (1824) a scaling-ladder, Ital. *scalada*, F. *escalade*. Only in LITHGOW, cf. next. See also SCALE *n.*[3]) — The ditch .. is mainly pallasaded with wooden stakes, for preuenting of suddain scallets; LITHGOW *Trav.* VIII 349. Their armes, a crosse, .. limbd like a scallet, trac'd with fleur du Luce; *Ib.* X 502.

Scalleting, *vbl. n.* [18th c. Eng. *scalade* (1729) to assault by scaling-ladder; SCALLET *n.*] The action of assaulting by scaling-ladder. — In case of scalleting they might have tumbled them [*sc.* little stones] over [the battlements] upon the assailants; 1645 LITHGOW *Siege Newcastle* 12.

Scallie broad, Skellie board, *n.* [? f. as SCALE *n.*[3] and BROD *n.*[2] Cf. 17th c. Eng. *scale* (1607) in this sense.] Appar. a measuring instrument prob. of printed paper pasted to thin board. — [In the testament of a skipper] Item ane litill paper broad pryce therof xii s. .. Item ane skellie broad pryce thairof xii s.; 1648 *Edinb. Test.* LXIII 315b. Accompt of the mathematicall instruments belonging to the Liberarie of Aberdein .. Ane planyshære vpon timber. Ane instrument called the theodolite. Ane litle timber quadrant. Ane other litle compas. Ane scallie broad. Ane great timber sphære, etc.; 16.. *Marischal Coll. Rec.* 222.

Scalling, var. of SCALDING *vbl. n.* **Scallit,** var. of SCALDIT *ppl. adj.* **Scallope,** var. of S(C)HALLOP *n.*

(Scalop,) Scaleb, Skalap, *n.* [North. ME *skalop* (?*a*1400), *scalap* (1401), ME *scalop* (v.r. *-ap*, Prompt. Parv.), aphetic f. OF *eschalope, escalope* (*c*1220 and mid. 13th c. in Larousse).] A scallop, the shell-fish. Also *attrib.* with S(C)HEL(L *n.* — Tres conche sculpte que wlgaliter dicuntur scalebs; 1382 *Makgill Charters* (Sc. Rec. Off.) No. 6 (1). — *attrib.* For all hir banis and hir skalap schellis .. Thay ar sa pretius into merchandise Thare is na man can sett thame to richt price; HAY *Alex.* 16144.

Scalour, *n.* [OF *squal(l)eur* (Godefroy), L. *squālor*, f. *squalēre* to be dry, rough, dirty, etc. Cf. e.m.E. *squalor* (1621).] Dirty or repulsive conditions; squalour. — All thair saulis salbe transformit Sum in bichis and sum in beiris .. Aganis the scalour [*M.* statour] that thay wer in For vengence of thair deidly sin; ROWLL *Cursing* 137 (B).

Scalp, Scaupe, *n.* Also: **skalp, scalpe, scawpe, skaup, skawp, scap(e, skap(e, skaipe; scaap; sclap, (sclop).** [North. ME and e.m.E. *scalp* (*a*1300), also *scawpe* (1541), ? of Scand. origin. Cf. ON *skálp-r* a sheath, Dan. dial. *skalp* shell, husk.]

1. The top, or crown, of the head.

With brokin skap [*Bann.* scalp] and bludie cheikis reid, This volff weipand on his wayis went; HENR. *Fab.* 1034. To hede of kyn, bot I wait nought Quis est ille, than I schrew my scawpe [*B.* skape, *M.* scaupe]; DUNB. (OUP) 116/52. Howbeid I se thy skalp skyr skawd; LYND. *Sat.* 2485 (B). Iis brek thy skaipe lytill foull aipe; MAXWALL *Commonpl. Bk.* fol. 20a. He saw him cast hir on the scalpe and saw na bluid; 1608 *Melrose Reg. Rec.* I 62. Saying that he should gair the scape of William McClellane; 1637 *Kirkcudbr. B. Rec.* II 550.

b. ? The crown (of the road).

Vnles he duell within the liberteis of this burgh & scalp of the calsay; 1578 (1659) *Dumfries Fleshers* 30 Sept.

2. ? A bare piece of rock standing out from surrounding vegetation; ? a small bare knoll.

Na gers nor herbys wer visibyll Bot skauppis brynt with blastis boryall; DOUG. *Pal. Hon.* 144 (L).

b. *attrib.* Bare, stony, rocky.

The saidis landis is a dry hard skalp ground; 1536-7 *Sc. Hist. Rev.* VII 360.

3. A bank on which has formed a colony of shell-fish, esp. mussels or oysters; an oyster or mussel-bed or colony.

For further instances see MUSSIL(L)-SCALP *n.* and OYSTER *n.*

(*a*) Mussil skawpis; 1496 *Forres Chart.* MS (see MUSSIL(L)-SCALP *n.*). That .. nan .. pull, gader nor away tak nane of the mussillis nor kokillis now begingin to gader one ane now skap .. for yeir and day, that the forsaid skap be perfectly growing; 1521 *Aberd. B. Rec.* I 98. The preserruation of the mussill scaap. .. The said James .. had revin wp the stringis of the mussillis & broken the scaap thairoff; 1589 *St. A. B. Ct.* 16 Dec. The linkis of Sanctandrois .. with the mussill scap; 1612 *Acts* IV 516/1. Infeft and seasit in .. skalpes within the sea mark wrak and weare; *c*1660 *Old Ross-shire* I 229. Dutch vesselis have repaired hither in so great abuondance that the furneissing them hath mutch brokin the verie scapes; 1664 *Edinb. B. Rec.* IX 340. Skippers [of two boats] .. inactit .. to pay .. for ther bygone transgressione done to the scakkes [? *erron. for* scappes]; 1666 *Old Ross-shire* II 83. Mussill scalpes; 1672 *Retours* I Inq. Spec. Elgin et Forres (128). In regard of the scarcity .. of oysters and their scalps; 1687 LAUDER *Notices Affairs* II 829. On Monday there cam three Morray boats to the tounes mussall sclaps; 1711 *Old Ross-shire* II 86.

(*b*) Nostrum burgum .. cum omnibus sclopis infra eundem et pertinentibus ad eundem; 1536 *Kirkwall Chart.* 8.

Scalve, var. of SKELF *n.*[1]

(Scaly,) Skailly, -ie, Skealie, *adj.* [e.m.E. *skaly* (1528); SCAL(E *n.*[2]] Covered with scales. — The skaillie [*1591* skealie] suimming race Quho uith thaire litle finnis Do cleue the frothie seas apace; JAMES VI *Poems* I 109/33. The skailly squadrones of the liquid lakes; MURE *Dido & Æneas* III 127.

Scalyie, var. of SCAILȜE *n.*

Scalyness, *n.* [17th c. Eng. *scalinesse* (Cotgrave); SCALY *adj.*] — *Sibbald's Orkn. & Shetl.* 53 (see SCAL(L *n.*).

Scalynge, var. of SCAILING *vbl. n.* **Scalyt,** var. of SCAILIT *ppl. adj.* and SCALIT *ppl. adj.* **Scalȝe,** var. of SCAILȜE *n.*

Scam, Skalm, *v.* [Later Sc. dial. *scam* to scorch, 19th c. north. Eng. dial. *scam* to stain, blotch, discolour (also *scaumy* blotched), Norw. *skjemma*, Sw. *skämma* to spoil, mar.] *tr.* **a.** To spoil, damage. **b.** To scorch. — **a.** Reffusing the said clayth .. becaus the same is skalmit be salt watter be the lek of the ships; 1628 *Stirling Merch. Guild* 47. — **b.** Ane fyrie cross of tymber, quhairof everie point of the cross wes scamit and brynt with fyre; SPALDING II 386.

Scamble, var. of SCAMEL *v.*

Scamel, Skemell, Schamill, *n.* Also: **scamle, skamel, -ill, -yll, -le, skammell; skemle, -il(l, skemmel(l, -il(l, -yll; skaymle; scambill, -ble, skambill, -ble; skembell, -ble; schamyll; schemill, shemle, shemmell; shamble, schambil; shemble, schembill, scheambill.** [ME and e.m.E. *schamel* (*a*1289), *-il* (*c*1305), *shamyll* (Rolle), *sheamble* (1477), etc., OE *sc(e)amel*, OS *skamel*, MLG *schemel*, MDu. *schamel, schemel*, ON *skemill*, L. *scamellum*, dimin. of *scamnum* bench.]

1. A bench.

Thai xxx^ty dayis his band thai durst nocht slaik, Quhill he was bundyn on a skamyll off ayk, With irn chenȝeis; *Wall.* XI 1352.

2. *specif.* A bench or table on which meat or fish was displayed and cut into portions for sale. Passing into 3 below.

Once, *pl.* with *sing.* concord.

(1) At thai sell flesch and coueris it and nocht opin in the window or the scamel; *Acts* I 333/2. Thai discharge the skamlys, and at na nytbor set thair stulys to mak merchandice, bot under thair awn stair; 1538 *Haddington B. Rec.* (Robb) 21 Sept. Accusit for the bying of slayne motone vpon the skamillis of this brucht fra vtheris flessouris and selling the samyn incontinent to the nechtboris; 1540-1 *Elgin Rec.* I 58. All fleschouris and the takismen of the skemlis tobe fre to brek fischis apon the commoun skemlis; 1549-50 *Stirling B. Rec.* I 58. Ane burd to cum no farther furth nor the owir stop of the said buith dur to lay his geir on daylie becaus the want of the said windok vnder the pane of birnyng of the skemmillis and burdis quhairon the burd salhappin to stand; 1551 *Edinb. B. Rec.* II 160. Quha belewit that Our Saluiour wald hef giffin his flesche to hef beine couponit and diuidit as flesche deuidit one the skamles; Q. KENNEDY *Breif Tract.* (ed.) 118/18. The fische skemmyllis sett to Robert Scott for vij merkis; 1562 *Stirling Chart.* 211. It is statut that no fische be bocht to be brokin on the skemlis be the flescheris; 1563 *Aberd. B. Rec.* I 356. The flescheris skamlis [*Maitl. Q. 226/19* skambillis] are gane dry The heiland men bringis in na ky; *Maitl. F.* 438/19. Fesche skammellis set to James Wolsone for iiij li. xl d.; 1595 *Stirling Chart.* 219. That the haill fleshe skemmellis be put behind the tolbuithe quhairthrow the flesche mercat may be thair altogidder; 1601 *Kirkcudbr. B. Rec.* I 401. And the flesh skambles at the last; 1649 *Old Dundee* I 421. The councill .. did [not] .. permitt any flesh stocks or skemells to be affixed on the grund of any pairt of the hie street; 1671 *Edinb. B. Rec.* X 103.

(*b*) That na schamylls be sett one the gait onder the pane of escheiting of thair paks; 1530 *Haddington B. Rec.* (Robb) 11 Oct.

(2) For makking of ane skaymlis of tre at the fysche cors, for laying of the quhyt fysche tharupoune; 1561 *Aberd. B. Rec.* I 334.

3. *coll. pl.* The meat or fish market of a town, within which animals or fish might be slaughtered and prepared for sale.

Like the MERCAT-CORS of a burgh, a place where public proclamations might be made, and public punishments inflicted.

(*a*) 1549-50 *Stirling B. Rec.* I 58 (see 2 (1) above). Skemmils; 1551 *Old Dundee* II 299. [The Council] dividit the burgh into four quarters .. the quarter frae the skemmels to Silvester Ramsay's in Argylesgait; 1553 *Ib.* 191. That na owtlandis flescher .. slay ony flesche at the skemils .. except on the tua mercat dayis; 1582-3 *Elgin Rec.* I 170. Ane proclamatioun to be maid at the mercat place callit the skemles be ane townis officier; 1591 *Ib.* II 18. Thatt quhatsumeuir persoun beis .. found to haif spoikin filthie and abhominable langage .. salbe .. convoyit .. to the tolbuyt .. and fra thence to the skameles and thair be phippit; 1592 *Ib.* 24. Thay marchit .. to Edinburgh, and plantit a gardehous at the common scambles thair; *Hist. Jas. VI* (1825) 119. Vpon the hie street .. betwixt the flesh skembells and the old toolbuith; 1597 (*c*1650) *Dundee B. Laws* 50.

(*b*) That the shemmells be theikit without delay; 1627 *Kirkcaldy B. Rec.* 159. Tack of the scheambills to James Winchester for ane yeir .. the raitts due to him to be .. each ox 2 s. 8 d. [etc.]; 1680 *Elgin Rec.* I 324. Roup of the schembillis; 1687 *Ib.* 343. Accompt of the comon good of .. Montrose. .. The flesh shemles; 1692 *Conv. Burghs* IV 596. The selling of all sort of unsufficient meat in the shambles or shops of any of the freemen; 1698 *Dumfries Fleshers.*

4. *Comb.* with *mail* MALE *n.*[1] and *stok.*

(*a*) The tollis and skemmill maill; 1548 *Elgin Rec.* I 94. For the half towis and skemill mailles of Witsundaye and Mertimes; 1553 Donaldson *Cramondiana* 53. Thay sa vnderstude .. the flesch as rent in pecis in a deid carion, or sauld on skemble stoks [L. *in macello*]; HAMILTON *Cath. Tr.* 47. Lyk flesche bocht at the scamble stokis; Id. *Facile Tr.* 376.

(*b*) The prouest etc. hes set the comond tollis with the schamill maill to Dauid Tailȝour; 1542 *Elgin Rec.* I 73. The haill tovn hes sett the haill tollis and schemill maill for thre ȝeiris; 1545 *Ib.* 86.

5. *fig.* **a.** A place of death, carnage or execution.

Bot caryit out as a corse, or rather drawin, as it wer, to the schambilis; BUCH. *Detect.* (1727) 65. [In March 1649] they set up their shembles in Edinburgh; FRASER *Polichron.* 338.

b. In such phr. as *to sell, lede, convoy* or *deliver* (a person) *to the scamels,* to betray to murder, to deliver to execution.

Sum of thame ar counsalouris of the kingis slauchter, sum convoyaris of him to the scamles; BUCH. *Wr.* 23. We sall avenge it on that clan, Ȝour freind that to the scambills sauld; 1572 *Sat. P.* XXXV 94. Syne sauld him to the skambillis lyik ane scheip; *Ib.* XXXVI 40. Till they be led to the skemles as wes the poor Erll of Mortoun; 1584 COLVILLE *Lett.* 49. He recompansit that loss with the selling of the noble Erle of Northumberland to the scamblis; *Hist. Jas. VI* (1825) 23. He [*sc.* Lethington] died in Leith before that the rest wer delyuerit to the skamles; MELVILLE *Mem.* 256.

(Scamel,) Skemmill, Scamble, Skambell, *v.* [e.m.E. *scamble* (1539); SCAMEL *n.*] *intr.* To sponge or scrounge. — Be his hois be pointtit vp and schone on his feit He gois to skemmill vp and doun To drynk he is evir meit; *Bann. MS* 158b/74. Povertie maks such sort of people turne cunny-catchers, .. skambell, beg and borrow of their betters; GRAHAME *Anat. Hum.* 8b. They goe scambling like butchers dogs in Lentron; *Ib.* 19.

(Scameone,) Skameone, *n.* [Aphetic, f. EXCAMBION *n.*] Exchange of lands or property. = EXCAMBION *n.* 1, EXCAMBIUM *n.* — Gewin .. to the said religios men of Newbottill in skameone of that pece land .. gewin to hym .. ane pec land [etc.]; 15.. *Newbattle Chart.* MS (Reg. H.).

Scamle, var. of SCAMEL *n.*

Scamler, Skamelar, *n.* Also: **skamlar; skemlar, -layr, -ler; scammyllair.** [SCAMEL *v.* Cf. Gael. *sgimilear.*] A scrounger, sponger or parasite; a sycophant, 'hanger-on'. — Skaldit skaitbird, and commoun skamelar [*M.* scammyllair]; KENNEDY *Flyt.* 37. On ȝour hienes follows eik .. Scaffaris, and scamleris; DUNB. (OUP) 130/45. The maist parte of the knichtis and men of armys .. war passand like skamlaris [L. *lixæ*] throw the cuntre; BELL. *Livy* II 153/15. Callit him commoun skemlar; 1563 *Inverness Rec.* I 107. Confessit he callit the baillie skemlayr; *Ib.* *Assecla*, a skemler, a page, a lackay; DUNCAN *App. Etym.*

(Scamling,) Skemlin, -land, *ppl. adj.* [e.m.E. *scambling* rapacious (Shakesp.); SCAMEL *v.*] Sponging, scrounging; in search of easy pickings. — In calling of hir ane skemlin witche karling, 1591 *Elgin Rec.* II 15. For calling Bessie Harper a .. skemland stable raker; 1654 *Dumfr. & Galloway Soc.* III 25.

Scammony, -ie, Scamonia, *n.* [ME and e.m.E. *scamoiene* (*c*1205), *scamoine* (*c*1275), *skamonye* (1436), *scamely* (*c*1475), L. *scammōnia, scammōnium,* f. the Gk.] Scammony, a gum-resin, 'used in medicine as a strong purgative' (OED). — Scammonie; 1597 *Bk. Rates* 10. Half ane pund of scamonia at xvj s. the vnce wecht; 1598 *Edinb. Test.* XXXI 350b. Drogues of dangerous quality as antimony, scammony, arsenic; 1633 *Anal. Scot.* II 122.

Scammyllair, var. of SCAMLER *n.* **Scance,** var. of SCANSE *v.*

Scancellar, var. of CHANCEL(L)AR *n.* — Andro Lorde Auandale, Scancellar of Scotlande; 1472 *Lennox Mun.* 92.

Scand, *v.* [e.m.E. *scand* (1495), var. of *scan*, L. *scandere*, F. *scander.*] *tr.* To scan, examine closely. — So doeth thy workes vnto this day declair for lyflie art who list thy vers to scand [: land]; *Maitl. Q.* 257/8.

Scandale, var. of SCANDAL(L *n.* **Scandaleis,** var. of SCANDALIZE *v.* **Scandaleizeing,** var. of SCANDALIZING *vbl. n.* **Scandaleous,** var. of SCANDAL(O)US *adj.*

Scandaler, *n.* [SCANDALL *v.*] A slanderer. — Either to give up witnesses .., or els to be discerned scandaler; 1649 *St. A. Presb.* 142.

Scandalese, -is(e, varr. of SCANDALIZ(E *v.* **Scandaleux,** var. of SCANDAL(O)US *adj.*

Scandalizar, var. of SCANDALIZER *n.*

Scandaliz(e, -is(e, -ese, -eis, *v.* Also: **scandil-, skandilize, scandelise.** [e.m.E. *scandalyze* (Caxton), OF *escandalizier, scandalizer* (1190 and 13th c. in Larousse), eccles. L. *scandalizāre.*] *tr.*

1. To discredit, bring shame upon.

This moste tryumphant court of Rome .. Quhilk hes bene heirschyp of all Christindome .. Off treuth, the hoill Christin religioun Throuch thame ar scandalizat and offendit; LYND. *Mon.* 4954.

2. *passive. To be scandalesit*, to be shocked or horrified.

Sum of Christis auin disciples left him, being scandalesit at his vordis; HAMILTON *Cath. Tr.* 59b. [The Lords were] much scandelised with itt [*sc.* that paper divulged by those admitted to treat with him in camp]; ?1639 *Rep. Maxwell-Stuart Mun.* in *9th Rep. Hist. MSS* App. II 260/2.

3. To attack verbally; to insult, abuse, speak ill of; to accuse falsely; to slander.

Thomas Mosie .. has unjustlie scandalized the magistrates and counsell, and will not give obedience to their auctorite; 1652 *Peebles B. Rec.* II 2. Walter Stewart .. [complained] that the said Laird Lamont did scandaliz his name odiously in sayeing that [etc.]; 1659 *Rothesay Par. Rec.* 21. [John Makie had] scandalized [the deacon by declaring that he had no right to give] any condiscendence to the toune of Edinburgh's impositioune upon 8 d. of the pynt of the malt; 1659 *Bk. Old Edinb. C.* XX 95. Kathrein Hernsoune his spous scandilizit the said Jon Walker deacone in calling him ane dam boger [etc.]; 1664 *Glasgow Bonnetmakers* 31. Fyv witnes .. did declare that the said Patrick Mare and John Mare did offer no violenc nor did skandilize his wyff so far as they did heir or sie; 1665 *Forbes Baron Ct.* 257. That no persoune presume to scandaleis or reproche his magistrats; 1666 *Rothesay B. Rec.* 124. Denys that he had scandalisd James M'Fie .. in saying that he had a leman amongst the faries; 1670 *Kingarth Par. Rec.* 58.

Scandalizer, -ar, *n.* [17th c. Eng. *scandalizer* (1680), f. prec.] One who scandalises, in sense 3 of prec. — A damnable scandalizer of the Church; LITHGOW *Trav.* IX 409. This he confest he had said becaus he heard it and therfor he was repute to be the scandalizar sinc [*sic*] he made it publick; 1670 *Kingarth Par. Rec.* 58.

Scandalizing, *vbl. n.* Also: **-eizeing, scandelizing; skandilisione.** [e.m.E. *scandalizing* (1594); SCANDALIZE *v.*] The action of scandalising, in sense 3 of SCANDALIZE *v.* — The compleint .. against Thomas M'Vrarthie for scandalizing the said Margaret calling hir a witche and a witches gait .. and uther ominous vords; 1658 *Rothesay B. Rec.* I 18. He hes bine fund giltie .. of scandelizing and reproching of his deacones name and utheris; 1672 *Glasgow Weavers* 71. [To answer for] skandilisione and abusing [of George Room]; ?1679 *Kirkcudbr. Sheriff Ct. Processes* No. 301. Persons .. fined be the .. session. .. Katherine .. for scandelizing, 5 lib.; 1680 *Old Ross-shire* I 41. For abuseing and scandaleizeing Margaret Smyths name, in calling her a theif; 1683 *Peebles B. Rec.* II 111.

Scandal(l, *n.* Also: **-dale, -del(l, -dill, -dle, skandall, -dell.** [ME *scandle, scha(u)ndle* (Ancr. R.), e.m.E. *scandale* (1581), *scandal* (1582), *scandall* (1590), OF *scandale, escand(e)le* (12th and 13th c. in Wartburg), MF *escandale*, F. *scandalle* (1360 and 16th c. in Larousse), eccles. L. *scandalum* (c1180 in Latham), f. the Gk.]

1. a. A cause or occasion of weakening religious belief, arising from the actions or omissions of believers. **b.** Disrepute brought upon religion by such actions or omissions.

a. Swa as we sall na mair be an skandall to uther nations, as we have hitherto bene for .. causing the Word .. to be evill spokin of; *Second Bk. Disc.* in Calderwood III 555. Gestures, rites, holy dayes, .. liturgies .. Mak glad the hearts of Romish brood, And thairby weak ones scandells find; 1639 *Fugitive Poetry* II xvi 4/9. Papists and quakers .. have their frequent meetings .. within this brughe, to the great .. scandall of the gospell, and of all trew Christians within this brughe; 1670 *Aberd. B. Rec.* IV 261. It is a scandal for presbyterians to give a bond that natively tends to the weakening and dividing of the presbyterian church of Scotland; MCWARD *Serm.* 58.

b. Quakers .. not any more to .. disturbe the inhabitants .. by ther eroneous principalls and preaches tending verie much to the scandall of the treuth; 1670 *Aberd. Council Lett.* V 7. 1670 *Aberd. B. Rec.* IV 261 (see above). Acts made against drunkenness and persones staying in allhouses misspending ther tyme to the dishonour of God and scandle of religione; 1700 *E. Loth. Antiq. Soc.* II 47.

2. a. Damage to (a person's or, more freq., a group or community's) reputation; (suspicion of) discreditable or offensive behaviour, freq. in relation to religious persons or groups.

Freq., *to the scandall of* (the party affected).

(1) Ane minister of Scotland Mr Andro Melvill .. had confessit himselff to be the auctour of certaine verses .. tending to the scandell and dischonour of the Kirk of England; MELVILL 681. (The Lord Chancellor) employed a painter to paint on his seat in Dunfermline Church a cross or crucifix to the] great scandle of the communitie; 1612 *Dunferm. Ann.* 272. Give .. pasturage ceas .. they wilbe forceit for want of maintenance to leive thair kirks, to the heavie reproche and scandill of this estait; 1623 *Aberd. Council Lett.* I 209. For taking of Robert Robsone .. of his clath and impignerating of the samin and drinking thareupon to the great offence of God and scandell of the incorporation; 1653 *Edinb. Surgeons* p. 250. To the .. great scandall and reproatch of the petitioners, who are stained .. thereby [*sc.* bad craftsmanship] by straingers .. who .. imputs the blaime to the petitioners; 1683 *Dundee B. Laws* 554. To the opprobrie of the said toune and common scandall of the kingdom; 1690 *Paisley B. Rec.* 120.

(2) The scandall of the crucifix payntit vpon my lord chancellor his dask .. that the causer as also the paynter of that idolatrous monument and the minister foirsaid have highlie offendet; 1612 *Dunferm. Ann.* 272. They are nott admitted to sit as members of the Assembly though there be no scandal nor exception propounded therupon till their carriage be first tryed and approven: muchless they are to be admitted [as members of the Assembly] when there is scandal and exception propounded thereupon; 1652 *Nullity of the Pretended Assembly* 15.

b. An imputation of bad or evil conduct; an (unfounded or unproven) report of such behaviour.

Cf. later Sc. legal use = 'a report circulated prejudicial to the character or reputation of another on which a legal action may be founded' (SND), as in 'a process of scandal' (1708).

Whether in respect of the scandall againe George Hall .. for wrang done to his faything he might yet sit in judgment, it is found he might till he be tryit and convictit; 1639 *Dumbarton B. Rec.* 57. The session .. thinkis fitt to lay him by as a scandalous impenitent person till both he clear himself of the said scandall in Edinburgh, and till some evidence of his repentance appear; 1664 *Cramond Kirk S.* II 6 Nov. Ther was a scandle delated upon David Donald of Shangie that he had miscaried in speaking against the doctrine taught be the minister in time of sermon; 1666 *Alyth Par. Ch.* 98. Anent ane scandell verifyed upon him for going to Edinburgh to .. ane sooth sayer; 1667 *Strathendrick* 10. Ther was a scandall delated upon Janat Red .. for recetting .. ane slight woman .. in her howse; 1668 *Alyth Par. Ch.* 97. After tryall and examination of parties and wittnesses in the said scandall of unwarrantably cureing of cattell [the brethren] did pass sentance; 1669 *Alford Rec.* 144. That he has not cleird himself of the scandal that he fell with his wife befor mariage; 1670 *Kingarth Par. Rec.* 57. Suspended .. because of a scandall of being drunken at the west port; 1690 *Cramond Kirk S.* III 25 Oct. [They] were obsolved from the scandell of fornication; 1695 *Minnigaff Par. Rec.* 5.

c. In the phr. *under* (*the* or *ane*) *scandal* (*of* a specified offence) in sense b above.

(1) Mungo Campbell .. this long tyme hes lyen under the scandall of murther; 1645 Edgar *Old Church Life* 250. The session being informed that Alexander Wilson is .. under another scandall .. thinkis fitt to lay him by .. till .. he clear himself; 1664 *Cramond Kirk S.* II 6 Nov. Under the scandal of fornication; 1698 *S. Ronaldshay* 72.

(2) Both being under scandall, and subject to kirk censure; 1647 *Moray Synod* 87. Nether any who ar wnder scandall quhill they have given sufficient satisfactioune; 1650 *Glasgow Wrights Acts* 9. Delated that Allekander M'Pherson .. should ferry Elspa Frisel being under scandel to the Lowlands on the sabbath day; 1666 *Kingarth Par. Rec.* 43.

3. A cause or occasion of disgrace or shame; a shameful or disgraceful circumstance.

[The Lords of Council considered it] a foule reproche and scandell to this oure nation that suche a .. byk of laules lymmaris shoulde be sa long sufferit [to have residence in .. the kingdom]; 1609 *Aboyne Rec.* 233. [They] sall .. creave God mercie and the congregatioun to quhom they ar an skandell; 1624 *Elgin Rec.* II 183. The insolenceis of proffest .. papistes .. give ane publict scandal to the kirk; 1628 *Aberd. Council Lett.* I 291.

b. Moral offence given by an action.

Ther cariage gave such ane scandall to them that he, his wiffe & servant left the house & they tuo stayed ther alone; 1652 *Cullen Kirk S.* 12 Sept.

4. Open or flagrant sin; a sin viewed as an offence to the community or congregation.

With a paper on her browe, shewing her nortorious scandall; 1653 *Dunferm. Ann.* 328. If any [elders] .. shall be found culpable .. by doing and acting any scandellis lykewayes mentioned in the same; 1655 *Mouswald Kirk S.* 9 Dec. Anent Mr. Thomas Irland, his scandell of Anabaptisme, he .. did judiciallie declair before the Synod that he renuncit Anabaptisme; 1657 *Dunkeld Presb.* II 416. Obeyed according as he was injoyned for his fornicatioun .. and was received and absolved from his scandale by the minister; 1664 *Kingarth Par. Rec.* 34. To .. stand ane hour besyde the joggs for hir uncivill words and scandalles committed be hir wpon Jonnet Lidderdaill; 1672 *Kirkcudbr. B. Rec.* MS 11 May. Since the elders .. would .. live verrie remote from these in the paroch comitted to there inspectione .. wherby scandales might escape all notice; 1678 *Edinb. B. Rec.* X 359. What penalty is taken for scandalls? 1696 *Penninghame Par. Rec.* I 3. Flagrant reports of his miscarriage and off the scandalls in his family; 1700 J. M. Beale *Fife Schools* 256.

(Scandall,) Scandell, *v.* [e.m.E. *scandall* (1592); SCANDAL(L *n.*] *tr.* To accuse on the basis of popular report or rumour. — Persons suspected or scandelled of treason .. [to] be taken and remain in firmance, and their goods under sure borrows; 1690 *Act of Privy Council* 3 Jan.

Scandal(l)ing, Skandelling, *vbl. n.* [SCANDALL *v.*] **a.** Bringing (a person) into disgrace; discrediting. **b.** The action of slandering (a person), or spreading malicious reports. — **a.** Their errour, who think they may both subscribe the Confession of Faith, and receive the Service Book, and Canons, which is not only a direct scandaling of them, but also a ready way to put a weapon in their hands against our selves; 1638 *Rec. Kirk Scotl.* 87. — **b.** For skandelling of Patrick Gordoun in saying he did steil his kaill; 1643 *Boharm Kirk S.* 6 Aug. For scandalling of James Dausone wrongously in calling him theiffe and knave; 1670 *Stitchill Baron Ct.* 58.

Scandal(o)us, *adj.* Also: **-eux, -owes, -eous, -allous, -olous, -el(l)ous; skandalous, -el(o)us, scandlous; scandaluous.** [e.m.E. *scandalouse* (1592), *-ous* (1603), F. *scandaleux*, med. L. *scandalosus* (*c*1332 in Latham).]

1. Of something said or written, also *transf.* of the tongue: Uttering false or malicious reports; defamatory.

It is not the maist part of my deutie to repres thy scandaleux tong; FOWLER II 31/21. Whairin ar set doun ane number of fals and scandalus speiches; 1632 *Justiciary Cases* I 219. By a scandaluous paper given in to the parliament; 1646 *Glasgow B. Rec.* II 104. Among the rest of [the] scandalous calumnies; 1650 *Strathbogie Presb.* 130. A scandellous pamphlet of James Grahames; *Ib.* 136. Concerning scandalous reports passing upon ministers; 1671 *Alford Rec.* 174. Because of some scandalous crying fame against John Fork; 1672 R. Brown *Paisley* I 321. To prevent the dispersing of prophane or scandelous papers; 1699 *Edinb. Paper-sellers.*

2. Of a person: Guilty of, or notorious for, shocking or disgraceful behaviour. Also *absol.*

Ane skandelus persone not meit to have societie within this parosche; 1615 *Elgin Rec.* II 146. Convict .. in .. holding in hir hous of anie scandalous whoorish drunken lowns; 1647 *Dunferm. Kirk S.* 23. The persons who wer scandalous in Thomas Eizat's house, to mak their repentance publickly; 1650 *Culross* I 239. Grissell Caillie ane scandolous woman; 1657 *Dumfries Council Min.* 20 June. Considering that Mr. Thomas Irland is a man so scandalous and infamous; 1661 *Dunkeld. Presb.* II 417. The haill penalties dew to be payit be all furnicatouris, and wther such scandelous persones; 1662 *Glasgow B. Rec.* II 491. Skandalous; 1662 *Forbes Baron Ct.* 231. An infamous, lewde and scandalous persone; 1664 *S. Ronaldshay* 45. That processe .. against Margret Guthrie, scandalous for many yeares for vitchcraft; 1666 *St. A. Presb.* 84. To enquyre .. if any of the masters etc. be scandalous or guilty of immoralities in their life and conversation; 1690 *Misc. Hist. Soc.* X 131. Scandalous persons .. to have their names read out publickly from the pulpit; 1692 *Edinb. Kirk S.* in *Sc. Ant.* XIII 74. A convenient seat to be placed in the middle of the church for scandalous faultie persons to sit in to receive theire reproofs; 1693 *Culross Kirk S.* in *Sc. Ant.* IV 82. Mr. Alexander connived at one Mr. Robert Denoon, a most scandalous fellow, to officiate as schoolmaster; 1698 *S. Ronaldshay* 71. He should be reput scandallous and unclean; 1700 *Minnigaff Par. Rec.* 34.

absol. Alexander M'Gillis appointed to .. cause make a sackgown for the scandalous to appear in; 1697 *Penninghame Par. Rec.* I 6.

3. Bringing ill repute upon the church or a religious congregation; giving rise to scandal in senses 1 and 2 of SCANDAL(L *n.* **a.** Of behaviour or doctrine. Passing into sense 4. **b.** Of members of the church. **c.** Of ministers.

There is some overlapping with sense 2.

a. Guiltie of .. using of charmes and inchantments and otheris devillishe practizeis, offensive to God, skandelous to the trew religioun; 1626 *Aberd. Council Lett.* I 253. [The] proces .. contained many blasphemous poynts, both of Arminianisme and Poperie .. besyde his scandelous lyfe. .. the whole Assembly voited to his deprivation; 1638 *Rec. Kirk Scotl.* 163. Some scandelous doctrine preached in Dysert be the said Mr. William; 1652 *Kirkcaldy Presb.* 381. Ther scandelous way of mariage, quhilk was vnorderly and not according to the actes of the church; 1657 *Lanark Presb.* 103. The .. Synod considering the great and scandalous abuse of taking oaths of privat persons upon suspicion of malefice done .. to ther person or goods by witchcraft; *c*1662-3 Bentinck *Dornoch* 236. [They] by theire presence had countenanced the .. disorderlie dispensing of the ordinance of baptism [etc.] .. to admonish the said heritors of their sinfull & scandalous carriadge; 1696 *Culross Kirk S.* in *Sc. Ant.* IV 84.

b. That speciall cair be taken of the election of elders and deacons .. and every one who is ignorant or scandalous [*Alyth Kirk S.* scandalowes], or doeth not perform familie duties, be removed from the Session; 1649 *Brechin Presb.* 13. That gross ignorants and scandelous persons be not admitted to the Sacraments; 1650 *Boharm Kirk S.* 13 Oct.

c. Exhorting all presbetryes to purge out scandalous .. ministers; 1691 *Cramond Kirk S.* III 14 May. For want of the powerfull ordinances of the gospell in purity and besides the present incumbent being scandalous and not edifieing to the people; 1691

Old Ross-shire I 42. There is one [minister] whom I can prove to be very atheistical, ignorant and scandalous; *Presb. Eloq.* (1692) 13.

4. Of behaviour or wrong-doing: Flagrant; shocking; disgraceful.

Ane penalty .. incurred be the said Janet .. for the scandalous conversing with ane David Williamson; HOPE *Major Pract.* I 138. If the adulterer or adulteress foirsaid hes not bein in tymes bygane of ane scandlous counversatione; 1640 *Mouswald Kirk S.* 1 July. That all extreordinare and scandolous drinking and gosseping upon the Lordis Day at any tyme be forborne; *Ib.* For thair scandalous behaviour in abusing of Euphame Robertsone in hir awin hous; 1646 *Dunferm. B. Rec.* II 201. Scandalous offences of all sorts, whither it be uncleanes, drunkenes, suearing, cursing, malignancie, or anie other; 1649 *Alyth Kirk S.* in *Sc. N. & Q.* 1 Ser. XII 75. The scandalous careage of Mr. Neill McLachlan, paedagogue to McLaine of Dowart; 1658 *Argyll Synod* II 195. Hir cariage was scandalous lyke with Alexander Wilson; 1659 *Cramond Kirk S.* I. Inquyrit gif he knew anything scandalous in the said Mr. John Schoolmaster; 1665 J. M. Beale *Fife Schools* 179. To confesse his scandaleous sinns .. and engadge himselfe to no mor so; 1666 *Dunkeld Presb.* II 68. Some flagrant reports of scandalous cariages of [two ministers]; 1669 *Alford Rec.* 141. For thaire scandallous language remitts them to the Sessione; 1680 *Kirkintilloch B. Ct.* 107. [The schoolmaster was dismissed for] gross & scandalous faults; 1693 J. M. Beale *Fife Schools* 171. What a horrid and scandalous sin charmeing was; 1706 *Rothesay Par. Rec.* 204.

Scandalously, -lie, *adv.* Also: **scandalsle, skandelouslie, (skanderouslie).** [e.m.E *scandalously* (1602); f. prec.] **a.** In a scandalous manner, disgracefully. **b.** Slanderously. — **a.** Jeane Blackwood .. skanderouslie behavit hir self in remaning in ane barne with Johne Wyllie all nicht; 1633 *Dundonald Par. Rec.* 369. That they suld performe the band of mariage within the ordinarie tyme, and, in mean tyme, abstaine and not live scandalouslie; 1647 *Carnock Kirk S.* in Row xxxviii. They were accused that they lived so scandalsle; 1653 *Culross Kirk S.* in *Sc. Ant.* IV 161. That Donald M'Couk sould have lyn in bed scandalously with Cirstine Rodger; 1674 *Kingarth Par. Rec.* 93. *c*1675 *Falkirk Par. Rec.* I 232. — **b.** That no craftismen of the said craft speik reproachfullie or skandelouslie ane against another; 1662 *Lanark B. Rec.* 190.

Scandalowes, -uous, -el(l)ous, -elus, -(o)lous, varr. of SCANDAL(O)US *adj.* **Scandelise, -ilize,** varr. of SCANDALIZ(E *v.* **Scandelizing,** var. of SCANDALIZING *vbl. n.* **Scandel(l, -ill, -le,** varr. of SCANDAL(L *n.* and SCANDALL *v.*

(Scanning,) Skanning, *vbl. n.* [Late ME and e.m.E. *scannynge* (Prompt. Parv.), *skannyng* (1560). Cf. SCAND *v.*, also SCANSE *v.* 2.) — I sie many over bold with God almightie, making to many subtill skanningis [*pr.* skaimingis] of his blissed will, as lawers do with humane testaments; 1585 *Warrender P.* (SHS) I 175.

Scanse, Scance, *v.* Also: **scaunce, skanse, skance.** [L. *scans-*, p.p. stem of *scandere* to climb. Cf. SCAND *v.*]

1. *intr.* To give an impression, have an appearance, (*of*).

Cf. later Sc. *(ill-)scansed* having the appearance of, (ill-)looking (SND).

Thir iiij [*sc.* Mars, Juno, Venus, Saturn] scansyte of diuers complexioun, Bataill, debaite, inwy and destructioun; *Wall.* III 347.

2. *tr.* **a.** To scrutinise, examine critically.

The forsaid arbetoris owrman witht consent of parteis hes scansit and consydderit the propirteis commoditeis of all and syndre the forsaid landis; 1562-3 *Rec. Earld. Orkney* 113. With sax honest men, to skance and stent the samin throche the parochin; 1604 *Shetland Sheriff Ct.* (ed.) 118. His depositioun .. wilbe curiouslie scansed be our laweris; 1608 *Haddington Corr.* 215.

b. To ponder, think deeply about, reflect on.

And panse not, nor skanse not, The perill nor the price; MONTG. *Ch. & Slae* 1357 (Wr.). Full oft this mater did I skance [: pance] Bot with my self befoir; *Philotus* 803. I find, Adrastus, when I deeply scaunce [*I.* when deepely I designe, *J.* when I have search'd and seene] Th'effectual motiues of this fatall crosse; 1604 W. ALEXANDER *Crœsus* 1475 (D, E). As for the king, I professe not to scanse his power; 1620 Calderwood VII 421.

c. To wonder. Const. *clause object.*

What high conceit with curious thoughts can scance [*J.* What minde can think though thoughts the same entrance How [etc.]]; *a*1614 W. ALEXANDER *Doomes-day* III 851 (G, H). How that can be forgote I greatlie scance; ADAMSON *Muses Thr.* 161. To make a day of dayes .. Whilst when they end it, what should come doth scance; DRUMMOND II 161/8.

d. *intr.* To reflect.

And at the ishew, fine, and end, Oft he conceats and skanses; GARDEN *Elphinstoun* 184.

3. *intr.* and *tr.* **a.** *To scanse in*, to ponder deeply on. **b.** *To scanse upoun* or *of*, to talk, or think, about; to discuss; to comment on.

a. The philosophers quha skansed in curiosities, quhen this wisdom came to them, they kend na mair in it nor bairnis or infants; ROLLOCK *Wks.* I 373.

b. I knaw that the motioun of my voyage at this tyme wilbe diverslie skansit upoun; 1589 *Reg. Privy C.* IV 427. He took him leisure from his vocation, which was to look to Jesus, and to follow him to scanse upon John; ROLLOCK *Wks.* II 640. To scanse of these things ouer far it is but vaine curiositie; *Rollock's Thess.* 28. How strangeres scanses of thinges that thay heir; 1608 *Lett. & St. P. Jas. VI* 149. Euen the wysest princes will not skance sufficiently vpon the parcialites and pretences of some of ther .. conseillouris; MELVILLE *Mem.* 121. Ther commyng and demandis was dyuersly skancit vpon, some supponing .. others thocht [etc.]; *Ib.* 336. Howbeit I cannot scance upon kings crowns; 1638 CANT *Serm.* (1741) 40.

Scansing, *vbl. n.* [SCANSE *v.*] The action of scanning (verses); scansion. — Be scansing of versis of dyuers kyndis and to the mesur and quantatie of tham; *Art of Music* 34b. Thinking the verse to haue a good trot, & sophing them rather by thair sounding, nor scansing them be iust measure; FOWLER II 29/2.

Scant, Skant, *n.* Also: **scante, (scanth), scaunt, skent; schant.** [ME and e.m.E *scant* (*a*1350); f. or f. as SCANT *adj.*] Short supply, or insufficient numbers, (*of* persons, material or non-material things); scarcity, dearth; insufficiency.

(1) The king buskit him sone with scant of squyary. Wachis and wardroparis all war away; *Rauf C.* 273. Albeit the strenth [*1533, Ruddim.* scanth] of men ȝhe set not by; DOUG. I viii 84. Ʒe may persaue That skant of men we wald not haue; 1573 DAVIDSON *Sat. P.* xlii 352. Seeing papistry abounds in the North be reason of the scant of qualified ministers therin; 1586 *Bk. Univ. Kirk* II 659.

(2) In tyme of scant of watter [at the mill]; 1510 (*c*1580) *Edinb. B. Rec.* I 130. For skant of vittal the cornys .. Thai grand and syne buke at the fyre; DOUG. I iv 39. Ane huge pray quhilk for skant of schippis he mycht nocht carye; *Boece* 94. Thai ete .. horsis, doggis [etc.] .. throw scante of viueris; *Ib.* 594. The skant of ledderis was in caus that thai clame nocht the barmkinis; *Ib.* 612. Ʒe ladyis that hes na skent of leddir; LYND. *Sat.* Proclam. 274. Of bran and geill thair wes na skant And ipocras he wald not want; Id. *Meldrum* 1563. The prowest, baillies .. considdering the skant of prisoun houssis [etc.]; 1560 *Edinb. B. Rec.* III 66. Quhairthrow than grew to him skant of riches; ROLLAND *Seven S.* 2797. For schant of monney compellis men to sell; 1564 Cochran-Patrick *Coinage* I 92. Hwngir and darthe that now is lyk to be Throw deid of beistis and skant of corne this ȝeir; MAITLAND *Maitl. F.* 40/8. The .. drowth and skant of watter; 1591-2 *Edinb. B. Rec.* V 59. For the raritie and fewtie or scant of sum of thame [*sc.* geese] in vthiris cuntries; DALR. I 59/24. Their is greate scant of monye on Spinola his syde; 1624-5 *Mar & Kellie MSS* Suppl. 225.

(*b*) For povurtie and scaunt of his leving; 1535 *St. P. Henry VIII* V 24.

(3) Thow sall haue fouth of sentence and not skant; DOUG. *Pal. Hon.* 1401. For skant of tyme thai mycht nocht mak duelling; KENNEDY *Pass. Christ* 1309. Gret scant of tyme .. Has maid my

wark mair subtell and obscur; DOUG. (STS) IV 194/17. These wonds .. Which skant of bontye mair and mair prepairs; FOWLER I 149/8. Of ony kynde of .. cheir, of gemm or play, of ony kynd of musik, was nother want or skant; DALR. II 68/18. But quhen he wes made canceller, his scant of clergie bewrayed him; 1626 *Ancram & Loth. Corr.* I 42. Meikle scant of life in the world .. before He should have died for want; RUTHERFORD *Cry from the Dead* (1775) 12. A vote, declaring the king, for no scant of faults, incapable to govern; 1646 *Rec. Kirk Scotl.* 459.

b. Without const.: Shortage; lack of the necessities of life.

Be battell, pest, .. Sum perrellous plague or pouertie Hounger and skant [etc.]; ROLLAND *Seven S.* 6331. In time of derth, penuritie and skant; *Ib.* 8542. Mony peipill diet with greit skant; *Diurn. Occurr.* 39. Triumphant kingis .. Who now lyes spoyled of thair pompe and skant dois plague thair pryd; FOWLER I 81/101. Thay .. broght into Edinburghe many ky and oxin .. for supplie of thair scant and hungar; *Hist. Jas. VI* (1825) 99. The cornis war universallie corruptit, sa that scant and dearth followit the nixt yeir; *Ib.* 217. With me thairs be na scant, Thairfoir some grace vnto me grant; *Philotus* 22. Nather may thair appeir ony .. tackynnis of penurie and scant amongis you; 1617 *Aberd. B. Rec.* II 350. Scant maks dearth; CARMICHAEL *Prov.* No. 1334. This strong strenth .. quhilk wes nather for scant nor want givin ower; SPALDING I 305.

Scant, Skant, *adj.* and *adv.* Also: **scante, schant, skaunt, skaynt.** [ME and e.m.E. *scant* adj. (*a*1400), adv. (*c*1400), ON *skamt*, neut. of *skamm-r* short, brief.]

A. *adj.* **1.** Short in supply; insufficient in quantity to meet an immediate demand, barely adequate in amount; scarce.

Of material things, esp. commodities, also of non-material things.

Chiefly *predic.*

(1) *predic.* To [*C.* Til] scant all rynnand ryverys were For to be drynk till his powere; WYNT. IV 989. Wictaill worth scant or August coud apper; *Wall.* III 13. Halcrekis and foregere was amang thame skant; *Boece* 562b. For lesurris to lyne the nek .. witht becaus the samin furring was skant; 1539 *Treas. Acc.* VII 172. Thocht the corne war never sa skant, The gudewyfis will not let freirs want; LYND. *Sat.* 753. Bot the .. expensis that he maid on corne .. gart the victualis be deir and skant, quhilk vas occasione that the pure pepil deit for hungir; *Compl.* 156/8. Quhair gold is skant, siluer mon vs content; ROLLAND *Seven S.* Prol. 174. Nocht onlie growis gude money skant .. bot als the prices .. incresis; 1562 *Reg. Privy C.* I 212. Ane derth, quhilk maid the vittel scant; *G. Ball.* 35. *Ib.* 194. The wedder wes vehement cald and the coillis wer verie scant; *Diurn. Occurr.* 289. This proclamatioun .. maid the drink scanter in the said toun; *Ib.* 302. The comowne gras off this burght .. is baith wery small and skant; 1582 *Burntisland B. Ct.* 1 May. I reid of nocht bot it is scant, Of cattell, clothing and of corne; MONTG. *Flyt.* 190 (T). Within my vaines the bloud is skant and cald; HUME 34/12. The cuntrey being exceiding pure, the moneyis skairsche and skant; 1657 NICOLL *Diary* 194.

(*b*) Fysh and flaische was skaynt and der; CULLEN *Chron. Aberd.* 47.

attrib. That in tyme of scant watter ilk nychtbour thairof sall haue facultie and priuelege to grynd four laid of quheit in his dew place; 1514 (*c*1580) *Edinb. B. Rec.* I 146. Twa schipis or ma lyand in ane heavin at skant wattir; BISSET II 246/15.

(2) Thocht in my translatioun eloquens skant is; DOUG. I Prol. 307. Quhil in thar dry throtis the aynd worth skant; *Ib.* V iv 80. Thy Godhed na rychar nor ȝit mar skant Nowthir now nor than; *Ib.* X Prol. 19. Thair wes no plesour to thame that tyme skant; STEWART 16884. Becaus strange termes I want I haue na dout bot heir thay sall be skant; ROLLAND *Seven S.* Prol. 161. Perchance thay sall find freindschip skant; SCOTT xvi 54. My skill is skant; GARDEN *Worthies* 177. And if my judgement be not scant Some lybel will be relevant; CLELAND 110.

b. Chiefly const. *infin.*: Difficult (*to* obtain), ? because of real or apparent shortage or high price.

(1) Fra tyme thow want that stufe is skant To borrow in the town; BALNAVIS *Maitl. F.* 357/55. Skowis for making of treyis .. be verray bair and skant to be gotten; 1597 *Conv. Burghs* II 12. Smeddie collis was risseyn to hiecher prycis and mair scant to be had thane they war of befor; 1604 *Urie Baron Ct.* 8.

(2) Madame .. the price it is to hie... The copie clene I quite it is sa skant ȝit nocht the les I sall compeir said he; ROLLAND *Ct. Venus* I 900.

c. *transf.* In *scant time*, a period when things are less plentiful.

Now is past by the skant time of the ȝeir; ROLLAND *Seven S.* 8532.

d. Of persons: Few, when compared with the number needed.

The tormentaris wer then so scant Chryist for to scurge scairis found wer sax; MAITLAND *Maitl. Q.* 43/33. Good servants now are scant, Their fees likewise are grown exorbitant; *Copie of a Baron's Court* (1821) 33.

2. Less in size, length, or amount than would normally be the case or than would suit the circumstances described. **b.** *specif.* Of a measure or measured amount.

Than can ȝe be no maner want Gold thocht ȝour pose wer neuer sa skant; LAUDER *Off. Kings* 260. Women .. man be buskit wp lyk bryddis .. Thair gownis schant to schaw thair skin; *Maitl. F.* 440/121. That littil god of loue .. So .. childlike, with bowe thrie quartars scante; MONTG. *Ch. & Slae* 110 (W). By reason of the skant and narrow passage many were killed; SPOTSW. *Hist.* (1677) v.

b. The myllar mythis the multyr with a met skant For droucht had drunkyn vp his dam in the dry ȝeir; DOUG. VIII Prol. 40. Ordines the balyeis to excers and vesy the halff furlattis one Fridy neixt to cum, .. for nane is vissis that hes mettis that ar skant; 1525-6 *Selkirk B. Ct.* (ed.) 79. Howbeit your elwand be to scant; LYND. *Sat.* 4061. A .. flekket of ane quart with ane scant chopine of rois walter in it; 1575 *Edinb. Test.* III 318.

3. Deficient in certain respects.

Quhar scant was Scottis, I had nane other choys Nocht for our tong is in the selwyn skant Bot for that I the fowth of langage want; DOUG. I Prol. 118, 119.

4. Of things and, chiefly, persons: *Scant of*, short of, lacking in, poorly supplied with (goods, etc.). Also *proverb.*

I bene .. full of merci .. In suffrance scant of mony grete offense; *Quare Jel.* 198. Gif he be scant of ky in hevin; *Duncan Laideus Test.* 163. Tending thairby to mak the town skant of breid; 1554-5 *Bk. Old Edinb. C.* XV 11. I vois skant off paper at the makin heiroff; ?1565 *Breadalbane Lett.* 13 Nov. Off houshald graythe sum rycht skant war; MAITLAND *Maitl. F.* 46/9. He was skant of hay and horse corne; PITSC. I 306/16. The said wellis .. becumes swa dry and scant of watter that the towne can nocht be seruit; 1580 *Edinb. B. Rec.* IV 161. He was scant of clathis; 1584 SEMPILL *Sat. P.* xlv 1016. In consideratioun that the toun is scant of ministers [etc.]; 1584-5 *Edinb. B. Rec.* IV 382. In this ile .. is .. na woodis nor mosses quhairthrow thai are scant of fire; *Descr. Isles* 436. This ile .. being skantly inhabited but by very few, and they as .. skant of ciuilitie as number; 1598 JAMES VI *Minor Prose* 70/12. Simple people scant of ony religion; MONRO *W. Isles* (1961) 87. Being prisoner .. and skant of .. servants; 1616 *Misc. Hist. Soc.* II 169. The goodwyff being skant of meit; 1620 *Grant Chart.* 325.

proverb. I am not sa scant of claith as to sole my hose with a doken; CARMICHAEL *Prov.* No. 831. They are scant of newes that tels their father was hanged; 1639 J. CORBET *Ungirding Scottish Armour* 52.

b. Without const.: Badly off; not having enough (money).

And onelie pwir maid by his welth and by abundance skant; FOWLER I 44/201. I haif send to ȝow my compt becaus I am skant, vtherwayes I wald not haif send it; 1608 *Breadalbane Lett.* 16 March.

B. *adv.* **1.** Scarcely, barely, hardly. *Scant weill*, not well, badly, inadequately.

The STEWART quot. in (1) below may properly belong to SCANT *n.* (3).

(1) Half the riches that he has Sall scant be worth his viffis clas; *Gud Wyf & D.* 62. All thair flesche wald scant be half ane fill .. vnto that gredie gled; HENR. *Fab.* 2905. For slydernes scant mycht he hald his fete; Id. *Orph.* 305. So sair the magryme dois me menȝie .. That scant I luik may on the licht; DUNB. (OUP) 72/5. Greit abbais grayth I nill to gather Bot ane kirk scant coverit with hadder; *Ib.* 121/86. Grantis quhen twelf ȝheris he had beyn diligent To study Virgill, skant knew quhat he ment; DOUG. I Prol. 130. Scottis and Pichtis .. war skant sa mekle be Romanis extemyt, as sufferit mak seruice like vile sklavis; *Boece* 163b. This nobill king that scant of sic thing wist .. Tha branit him; STEWART 61109. Skant was he vp quhen at the eird was he; ROLLAND *Ct. Venus* II 648. Ane kynnisman skant may trow ane vdir; MAITLAND *Maitl. F.* 317/16.

(*b*) Pane eternal in body and saule that natur of creatur skaunt may sustene; IRLAND *Mir.* II 59/2.

(2) Says nocht ȝour sentens thus, skant worth a fas, 'Quhat honeste or renoun is tobe dram?' DOUG. IV Prol. 155. In it he left nocht scant worth ane kaill leif; STEWART 56288. Peter Tomsone sayd to the Abbat of Pasley in consell at the greyt men of Yngland was skant content at suber men had the realm in gyding; 1547 *Cal. Sc. P.* I 5.

(3) Latyn wordis .. That in our leyd ganand translatioun has nane, Les than we mynys thar sentens and grauyte, And ȝit scant weill exponyt; DOUG. I Prol. 366. Thay wis thair will that skant weill wenys; SCOTT xxvi 34.

b. Of a numeral, or a measured amount, distance, etc., stated in terms of numbers and fractions. Cf. A 2 above.

He had not with him bot skant ten; BARB. (1571) XX 444. Ane half yhere skant he wes byschape; WYNT. VII 3248. Skant ane thousand chaist all thare cumpaney; HAY *Alex.* 3383. The breid of the erth fra south till north .. is scant half samekle as the lenth; *Asl. MS* I 153/29. Thy fulies lust lestis scant ane maij; KENNEDY *Maitl. F.* 234/10. Of all our floyt .. skant Sevin .. salue remanys with me; DOUG. I vi 134. That secund Parys .. With his onworthy sort, skant half men beyn; *Ib.* IV v 77. He wes skant thre ȝere quhen he wes crownit; ABELL 116a. Pure claggokis .. Quhilk hes skant twa markis for thare feis; LYND. *Syde Taillis* 63. Scant twa houris afore the meting of the wardanis; 1553 *Reg. Privy C.* I 149. Palexis .. and Amadur .. War scant two mylis the kingis castell fro; *Clar.* I 1346. It is scant ane twelf month gane; *Maitl. F.* 244/7. The breadth of the island is .. scant thrittie mylles; PITSC. (1814) xxi. If we ha' scant a mark o' gear; G. STUART *Joco-Ser. Disc.* 11.

2. Scantily. = SCANTLY *adv.* 2.

The wynd swa skant wes than blawand; WYNT. VI 2494. A heart couragious never breathed scant; BOYD *Zion's Fl.* 153.

Scant, *v.* Also: **skant, skent.** [? ME (once, 1436, if reading correct) and e.m.E. *scant* (*c*1553); SCANT *adj.*]

1. *intr.* To become scarce; to run short.

Fra fude scantyt na plesance was that tid; *Wall.* VIII 980. In Wallace ost so scantyt the wictaill Thai mycht nocht bid na langar till assaill; *Ib.* XI 35.

2. *tr.* **a.** To give out (something) sparingly; to limit (a supply).

Had nocht ȝour wyne bene scantit [etc.]; *Bk. Chess* 345. She, being naturallie needie, skanted the victualls; CALDERWOOD III 283.

b. *To scant* (something) *on* (a person), to bestow in a niggardly manner.

No poynt of bewtie nature on hir scantit For scho hir paintit as goddes devine; *Clar.* IV 958.

c. *passive. To be scanted of* (something), to be inadequately supplied with, to be kept short of.

They fand themselves extremly scanted of powder and ball; P. GORDON *Brit. Dist.* 120.

3. Of the wind: To hinder (a ship) by not blowing favourably.

Their purposs was for Deepe, but the wind did skant them soe that thaye warr forced to land at .. Bullone; 1622-3 *Mar & Kellie MSS* Suppl. 152.

Scant(e, varr. of SANCT *adj.* and *n.* and SCANT *n., adj.* and *adv.*
Scant-frie, see SCOT-FRIE *adj.* **Scanth,** var. of SCANT *n.*

Scantly, -lie, *adv.* Also: **skant(e)-; schant-.** [Late ME and e.m.E. *scantly* (*c*1440); SCANT *adj.*]

Syntactically, freq. operating as a negative adv., see the notes to 1 below.

1. Scarcely, hardly, barely.

Freq., in early use chiefly, with inversion of following subject and verb.

Sometimes with appar. superfluous negative.

When followed by a temporal clause introduced by *quhen*, sometimes virtually = no sooner .. (than).

(1) And skantly had ilkane of tha A singill clath; *Leg. S.* V 151. That skantly had he the lyf-lad; *Ib.* xxvi 79. Scantly may a man trowe the grete .. riches .. of that toun; HAY I 39/11. Scantlie may he purches by his maill To leue vpon dry breid; HENR. *Fab.* 2754. Scantly was left in him his life; *Sir Eger* 2026. Of sa gret antiquite that scantly is thair memour of the lignie quhar of he descendit; *Loutfut MS* 8b. That scantly had he ane serwand to mak seruice to him; ABELL 9b. That scantlie is left ony of thair posterite; MAITLAND *Ho. Seytoun* xiii. Now skantlie in it may be sene Twelff gait glydis deir of a preine; *Maitl. F.* 439/55. Four parische kirkis till ane preicheir Quhen skantlie may ane man gyde ane; 1573 DAVIDSON *Sat. P.* xlii 909. Because the cuntrie .. is sa parrallous, that skantlie may ane schip saill thereby .. but cuming in dangeare [etc.]; BISSET II 233/8.

(2) For he for old scantlie mycht stand; *Troy-bk.* II 2374. *Ib.* 1633. He .. scantly had bene thare ane houre Quhen the lord callit at the ȝet; *Seven S.* 1930. Britouns skantlie weill walkynnyt fra slepe; *Boece* 49b. Ȝit we myct skantly spair ane houre .. to the audience of the same; 1570 *Cal. Sc. P.* III 483. For thai that had gude hors .. Hes skantlie now ane crukit meir; MAITLAND *Maitl. F.* 332/39. *Ib.* 432/7. Skantlie knawand in quhat pairt of the warld he was in; PITSC. I 294/32. The plenessing .. quhilk my ladie desyiris .. quhilk extendis skantlie to the half; 1580 *Coll. Aberd. & B.* 391. Skantelie; JAMES VI *Poems* I 143/266. In all this scho will skantlie get hir awin scho brocht with hir; 1605 *Edinb. Test.* XL 15b.

(*b*) He answert in extremis quhen schantlie he mycht profer the word 'Hang thame all on a gallos'; ABELL 97b.

(3) It that I tuke of ȝouris is scantly the costis of the men that I led with me; HAY I 166/32. Wntymys thou art, for it is scantly day; *Wall.* IX 1630. Fra the cite thai war nocht gane Twa legis bot scantly ane; *Seven S.* 1806. Quhilk war .. Scantlie the fourt part that tha brocht of toun; STEWART 37060. Thrie of the foirsaid tokins veyand skantlie [*pr.* skanthe] ane drope vecht; 1631 Cochran-Patrick *Coinage* II 77.

(4) That he durst nother scantlie speik nor luke; STEWART 33130. Thay will do nocht .. Nor skantlie speik without he giue hir leif; ROLLAND *Seven S.* 8011. For butt ane staff thow may nocht skantlie stand; WEDDIRBURNE *Bann. MS* 288b/95.

(5) Skantly lenyt don he was, Quhen the woyce on hym can cry; *Leg. S.* xix 228. Scantly sad this word he had Quhene twa feyndis apperit thare; *Ib.* xxxix 136. Scantlie had thay drunkin anis or twyse, Quhen in come .. our iolie cat; HENR. *Fab.* 325. Scantlie had Age restit him thair ane quhyle Quhen Conscience come cryand [etc.]; *K. Hart* 521. Skantlie had Nothat twa ȝeris regnit quhen be sa vile seruitute he supprisit the pepill; *Boece* 65. Skantlie had the herald .. rehersit this charge quhen skurreouris schew [etc.]; *Ib.* 127. Scantly war thir wourdis endit quhen on athir party the schout was rasit; *Ib.* 308. Skantlie was he turned quhene baith the airmieis

was redy to joyne and thyk togidder; PITSC. I 54/31. Skantlie was he weill come hame quhen he did persaif the craftie subtilltie; *Ib.* 139/16.

Passing into: **b.** Not easily, with difficulty.

Scantly the todir day Fand thai quhare the body lay; *Leg. S.* xvii 111. His nawyne nere tuk thai That scantly wan hym-self away; WYNT. IV 1356. Thar [*sc.* wise men's] smylyng scantly may men ken; *Thewis Wysmen* 104. Scantlie fra the cauld I micht defend; HENR. *Test. Cress.* 7. His wyff come out Scantlie scho micht our hy him; *Peblis to the Play* 146. We .. now scantly within our wallit cite The hope and weill of Italy defendis; DOUG. XII i 86. And scantlie [*P.* vnneth; L. *vix*] he gais away al to drawand him; NISBET *Luke* ix 39. And thai .. scantlie [*P.* vnnethis] swaget the pepile; Id. *Acts* xiv 17. vij wretaris successiue mycht skantlie .. write with sic expedicioun as he wald indite; *Boece* 183b. The sand .. blindit thame that tha mycht scantlie ken Quha wes thair freind or .. fa; STEWART 37820. Bot atheisme and superstitioun Hes sa me now environit about That scantlie can I find quhar to get out; ARBUTHNOT *Maitl. F.* 49/19. That scantlie hagbutis mycht schute for rainyn; *Diurn. Occurr.* 218. So fast the deir ran to his den His coulour I cud skantlie ken; BUREL *Pilgr.* I 75.

c. Only just; immediately before that.

As he passit in the hous the vyndois wer oppin the candillis scantlie lychtit; BUCH. *Indict.* 44. This message scantlie endit Erle of Douglas .. ansueris him againe; PITSC. I 21/6.

2. Scantily; inadequately. **b.** Thinly or sparsely (inhabited).

For all the proverbs ȝe pervsit ȝe thocht vs skantly skillit; MONTG. *Ch. & Slae* 918 (W).

b. 1598 JAMES VI *Minor Prose* 70/11 (see SCANT *adj.* 4).

Scantnes, Skantnes, Skantines, *n.* Also: **skantnesse, skanthness.** [ME and e.m.E. *scantnesse* (Chaucer; v.r. *skarsenesse*), *scantnesse* (*c*1412), *skantnes* (1574), *scantiness* (*a*1695); SCANT *adj.*]

1. The condition of being insufficient in number *of* people or things or of quantity *of* material things; shortage.

(1) Bot gif .. the iuge thruch wayknes of the court or skantnes of soytouris may ferrar contynew the daye; *Acts* I 30/2. For waiknes of soytouris of the court & scantnes of consale & vthiris; *Harl. MS* 4700 279b. Preistis of quham grete skantnes was for the tyme; *Boece* 63b.

(2) Fertill of folk with gret scantnes of fud; *Regim. Princ.* 180 (Fairf.). The gret skantnes that is now of money throw al the land; 1475 *Acts* II 112/1. Bettyr is to sustene .. grest skantnes .. batht of drink & met Na [etc.]; *Bernardus* 288. Mony .. levis in greit povertie and skantnes of thair necessare sustentation; HAMILTON *Cat.* 35. Throw penurite and scantnes off halfpennyes and other small money; 1554 Cochran-Patrick *Coinage* I 76. Maist exhorbitant derth and scantnes of fewall; 1563 *Acts* II 543/2. Thair wes ane great darth and scantnes of victuallis in Edinburgh; *Diurn. Occurr.* 306. For skantnes of paper I wrait to yow in hest; 1578-9 *Waus Corr.* 205. Skantnesse; JAMES VI *Basil. Doron* 90/3. The greit skantnes of .. barkit ledder; 1604 *Inverness Rec.* II 27. Skanthness; 1606 *Ib.* 47. There is great raritie and skantnes within the realme .. of silver; SKENE *Reg. Maj.* II 45.

(*b*) Dartht and skantines of wittuallis; 1595 *Banff Ann.* II 403.

b. Without const.: Insufficiency, or shortage, of the necessities of life.

(*a*) Sic skantnes and hungir followit; *Boece* 431. So greit scantnes ouir all Scotland thair fell; STEWART 44418.

(*b*) Considdering that victualis ar cariit furth of this realme be marchantis and utheris .. to the greit incres of derth and skantines; 1567 *Reg. Privy C.* I 571.

c. With possess. pron.: Lack of a sufficient supply (to meet one's needs).

The supplicatiouns givine in be the .. inhabitantis .. schawing thair skantnes of water; 1654 *Glasgow B. Rec.* II 294.

d. Inadequacy; lack of knowledge.

Or gife my langage be not lyik the laife For mekle neid and skantnes that I haife Of mother toung quhilk garris me seik and borrow; HAY *Alex.* 19308.

2. Parsimony.

It is degradyng till a kyngis croun .. To leire to conquest guddis with scantnes; *Regim. Princ.* 102 (Fairf.).

Scanȝe, var. of SKENȜE *n.* **Scap,** var. of SCALP *n.*

Scap(e, Scaip(e, Schap(e, *v.* Also: **skape, skaip(e; sckip-; schaip; schepe;** CHAPE *v.*[1]. [Early ME *scapien* (Layamon), ME and e.m.E. *scap(e* (Cursor M.), *skape(n* (Manning), *schape* (*c*1450), aphetic var. of ESCAPE *v.*]

1. *intr.* To escape (*from* (*out of*) captivity, prison, servitude, etc.); to gain liberty. = ESCHAPE *v.* 1.

(1) Wenand sum man .. His son .. Had lattin scape [*pr.* stape] out of presoun; *Troy-bk.* II 2947. He sall pay him ane hundreth thousand frankis or he scape out of his handis; HAY I 227/8. Where from thy spirit shall I scaipe? MURE *Psalmes* cxxxix 7.

(*b*) To remayne in suer ward .. and not to schap forth of the same; 1590 *Cal. Sc. P.* X 323.

(2) [He] gert put to dede a chiftane .. becaus and he had scapit, he wald nocht have cessit to mak him starkare were than before; HAY I 121/10. That the three lords were scaped thus; *Rosw. & Lillian* 100. Thow scapis in France to be a knycht of the felde; KENNEDY *Flyt.* 430. Think not to scape for .. Thou there shall be condemn'd as well as I; Fowler I 390/5.

(*b*) The juge .. dange hym in a dongeone depe That he na schapit; *Leg. S.* iii 216.

2. To escape from (*fra, of*) danger. = ESCHAPE *v.* 2.

(*a*) The theif sayis that ewer I scaip Na ane stark widdie gar me gaip; DUNB. (OUP) 169/71 (M). Now am I scapit fra that effray; LYND. *Sat.* 627 (Ch.). Bot he skaped by the nights [*M.* michtis] of Marie; *Christis Kirk* 88 (Laing MS). Him .. That so had skaipit betuix the bow and string; *Clar.* II 325. Young Henrie skaipit home is hurt A souldier shot him with a bow; *Reid Swire* 129. ȝeit he sckipit at that tyme; PITSC. II 54/4. He beand scapit of that danger, Hame through he past; 1584 SEMPILL *Sat. P.* xlv 726. For her pryde but punishment skaips free; CRAIG III 18. Faine thay uaire to skaipe thaime selfis; JAMES VI *Poems* I 245/404.

(*b*) And tocht gif he sould schape it hed bene schame; HAY *Alex.* 3852. The husband .. wes fane he schaippit from thair ill; HENR. *Fab.* 2373. In the gretest danger that eny man mycht be in and schape with his lyff; *c*1515 WILLIAMSON in Doug. (Sm.) I p. xxvi. This lady .. Full pitiouslie did weipe .. Beseikand God to send hir help .. To schape out of that dreidfull wildernes; *Clar.* III 712. Thair schaipit nane of all that greit nauy; *Wall.* (1570) VII 1070. Quhen quene and court could not get him conuict .. Thay fryit in furie that he schaipit quick; 1573 DAVIDSON *Sat. P.* xl 163.

3. *tr.* To escape (danger, suffering, etc.); to get safely away from.

And thus the scorne and the scaith scapit he nothir; DUNB. *Tua Mar. W.* 358. To scape the fray he was so fane [etc.]; 1584 SEMPILL *Sat. P.* xlv 705. My mortall foes new scaip't the Graecian sword; MURE *Dido & Æneas* I 225. And you .. schuld skaipe bothe gutt and wther diseissis; 1622 Innes *Sketches* 517. We, who scapt those grievous crosses, Did hide our selves in bogs and mosses; COLVIL *Whig's Suppl.* (1681) I 56.

(*b*) I am .. sik .. and tinkis nocht to schepe parrall of my lyf; 1560 *Breadalbane Letters* MS No. 4. He prayit .. that ȝe micht turne That ȝe micht schaip eternall schame; 1573 DAVIDSON *Sat. P.* 134.

b. Of a thing: To escape (a person's) notice.

My firmer substance scaipt thee not In secreet when forth broght By thee; MURE *Psalmes* cxxxix 15.

Scape, var. of SCALP *n.* and SKEP(E *n.*

Scapelry, Skaplarie, *n.* Also: **skaiplar(i)e, -y, skapellarye; skapular.** [ME and e.m.E. *scapelori* (Ancr.

R.), *scapelarie* (1402), *skaplorie* (*c*1440), *scapelerey* (*c*1474), *scapulare* (Cath. Angl.), *scapelar* (1499), *scapuler* (1546), OF *scapel-, scapulaire* (early 13th c. and *c*1380 in Larousse), med. L. *scapulare* (*c*1220 in Latham), L. *scapulārium*. Cf. SCABELL *n*.] A (short) cloak worn by certain religious, a scapular.

To tak the skapellarye, to join a religious order.

(1) With that scho kest the cule away & scapelry but delay; *Leg. S.* xxxi 672. With quhom my doingis may be kepit maist secrete becaus of thare wyd cloikis & skaplaries that hingis befor thare wame; MYLL *Spect*. 292/21. And ȝe sell se ws sone degraid thame Of cowle and skaiplarie; LYND. *Sat*. 3617 (B) (see *CHAPLARIE). I knaw that cowle and skaplarie [*B*. skaiplary] Genners mair hait nor charitie; *Ib*. 4260 (Ch.). Foure elnis lylis worsate to be ane skaiplare and doublett; 1546 *Treas. Acc.* IX 51.

(*b*) Fower syed skapulars & twa schort; 1530 *Balmerino and L. Chart.* II 32.

(2) Or gyf ȝe tuke the skapellarye, That ȝe mycht leif more plesandlye; LYND. *Mon*. 5858.

Scap(e)thrift, *n*. [SCAPE *v*. and *thrift*.] A spendthrift; a ne'er-do-well. — He gatherit a power of wickit scapthriftis, alias reivars and sorneris, and with the same . . beseigit the castell; *Breve Cron. Earlis Ross* 11. Accompanied with a number of scapethrifts and rebells, as the historie calleth them; GORDON *Geneal. Hist.* 27.

(Scaping,) Skaping, *ppl. adj*. [SCAP(E *v*.] Describing a place to which to escape, providing refuge. — No skaping place is in the seas thoch men uoulde Mars refuse; JAMES VI *Poems* I 239/347.

Scapperar, *n*. (?) — By this soverane salve and medecin may this blote onelie be takin away meit for scapperars, even by confession of a fault in the awin degrie; 1585-6 CARMICHAEL in *Misc. Wodrow Soc.* 443.

Scapthrift, var. of SCAP(E)THRIFT *n*.

(Scar,) Scair, Skar(r, *n*.[1] [ME and e.m.E. *scar* (Wyclif), *scarre* (1530), ? aphetic f. F. *escarre, eschar(r)e* (15th c. in Larousse).] *fig*. **a.** A scar. **b.** ? A blemish or fault. (But perh. the text is corrupt.) **c.** *comb*. ? Cleft or split with the scars of conflict. — **a.** It is thought that the skarr of ther greiffis doth still stick in ther hartis that awayte opportunitie for revenge; 1582 *Cal. Sc. P.* VI 221. — **b.** The pickman . . being persewit be Stewin Forbis for spilling of certane cornis be grinding thairof . . is ordaneit . . to pey ten pundis toties quoties he happin to fall in the lyk scair; 1615 *Urie Baron Ct.* 14. — **c.** Their wiues and children cry when will they come [*sc*. back to Scotland]? Yea, yea they come but with an empty hand . . Wherein I vow that England turnes a curse To . . my spent gentry and their purse . . And why? cause in a Rodomunto, they Play the Orlando Furioso aye: As well in humours, as in lavish charges, Which makes most femals weare such skar-clift targes Where deepest strokes in strugling force are given Till both mens ribs and rigs are backward driven; 1633 LITHGOW *Poet. Remains* 92.

Scar, Skar, Skair, Skyre, *n*.[2] [e.m.E. *scarre* (1530), *scare* (*a*1548); SCAR *v*.[1]] **a.** A (sudden) fright. **b.** Fear. **c.** ? An alarm signal (cf. OED *Scare* n.[2] 3). — **a.** This I said to geue hir a litle skar to mary, be the way; MELVILLE *Mem*. 159. For of his springs I gat a skair [: sair]; 16.. *Sempill P.* 44/84. — **b.** For his misbehauiour in sending sic cartellis, in scar and terrour to others; 1614 *Lett. & St. P. Jas. VI* 226. — **c.** Chargeing all and sindrie oure soverane ladeis liegis . . that thai cum and concur witht the wardanis in the defence of the cuntre and resisting of our auld innimeis as thai sal be adverteist owthir be skyris, balis or schott of gune furtht of Hume; 1558 *Treas. Acc.* X 384.

(Scar,) Scare, Skar, Skair, Skarr-, *n*.[3] [ME and e.m.E. *skerre* (14th c.), *skarre* (1387), ON *sker* a rock, reef.] **a.** A sea-cliff or crag. **b.** A rock or reef in the sea; a small rocky island. **c.** Only in the south-west: A bank of gravel and stones. Cf. SND *Scaur* n.[2] (2) 'on the upper Solway Firth: a bank of gravel and stones running out from the shore'. — **a.** Ane fair castell standand on the se skar . . the castell of Dumbar; STEWART 32723. — **b.** Dimidietate terrarum nuncupatarum Cuthilsyid et the Skair de Littill Segieden prout in carta; 1622 *Reg. Great S.* 124/1. For your shippe cannot approach nere to the land, since you say that these scares are many rockes covered with a shallow water; BLAKHALL *Narr*. 152. — **c.** Piscariam salmonum [etc.] . . de lie skaris et cowpeis de Cummertreis . . et lie Brewingskare ad dominum de Thorthorwald . . Hallgarthskar, Littillhallskar, Dirchallskar, Powskar, Newskar et Waitskar, cum ceteris lie skarris, draughtis, haldis, laikis et nettis infra dictas bondas; 1607 *Reg. Great S.* 672/2. 1637 *Retours* I Inq. Spec. Dumfries (162).

(Scar,) Skar(r, *adj*. Also: **scarr, scare, sker, scer.** [Late north. ME *skar* ? 'easily provoked' (OED) (*c*1460), ON *skjarr* (of animals) shy, timid, whence *skirra* SCAR *v*.[1]]

Also in the later dial. of north-west England.

1. Shy, timid, backward; hesitant.

Mesapus . . Dovn from hys stowt cursor, na thyng skar, Smait hym a grewos wond; DOUG. XII v 186. Prent the wordis . . Quhilkis ar nocht skar to bar on far fra bawrdis; SCOTT i 211. Thocht ȝe wes skar, Me think that now ȝe cum ouir nar; 1573 DAVIDSON *Sat. P.* xlii 62. Ȝe se thairfoir that ȝe ar mair skar nor vas S. Hierom; BURNE *Disput*. 132. Sen thair is nane that darre Æneas said bot all seamis uery skarr; JAMES VI *Poems* II 125/45.

2. a. Of a horse: Easily scared, restive.

The cappill . . is nought skeich, na ȝit sker [*M*. scer], na scippis nought one syd; DUNB. *Tua Mar. W.* 357. Bewar with this hors for he is werray skar & subtil . . in the end he castis the riddar oft tymes in the myre; ABELL 55b. The horse being scare, he twice threw him of; 1679 LAUDER *Notices Affairs* I 238.

b. Of a bird: Alarmed, terrified.

For quhen the bird is lampit in the lyme It helpis nocht that tyme thocht scho be skar, Becaus in tyme that scho culd not bewar; STEWART 4493.

c. Only in Shetl., of sheep: Wild.

[There is an ancient act] that none keep scar sheep except it be in holms of nesses diked in [etc.]; ? Hibbert *Shetland Islands* (1891) 185. Here are good store of sheep, many whereof are scarr or wild; 1684 *Descr. Zetland* 79.

Scar, Skar(r, *v*.[1] Also: **scarr(e, skarre; scaur; scare, skair; sker(r, skere; (sakir).** P.t. and p.p. also **skard, (scart).** [ME and e.m.E. *skerrenn* (Orm), *scarr(en* (14th c.), *skayre* (north., *a*1400), *skere* (*a*1450), *skare(n* (*c*1450), *skeare* (1558), *scarre* (1568), *scare* (1573), rare ON *skirra*. Cf. Norw. *skjerra*, Sw. dial. *skjarra* to scare, f. ON *skjarr* shy, timid, startled.]

1. *tr*. To make (a person, also an animal) afraid; to frighten, terrify.

(*a*) For althir-mast That takine [*sc*. the cross] the feyndis sakiris [*ed*. sckiris] & of thar mycht & purpos merrys; *Leg. S.* xxv 595.

(*b*) Ȝow wil he neuer skar; *Prestis of Peblis* 882. With that they raid away as thay war skarrit; DOUG. *Pal. Hon.* 298. Cum thow agane to skar us with thy strais We [etc.]; DUNB. *Flyt*. 214. And maid . . sic ratling that be noyis thairof thai fleyit and skarrit all the hors . . sua that . . thai mycht nocht be haldin; *Boece* 606b. Quhar garris ȝow cry me for to skar; *Bann. MS* 134a/25. Al your speeches and hard conditions shal not skarre vs; 1598 JAMES VI *Minor Prose* 66/35. [They have] sa scarrit and terrifeit the said Robert Jowssie [that he dare not return within this realm for fear of the like rigour]; 1599 *Reg. Privy C.* VI 60. I can not tell quhat hes him skarrit, Or hes the man amazed; *Philotus* 1183. Vtheris ar scarrit to cum thairvnto; 1604 *Aberd. B. Rec.* II 263. And I think that whilk scarres some most is feare to offend the kings majestie; 1638 *Rec. Kirk Scotl.* 167/1. All these commands . . will sufficiently scar all the fearers of God to join; SHIELDS *Hind Let Loose* 553.

(*c*) They [*sc*. witches] daunsed a whyle togidder till they were skaired by some people coming by; 1661 *Soc. Ant.* XXII 253.

b. ? To surprise, come upon suddenly; to give a fright to (someone).

Scho had severall meetingis with all tham . . somtymis at Androw Erskins litle hous quhair they war skard on night with James Moreis; 1658 *Sc. Ant.* IX 51.

c. ? To fear (a possible consequence).

For mene suld skere That thai ware to be il or ware; *Leg. S.* xxv 35.

2. To frighten *away* (a person or animal); to alienate by arousing fear or apprehension; to drive off *fra (fro, (furth) of)* a person or place. Also *fig.*

(1) Sa smartly thair thay can thame skar That the fecht halely warpit thay; *Alex.* I 1656. The spensar had na laser .. to skar nor chace; HENR. *Fab.* 304 (H). Yone knicht to scar with skaitht ye chaip nocht but scorne; *Gol. & Gaw.* 279. Howbeid gud counsale hestaly be not hard, With ȝung princis, ȝet sowld thay not be skard; LYND. *Sat.* 981. Sir hald ȝou stil & skar them nocht Till ȝe .. se quhat men them leids; *Ib.* 2340 (Ch.). *Ib.* 4368 (B). Grit skayth wesd to haif skard him; *Christis Kirk* 65 (B). Ay lyand lyk ane ruik, if na man wald scar thé; MONTG. *Flyt.* 108 (T). All the kye in the cuntrey they skarred & chased; *Ib.* 487 (H). That the deir and rae in Beinmoir is skarrit; 1614 *Keeping of Deer Forests* in *Argyll Mun.* (*Misc. 17th c.*) 2 July. Gif Patrik Pitcarne hed nocht skarrit him he waild haif presentit his fisch to the marcat to sell; 1621 *Perth Kirk S.* MS 2 July. Ther minister, his onkyndlie vsage of thame, haid scarred thame that they culd nocht communicat with him; 1631 *Strathbogie Presb.* 2. They will scarr and flie and missuse the gentilwoman; 1642 *Thanes of Cawdor* 287.

(2) That he mot skar thay felloun fleis away; *Prestis of Peblis* 516. The grettest gun .. will nocht scar thame nor chais thame away; DALR. I 25/15. [Persons who use the said weapons for shooting wild fowl, so that they are] scarrit and chaissit away; 1601-2 *Reg. Privy C.* VI 353. For laying on of the net to skar away the kees; 1638 *Peebles Gleanings* 319. Bot this schip lay still .. and skarrit away the salmound fishes; SPALDING II 242.

fig. The more to skar away Christian contemplation ye will finde for object to your eye [etc.]; BIRNIE *Kirk-b.* xvii.

(3) Thus wer thay fro the meit skerrit [: merrit]; *Colk. Sow* I 193. To .. skere thaim quyte Fra me that thai cum neuer agane; *Seven S.* 2622. Na dog durst fra my denner sker me Quhen I was tender with the king; LYND. *Compl. Bagsche* 115. I thocht I wald nocht sker thame of that place; LICHTOUN *Dreme* 43 (M). Be daylie hanting reparing [etc.] .. within our said forrest scarris fleis and banesis the .. wyld bestes furth of the samin; 1578 *Glenartney Doc.* I being .. skarrit and chaisit fra my familie; 1651 NICOLL *Diary* 41.

b. To deter (*from* (*fra*) a place, action or proposed action) by arousing fear of danger, suffering, etc.; to daunt.

(1) Is neyther castell, torret nor ȝit tour May scar him; *Prestis of Peblis* 1254.

(2) ȝe skar the wayklings from the wourd receauyng, Throw ȝour vngodlie and vicius behauyng; LAUDER *Minor P.* I 414. [Such actions] wald dasche and skar his ennemys fra all ther desperat enterpryses; MELVILLE *Mem.* 221. The liegis wer scarrit fra expeiding of thair richtis and securiteis by reasoun of the extraordinarie extortioun foirsaid; 1612 *Reg. Privy C.* IX 323. Albeit skarrit and hindreit, throw the fidelitie of ȝour keiperis, fra doing thairof; 1615 *Crim. Trials* III 312. Thes .. are now upoun this preconsaivit fear scarrit thairfra [*sc.* buying land]; 1623 *Aberd. Council Lett.* I 211. If they came, it might scarr people from hearing of them; 1638 *Rec. Kirk Scotl.* 148. Thir .. would scarre any man from imbraceing your violent and tumultuarie invitation; 1639 BAILLIE I 229. The third great end is to deter and scar people from sin; W. GUTHRIE *Christians Gt. Interest* (1776) 103. No fear of danger could him ever scar From diligence; 1688 *Elegie Death James Renwick* 5. By which means the people were scarred from the ordinances; 1701 *Strathendrick* 109.

(3) The Lyon and the Leopard From louping and scouping war skard; BUREL *Pilgr.* I 44. Some indirect course was taken to scarr the fishes from coming that way; 1664 *Reg. Privy C.* 3 Ser. I 573.

3. *intr.* To be frightened; to take fright; to feel fear, terror, etc.; to fear (*to* do something).

(1) The sowtaris hors scarrit [*B.* start *or* scart, *M.* stert] with the rattill; DUNB. (OUP) 156/194 (Asl.). With him [*sc.* Saint Francis] and with his abbeit bayth I skarrit; *Ib.* 165/6. This Calcas .. Skars at the last, throu .. slycht Of Vlixes constrenyt; DOUG. II ii 126. Quoth Courage, 'Scourge him first that skars'; MONTG. *Ch. & Slae* 1415 (Wr.). First I employed Duncan Campbell [to carry out a murder] quho scarred with it and refusit it; 1596 *Highland P.* I 181. So neir in skaith, suppose thou skantly skarris; MONTG. *Sonn.* xlviii 8. Scarre, scarre o my soule, with sinne; D. LINDESEY *Heavenly Chariot* 11. Many brethren who had assented to the maine point begann to scarr; SCOT *Narr.* 105. Let none scaur because they see some fall away; DICKSON *Wr.* 95. But it is the strangeness & unusualness of these apparitions maks you scare; 1674-6 KIRK in *Il Cappellano delle Fate* 159.

(2) Quhat neids ye skar .. To puneis sic as proudly dois rebell? 1570 *Sat. P.* xxiii 121. Then prayers, almesdeids, and tearis, Vhilks ȝit to skorne ȝee skantly skar, Sall mair availl than jaks and spearis; MONTG. *Devot. P.* vi 62. For in time past I never scarred, without scruple to wrap myself .. in the involucre of your highness' will; 1608 Craven *Argyll Diocese* 55. And that thair is no cause to skar to deale with him; 1613 *Stirlings of Keir* 481. Thou must not skarre vpon thy soares to looke; MURE *True Crucifixe* 1775. How farr their hearts did scarr to convict; FORBES *Rec.* 492. I find that great men, especially old friends, scaur to speak for me; 1637 RUTHERFORD *Lett.* (1894) 240. Heartless fears .. scaring to apply Christ and to make use of the promises; 1650 *Comm. Gen. Assembly* III 149. To be very loath, shy and scarring to meet with the Cross; DURHAM *Subtile Self* 117. Many of you now skars to hear tell of Christ's coming to judgement; Id. *Comm. Rev.* 25. Our dry bones starts throw the skin of it, that men and women scarrs to own it; M. BRUCE *Rattling Dry Bones* 35.

b. To draw back, as in fear, *from* an action or activity; to withdraw in alarm.

Than of ȝow I haid sik dout, That from my sleip soon did I skar; J. STEWART 149 § 9. Even the deboashedest amongst us, might learne to scarre from attempting .. vanitie; D. LINDESEY *Heavenly Chariot* 22. The bishope forbade the people to scarre from communicating for the offering that they were to give, seing these that offered little were notwithstanding accepted; CALDERWOOD VII 298.

4. *To scar at*, in the above senses: To take fright at, feel fear of, be alarmed by; to turn away from, as in apprehension, disgust or rejection; to be deterred from or put off by.

(*a*) 'Schir,' said the foxe,' .. at my figure beist and bird will skar'; HENR. *Fab.* 1971. For sa mekil as thai at the sey monstreis skarrit; DOUG. VII xii 180. Yit be nocht discouragit nor skar nocht thairat; 1568-9 *Hist. Fam. Kennedy* App. 20. Bot gif that thay grow proud and heich And skar at ȝow as thay wer skeich And on na wayis will bow thame; 1572 *Sat. P.* xxxi 218. ȝe suld nocht skar at the mater, becaus ȝe haue ane scho Paip besyd ȝou for head of the Kirk of Ingland; BURNE *Disput.* 96b. You skarr at them whom you were blyth to see; 1607 W. COWPER in Scot *Narr.* 195. So ye deuyd the losse betuix yow .. But if my lord sal scar at this your lordschip may wrytt that [etc.]; 1627 *Misc. Hist. Soc.* I 101. Scar not at suffering for Christ; 1637 RUTHERFORD *Lett.* (1671) 230. Skarre; HENDERSON *Serm.* 30. But scars at gnatts, least swallowing those, Be tymes I greatest camels eat; 1638 *Fugitive Poetry* II xvi 4/11. Finding the people there to scar at my company, so that none would give me a lesson; 1661 R. Brown *Paisley* I 353. Be not afraid .. neither scar ye at His cross; 1666 in *Naphtali* (1693) 341. Mr. Jhon Stewart preachd and baptizd two children .. yesterday. I scarr at thes things; 1678 BRODIE *Diary* 393. We should not scarr nor weary at the sufferings for Christ; 1688 A. SHIELDS *Some Notes or Heads of a Preface, Lecture and Sermon preached at the Lothers in Crafoord Moor* 16. Scarr; 1691 *Sel. Biog.* I 502. What ails poor harlot Scotland to scar so much at that noble gift? 1710 in Calderwood *Dying Testim.* (1806) 157.

(*b*) Therefore let none of you scare at the greatness of your sins; WELSH *Forty-eight Serm.* 59. When folks scares at Christ, and discourages and faints through their mistaking of Him [etc.]; DURHAM *Comm. Rev.* 44. Men should not scare at the sacrament because of the profanity of others; Id. *Scandal* (1659) 138. Every parish will scare at thee as a monster of men; 1699 T. BOSTON *Art of Man-fishing* (1900) 52.

(*c*) Quhy skerris thow at this holy weid? DUNB. (OUP) 165/11 (M).

Scar, *v.*[2] (e.m.E. *scar* to disfigure by inflicting a wound (1555).) *tr.* — Lik a madman that laughs whilst he is scaring his oune flesh; 1659 A. HAY *Diary* 19.

Scar, Skar(re, Sker. [ME *sker,* once, in the phr. *o sker* (Manning), of unknown origin. Cf. SCAR *v.*[1]] *A(ne (one) scar,* aside, apart, at or from a distance. — Neptunus hid hym lyke one sker; LYND. *Test. Pap.* 126. Thay play bokeik evin as I war a skar [*Ch.* ane skar]; Id. *Sat.* 1633. Standing a scar from the reiding and in feare from the pervsing the sam; FOWLER I 303. My friends .. turn aside, and glance on me a skarre And some, like barking dogs, begin to gnarre; *c*1611 MELVILL in *Fugitive Poetry* I ii 10/1.

Scarce, var. of SCARS(E *adj.* and *adv.* **Scarc(e)lie, -ely,** varr. of SCARSLY *adv.* **Scarcenes,** var. of SCARSNES *n.* **Scarcht,** var. of SCARTH *n.*[2] **Scarcitie, -y,** varr. of SCARSITY *n.*

Scar-crow, Skar-craw, *n.* (e.m.E. *scarre crow* (1553), *scar crow* (1592) a person, or device, to frighten birds, also (1589) *fig.* = 'something (not really formidable) that frightens or is intended to frighten' (OED).) — This was but a skarcraw to putt them in feare, where they needed not feare; CALDERWOOD VII 97. It may serve for ane beacon and scar-crow to the descendants to hold of thesse rocks and shelves; 1691 LAUDER *Jrnl.* 301.

Scar(e, Schare, *n.*[1] Also: **skar(e, scair(e, skair(e, scayre, skayr; (skiar); s(c)hair, shaire, s(c)hayr(e, share, shear, sheaire; saire.** [ME and e.m.E. *share* (1372), OE *scearu,* also MDu. *scare, scaer* a share in common grazing ground, MLG *schare* a share, troop of soldiers, ON *skera* to cut.]

1. The part allocated or belonging to an individual when a whole is distributed among a number of persons; one's portion (*of* something); a share in a non-material benefit or loss; also the portion contributed by an individual.

(1) The half [*sc.* of the treasure] ȝe hecht ws till our scaire; *Seven S.* 1726. The emprioure .. Gaif thaim the scair that to thaim fell; *Ib.* 1748. The scho ravyn now cravis the bird And sais scho aucht to be his hird The hie rawyn says scho aucht na scaire; *Ib.* 2665. Quhen all is drunkin I got to skair The likkyngis of the cop; LYND. *Sat.* 4382. The foulis that fleis in the air The doggis leuingis to tak vp to thair skair; ROLLAND *Seven S.* 10700. To refound .. to schir James Duff .. iij firlottis quheitt greit aittis comprisit on thair gudis of this last crope of saxtie tuay yeiris, euerilk man thair skayr pairt and pairt lyke; 1562-3 *Inverness Rec.* I 101. I ordine the barne that my wyf is with now instantlie .. gif it be ane madin barne bot haif hir playne skair as the rest gettis; 1583 *Edinb. Test.* XII 199. To the end that sho may haue the bettir skaire Of oure forfather tyed to bedd; JAMES VI *Poems* I 149/337. I leif the kyndnes of my mailling to .. my sonne .. Togidder with .. pleucht and pleucht graytht by skair and daill; 1589 *Edinb. Test.* XXI 93b. Schiris play all fair .. Ilke man his skair, & syne taks rest; MAXWALL *Commonpl. Bk.* fol. 5b. The said Robert Moir gat twelf scheip for his skair; 1635 *Justiciary Cases* II 322. The gentlemen and others .. resolved .. to repair and big up the said kirkyaird dyke, alloting to everie pleuche in the said parish suche a proportion of the said dyke as sould fall to thair scair; 1637 *Reg. Privy C.* 2 Ser. VI 391. When the Lord shows mercy to the land, ye sall find no skair in it; HENDERSON *Serm.* 402.

(2) My wyf to haif hir skair of the seid corne; 1583 *Edinb. Test.* XII 199. I ordine the barne .. to haif the .. seid corne that lyis vpoun tredaill by his skair that he gettis of the rest; *Ib.* To haue thair skair of timber conforme to vse and vont; 1587 *Aberd. Chart.* 347. And hes the vyt that rair Men dois declair thair happie skair of lair; J. STEWART 148/9. Ȝea and ilk ane of ȝow gat ȝour skair of the saidis robit gudis; 1610 *Crim. Trials* III 106. And allegit that he vantit his skair of the vining of the mairt ky sauld be the saidis Robert; 1629 *Monteith Stewartry Ct.* 7 May.

(*b*) [He is] to pey such shaire of any taxatioun as the rest of the tennentes within the said parochin; 1667 *Galloway P.* 7 Feb.

(*c*) The said Issobell .. constitut Mr. Gilbert Annand hir brether .. hir lawfull procuratour and commissioner to uptak and resseave her saire of quhatsumever guds geir dettis [etc.] .. perteins .. to hir .. be deceis of the said .. Robert; 1601 *Misc. 3 Spald. C.* II 57.

b. quasi-*adv.* *Skar skarlik,* on the basis of fair or equal shares, each having or contributing a like share; 'share and share alike'.

To mak cumpt reknyng & payment till Sande Sympsoune of xvij barellis of vnȝonis skar skarnlik [*sic in transcr.*] as thai war sald for in Leith; 1522 *Dundee B. Ct.* I fol. 101b (4 Aug.). That beand previt the pertinaris to pay that sowm skar skar lik; 1523 *Ib.* fol. 140b (24 April).

2. A portion of the cargo of a ship, a definite portion of a property owned or invested in by a number of persons in common, esp. the freemen or indwellers of a burgh, passing into a share in the profit or loss of a venture. Passing into **b.** A 'share' in a joint-stock company.

(*a*) Thom Wilȝamson [etc.] .. to by al gudis cumand in be se .. And ilk quartarismaister sal haf tyl his fe of ilk bate a skare tyl his fe togider with his awin skar and the refus of the gudis sal be bocht to the profetis of the fremen and na man out dwelland without the town sal haf a skar; 1452-3 *Ayr B. Ct.* 12 March. That wrangwisly he .. held fra him certane gudis anentis his scowt scare in fische & siluir as efferis tilhis part & wald nocht cum to compt & rekinyng; 1490 *Prestwick B. Rec.* 33. The haill consell .. of Wigtoun hais maid Gilbert erle of Cassillis .. burges wyth the consent of the haile .. comunte pleg to keip scot & lot to the toun & to tak gret skair vyne irn & salt & wdir merchantdice Patrik McKe; 1518 *Wigtown B. Ct.* fol. 78b. Lord Maxwell hais tane on hand to be in the grete skair pleg to keip scot & loit to the toun Thomas Hannay; *Ib.* Thir ar the gret skaris Gilbert erle of Cassillis plege, Patrik Mwre plege [etc.] .. the myd skair the commissar plege Jo[es] M[c]crekan Michell M[c]Connill plege [etc.] .. The small skaris .. Gibbe M[c]Ke plege [etc.] .. The samin day thir skarismen abone writtin hays [*MS torn*] thame & thair plegis costand at thair skaris in all tyme tocum as thai ar fore writtin. And it is statuit .. be the consent of the prouest .. & the said skarismen that quhen it happinnis ony of thame to refuys & takis nocht thair skair it salbe lefull to the .. officiaris of the town to geir on the rafuserris bring thair skaris to the dur & streng & tak the redrest geir at thair plegis hais & pay the merchand for thair skair and at nay skarisman in ony tyme tocum geif thair skair till ony vplandman quhill it be fyrst cullarit with himself vnder the pane of tynyng of thair fredome; 1518 *Wigtown B. Ct.* fol. 252a, 252b. The .. bailȝeis has coft ane schip of salt frome France man callit Geliame for xvi crounis vi d. for the boll of vattir met the thrid part hydis .. thrid part clayth & thrid part siluere to be pait within viii dais .. The .. bailȝeis & scayrmes [*erron. for* scarismen] ilkman for his awin scayre has tane the charge of the salt quhill scho enter in the hawin pleg ilkman for him self as thai ar writtin & scarit; 1525 *Ib.* fol. 163b. The alderman of Wigtoun Patrick M[c]Kee hais maid James Lang burges and the said James hais fund Adam Ahannay pleg to keip scot & loit to the toun & to tak his skair of schippis & wthir necessaris; 1530 *Ib.* fol. 233b. Of the gift of .. sowmes of money, skaris of schippis, gudis and prisis thairof [etc.]; 1546 *Reg. Privy S.* III 318/1. Comforme to his scayre of the said schip; 1551-2 *Dundee B. Ct.* II fol. 129 (29 Jan.). The nychtbouris of the brucht hes admittit the said William .. nychtbour and indwellar and also to haue his skaris of all necessaris that cummis to the towne; 1561 *Inverness Rec.* I 66. Comperit Petyr Waus and nayis him to haif bot the thred skayr of the said schip; 1572 *Ib.* 223. [The council] ordaines the nighbours to have their skair of the said tobacco; 1643 *Edinb. B. Rec.* VIII 31.

(*b*) Patriyk Mvre alderman callyt to the .. commvnyte challans & the marchiand of the schyp fore the vranguis vythaldyng of ane hoghed of his avyne schare .. the alderman sall paye the tua hogheddis .. to the marchyand of the schyp ane beand alovyt to hyme fore the smal covstomys; 1526 *Wigtown B. Ct.* fol. 183a. Patrick Mvr [etc.] .. hais coft ane schip of gascone vyne gud & marchiand .. & pleg fore the pament the forsaid viii personys .. euere ane fore thare awyne schayre; 1531 *Ib.* fol. 250b. He had tane Walter Hudsoun dettour for the saidis scharis upoun quhais confessioun the juge decernis .. Walter to pay to the said Johnne his scharis; 1558 *Admir. Ct. Bk.* (St. S.) 80. No nutrel man shall have any shair of my busenes or the profets arising therefrom; 15.. *Stirling Archæol. Soc.*, 27-8, p. 24. I did enter in the corporation of the fisherie trade .. and I to have my peart as other in the benifitte or my sheaire in the losse of the wenture; 1670 WEMYSS in *Sc. Diaries* 128. That evry persone of the company gett ane accomptt what he owes of stock, rent [etc.] .. That they who owe nothing gett ther annual-rent allowed them out of the first cloath falls to ther shear; 1684 *New Mills Manuf.* 65.

b. Payed in be the defunct to the peaper manufactorie for his shairs tharin iij^c lib. money forsaid; 1699 *Edinb. Test.* LXXX 331. When we came to London .. We .. ran up and down In rising stocks to buy a skair; W. Thomson *Orpheus Caledonius* (1733) II 20.

3. A part taken in an activity, etc.

And Christ wer heir, thou wald crucifie him agane .. gif sic ane turne wer in handis thou wald have an scair in it; ROLLOCK *Wks.* I 375. In the ensueing parliament or in any of your maiesties judicatories whair .. wee ar to have ane schair; 1660 *Conv. Burghs* III 524. I think he is not pleas'd that I am in this statione, wherof perchance he designed himself a shayr; 1667 *Laing MSS* 373. And have the halfe shaire in the government; 1682 Insh *Colonial Schemes* 194.

b. One's part or share; one's lot or allocated portion; lifespan.

That ressoun is bot vane To say a man may do na mair But serue a kirk vntil his skair .. Paull himself .. At a kirk did not ay remane; 1573 DAVIDSON *Sat. P.* xlii 556. Six hundreth yeiris & fourtie uas hir scaire Quhilk nature ordanid hir for to fulfill; JAMES VI *Poems* I 45/45.

4. A part, piece or portion (*of* something).

Some quots. may belong in 1 above.

(*a*) The froit .. He plukit and kest dovne gud scare; *Seven S.* 553. The wrangus withhalding fra him of ane quarter skair of the corne; 1511 *Alloway Baron Ct.* 18 May. ii cronys .. for the price of ane skair of beif; 1518 *Wigtown B. Ct.* fol. 82a. The inqueist will nocht grant to geif to ony owt townis burgessis that are preistis ony skaris of ony fredome; 1550 *Ayr B. Ct. & Council Bk.* 17 Nov. The saidis fischearis .. was in wse to bring ane large skayr of all keynd of thair fischis to the said castell; 1564-5 *Inverness Rec.* I 122. Payit for .. ane skair of a daill; 1599 *Edinb. D. Guild Acc.* 713.

(*b*) Lines which sphears in equall shares divyde; MURE *Sonn.* 2 Ser. viii 1. That they [*sc.* tanners] sall nather steip thair hydis in the burne nor wasche thair shair thairintill; 1656 *Glasgow B. Rec.* II 337.

5. *specif.* Chiefly in Kirkcudbright: A portion of the common land of the town let or feued to one of its inhabitants. Also *attrib.* and *comb.*

Quots. of 1584 and 1595 in (3) may belong in (2).

(1) That the sandis .. the quhilk contenis .. xxiiii akeris be set for the termys of xix ȝeris .. to the personis wnder writtyn .. And it is statute .. that all thir personis sal bruk & jois the said tak .. & within thre of the first ȝeris .. ilka man of his awn part & skare wil mak the town skathles of thai sandis .. and at ilke nechboure sal haf full power to set his skare .. til ony .. indwellar within the town .. and at na skare be set but consent of the alderman & the bailȝeis; 1476-7 *Ayr B. Ct.* 9 Feb. Ane fens maid .. vpoun Johnne Dungalsoun fischearis skair in the Borland; 1576 *Kirkcudbr. B. Rec.* I 7. Thomas McClellan prowest is becumin actit .. to set to the burgesses of the burgh the teynd schewis of the lands of the Bordlands viz. euery man his awin scair teynd .. peyand thairfoir yeirlie ilk grete scair xv s. vj d. being in the haill [£38 15 s.] yeirly; 1577 *Ib.* 57. It salbe lesum to euery man to tak his scairis of land in few for the payment of viij d. yeirly augmentatioun of ilk scair alsweill the Borland as vther land beneith the town .. euery man sall haif thairvpoun ane nyntein yeir tak of his awin scairis for the said peyment yeirly .. quhairby the saids scairis salbe specialie estemeit to thame as kyndele; *Ib.* 57, 58. For the maill of thair skair; 1580 *Ib.* 105. Reserwand alwais the tollerance and occupatioun of the said third part skaris in the Mylnflat to the said Thomas; *Ib.* 130. Awand to Johnne Meikill thair commoun clerk iiij lib. money for the maill aucht be the toun to him for the clerks skair of the yeir bygane and thairfoir ordanis the thesaurar to pay him the same and lykewais .. to pay to him yeirlie heireftir quhill he be enterit to the clerks skair of land iiij li. money; 1581 *Ib.* 152. It is statute .. that the possessours of ilk skair land beneth the toun saw ane auchlot peis yeirlie vpoun thair skairs in tyme cuming that the land may be the bettir manurit; 1584 *Ib.* 197. The possessouris of the Boirland skairis; 1595 *Ib.* 323. The landis of Meikill Kirkland .. ar dewydit in fyiftie skairis and set in few ferme to the burgesses .. for peyment yeirlie of ij s. money for ilk skair .. the Mylnflat .. contenis xxv skairis and ar set in few ferme to the inhabitantis of the said burgh for .. ij s. of few maill [yearly] for ilk skair [etc.]; 1597 *Ib.* 339. Particatam lie skair jacentem in Mylneflat .. inter lie skair olim ad Joannem Dungalsoun piscatorem pertinen. [etc.]; 1603 *Reg. Great S.* 499/1. Thair is na commoun lands pertening to [Kirkcudbright] .. but onlie the landis callit the Boirlands .. ; ilk halfe of the saids Boirlands is devydit in fiftie skairs or pecis of land quhilks war sett in tak to certain inhabitants .. for nyntein ȝeirs and is of long tyme now outrune, and is possessit presentlie bot be tolleratione quhill new takis .. be maid thairof; 1612 *Conv. Burghs* II 346. [The skair called the] ill shapen cloak; 1672 *Kirkcudbr. Sheriff Ct. Deeds* I 396. William Reid thrie skiars and ane two pairt; 1673 *Kirkcudbr. B. Rec.* MS 3 April. Umquhill proveist Fullartouns skair under the Laighe Gate; *Ib.* 12 or 17 Oct.

(2) The .. covnsall present assignit to Thomas McCrone burges the half skair of land last occupeit be vmquhill Stewin McCaill now waikand in thair hands; 1576 *Kirkcudbr. B. Rec.* I 10. [They] fyndand the petition .. ressonable in seiking of ane skair of thair commone burrow lands hes grantit to the said Patrik the first waikand skair and land that [etc.]; 1578 *Ib.* 85. Ane skair of thair commoun landis of Mylnflat; 1580 *Ib.* 111. 1581 *Ib.* 152 (see (1) above). That ilk burges taxman and possessour of the skairis of land of the Bordland Mylnflatt and lands beneth the toun and ruds sall pey [etc.]; 1585 *Ib.* 206. [The] Boirland hird .. ij s. for keiping ilk skair of corne to be peyit ȝeirle at [etc.]; 1606 *Ib.* II 10. Ther common landis .. quhilk is devydit in fyftie skaires of land quherof quhen the ane halff is vnder crop the vther halff lyes ley; 1627 *Ib.* 342. The burgh of Kirkcudbright to send with thair commissioneris ane perfyte rentall of the skaires of the commoun lands callit the borelands; 1639 *Aberd. Council Lett.* II 120. Of the skairis halff skairis particatis of landis and just half pairt of the thrie croftis; 1655 *Kirkcudbr. B. Rec.* II 987. And litle skare of beir land at the Kittroch hill a little skare of beir land callit the Old Ship skair; 1669 *Kirkcudbr. Sheriff Ct. Deeds* I 290. With certain skairs and riggs of beir land; 1674 *Kirkcudbr. B. Rec.* MS 13 July. To mak payment .. for ane skair of the ruids possest be them; 1679 *Ib.* 18 June. [His whole tenements of houses, barns, lands, or skairs or rigs of land belonging to him]; 1695 *Kirkcudbr. Sheriff Ct. Deeds* II 826 (16 Oct.). — Tua skairis beirland and ane skair ait land in the Boirland; 1599 *Kirkcudbr. B. Rec.* I 372.

(3) *attrib.* and *comb.* 1584 *Kirkcudbr. B. Rec.* I 197 (see (1) above). He leuis to Jonet Carsin .. hir half skair land in the Boirdland; 1595 *Edinb. Test.* XXVIII 30b. For selling his tua pairt skairland in the Boirland; 1603 *Kirkcudbr. B. Rec.* I 438. Seasing maid to him be vmquhill Thomas McCrone of his skairland in the Mylnflat; 1611 *Ib.* II 97. — 1577 *Kirkcudbr. B. Rec.* I 57 (see (1) above).

Scare, *n.*[2] ? Abbrev. form of SCAUBERT *n.* — For ane scare to William's sword 18 s.; 1667 *Old Ross-shire* I 60.

(Scare,) Skar(e, Scair, S(c)hair, *v.* [e.m.E. *share* (1586), *shaire* (1590); SCAR(E *n.*[1]] **a.** *tr.* To divide into shares or portions. **b.** *intr.* To have a share, participate (*with* someone) (*in* something). — **a.**

Quo die the inquest has ordanit thir men quartarmaisteris of the watter, & to skar the gudis the alderman [etc.]; 1462 *Ayr B. Ct.* 26 April. Ilkman for his awin scayre has tane the charge of the salt quhill scho enter in the hawin pleg ilkman for him self as thai ar writtin & scarit; 1525 *Wigtown B. Ct.* fol. 163b. Quhen ony blok of ky beis bocht in meanis amangst tham, that nain of tham be taikin and slain .. quhill thay be skarit and devydit; 1612 *Glasgow B. Rec.* I 332. Their land .. Hee 'mongst his people shaird; MURE *Psalmes* cxxxv 12. — **b.** Quhairof they scair in the proffeit; 1625 *Conv. Burghs* III 181. And I schall schair with yow in everie conditione it sall please God to bring us in; 1640 *Laing MSS* 205.

Scare, var. of SCAR *n.*[3], *adj.* and *v.*[1] **Scaresity,** var. of SCARSITY *n.*

Scarf, *n.* Also: **scarfe, scarff, skarf(f, skairf, skerf(f, skerffe; skraf, skref; (skaf).** [e.m.E. *scarf(f)e* (1555), *scraf* (1562), *skarfe* (1566), *scarf* (1624), OF *escrepe, echerpe, escharpe* (12th c. in Larousse), MDu. *scharpe.*] **a.** A scarf, a broad strip of fine stuff, worn loosely over the shoulders as an ornamental accessory to one's dress. In quot. *Waus Corr., scarf* may be erron. for SCART *n.* **b.** A (scarf used as a) sling for a wounded arm. **c.** A military scarf or sash. **d.** A black sash worn over the shoulder as a token of mourning at a funeral.

a. And now quhen I haif forsakin all wthiris that offerit me siluir .. I haif resauit ane scarf for my raward; 1576-7 *Waus Corr.* 139. Sax yards of tefenie to be two skrafis; 1603 *Montgomery Mem.* II 246. Skrefis; *Ib.* 250. Tuelf .. ourlairis with ane skaf, ane ring, ane new antefute of veluot [etc.]; 1607 *Edinb. Test.* XLIII 152. One preinȝie of silver to the skarf; 1608 *Tailor's Acc. Bk.* A 70. Tua riche skarffis imbroderit at xxx lib. the peice; 1618 *Edinb. Test.* L 164b. Fyve imbroderit skairfis at xx lib. the peice; 1620 *Ib.* 352. 3 ellis of taffitie half breid for a scarfe; 1666 *Household Bks. Archb. Sharp* in *Misc. Maitl. C.* II 540. Skerffe and hoods; 1667 *Old Ross-shire* I 60. Lutstring for skarfs; *c*1680 *Bk. Dunvegan* 194. For an alamod skerf, £20 10 s.; 1696 *Household Bk. Gr. Baillie* 192. 5 ells lutstring .. for hood and skerff; 1701 *Fam. Rose* 390.

b. The wound [in his arm] was haillit and he went that nycht to Tarsappie his scarff being cassin away; 1625 *Justiciary Cases* I 23.

c. Whosoevir sould ding the samyn [*sc.* papingo] doun sould be capitan, and have ane benne or scarff consisting of the value of twelff pundis Scotts or thairby; 1665 *Irvine Mun.* II 91. Some ask'd .. Why .. plaids for scarff about his middle; COLVIL *Whig's Suppl.* (1681) II 16.

d. Alamod, narrow for a murning skarf; *c*1680 *Bk. Dunvegan* 190.

(Scarf,) Skarfe, *v.* [e.m.E. *scarfe* (1598), *skarfe* (1605); f. prec.] *tr.* To blindfold with, or as with, a scarf. — She shot Loue, that he was faine To skarfe his eyes and begge the bow againe; CRAIG II 26.

Scarf, var. of SCARTH *n.*[1] **Scarfe, Scarff,** varr. of SCARF *n.*

Scarificatioun, *n.* [ME and e.m.E. *scarificacioun* (*c*1400), *scarification* (1533), L. *scarīficātiōn-* the action of *scarīficāre.*] The action of scarifying. — As mair extreme cure sic as ventosis, scarificatioun, blude drawing; SKEYNE *Descr. Pest* 44.

(Scarisman,) Skarisman, Share(s)man, *n.* Also: **(scayrmes).** [Cf. 19th c. Eng. *sharesman* (1895); SCAR(E *n.*[1] 2] **a.** In Wigtown, a shareholder in a venture. — **b.** One who has a share of something; a partaker. = S(C)HARER *n.* — **a.** 1518 *Wigtown B. Ct.* fol. 252b (see SCAR(E *n.*[1] 2 (*a*)). Scayrmes [*erron. for* scarismen]; 1525 *Ib.* 163b (see SCAR(E *n.*[1] 2 (*a*)). — **b.** Hee is part-taker of flesh and blood, fellow-brethren, and coheyres, shares-men in all the Fathers goods with him; DICKSON *Hebrews* 14. Yea, though he should undo them, He's sharesman of the harme; 1640 MURE *Counter-Buff* 245. Sharemen and partakers of all that is given unto him; DICKSON *Psalms* (1653) 308.

Scarket, *n.* (?) — I haue payet my part of iii honderis poundis and my part of the scarket that was coft; 1589-1600 *Skipper's Acc. (Morton)* 65a.

Scarlating, var. of SCARLETTING *adj.*

Scarlet(t, -lat(t, *n.* and *adj.* Also: **-leth, -late, -lait, -lo(i)t; skarlet(t, -lete, -lat(t, -late, -lot(t, -lote, -loth, -lit(t, -led; skerlet, -latt, -lot; skerllat.** [ME and e.m.E. *skarlet* (*c*1250), *skerlat* (*a*1533), OF *escarlate* (*c*1160 in Larousse), med. L. *scarlata* (*c*1155 in Latham), Pers. *saqirlāt*, anc. Arabic *siquillāt*, MGk. **sigillatos*, L. *sigillatus*, ornamented with little figures. Cf. ON *skarlat.*]

A. *noun.* **1. a.** A rich cloth, of various colours, freq. bright red. Passing into **b.** Cloth, or clothing, of scarlet colour. **c.** *pl.* Various sorts of scarlet cloth as in a or b above.

[Vna vlna scarleti; 1230 *Reg. Episc. Aberd.* II 267. 1373 *Exch. R.* II 439. Quatuor ulnis de scarleto; 1429 *Ib.* IV 499.]

a., b. (1) Scarlot furrit with riche furring; *Alex.* I 2522. Skarlete .. for ij cotis; 1473 *Treas. Acc.* I 41. His hude off scarlet [*H.* skarlate] bordowrit weill with silk; HENR. *Fab.* 1351. Scarloit; 1495 *Treas. Acc.* I 225. Scarlate to hing the kingis inner chamir in Strivelin; 1501 *Ib.* II 23. Scarlait; 1502 *Ib.* 150. For doublattis to the saidis servitouris ilk man haffand x quartaris skarlot; 1516 *Ib.* V 96. Scarlett; 1517 *Lanark & R.* 196. Skarlote and crammosie, With meneuer, martrik, grice, & ryche armyne; LYND. *Test. Pap.* 1047. The honest burges .. Sum in scarlot, and sum in claith of grane, For till haue met thare lady souerane; Id. *Depl. Magd.* 118. Threttie he put in his luferay, In scarlot fyne, and of hew grene; Id. *Meldrum* 1167. A stand of red scarlet; 1559 *Reg. Episc. Aberd.* I xc. Skarlote; 1565-6 *Treas. Acc.* XI 467. Preistis, leif ȝour pryde, Ȝour skarlet and ȝour veluote soft; *G. Ball.* 196. The lord Hammilltoun .. tuik his clok of skarlat and cust it wpoun him; PITSC. I 320/15. Frensche reid stamyn scarlot at iij li. the eln; 1581 *Edinb. Test.* X 76. The scaffeld .. was covered with linin cloth, then sand, next cloth, then scarleth; 1584 *Gowrie P.* 36. For skarlatt to be ane sedill cuffering; 1608 *Argyll Acc.* 17 Nov. Ane cloak of reid skarlatt; 1629 *Dumfries Test.* IA 215. For .. skarlet .. to be my ladies petticoat; 1666 *Household Bks. Archb. Sharp* in *Misc. Maitl. C.* II 539.

(*b*) The habit of ane erll .. salbe a mantillat of skerlatt opinnyt before; *Loutfut MS* 120b. v elnis skerlet to be the king ane schip sclop with ane hud; 1511 *Treas. Acc.* IV 193. ij pair of hois of skerllat xl s.; 1519-20 *Stirling B. Rec.* I 2. Ane pair of hois of skerlot price iiij li.; 1539 *Ex. Processes* No. x (Scott v. Scott & McDowall).

(2) For v quarteris Londone scarlat to lyne the samyn [doublet] lx s.; 1501 *Treas. Acc.* II 30. *Ib.* 35. Scarlet of Ingland; 1502 *Ib.* 295. Inglis scarlet; 1506 *Ib.* III 50.

(3) Narrow skarlot; 1525 *Treas. Acc.* V 253.

c. That na man .. that levys be merchandice bocht .. and thare wifis weire clathis of silk nor costly scarlatis in gownys; 1457 *Acts* II 49/2. Of this woll is maid the fine skarlettis, with mony uthir granit and deligat clathis; BELL. *Boece* I xxiv.

2. Scarlet colour, bright red.

The litstar that will put the maist noble colour of scarlet in a claith first he inducis and causis vthir colouris in the claith and disponis it to the recepcioun of the maist noble colour; IRLAND *Mir.* III 87/18. All litting by weight is discharged after the above writen deat, except bodayes and ingrand collers, such as scarlet; 1697 *Dundee B. Laws* 561.

B. *adj.* **1. a.** ? Made from scarlet, the cloth, see A 1 above. (? Only in early examples, and even in some, or all, of these, perh. properly belonging to b below.) **b.** Scarlet in colour.

a., b. [A] rede skarlet cape lynit wyth sylk and furret wyth ermyn [and] a new other skarlet cape off Florence skarlet broune cugnyt lynit wyth sylk and furret wyth menyuer; 1453-4 *Coll. St. Salvator* 153 n. 1506 *Treas. Acc.* III 351. Grene dames to breke the hale scarlet hos; 1506-7 *Ib.* 252. Crysp scarlet levis sum scheddand, .. Kest fragrant smell; DOUG. XII Prol. 125. As quha byspark wald the quhite evor Indane With scarlet droppis; *Ib.* ii 38. Quhat may ȝone fell freik be, With the quhite berd and scarlat ene dois bleir; ROLLAND *Ct. Venus* IV 667. A skarlet huid; *Bann. MS* 257b/37.

Scarlett; 1628 *Wemyss Corr.* 46. Send me .. skarlet or fyin gray Lundon cloith [etc.]; 1636 *Sutherland Bk.* II 168. iii ell of reid skarlett freise; 1638 *Black Bk. Taymouth* xxiii. What meanes .. this short skarlet cloake, and all this gay graith of yours? Row 462. Skarlitt; 1649 *Wemyss Chart.* 232. 8 ell of greine and skarlett rubene; 1650-1 *Fam. Rose* 349. 2 yeardis and ¼ of skarled bais; 1663-6 *Household Bks. Archb. Sharp* MS 42b.

(*b*) A gon and a hud and a bonat scarlat; 1457 *Peebles B. Rec.* I 119. ʒe Ismalitis, with scarlat hat and gowne, ʒour bludie boist na syith can satisfie; *G. Ball.* 104. This foule .. Whose taill of coulour was celestiall blew, With skarlat pennis that through it mixed grew; JAMES VI *Poems* I 44/40. The justice-generall .. sall weare a reid scarlatt gowne .. and the justice-depute and justice-clerk principall sall weare blak gownis; 1610 *Reg. Privy C.* VIII 614.

(*c*) Mercurius .. Doctour in phisick, cled in ane skarlot goun, And furrit weill, as sic ane aucht to be; HENR. *Test. Cress.* 250. Purpour colour, punik and skarlote hewis; DOUG. *Pal. Hon.* 537. Ane reid skerlot bonet; 1512 *Treas. Acc.* IV 370. Weluote to lyne ane skarlote cloke; 1565 *Ib.* XI 418. Skarloth; 1567 *Ib.* XII 51. A wrache to weir an nobill skarlot goun; *Maitl. F.* 196/33. Ane vyllicot skarlot cuminge hame from Lundoun that I send thair to lit; 1627 *Brechin Test.* IV 292.

(*d*) Half ane quarter of skarlit stemyng, x s.; 1597 *Tailor's Acc. Bk.* A 12b. Ane rid skarlott ryding coat; 1638 *Household Bk. M. Stewart* 24.

2. *comb.* Qualifying the name of another colour, esp. *rede* (red).

In 1453-4 quot. it is not clear whether a comb. is intended.

Riche was his gownis .. For Sonday silk for ilk day grene & gray His wyf was cumly cled in scarlet reid; *Prestis of Peblis* 211. In rob royall wes all of scarlat reid; STEWART 38147. Hir kirtill was of scarlot reid; LYND. *Meldrum* 121. Ane new skarlet rid wylicoit; 1612 *Edinb. Test.* XLVII 87. — Scarlet broune; 1453-4 *Coll. St. Salvator* 153 n. (see B 1 above). Scarlot gray stamyn; 1584 *Edinb. Test.* XIV 132.

(Scarletting,) Skarlet(t)ing, *adj.* Also: **scarlating.** [? SCARLET(T *n.* Cf. e.m.E. *scarlet* v. to colour scarlet (1641), clothe in scarlet (1553).] = SCARLET(T *adj.* 1. — Reid skarletting serge; 1569 *Treas. Acc.* XII 175. ij elnis reid scarlating stemmyng to cover twa sadillis; 1572 *Ib.* 317. Reid skarleting stemmyng; 1586 *Ib.* MS 122.

(Scarlik,) Skarlik. See SCAR(E *n.*[1] 1 b.

Scarlo(i)t, varr. of SCARLET(T *n.* and *adj.*

Scarlote. ? Erron. for *scariole* broad leaved endive. — Applicand the samin hait vpon the partis pectoralis with ane lytill scarlote trampit in the decoctioun; SKEYNE *Descr. Pest* 35.

Scarman, var. of SQUAREMAN *n.* **Scarmeshing, -ishing,** varr. of SCARMIS(C)HING *vbl. n.* **Scarmis,** var. of SCARMISCHE *v.*

Scarmis(c)h(e, Skirmis(c)h(e, *n.* Also: **skar-** and **-mis, -mys(che; mus(c)h(e, -mwsch, -mouche, -mus; -meis(h, -mes; sharmes; scremouche; scirmish; sky(i)r-** and **-mise, -mist, -me(i)s, -mish, -mysshe, -musche; skirmage; scrymishe; scrymmage; skyrmyng.** [ME and e.m.E. *skarmoch* (14th c.), *skarmyssh, -much* (both Chaucer), *skarmich* (*c*1400), *skrimishe* (1505), *skyrmosh* (1514), *scrymage* (1549), *scresmys* (1557), *skyrmush, skermyche* (both 1560), *scaramoche, screamishe, scrimishe* (all 1562), *scirmage* (1567), *scirmige* (1581), OF, MF *escremie* (12th c. in Larousse), *eschermie, escarmie* (12th-15th c. in Wartburg), MF *escarmuche* (14th c. in Wartburg); SCARMISCHE *v.* See also CARMUSCHE *n.* and ESCHARMOUSCH *n.*]

1. A skirmish, or skirmishing, an irregular encounter or minor conflict, esp. between small groups of opposing forces.

(1) Ane wond that he gatt at ane scarmus; BELL. *Boece* II 287. At the first skarmusche, Schir Johne Claremont, merschal of France, was slane; *Ib.* 447. Thir brethir rais ilk day in grete haterent, persewing vthir with maist cruelte and slauchter. At last thai mett togiddir at ane skarmusche in quhilk Remus alwayis was slane; Id. *Livy* I 22/6. Thair wes dayly raidis, skarmisheis and invasionis; LESLIE 48. Thair was mony scarmushis and invasionis; *Ib.* 186. A skarmish in Skotland, betwix Edinburgh and Leith, betuix Huntly .. against Morton; 1571 *Misc. Bann. C.* III 126. Some skarmessis thare war betuix the one and the other; KNOX I 221. At ane scarmische Williame Seytoun .. followit sa fast that he was takin prisonare; MAITLAND *Ho. Seytoun* 22. Thair was ane skarmys betuix my Lord Lyndsay and the Lard Wyrmyston; CULLEN *Chron. Aberd.* 38. He assalyeit thame in the nyght. The scarmish was great on both sydis dyvers slayne sindrie hurt and many preasoners tayne; *Hist. Jas. VI* (1825) 139. In the quhilkis carmeissis, M[c]Leane .. sleyis the Erlle of Huntlyis fader brother, .. thar being ane gritt skarmeis; *Chron. Kings* 152.

(*b*) Thei wold go furth to draw on a skyrmise; BANN. *Memor.* 237. At all skirmishes and vther tymes of pursuite of the king and his faithfull subjectis; 1572-3 *Ib.* 303. Skyrmishis; *Ib.* 177. This skyrmis conteanewed fra 7 houris .. till 5 houris at ewin; Id. *Trans.* 177. In this haill skyirmis, thair wes ane suddart slane; *Diurn. Occurr.* 219. Skirmist; *Ib.* 58. Skirmeis; *Ib.* 246. Lang tyme efterwart this skirmische was callit be the commons .. 'clenʒe calsay'; PITSC. I 283/29. And making freindship with sa rude a people thay thocht that eftir sum lycht skirmishes thay mycht grow into ane people; DALR. I 75/17. At the fecht or skirmisch of Bintoiche; 1604 *Hist. Clan Gregor* I 378. Skirmeissis; *Chron. Kings* 34. Skirmesis betuix the Englisches and Scottis; 1654 NICOLL *Diary* 121. Glenfroon famous for the scirmish betwixt the M[c]Gregors who ravadged this countrey in 1603 where the Colquhouns .. were defeat and many of the gentry .. killed; 16.. *Macfarlane's Geog. Coll.* II 199. — Everilke skirmage and new debait Of Cyprianis and of Saraseinis; *Clar.* III 321.

(*c*) Ane Inglisman, on the gait, saw he play At the scrymmagis [*M[c]D.* scrymmage], a bukler on his hand; *Wall.* III 359. Then Longaweill .. With Wallace past, as ane to that scrymmage. Thir four hundreth rycht wondyr weyll arayit Befor the toun [etc.]; *Ib.* IX 458.

(2) With skirmische and conteryng ilk da; STEWART 18907. Skirmusche; *Ib.* 10278. That they shuld burne the barnes about the house ande giff scrymishe thereunto; 1547-8 *Cal. Sc. P.* I 82. [The earl .. sent for his horses, and pursued them] in skyrmysshe [to Craigmyller]; 1571 *Ib.* III 595.

b. A raid or incursion.

(*a*) Thai come aganis the Romanis with ane haisty skarmusch; BELL. *Livy* I 86/27. Be licht iupertyis and skarmwschis all the somer thai suld impesch Romanis; *Boece* 147b. [The inhabitants provoked by] sum lycht scarmysche [rashly] mellit [with the enemy]; 1559 *Cal. Sc. P.* I 282.

(*b*) Thay wald nocht pas in Ingland at that tyme to invaid it with na sharmes; PITSC. II 119/23.

c. A sudden assault *on* or *gainst* a person or an enemy force.

Quhen that thai trowyt the Scottis was all at rest .. With v thowsand welle garnest and sawage Apon the Scottis thai thocht to mak scrymmage [*1570* skirmage] And at [the] ʒet wschyt owt; *Wall.* VIII 814. To mak ane scarmouche on the Romans; *Compl.* 114/13. Thair is ane slauchter committit in the scremouche wpoun George Drummond; 1554 *Corr. M. Lorraine* 384. For to yokke a skarmushe hotte .. Gainst Adamis head; JAMES VI *Poems* I 147/308.

d. A series of raids intended to engage enemy forces over a wide area; skirmishing as a strategy.

Congall .. aduertist the king of Pichtis he suld be of conforte and with .. skarmusch suld defend his landis quhil his cuming; *Ib.* 262b. King Malcolme .. contracting ane armye passit aganis inemyis devising be skarmusch alanerlie refrene thare audacite; *Boece* 449.

e. *fig.* A quick darting movement (? forward then back again).

The iugler fyne .. can no uayes makke to dance Makke skarmushes rinn & retyre & syne againe auance His littill marmousettis; JAMES VI *Poems* I 133/170.

2. A mock battle. See also SCARMIS(C)HING *vbl. n.* b.

(*a*) Scarmis; BANN. *Trans.* 98 (see SCARMISCHE *v.* 2.). It is the kyngis maiesteis will that the young men .. mak his hienes ane scarmeish their, quhairfor thei ordane James Ros to send tham .. fourtie pound wycht of pulder; 1578 *Edinb. B. Rec.* IV 79.

(*b*) Skyrmeis; BANN. *Memor.* 91 (see SCARMISCHE *v.* 2.).

3. ? Some sort of firelock.

A hackbuit with a snap, .. a hackbuit with a skarmische; 1589 *Wedderb. Compt Bk.* 182. [His say viz.] ane skarmisch of ane hagbute; 1627 Whitelaw *Sc. Arms Makers* 198. [David Fender gunmaker .. his sey] ane scarmisch work to be made; *Ib.*

(Scarmische,) Skarmisch, Skirmis, Skirmis(c)h, *v.* Also: **scarmis; -mysch; skyrmise; scirmishe, scrimische, scrimmish.** [ME and e.m.E. *skyrmysshe(n* (1385), *scarmysshe* (*c*1420), *scarmush* (*c*1470), *skyrmysh* (*a*1513), *skrymysshe* (1523), *skryssmys* (1554), *scaramosh* (1562), OF *eskirmir* (1080 in Larousse), MF *escarmucher* (*c*1350 in Larousse), OHG *skirm-, scirman,* Gmc. **skirmjan* to defend.]

1. *intr.* To engage in irregular, light or desultory fighting; to skirmish. Also *fig.* **b.** *To skirmish upon* (a place), to carry out a sudden raid or attack on.

(*a*) And skarmischand with him [*sc.* Huntly], thaj send word to .. Murray to come with the Louthiane men; *Diurn. Occurr.* 74. Hou the congregatioun .. skarmisched with Leith; PITSC. II 162 h. of ch.

(*b*) Our suddartis, seing the enemie, foulishlie wold gae vpon thame and skyrmise; BANN. *Memor.* 213. He continwallie skirmischit with thame; *Diurn. Occurr.* 170. [They] skirmischit witht the Scottis the space of ane moneth; PITSC. I 397/1. Wher themperour send doun some harguletis on horsbak to skirmysch; MELVILLE *Mem.* 25. Thai skirmisched about thrie houres; 1650 *Rec. Kirk Scotl.* 614.

(*c*) Iff we can itt is thoughtt we will feaght or att leist scrimmish this nightt; 1650 *Milne-Home MSS* 101.

fig. How that with thame I skirmisht not, nor with theme wold anes mell; FOWLER I 293/65.

b. The hagbuttaris also skirmishing vpoun the croftis the space of ane hore and more; BANN. *Memor.* 124.

2. To take part in a mock battle.

Ther was .. a foulishe vanitie devysed in the castell. The one pair of the captanes shuldieoris touke vpoun thaim to scirmishe [BANN. *Trans.* 98, scarmis] in maner of ane assault to the castle; the vther pairt .. tuike the defence .. of the castle. The skyrmeis [BANN. *Trans.* 98, scarmis] begouth about 8 horis at nycht... About the end of the scirmiscing [BANN. *Trans.* 99, scarmusching] the castle began to discharge cannones .. and so shortlie the counterfout assaultaris touke the flicht; BANN. *Memor.* 91.

3. *tr.* To engage (an enemy) in a skirmish; to harass with skirmishing.

[We lie nightly in our arms, most part on horseback,] .. to scrimische them; 1559 *Cal. Sc. P.* I 285. The Inglismen skarmisched thame oft; *Diurn. Occurr.* 47. They .. fell to the Inglischemen and skirmischit thame so haitlie that they caussit thame reteir; PITSC. II 109/3.

4. *It wes .. skirmissit*, skirmishing took place.

Thaj remanit .. abyding the erle of Mortonis furthcuming quha wes than in Dalkeith with my lord regentis men of weare. And at his furth cuming, it wes crwellie skirmissit betuix the saids pairties; *Diurn. Occurr.* 218.

(Scarmischar,) Skyrmissar, *n.* Also: **skirmisher, -meisher.** [e.m.E. *skirmishar* (1565); f. prec.] A member of a small party of soldiers sent ahead of the main body to engage the enemy in skirmishing; a skirmisher. — The Frensch .. had send furth thair skyrmissaris quha .. contyneuit skyrmissing till [etc.]; KNOX II 59. *Veles,* a skirmisher; CARMICHAEL *Etym.* 19. — A litill pincke .. called the skirmeisher [BANN. *Trans.* 35, skyrmisher]; BANN. *Memor.* 51.

Scarmis(c)hing, Skarmus(c)hing, Skirmissing, *vbl. n.* Also: **scarmising, -mus(c)hing, -mousching, -meshing, skarmisching, -missing; -mus(s)ing, -musyng, -musheing(e; skyrmissing, -mishing, scirmiscing; scrimmiging.** [ME and e.m.E. *skarmysshynge* (v.r. *scarmuchyng*; Chaucer), *skirmishing* (1592); SCARMISCHE *v.* See also CARMUS(C)HING *vbl. n.*] **a.** The action of skirmishing, in sense 1 of SCARMISCHE *v.*; an instance of skirmishing, a skirmish in sense 1 of SCARMIS(C)H(E *n.*

(1) The nixt summyr wes frequent and continewall skarmusching; BELL. *Boece* (M) I 105. To stop the Romanis fra invasioun .. be frequent skarmusyng [*1821* incursionis] erar than playn batell; *Ib.* 126. Followit frequent preiking, skarmussing and singular battallis on athir side; *Ib.* (1821) II 55. Skarmussing [*v.r.* scarmusching]; Id. *Livy* I 78/7. Thei maid hott skarmisching; KNOX I 86. At the sea-coast began the skarmissing; *Ib.* II 7. And sua held thame that day occupeit skarmusheinge; LESLIE 181. The parteis .. suddantlie entered in skarmushing, first with bowis and arrous; *Ib.* 184. Dyvers companeis .. assailyeit be scarmushing so neir the Inglis campt, that [etc.]; *Ib.* 201. Thair was .. hat scarmishing .. with harcabushers, culveringis and pistolettis; *Ib.* 282. Skarmusheing; *Ib.* 197. Sum captiues quhilk war taine in the skarmising befoir the chok of the battell; PITSC. I 127/32. Edwart .. fell in skarmisching with the Scoittis; *Ib.* 157/11.

(*b*) Thair wes daylie skirmishing Quhair men of armis brak monie sting; LYND. *Meldrum* 253. He wald not gett him in the skirmissing becauis he knew he was bot a cowart; KNOX II 10. Skyrmissing; *Ib.* 69. Whair was thicke skyrmishing and grit shutting fourth of the castle; BANN. *Memor.* 234. Thair was heat skirmishing betuix the congregatioun of Edinburgh and the Frenchemen of Leytht; PITSC. II 162/3.

(*c*) The ennymie now lays in Heddingtone .. yeisterd ther was soum los bott nott many presed be God this scrimmiging sowm .. thinks he is retyring bot otheris thinkis .. he will nott retir till he feaghtt; 1650 *Milne-Home MSS* 101.

(2) *sing.* He fell in the hands of the said capten efter a light scarmeshing; *Hist. Jas. VI* (1825) 128.

pl. The Romanis .. wer be ythand skarmussingis sa invadit .. that [etc.]; BELL. *Boece* (M) I 109. Than followit continewall skarmusingis with lycht hors; *Ib.* 158. The fulich Gabynis .. straik sindri small skarmischingis and batallis aganis the Romanis; Id. *Livy* I 116/2.

b. ? Mock battle, military exercises or simulated battle.

Scirmiscing [*v.r.* scarmusching]; BANN. *Memor.* 91 (see SCARMISCHE *v.* 2). The four quarter maisteris to be delt in tua quarteris at scarmousching or wapinschawing with anseinyeis; 1601 *Glasgow B. Rec.* I 218.

Scarmit, var. of SCHAMARRED *ppl. adj.* **Scarm(o)us(c)hing,** var. of SCARMIS(C)HING *vbl. n.* **Scarmush(e, -ysche,** varr. of SCARMIS(C)H(E *n.*

Scarpen, -ing, Skarpin, Scrippion, *n.* [17th c. Eng. (1611 Florio), gl. Ital.: *scarpini* scarpines, pumps or sockes, F. *escarpin.*] ? A light shoe or sock. — Thair dry scarpenis [*Maitl. F.* 440/95 scrippionis] baith trim and meit Thair muillis glitterand on thair feit; *Maitl. Q.* 229/95. Twa pair of mullis with thair scarpingis, .. ain pair of marikine mullis with ane pair of marikine schuis; 1583 *Cal. Sc. P.* VI 690. Syid gounes and larg breikks, mules and skarpins ver not in vse in th' Apostles dayis; COLVILLE *Paraenese* 116.

Scarping, *vbl. n.* (17th c. Eng. *scarping* (1639) a steep slope, the rocky face of a hill, also 19th c. Eng. *scarp* (1803) to cut to a steep face, to slope.) — Quhilk with the scarping and cutting of peeres of the rock .. may cost [etc.]; 1690 *Fraser P.* 269.

Scarping, var. of SCARPEN *n.* **Scarr,** var. of SCAR *adj.* and *v.*[1] **Scarre,** var. of SCAR *v.*[1]

Scarrell. [e.m.E. *scurrill* (1567), *scurrile* (1586), MF *scurrile* (1495 in Larousse), L. *scurrīlis.*] Scurrilous. — They vpbraide the spirit of grace .. with base and scarrell jests; BOYD *Last B.* 99.

Scarring, *vbl. n.* Also: **skarr-.** [SCAR *v.*[1]] **a.** *lit.* and *fig.* The action of frightening away, in the senses of SCAR *v.*[1] 2. **b.** Fear, or hesitation, (*to* do something), in the sense of SCAR *v.*[1] 3. **c.** The action of turning away from, or rejecting, in the sense of SCAR *v.*[1] 4.

a. He pretendis his said new fischeing to be betuix thame and the sey and in the sey forganis the watter mouth of Don thair scarring and debarring of the salmound fischeis with cobillis, airis, and nettis, stopping thame to entir in the watter; 1580 *Reg. Privy C.* III 295. Quhat tryall we twik of the beirar annent the skarring of ȝour deir; *c*1600 *Breadalbane Lett.* Lastly, consider for scarring from such games, these two things; DURHAM *Commandments* (1675) 111. By thus skarring people from serious considering the state and danger of their souls; *Just & Modest Reproof* (1693) 15.

b. Pusillanimity is a scarring to reach to, and adventure upon, what a man is able for; DURHAM *Commandments* (1675) 199.

c. With a sharp rebuke .. from .. Canterbury, calling our scarring at them scandalum acceptum, non datum; 1617 *Sel. Biog.* I 95.

Scarrs, var. of SCARS(E *adj.* and *adv.*

Scarrsbrugh warning. Also: **Skairs-; Scairsbrough** and **wairning.** (e.m.E. *Scarbrough warnyng* (1546) very short notice, or no notice at all; a surprise.) — This morning Sir George Home come to me with verie Scairsbrough wairning frome his majestie; 1592 *Cal. Sc. P.* X 710. That is Scarrsbrugh warning; Ferg. *Prov.* MS No. 1412. Ye giv me Skairsbrugh warning; *Ib.* No. 1646.

Scars(e, Scairs(e, *adj.* and *adv.* Also: **scarce, skars(e, skarce, scarrs, scarsse, skarsh, scharce; skairs(e, skairce, scairis, skairis, scairch, skairch, (scairth), skairsche, schairs, schairce; skers(e, skeris, skerche; sciers.** [ME and e.m.E. *skars* (1290), *scarse* (1340), OF *eschars, escars* (12-13th c. in Godefroy), late L. **scarsus,* f. p.p. of **excarpere* = L. *excerpere* to pick out.]

A. *adj.* **1.** Ungenerous in giving or spending; parsimonious, stingy, mean.

Ȝon couetous, ȝon skarce lymmar; *Alex.* II 206. Bot couetous, scarce and quyed He was sa, that [etc.]; *Ib.* 7149. For men sould scarce men, hard and proud And couetous, alwayes despyse; *Ib.* 9592. Scars wrechitnes, that neuir wan land; *Ratis R.* 318. Skars keping And auaritious wynnyng; *Colk. Sow* I 29. Scars spending, skathis gentris; *Ib.* 486. Ane fre liberale hert quharein noblenes inhabitis suld nocht be scars & haldand; *Porteous Noblenes* 182/18 (A). Be nocht a wreche nor skerche in ȝour spending; DUNB. (OUP) 38/5. Be not ouir skers nor ȝit ouir lairge; *Bann. MS* I p. 25/41. His table was nocht .. dilitious as diligat personis requiret, nouther was he ouer skairs, narraw or gredie; DALR. II 261/11.

2. a. Of an amount or quantity: Only just (the amount or quantity stated) and no more. **b.** Of a length of time: Short; restricted. **c.** Of a meal: Sparing; barely adequate.

a. Of xxiiij unce skars weit; 1511 *Treas. Acc.* IV 533. Ane skers elne of .. taffitie; 1593 *Edinb. Test.* XXV 191.

b. Providing sic defence as for the skars tyme thai mycht; *Boece* 223.

c. At evin only thay first prepaired the table, and that verie scharpe and skairs; DALR. I 90/10.

3. Of food or money: In short supply; barely adequate to meet ordinary needs; sparse. **b.** Applied to a period of scarcity, (a time) of dearth.

The pepill leiffit with sa skars and naturall fude that thay .. war preservit fra all maner of maledyis; BELL. *Boece* I 51. Ewin, to stabil his realme in virtew, commandit the young children of his realme to be nurist with skars and hard fude; *Ib.* 60. He levit .. ane hevinly life of skars fude and sleip; *Ib.* II 182. The Romanis be skars vittalis or sum vther displesouris wer constranit to pas oute of Brigance; *Ib.* (M) I 153. Quhilk victuallis ar sa skarce and evill to be had that [etc.]; 1604 *Shetland Sheriff Ct.* (ed.) 123. The cuntrey being exceiding pure, the moneyis skairsche and skant; 1657 NICOLL *Diary* 194. If I could get him to louse his purse upone my land, when mony is so scarse; 1667 *Argyll-Lauderdale Lett.* 71.

b. Well provyded in hors meat, which is a great consideratione this scairs year; 1682 in Turner *Mem.* App. 284.

4. *Scarse of* (*in*) (something). **a.** Poorly or inadequately supplied with; short of. **b.** *specif.* Sparing in regard to something.

He wes in purite of life maist clene .. To keip the sam he wes wera sobir in meit and drink & skars in sleip; ABELL 65b. Thame of Leith war verray skairce of victualls; KNOX II 72. For the captane of that house had made it scairse of victuallis; BANN. *Memor.* 297. Quhairby this contrey being sa plentifullie furneist of befoir is becum altogidder skairch of sic wairis [*supra* wyld foull]; 1599 *Acts* IV 180/2. Scairth; 1600 *Ib.* 236/2. The mercet is skars of meill; 1637 *Kirkcudbr. B. Rec.* II 562. Being skarce of elding, a boy .. broght in some old turfes; 1647 HOPE *Diary* (1919) 135. When they wer skairce of anything .. shee consulted with that spirit; 1650 *Brechin Presb.* 37. Many ar scars of cornis altho plentifull in pasture; *c*1641-54 J. GORDON in *Macfarlane's Geog. Coll.* II 577. If ȝow be scarce of [mercury] [etc.]; 16.. *Alchem. MSS* V

B. *adv.* **1.** Barely, hardly; only just.

See the notes to SCARSLY *adv.*, which also apply here.

Scars .. quhen, no sooner .. than (see the note to SCARSLY *adv.* 1 b and to SCANTLY *adv.* 1).

(1) Is he sa ganand to corrupt ȝow, that has skars [*v.r.* skers] power to corrupt the servandis? BELL. *Livy* I 266/27. And ȝit ȝe giue skarse credence in that cace; ROLLAND *Seven S.* 7440. Ther are yet but two byshopes arrived .. but scharce dare put their noses owte of their doores; 1561 Knox VI 128. The corbeis .. May scarce the heate abide; HUME 30/154. Thir fascheries war skarse weill sattelit, when out braks a graitter contradiction; MELVILL 123. Sa dashit .. with terror .. that he could skarse sitt; *Ib.* 246. Sanders, and the laif .. Who scarrs amongs our pastours trew war kend; 1611 Id. in McCrie *Melville* 446. Fishes, fleschis, fowllis, .. scarce gettabill in Abirdene; SPALDING II 241. Scarse were the cornes inned .. at Hallowmas; CALDERWOOD VII 514. Scarsse; BALFOUR *Ann.* I 120. Scarce there was any rain from March to neir the end of June; 1681 LAUDER *Observes* 42.

(*b*) Skers; BELL. *Livy* I 266/27 (see (1) above). That skers hes left bot skin and bon; SCOTT xxi 6. [She] anserit that scho serueit in this town ane ȝeir imediatlie befoir scho enterit his seruice quhilk he skers culd qualifie; 1619 *Perth Kirk S.* MS 18 May.

(*c*) Skairis aneuch to sustene thir thair realme; 1565 *Reg. Privy C.* I 402. Schairs durst scho sob; J. STEWART 19/160. Thocht skairse it vey the valeur of ane fas; *Ib.* 65/12. That skairs war thay able to reteine thair awne ony langre; DALR. I 159/20. Our folkes .. skairse mycht .. byd the firste brasche; *Ib.* 195/16. That skairs culde thay draw thair ende; *Ib.* II 299/12. Quhiche insolence is .. skairse excussabill, 1608 *Crim. Trials* III 49. A basse-borne passioune schairce rype till rottin; MURE *Early Misc. P.* i 111. I do skairse think that it can be .. altered to the bettir; 1613 *Lett. & St. P. Jas. VI* 213. Quhan the mylne hors war skairs able to cors the watter; 1629 *Justiciary Cases* I 97. As thair pay .. would scairch intertein them; 1643 *Acts* VI I 15/2.

(*d*) I fear you shall [s]ciers fynd tym to ried this so I will say no mor; *c*1666 *Laing MSS* 351.

(2) Scars this wes said quhen .. The sessonabil ayr pipis vp; DOUG. III viii 47. Scars had Juno thir wordis brocht to end, Quhen [etc.]; *Ib.* XII iii 90. With painfull paine thir wordes skairs birst out Quhen I wes forcit to turne my face away; *Maitl. Q.* 206/153. Skairs vas vpsprung as ȝit the chirming lark, Quhan .. I did mark [etc.]; J. STEWART 215 § 63. Lyon harat skairs til Edinburgh was cum quhen the king selfe was vpon the way; DALR. II 216/28. And skarsh wes stopt the torrent of that tide, When it [etc.]; GARDEN *Kings* 68.

b. Qualifying a numeral, measurement or (period of) time.

That it was scars mydnycht; DOUG. III viii 14. It is scarce ȝit nyne houris gane of the clock; ROLLAND *Seven S.* 4570. A counterfuted dule quhilk .. lastit scars aucht dayis; 1568 Hosack *Mary Q. of Scots* I 542. Thair is skarse twentie .. Sufficient men that I can call; 1573 DAVIDSON *Sat. P.* xlii 263. The tormentaris wer then so scant .. scairis found wer sax; MAITLAND *Maitl. Q.* 43/34. Skairse the lenth of ane buird; 1591 *Crim. Trials* I II 245. Skairce; DALR. I 56/19. Ane fyne bed .. viii fet skeris of lenthe; 1598 *Skipper's Acc. (Morton)* 36b. Ane rounde, skairs sax feit breid, and sax feit weyd; 1600 *Misc. Bann. C.* I 144. Their was skers twa vnce betwix his deid and his lyf; *Ib.* 145. Ther was skerse 4 agreid together; 1627 *Haddington Corr.* 154. A small iland scarce of a myl of lenth; c1641-54 J. GORDON in *Macfarlane's Geog. Coll.* II 516. He and his ladie cometh scarce to the kirk once in a year; 1656 *Lanark Presb.* 100. There were .. scarse fourty healthfull to doe duty; *Siege Castle Edinb.* 70.

2. a. In DOUG.: Rendering L. *vix*, sometime appar. without semantic force: Just. **b.** With difficulty.

a. Calcas held his tong ten days .. Skars at the last .. the laith word furth braid; DOUG. II ii 126. A few wordis skars as I mycht furth bryng For to comfort that maist lamentabill wight; *Ib.* III v 63. Skars on the ferd day at morn dyd I spy, Heich from the wallys croppis, Italy; *Ib.* VI v 125.

b. The myst was so thick .. that skairse mycht any man espy ane other; KNOX II 269. Skairse could he conteine him selfe .. bot behouet to dissimule the mater; DALR. II 7/12.

3. Sparsely, meagrely.

The onlie medcin for to .. presarue alyue Is skers by dyet wsd on homlie cheir to fead; Fowler I 394/15.

4. Rarely; hardly ever.

The turtle dwe, the feldifare, the nichtingale, with vthiris natiounis ar frequent, bot skairs with us ar fund; DALR. I 40/13. I haive sent ȝow .. thrie hunder wanettes, which are ȝit scairs to be gottine; 1637 *Breadalbane Doc.* No. 700.

Scarse, *v.* [ME and e.m.E. *skarse* (Gower) to become less, diminish, *scarsyn* (Prompt. Parv.), *scarse* (a1500), SCARS(E *adj.*] *tr.* To furnish with scarcements (SCARSMENT *n.*). — The said Villiame Robertsone hes scarsit the ane half of the said gawill nixt his [? seid] land to inlarge and mak rowm his loft; 1581-2 *Perth B. Ct.* 20 March.

Scarsetie, -ye, varr. of SCARSITY *n.* **Scarshes,** see SCARSNES *n.*

(Scarsit,) Scairced, *ppl. adj.* [Cf. SCARSE *v.*] Inadequately supplied (with (*of*) victuals). — The marques .. desyred, that in regard of the season .. and scairced of victuals, a part of the foot might only be given to him; 1646 BAILLIE II 417*.

(Scarsity,) Scarsitie, Scairsitie, -tie, *n.* Also: **scarc-, scarss-, skars-, scares-; scairc-, skairs-, skairc-, skairsh-, skairch-, skairth-; skears-, skearc-, skers-** and **-ité, -ity, -etie, -etye, -tye.** [ME and e.m.E. *skarsete* (a1370), *scarsetee* (Chaucer), *scarste* (Wyclif), OF *escharsetez*, *escarceté* (c1250 in Godefroy); SCARS(E *adj.*]

1. Niggardliness, meanness, stinginess. = SCARSNES *n.* 1.

Pride, inuy and skarsite .. Hes brocht thé now to thyne ending day; *Alex.* II 9606.

2. Smallness of amount, quantity or number, when compared with what is needed or required. **a.** Shortage *of* something, esp. of food or money; insufficiency of supply.

(*a*) Great skairstie of bread was in the haill citie; 1554 KNOX III 189. The Englis army, for skarstye of victualles .. retearis thame owir Twead; *Ib.* I 80. Skarsetye; LESLIE 286. The present hunger, derth and scarcitie of viveris within the burgh of Edinburgh; 1572 *Reg. Privy C.* II 148. Hagbuttis and murrionis, quhilkis, be ressoun of the scarsitie of thame, may not gudlie be sparit; 1581 *Ib.* III 398. The skairsitie of buikis and volumis of all sortis; 1589-90 *Trans. Bibl. Soc.* VIII (1927) 182. Be the grite penuritie and scairstie of the said siluer and allayed money; 1593 *Acts* IV 48/2. Be the grite scairstie of small money presentlie current within this realme; 1594 *Reg. Privy C.* V 190. The skairthtie of the saidis partridgis and murefoull; 1599 *Acts* IV 181/1. Scarstie; 1600 *Ib.* 236/2. Skairstie; MOYSIE 6. For scairstie of chamberis .. to serve the haill studentis within the said colledge; 1633 *Aberd. B. Rec.* III 63. Thair wes great skairstie of wictuall and elding; 1635 *Chron. Perth* 34. Gryt skarsitie of white fishes .. to the hurt and hunger of the poor; SPALDING II 154. The great skairchtie of provisiounes for horses; 1650 NICOLL *Diary* 9. This tyme .. thair wes great skairshtie of wynes; 1653 *Ib.* 105. Excuse the scarsitie of paper; 1676 *Old Ross-shire* I 300.

(*b*) The skearsitie of armour in this burght; 1626 *Aberd. B. Rec.* III 6. The dearth and skersitie of malt; 1649 *Aberd. Council Lett.* III 143. Skearcitie; 1666 *Old Ross-shire* II 84.

b. Shortage of persons necessary to fill offices, carry out duties, etc.

The scarsetye, yea nuilletie, of cardnalls partakers which wer and ar the fontane .. of al altercations amangs theme; FOWLER II 94/8. The scaircetie of skilled and weill affected assysoures in thir caisses; 1606 *Melrose P.* 12. The ressone of this scarsetie wes becaus the covenanteris .. causit the haill seruandis .. flie out of the countrie for feir; SPALDING I 262. In respect of the scarsitie of ministers in Cathnes and Orknay; 1651 *Comm. Gen. Assembly* III 189.

c. Of non-material things.

To impair ȝour lordschippis praises by scarcitie of words; 1581 FOWLER II 10/14. The scarcity of faith in the earth; 1637 RUTHERFORD *Lett.* (1894) 506.

d. Deficiency; shortage (as opposed to abundance).

Ane adjectiue and a verbe signifeing plentie or scairsitie governes the genitiue or, ablatiue; *Rudiments* fol. 38a.

3. Shortage, in a community, of the necessities of life, esp. of food; dearth.

It hes plesit God to visie thame with the lyke derth and scarssitie; 1576-7 *Reg. Privy C.* II 589. The saids factouris in tyme of scaircitie does buye victuall .. before the hand and vtters the samin agayne at ane hier rate; 1624 *Conv. Burghs* III 165. Since that dearth [of corn] is ane foirrynner of scaircitie; *Ib.* 170. And that the towne wold be pleased in regaird of the scairctie of the ȝeire to augment his stipend; 1650 *Inverness Rec.* II 203. Thair wes such great skairshetie in Edinburghe, that all sorts of viveris, meit and drink, could hardlie be haid for money; 1650 *Rec. Kirk Scot.* 615. Skairshitie; *Ib.* That some plan be devised for distributing the money collected for the poor in this time of scaresity and dearth; 1696 *Edinb. Kirk S.* in *Sc. Ant.* XIII 77.

4. a. A spartan régime, under which rejection of luxury is regarded as a virtue; simplicity of life.

This king commandet to bring vpe the ȝouth with al hardnes and skaircetie; DALR. I 145/12.

b. Poverty; penury.

Reduced to such penarie and skairstie that they are note able to subsiste any longer; 1675 *Kirkcaldy B. Rec.* MS 23 Aug.

Scarsly, Scairslie, *adv.* Also: **scarc(e)-, skars(e)-, skarce-, schars-; skairs(e)-, skairc-, skayirs-; skers-, scers-; skirs-** and **-lye.** [ME and e.m.E. *skarslich(e* (14th c.), *scarsly* (Chaucer); SCARS(E *adj.*]

Cf. SCARS(E *adv.*

1. Barely; only just; not quite. Passing into 3 below.

Modifying (chiefly) verbs, also numerals and, rarely, adjectives. Freq. disjoined from the word, or word group, modified.

Sometimes with inversion of subject and verb.

(1) The king Robertis sone Dawy .. than bot fyve ȝer had scarsly; BARB. XX 38. This prince rang skarsly two yeris; BELL. *Boece* I 40. Ane word to me skarslie scho had to speik; ROLLAND *Seven S.* 2360. When skairslie coulde ten be founde into ane contrie, that rychtlie knew God; KNOX II 444. Thair was scarslie the ane halff of the assyis cuntriemen; 1576 *Orkney Oppress.* 44. Skarslie the half of the materialles, lyme and stean; 1600 MELVILL

6. Skersly consisting two hundreth and fortie communicants; 1649 *Laing MSS* 248.

(2) Quhilk infeftment is weill auld, the letteris .. consumit, and be schort spaice the same wil becum skairclie legiable; 1407 *Montgomery Mem.* 20. His vissage semyt scarsly blith, .. He was .. no thing lycht; DOUG. VI xv 39.

(3) Sa paysand was this cote that scarsly mycht Phegeus and Sagaris .. Bair it on thar nek; DOUG. V v 37. The lift micht skarslie be sene above thair heidis; BELL. *Boece* I 268. Sa wery that he mycht skarsly draw his aynd; Id. *Livy* I 250/23. The quene .. skarselie wold geve a good worde .. to any .. favoraris of the Erle of Murray; KNOX II 358. And skarslie in so lang tyme is the air mouit and reneuit; SKEYNE *Descr. Pest* 12. Rome .. sal skarslie haue the maiestie of ane impyre; BURNE *Disput.* 134a (Jam., s.v. *Imminish*). I have evin now gottin it, and skarslie read it over digestlie; 1585 Calderwood IV 310. The sea so calme as scarcelie ebbs or flowes; JAMES VI *Poems* II 80/18. Sa of a noble and ritche prince as ever governde in Scotland, he [*sc.* the Earl of Morton] becam a pure prisoner, wha skarslie haid to furnise his necessitie; MELVILL 116. My wound, skarslie stemit, beguid to blude apace; *Ib.* 145. And skarslie .. gott sa mikle as a fear answer; *Ib.* 435. In poore menis names skarslie knowne in Scotland or England; 1611 *Reg. Privy C.* IX 267. Ten thousand punds will scarcelie mack conquese above five hundreth punds of constant yeirlie rent; 1630 *Aberd. Council Lett.* I 331. Scarcely; 1648 *Sc. Hist. Rev.* XIII 384. Scarclie; 1679 *Laing MSS* 419.

(*b*) Mony vtheris small wardis and mariagis, quhilkis skairsly may be nowmerit; *c*1522 DOUG. in Doug. (Sm.) I p. cix. Skairslye; BELL. *Boece* (M) II 325. Thair gouernour skairslie know gif he had to the foir aucht hunder futmen; DALR. II 325/24. Skairsely; MELVILLE *Mem.* 45.

(*c*) Sum vtheris tounes and dorpes, quhilkes a hail ȝeir skirslie culd be clinsed; DALR. II 294/4.

(4) Thai .. mett him with sik properant fury that skairsly had ony of the sydis lasar to cast thair darttis; BELL. *Boece* II 128/23 (M). Scarslie wait I .. Quhat till eschew or quhat for till embrace; ARBUTHNOT *Maitl. F.* 49/24. As no reason could satisfie him, so skarcelie could he haue patience euer to heare it offerid; 1594 *Lett. Jas. VI to Eliz.* 102.

b. Preceding a temporal clause introduced by *quhen*, virtually = no sooner .. (than).

Skarsly [*Sm.* scharslie, *Ruddim.* skarslye] the statw was in thar tempill vpset Quhen all hir membris bittir terys swet; DOUG. II iii 53. Scarsly the auld thir wordis had warpit owt, Quhen sone the ayr begouth to .. rowt; *Ib.* xi 23. *Ib.* V xiv 57. Flanders, .. wher he had scarslie remained tuo yeirs, when he wes agane recalled; GORDON *Geneal. Hist.* 144.

c. *Scarsly gif,* ? probably .. (not). Appar. indicating the probable non-fulfilment of the action of the following verb(s).

And scarslie gif the said Hector of Coill dar meyn ta the quenis majestie or persew Hector [of Dowart] thairfoir .. becaus of his creweltie and tyranne, quhilk he will use aganis .. Hector of Coill. .. And siclike, because the said Hector of Coill dar nocht persew .. the said Hector [etc.]; 1564-5 *Reg. Privy C.* I 312. The wreit .. quhilk is not now vsit; and skairslie gif ony man can reid the samyn; 1591 *Aberd. B. Rec.* I ix.

2. a. ? In a small degree. **b.** Scantily; inadequately; sparingly, parsimoniously.

a. Gif for him self euery man scarsly [*Arund.* scersly, *Harl.* skerslie] May for his faltis schawe sure sufficians; *Contempl. Sinn.* 771 (Asl.).

b. He that sawis scarslie sal alsa schere scarslie; NISBET *2 Cor.* ix 6. He .. on his veyage past Fameist for fude and richt skarslie repast; ROLLAND *Ct. Venus* II 473. Skayirslye; WINȜET I 122/24. Thai sowped skarslie becaus thair schippis war takein in the quhilk was thair victuellis; KNOX II 13.

3. With difficulty.

There may be some overlap with 1 above.

Quhen from the scharp rolk, scarsly, with gret slycht, Sergestus gan vpwreil hys schyp; DOUG. V v 51. Skarslie als he chaipit with his lyfe, He wes so straitlie sted; STEWART 21441. Thir twa ȝoung men .. war sa wonder like .. Skarslie micht nane knaw the ane be the vther; ROLLAND *Seven S.* 8745. I can skarslie weill abyd the tearis of my awin boyes whome my awin hand correctis; KNOX II 389. Quhen ane man hes vnder cure Sa mony thousandis, .. Skarsly will he ken ilkane Quhen twentie ȝeiris ar cum and gane; 1573 DAVIDSON *Sat. P.* xlii 489. Men culd skairslie contene thameselffis from teiris and murning; PITSC. I 110/17. FOWLER I 101/119. Bot .. the puiris bene far fra the cunyehous can scairslie gett for the xvj s. pece xij s. or thairby; ?*a*1600 Cochran-Patrik *Coinage* I 104. He is scairslie able to draw his end or to speik; 1600 *Reg. Privy C.* VI 112. A sweik thik mist .. wherby we might bot skarslie gis at the sight of the land; MELVILL 170.

Scarsment, Scairsment, *n.* Also: **skars(e)-, skaris-; skairs-; scerce-.** [Med. L. (north. Eng.) *scarcementum* (1398 in OED); SCARSE *v.*] A ledge or shelf of stone in a wall; *specif.* in the foundation of a building serving as the base for a wall, or in a wall as a support for cross-beams, etc.

Pinnakillis, fyellis, turnpekkis mony one, .. Skarsment [*L.* skarsement], reprise, corbell and batellingis [etc.]; DOUG. *Pal. Hon.* 1434. And that the said George hes ane skarisment with tymmer table and watter table that [etc.]; 1529 *Edinb. Guild Ct.* 29 April. That the said George Mayne sall leif still the scarsment on the north of the said gavill; 1532 *Ib.* 19 Nov. To that effect to big and upput ane upricht gawill fra the grund of the skarsement of ane auld gavill alreddie foundit in the said Thomas land; 1569 *Canongate Ct. Bk.* 44. For poynting of scercementes and for poynting of wondokes; 1611 *M. Works Acc.* (ed.) I 336. That ane skairsment fra the ground of asler wark be biggit at the north side of the tobuith stair of ane ressounable hight above the steps for saifty of persons repairing up & doun; 1620 *Dunferm. B. Rec.* II 124. The magistrattes .. grantes licence .. to .. John Crichtoun to found the said gavill upon the saids .. scairsments of the said putt joynd to the steiple, the said John Crichtoun not extending .. the said gavill beyond the saids tua first scairsments; 1664 *Peebles B. Rec.* II 61.

Scarsnes, Scaircenes, Skairnes, *n.* Also: **scarce-, scarss-, skars-, skarsch-; skairs-, skairce-; skears; skarsines; skernes.** [ME and e.m.E. *scarsnes* (Cursor M.), *scarnes* (*c*1400), *charsnes* (1508); SCARS(E *adj.*]

1. Niggardliness, meanness, stinginess.

For scarsnes in ȝow .. sa rutit is That thair is na thing that ȝe leif On ony syde; *Alex.* II 6654. *Ib.* 7007. Larges relevis & succouris a man and scarsnes interdytis nobilite; *Porteous Noblenes* 182/20 (A). Away was falsit, .. Away was nigardie and all skarschnes; *Clar.* V 1832.

2. Shortage of the necessities of life; dearth. Cf. SCARSITY *n.* 3.

Na king suld .. wit of scarsnes nor of wrecchit thing That micht his hart to dreid of mister draw; *Regim. Princ.* 93 (Maitl.). Wastand their lands .. with multitude of scheip .., thereby trubland Gods people with skarsnes, povertie, and outer hairschip; SKENE *Reg. Maj.* II 8. [Not able to harvest his crop] be reasone of the skearsnes this present yeir, [Ord sold his crop to the landlord]; 1623 *Aberd. Sheriff Ct.* II 272.

3. Deficient supply (*of* persons or material things needed for a particular purpose, also *lit.* and *fig.* of a non-material thing). Cf. SCARSITY *n.* 2.

(1) The want and laik of cunning men, raritie and skarsines of thame to teche and reid within oure realme [etc.]; 1555-6 *Reg. Privy S.* IV 555/2. Thai wantit servandis and could get nane to fee .. throw derth and scarssnes thairoff; 1576 *Orkney Oppress.* 64. To the end the straingeris .. may not find skaircenes of men alsueill as meitt; 1617 *Dundas Fam. P.* No. 50.

(2) It is fund that the lattin furth thairof [*sc.* the burgh loch] is the principall caus of the scarcenes of watter .. in the drowth and

somer seasoun; 1580 *Edinb. B. Rec.* IV 161. For skairsnes of tymer in this west cuntrey; 1597 *Conv. Burghs* II 12. The not selling of barkit hydes quhilk occasiones the skearsnes of schewes; 1666 *Old Ross-shire* II 82. Thomas Robertsone .. [complains] upon the scarshes [*for* scarsnes] of the watter in the windmill; 1670 *Edinb. B. Rec.* X 83.

(*b*) Throw the greit dartht, penuritie and skernes of victuellis that is presentlie in the cuntrey of Orknay; 1578 *Rec. Earld. Orkney* 354. Becaus of penuritie and skairnes of tymer; 1590 *Glasgow B. Rec.* I 153.

(3) Penuratie of and scarsnes of nottis in ane part of ane sang; *Art of Music* 26b. They will find greit inlaik and scaircenes of this spreit; ROLLOCK *Wks.* I 447.

Scarsse, var. of SCARS(E *adj.* and *adv.* **Scarssitie, Scarstie,** varr. of SCARSITY *n.*

Scart, Scrat, *n.* Also: **skart, scarth, skarth.** [e.m.E. *scrat* (1542); SCART *v.* With the common Sc. metathesis of *r* + vowel (cf. GERS *n.*, GRES *n.*, etc.)] A scratch. Also, *scart free, hail(l scart(h,* without a scratch, uninjured, unhurt. See HALESKARTH *adj.* for further instances.

(1) With scarts and scores athort his frozin fronnt; POLWART *Flyt.* 542 (H). Ane scart on the left neis; 1623 *Orkney & Shetl. Test.* I 94. All the children safe without a scart; FRASER *Polichron.* 455.

(*b*) We will haw the byttis, and rywis, and scrattis in owr bodies; 1662 *Crim. Trials* III 611. *Ib.*

(2) All whom the lawyers do advise, Gets not off scart free; CLELAND 110.

(3) Thy slycht and wylis sal thé nocht beir away, Nor haylscarth [*Ruddim.* hail skarth] hyne do turs thé hame fra ws; DOUG. XI xiii 155. *Ib.* XV 69. Haill skart; PITSC. II 103/33 (see HALESKARTH *adj.*).

Scart, Skart, Skrat, *v.* [ME and e.m.E. *scratte(n* (*a*1225), *skratt(e* (*c*1440), *scrat* (*c*1500), MSw. *kratta,* OF *gratter* (1155 (Wace) in Larousse), *esgrater* (Godefroy), f. Gmc. **kratton.*]

1. *tr.* To scratch, lacerate or mark (a person, his (one's own) face, etc.) with the nails, etc. Also in *fig.* context. **b.** *To be skarted* (*in* the leg), to be lightly wounded.

(*a*) With hyre handis [she] skartyt hir face; *Leg. S.* xxvii 249. To se him scart his awin skyn grit scunner I think; DUNB. *Tua Mar. W.* 93. Hyr rosy chekis to tor and scartis sche; DOUG. XII x 129. Scho .. Raif doun hir hair and skartit hir visage; ROLLAND *Seven S.* 1171. Scho .. began to skart and rife hir skin; *Ib.* 2726. The said Elspet and Mareoun .. drew her hayr, scartit hyr face [etc.]; 1566 *Inverness Rec.* I 138. I sall ȝow skart quhill that ȝe bleid; *Bann. MS* 134a/39. The said Bessie .. hes skartit me with hir naillis; ?1586 *Glasgow Prot.* IX 155. His hands outragius did his visage skart Maist horribile; J. STEWART 94/524. The said Robert .. flew on his face and .. skarted him; 1629 *Reg. Privy C.* 2 Ser. III 231. Scartavit prælustre caput; *Polemo-Mid.* 20. Formosas scartabo facies & vos wirriabo; *Ib.* 161. They .. beat, abuse and scart us with their hands; 1670 *Old Ross-shire* I 308. Then up with plaids and scarts her thighs; CLELAND 34.

(*b*) Hyr face scho skratit; *Leg. S.* xxxi 365. Mad hyr sa bludy & bla, & skratit hir [etc.]; *Ib.* 523.

fig. Some eminent divines Makes their assertions here to thwart, And one anothers cheeks to scart; CLELAND 89.

absol. He seikis for his sark Ay to skart and to claw is his first wark; *Bann. MS* 158b/86. It skittered, it scarted; they skirlde ilk ane; MONTG. *Flyt.* 486 (H).

b. Cromwel skarted in his legg; Col. Hacker killed; 1650 JOHNSTON *Diary* II 6.

2. To scratch aside (earth) in the process of digging; to tear (a hole) *in* (something) by scratching.

The modiwart .. Quho fast the eirth culd scraip and scart; BUREL *Pilgr.* I 191. A bustle will claw his womb long before he skart a hole in it; 1658 R. MORAY *Lett.* fol. 219.

b. *To scart* (a person) *out of eird,* to dig up by scratching or scraping with the hands or nails.

Quhen he [*sc.* your son] is deid ȝe .. wald be fane to scart him out of eird; ROLLAND *Seven S.* 5737.

c. *intr.* Of a hen: to scratch, scrape at a surface (in looking for food).

For his hen scarting on the baillies house; 1660 *Lanark B. Rec.* 181.

3. *intr.* To gather or scrape (items) *togidder* with difficulty. **b.** *intr.* To save parsimoniously.

Certean articles, (skartit togidder be him,) or rather blottes of that .. ordour quhilk sould be in Chryst's Kirk; 1583 Melvill 159.

b. If loue of money .. Moue thee to scrape, to scart, to pinch, to spare [etc.]; MURE *True Crucifixe* 2573.

Scart, p.t. and p.p. of SCAR *v.*[1]

Scarth, Scart, Scarf, Scrath, *n.*[1] Also: **skarth, skart.** [17th c. Eng. *scarfe* (1668), ON *skarfr.* Also in the later dial. of Orkney and Shetland.] A cormorant.

(*a*) The scarth a fische fangar, And that a perfyte; *Howlat* 181 (A). Pyssacone, After his deith, .. Intill ane skarth transformit was anone; DOUG. *Pal. Hon.* 1206. Quhar skarthis [*Ruddim.* skartis] with thar bekis .. glaidly thame pronȝe and bekis; Id. *Æn.* V iii 49. Yielding no commodity to the neighbouring inhabitants, but skarts and solane gees; *Sibbald's Orkn. & Shetl.* (1845) 5. The other birds which nestle in the Basse are these; the scout, .. the cattiwake, .. the scart [etc.]; RAY *Remains* 192. It [*sc.* the Bass Rock] has abundance of common doves, turtle-doves, skouts, skarts, and all sorts of sea fowls; 16.. *Macfarlane's Geog. Coll.* III 113.

(*b*) The sussing of the kitnings, clamring of scarfes, whimpring of fullmarts; URQUHART *Rabelais* III xiii 107.

(*c*) The scrath, the budoch, are two great black fowls; .. these .. dive under water; 1683 *Coll. Aberd. & B.* 100.

Scarth, Skarth, *n.*[2] Also: **scarcht, skartht; skart; scratche; karthe.** [ME and e.m.E. *skratt* (15th c.), *scrate* (*c*1475), *scrette* (Caxton), *scrayte* (1552), 19th c. north. Eng. dial *scrat, scart, scratch* (1878) ?OE **scratta* (?miswritten *scritta*), ON *skratte* wizard, goblin, monster.] **a.** A monster. **b.** A hermaphrodite. — **a.** Scarth fra scorpione, scaldit in scurrilitie; DUNB. *Flyt.* 58. *Ib.* 194. Ane scabbit skarth, ane scorpioun; Id. *Tua Mar. W.* 92. Ane of thame amang the Pichtis wes borne callit a skart conuenand with tua naturis with ane swynis heid fra the breist with gwis feit bot mannis leggis; ABELL 67a. — **b.** Of the skartht that was born of baith the kyndis maile and female; PITSC. I 145 heading. Ane bairne .. quhilk had baitht the kyndis of maill and famell callit in our langage karthe [*v.rr.* scarcht, scratche]; *Ib.* 145/20.

Scarth, var. of SCART *n.*

Scarting, Skarting, Scrating, -yng, *vbl. n.* [ME and e.m.E. *skratting* (Cursor M.), *scrattinge* (1555); SCART *v.*] The action of scratching, in sense 1 of SCART *v.* **b.** *specif.* The action of scratching so as to relieve an itch.

(*a*) Scho .. With skarting causit hir face to bleid; ROLLAND *Seven S.* 1203. For trublance .. in stryking of hir, scartyng of hir and dynging hir to the erd; 1574 *Glasgow B. Rec.* I 7. Byting and scarting is Scots folks wooing; FERG. *Prov.* No. 204. In stryking hir on the face and skarting thairof to the effusion of hir bluid; 1607 *Dunferm B. Rec.* II 34. Scarting and byting is Scots mens wowing; CARMICHAEL *Prov.* No. 1344.

(*b*) Convict .. for .. scrating and ryveing of wtheris faces; 1641 *Aberd. B. Rec.* III 256.

b. As the maister wold .. teach his schollers .. guid maners .. and that they be taught to abandon all wnciveill gesturs, as skarting of heid, armes, etc.; 1640 *Dundonald Par. Rec.* 467.

Scat, var. of SCAT(E *n.* and SCAT(T *n.* and *v.* **Scatald,** var. of SCAT(T)ALD *n.*

(Scatch,) Skaitch, Skaitge, Skutch; Skack, *n.* [Late ME and e.m.E. *scache* (1420), OF *esc(h)ache* (12th c. in Wartburg).] A pole used in construction, ? a scaffolding-pole, also, the uprights of a ladder. **b.** A stilt. — For i[c] firryn sparris for the skaitgis to fut gangis; 1530 *M. Works Acc.* (ed.) I 38. Planscheour naill and .. garrow naill to eik the ledder with and naill on the skaitchis; 1563-4 *Edinb. Old Acc.* I 469. For skutches to the futgang, ij s.; 1578-9 *Haddington Treas. Acc.* Ane aikkan rufe spar to be roungis & skackis to the saidis ledderis; 1583 *Edinb. D. Guild Acc.* 178. — **b.** Others grew in the legs and to see them you would have said they had been .. men walking upon stilts or scatches; URQUHART *Rabelais* II i.

Scat(e, Scait, *n.* Also: **skat; skait(t, skaite, skay(i)t; sket(t; skeitt, skeite, skeat; schayt.** [ME and e.m.E. *schat* (c1340), *scate* (c1375), *schate* (c1475), *skete* (1538), ON *skata.*] A skate, the large flat-fish; skate's flesh, for use as food. In *pl.*, inflected and, chiefly, uninflected. Also *attrib.* and *comb.*

sing. and *uninfl. pl.* Ane hundreth herd scat; 1516 *Fam. Rose* 189. Ane hundreth scat, price iij lib.; 1517 *Misc. Spald. C.* II 79. [Alexander Innes is entitled to] a pennyworth of fish of every fish boat thereof viz. 12 small fish for a penny or scate for a plack; 1556 *Old Ross-shire* I 15. In the sea they catch .. whitting, mackrel, turbot, scate [etc.]; WALLACE *Orkney* 17.

(*b*) And sum with rumpillis lyk ane skayit And geis and capounis rostit hait; ROWLL *Cursing* 119 (M). i skait vi d.; 1512 *Household Bk. Jas. IV* 21b. x dry scait; 1524-5 *Wigtownshire Chart.* 64. Pro uno lie skait 2 denarios; 1561 *Reg. Episc. Morav.* App. 404. Greit fische, sic as skait, keiling or lyng; 1564-5 *Reg. Privy S.* V I 558/1. Lie kylling, lyng, halecum, lie skait, saithis; 1571 *Ib.* VI 227/2. Skayt; 1572 *Inverness Rec.* I 223. That the flescheouris .. sall have for breking of ilk kylling, lyng or skaitt, tua penneis [etc.]; 1580-1 *Edinb. B. Rec.* IV 198. Ane dossane of keilling and ane dossane of skaitt merchand wair; 1594 *Orkney & Shetl. Rec.* 218. This cuntrey sindrie fresche water lochis hes that abundes in mony kyndes of fische, cheiflie in thrie, killine, skait, and makrell; DALR. I 13/20. Gryt lyne fische, sic as leing, turbat, keling, & skaitt; a1600 *Aberd. B. Rec.* (Jam.). The stealing of the yarne, salmond skait and long fisch; 1671 *Rec. Old Aberd.* I 121. Further down the sands, neer the sea they take keilling and skait by hooks baited and laid upon the sands, which they get at low water; 1692 *Macfarlane's Geog. Coll.* II 116. [Bill of fare]: Frydd skaite; 1698 *Marchmont* 185.

(*c*) Ane dosoun of dry sket; 1570 *Crail B. Ct.* MS 25 April. For a littil fresche sket to the servandis denners & supers; 1631 *Buccleuch Household Bk.* 3 Oct. Skett; *Ib.* *Squatina,* a sket; WEDDERBURN *Voc.* (1709) 11.

(*d*) Ane fresche skeitt; 1599 *Household Bks. Jas. VI* 22 June. Nyne dussoun skeite; 1602 *Shetland Sheriff Ct.* (ed.) 49.

(*e*) Fourtpairt peis prickettis coillis vinagir skeat; 1597 *Household Bks. Jas. VI* 27 March. *Sibbald's Orkn. & Shetl.* (1845) 52.

infl. pl. For stink of haddockis and of scaittis; DUNB. (OUP) 201/9. Hard skatis; 1515 *Treas. Acc.* V 21. Of dry fische keling ling & skaittis sax hundreth price of the dosone ourheid tua merkis; 1588 *Edinb. Test.* XIX 8. ix skaittis .. ij dray skaittis; 1598 *Stirling Ant.* III 308. An elephant .. a great beast .. lowged like two skats lying closs to its heid; LAW *Memor.* 176.

(*b*) Fleucca & sketta pererrant Per costam; *Polemo-Mid.* 4.

(*c*) Thair is gottin amang the roch stanes of the dyke foirsaid flewkis, podlokis, schaytis and little hering stikand fast amang the stanes; MONRO *W. Isles* (1961) 77.

attrib. and *comb.* Thay dang him with skait rumpillis; BELL. *Boece* II 98. xxv skaitt nettis price of the pece ourheid tua merkis; 1596 *Edinb. Test.* XXIX 362.

b. As a place-name element, in *Skait-raw(e.*

ij rudis in the Skaytrawe [in Nairn]; 1506 *Thanes of Cawdor* 118. The siemen of Fischeraw .. Dunbar Skaitraw Haymouth and vtheris sey townis on the south syd of Forth; 1592 *Acts* III 614/1.

(Scate-bread,) Skatebread, *n.* [? f. or f. as prec.] Some sort of small fish. — Catalogum piscium .. qui in regionibus Scotiæ austro et occidenti reperiuntur, .. pillock, skatebread [etc.]; SIBBALD *Scot. Illustr.* III 37. Skatebread, which is a little small fish an inch and a half long, shining clear as silver; 16.. *Macfarlane's Geog. Coll.* III 191.

Scatell, var. of SCAT(T)ALD *n.* **Scater,** var. of SCATTER *v.* **Scatering,** var. of SCATTERING *vbl. n.* **Scatfa, -fall,** varr. of SCATTFALL *n.* **Scatfalling,** var. of SCATTFALLING *vbl. n.* **Scat-free,** var. of SCATT-FRE *adj.* **Scat-frie,** var. of SCOT-FRIE *adj.* **Scathald,** var. of SCAT(T)ALD *n.*

Scath(e, Scaith, Schath(e, Schaith, *n.* Also: **scatht, sckath, skath(e, skatht, skat; scaitht, scaythe, sc(k)ayth(t, scayt, sckait, skaith(e, skaitht, skaiethe, skayth(e, skaytht, skaythth, skaythit, skait, skayt; skeith(e, skeit, skeyth; skeath(e; sketh, sket; skyth; scaid; schatht, schat(e, schacht, schatth, chad; schaith, s(c)hayth, schait(e; schaid, schayd; schecht.** [ME and e.m.E. *scathe* (Layamon), *scath, skath* (Cursor M.), *schathe* (c1350), OE *sc(e)aða* one who injures; hurt, injury, ON *skaði* harm, damage, OFris. *skatha, skada* injury.]

Freq. in collocation, esp. in sense 1, with HARM *n.* and HURT *n.*

1. Harm, damage; hurt suffered, threatened or perpetrated. Freq. const. *to* (*till*) the person or body of persons suffering harm. Also a particular instance of harm, etc.

As: physical damage or injury, damage to (one's) interests or property, defeat or loss of men in battle, spiritual or moral harm, etc.

See also *ILL-SCATH.

Some quots., esp. with possessives ((3) below), refer to harm threatened or perpetrated in contradistinction to the more frequent sense of harm suffered.

(1) Gif ony .. wittis or persavis ony maner of grefe skath or perele apperand til othir he sal .. revele it; 1409 Slater *Early Sc. Texts* No. 72. For fault of .. provisioun men wald purvay thame that thai suld pas with lesse cost and scathe; HAY II 112/24. In .. lessing of the saide fare hurt .. to oure said burgh and skath to the cuntre; 1459 *Ayr Chart.* 31. Slewthe and delay oft causis mekyl skatht; *Bernardus* 48. To thé it is bot dowble skath; *Bann. MS* 80b/46.

(*b*) Gyff ony of thaim thar-at war wrath Thai watyt hym wele with gret scaith; BARB. I 202. That cours had turnit to great skaith To thame; *Alex.* I 2623. That thow may with thin eris here .. To knaw profyt and eschew scaith; *Ratis R.* 145. For till eschew apperand greitt skaith; 1436-7 (c1580) *Edinb. B. Rec.* I 5. For litill tribut littill scaith will mak; HAY *Alex.* 2611. Dredand for scaith sche fled quhen scho him saw; HENR. *Orph.* 103. Throw wer sic skaith, at labour was away; *Wall.* III 12. To the eschewing of dampnage and skaith that mycht happin betuix thaim; 1508 *Reg. Privy S.* I 258/2. To me this was fyrst apperans of skaith; DOUG. II ii 66. To set apon thame and await with skaith The place surly we haue espyit baith; *Ib.* IX v 29. Causis quhar occultacioune of the verite may gener gret dampnaige hurt or skaitht to the innocent heir; 1525 *Dunferm. B. Rec.* I 279. And the mater was nocht fer fra appering skaith [L. *non procul ab iniuria*]; *Boece* 37b. Traist nocht .. of me That I .. cruell be, .. The quhilk to me war ouir greit skayth and lak; STEWART 36720. Robert the Bruce for dampnage skayth and pand That he had tane out of Northumberland Threttie thousand [pounds] .. To King Edward for recompans sould pa; *Ib.* 51361. It behuifit to set ane pairt of thair commoun mvre quhilk was maist proffitable and leist skaith to the tovne; 1536-7 (c1580) *Edinb. B. Rec.* II 84. Consideratioun of gret skaith that cummis of ane evil toung; HAMILTON *Cat.* 11. And the said Schyr Jhon Borthuik not to hav falyet tharintyll, nor to have deservit infamite, punischement, nor skayth tharfor; 1561 *St. A. Kirk S.* 103. Ane freynd .. Quha stall away thair styngis bath clair And hid in secreit wayis For skaith; SCOTT ii 59. I sall wey bayth in ane ballance Wynning and skayth I sall we bayth; *Maitl. F.* 368/10. Quhair of come greit skaith on this realme; *Diurn. Occurr.* 4. Sinne bringes sua greate skaithe to the saule; KING *Cat.* 14. That thair wald be grit skayth baith be see and land; 1590-1 *Crim. Trials* I II 232. Weiris ar begun with gret slauchtir on baith sydes, disproffet and skaith; DALR. I 330/31. Summa ius, summa injuria; there is skaith aneugh alreddie; 1626 *Ancram & Loth. Corr.* I 41. A lantrone .. for preserving the candle or light thei sie with for eshewing of skaith may cume therby; 1669 *Glasgow B. Rec.* III 115. Frie .. from future personall and uthir hurtis and skaith threatned and minaced be the said Georg; 1706 *Corshill Baron Ct.* 217.

(*c*) Be the quhilk hurt or skeath may follow or be done to .. our gude friends; SKENE *Reg. Maj.* II 26.

(*d*) [We] annullis .. & brekis al & sindry charteres .. that may mak derogacioun schaith hurt .. or hendryng to the said lord; 1474 *Reg. Morton* II 223.

(*e*) That thrw the hyding of suthfastnes schate or preiudice may be engenerit till ony innocent; 1433 *Swintons* App. xxxvi.

(2) And thai couth nocht persawe the skaith That towart thaim wes apperand... Thai trowyt that he [*sc.* the English king] .. Wald hawe iugyt in lawte; BARB. I 82. Sen this fule was causar of this scaith Richt so [etc.]; *Fyve Bestes* 23. And the said Nicholas skipar protestit quhat skait hapint to cum or his said schip; 1532 Kennedy *Aberd. Ann.* II 478. Acknowledged payment .. for all the skayth and heyrschip of gudis taken by the said James servants from Holhouse; 1544 *Prot. Bk. Hew Rig* 32a. And sall radund the skaith again gef ony hapinnis to be; 1571 *Lanark B. Rec.* 55. I wald first apne vp the hail mater to ȝour hienes of the reife, skaith and our silens hithirto; DALR. II 132/16. For satisfactioun and assythment of the skaith hurting bluiding and wounding of .. John Chalmer; 1606 *Antiq. Aberd. & B.* IV 58. The skaith of such a sinful neglect; RENWICK *Serm.* 458.

(*b*) Bot the skeath fell vpon thameselvis for they slew thair awin man; BANN. *Trans.* 173. Thankand his God .. For his delyvering of that skeath; 1584 SEMPILL *Sat. P.* xlv 632.

(*c*) The skyth is noicht greitt, onlye ane man deid; 1606 *Lett. & St. P. Jas. VI* 84.

(*d*) Nocht-than thai ordenyt gret landis to bet The schath; *Leg. S.* xxxi 470.

(*e*) For the vmcheving of the apperand schayth and damagh of the haill commond veill of the said brucht; 1547 *Elgin Rec.* I 91.

(3) Quhill he be vengit of his skaith; *Alex.* II 10452. Foralsmykle as we ar informit .. that thai dreid the evil and skath of oure enemies of England; 1450 *Edinb. Chart.* 71. Be quhat resoune than suld he consent .. till his awin scathe of destructioun of his contree, unmaid sekir to be amendit? HAY I 185/29. Of vtheris menis scaith thai ar ay fayn; *Thewis Wysmen* 405. John Clatter .. askyt .. lawborowis of Mechel of Forest .. and than the balyais askyt at the sayd John gyf he doutyt his skath and than the sayd John suor apon a buk the gret ayth that he doutyt his skath .. than the said Mechel com law borch for hym self .. that the sayd John Skatter suld be skathlas of hym; 1470 *Peebles B. Rec.* I 163. Than had bene honest tyme .. Till haue previdit for thy frendis skaith; DOUG. X ii 92. Quho maist to my skaith consentit, Within few ȝeris full sore repentit; LYND. *Complaynt* 295. Thay that stoppis nocht thair nichtburs skaith .. or warnis noth thaime thair of befor; GAU 17/13. Sum aysyer way to prowyid ȝour bayrnis .. nor be my skayth and dishonour; 1562 *Reg. Morton* I 10. He was herreat, Lamenting sair his losse and skayth; 1584 SEMPILL *Sat. P.* xlv 75. He .. lamentis his kais, skaith and diffett, quhilke of the Scottis he had receiuet; DALR. I 231/23.

(4) Al man that has back yettes close thaim swa that thair cum na skath throw thaim to the toune; 1442 *Aberd. B. Rec.* I 9. The quhilk bill .. findand na skaith nor hurt thairin we have approveit; 1473-4 *Edinb. B. Rec.* I 28. That na scaith nor infectioune sall cum within his house; 1529 *Ib.* II 17. And als protestit that in cais ony skayth or danger come to the pannellis .. that the falt thairof be nocht input to him; 1556 *Ib.* 254. Thair wes na vther skaith; 1570 *Sat. P.* x 286. No skaith to Scotts .. Can come be these thy windie wants vnwise; GARDEN *Worthies* 46. Thare was thrie of thame within the hous slane and the same win without ony mair skayth; BANN. *Trans.* 123.

(*b*) Gif ony keip or persayvis ony schath aperand to the tovne; 1412 *Aberd. B. Rec.* I 389.

(5) Gyve ony thareat war wrath, Wayt hym welle wyth a gret skath; WYNT. VIII 2744. A weilfair and a skaith Salbe common .. to ws baith; DOUG. II xi 53. Ane general thraitning to do ane evill turne .. can haif na coherence with ane speciall skaith following thairaftir; 1622 *Crim. Trials* III 517.

(6) The towne At he defendit had fra skath .. and fra ruyne bath; *Leg. S.* i 553. Bot ye do it, fra scaith ye may nocht scheyff; *Wall.* III 264. For thire ar wordis of .. wertew .. Baith saull & body fra scaith to reskew; *Contempl. Sinn.* 1024 (Asl.). Sa mot thou Troy quham I sal salue fra skath, Kepe me thy promys; DOUG. II iii 31. Lufe weil and keip fra skath Our a ȝong son; *Ib.* xii 55. In time cuming for to keip ȝow fra skaith This I ȝow say [etc.]; ROLLAND *Seven S.* 9072. Ȝour grit necgligence That makis na defence Fra scaythe it [*sc.* this realm] for to keip; MAITLAND *Maitl. F.* 309/7. My Lord Athoill .. sall .. keip his auin cuntrie fra skaith; 1576 *Warrender P.* (SHS) I 134. As they might saiff thair contrie fre skaitht and boundage; PITSC. II 103/28. It hes ane cradle about it of clay to preserue it from skaith; 1662 *Crim. Trials* III 612. Our cornes not being fenced about therefor it gav us warie much adoe to keip theme frome skaith; 1674 *Kirkcudbr. Sheriff Ct. Processes* No. 206. — For to kepe Machore of skath; *Leg. S.* xxvii 640.

(*b*) Put her shell run in a closs head, To keep her shell frea skeit; *c*1700 *Fugitive Poetry* II xli 4/98.

(*c*) Ilke man .. sal stand gaird .. to kepe .. thair nychtbouris als fra schaith and susprice; 1412 *Aberd. B. Rec.* I 389.

(7) *pl.* The qwhilk skathis our lege lorde the kynge and his liegis has paciently tholyte; 1405 *Douglas Corr.* 64. That he arest sic folkis [*sc.* trespassers] and chalange thame as brekaris of the kyngis pece taxand the kingis scathis apone thame; 1424 *Acts* II 4/1. Gude justice has .. to mak the scathis [*Fairf.* schathis] hale; *Regim. Princ.* 234 (Marchm.). Decepcioun composicioun and mony vthir skaithis; IRLAND *Mir.* III 148/10. Sanct Bernard .. Desyrand fra thar [*sc.* fiends'] scaithis to reskewe .. Synnaris; *Contempl. Sinn.* 1331 (Asl.). And neuir to here his or tharis skaithis nor se it bot sall leit it [etc.]; 1506 *Charter* (Reg. H.) No. 693. Quhat skaithis and offence That women dois with cullourit eloquence; DUNB. (STS) lxxxiv 27. How mony skathtis followit apoun licht and flowand myndis; *Boece* 580b. France and Ingland faucht with baith thare skathis; ABELL 96b. And for sythement to be maid be hym for the scathes forsaid; 1552 *Breadalbane Doc.* No. 82. The deacon of the flesheors crafts .. sall appoynt two masters of thair craft vpon ilk mercat day .. to appryse all skaithes of all flesh in the land mercat; 1564 (*c*1650) *Dundee B. Laws* 30. To tak afald part .. in resisting .. of all skayththis; 1571 *Grant Chart.* 146. Command .. to refound the skaythtis and iniurieis done be thame of befoir; PITSC. I 82/26. That ane kinde of gear suffer nocht twa skaithes [*i.e.* of being spoiled and being cast overboard]; WELWOD 69. The foldis arestment maid [anent] certane sketis of his; 1602 *Shetland Sheriff Ct.* (ed.) 16.

b. *But (forouten, without(in) (ony) scathe*, without suffering harm; without causing, or threatening, harm.

Appar. sometimes a mere rhyming tag.

(*a*) This nuryse .. Forȝetil of the mykil wath That tyd the barne for-outyn skath; *Leg. S.* xxvii 214. He feld twa stowt fechteris And but skath went till his feris; WYNT. IX 690. Bot free passage but scathe to the contree suld nocht be denyit; HAY I 185/19. The leym of fyre and flambe, but ony skath, In his haris .. Kyndyllis vp brycht; DOUG. II xi 9.

(*b*) He .. Far by passit withoutin mare skaith; *Alex.* II 1231. And syne be tane but harme or skaith; *Ib.* 4928. He said the barne will ȝe gar bring Agane quhill I haf tald but scaith Syne do ȝour plesans of ws baith; *Seven S.* 1877. Gif he maid the samyn myln .. nocht sufficient and without hurt or skayth to the rest of the townis mylnis; 1556 *Edinb. B. Rec.* II 249. And now withouttin skaith I will obtene the blyssing of ȝow baith; ROLLAND *Seven S.* 4331. That the kingis lieges might lief but harme or skaith; LESLIE 15. Craigmillar .. was haistilie gevin to thame, promesand to keip the samyne without skaith; quhilk promes thai break; *Diurn. Occurr.* 32. On the morne the Scottis without any skaith fled misordourlie;

Ib. 36. And in comes Johnne Moat, and sa they wer red without skaith; *c*1610 *Lanark B. Rec.* 121. The assaillantis fynding the place .. vn-wynnabill without gryte skaith; Spalding I 291. For bringing it home ag[ain] safe without skaith; 1675 *Cullen Kirk S.* 25 Nov.

(*c*) Certane of his adversaris .. maid a litle pricking without skeathe; Moysie 15.

(*d*) That serpent .. he in the fyre but schathe Fra hym keste; *Leg. S.* ii 41.

c. *To the scathe of* or *to* (a person's) *scathe, in scathe of* or *in* (a person's) *scathe*, to the hurt of, causing harm to; (to be) *at scathe of*, of danger to, a cause of damage to.

(1) It shal not turn you to your skaith; *Sir Eger* 2235. Our said schiref .. hes forebydding his clerk of court to gif to the said James the coppy of the said proces .. to his grete skaytht; 1522 *Fife Sheriff Ct.* 175. To the gret hurt & scaith of our said souerane lord his liegis and subditis; 1524 *Acts* II 286/1. To the innomerable scaith of our soverane lordis liegis; 1528 *Acta Conc. Publ. Aff.* 295. To the gret skaith and herschype of thaim quhilk hes ane rycht actioun of the pley; Hamilton *Cat.* 100. The cummeris and expenssis of the reductioun tharof to his hevy dampnage and skaytht; 1557 *Aberd. Sheriff Ct.* I 165. It wes done to the skayth and sklander of the toune; 1562 *Crim. Trials* I i 421. To the greit skayth and hendering of my barne and cornis thairin; 1565 *Inverness Rec.* I 129. All contractis .. maid be minoris in thair les age, to thair enorm hurt and skaith, ar of nane avail; Balfour *Pract.* 179. To the greit skayth and violatioun of the actis; 1579 *Inverness Rec.* I 266. Certaine persones .. did tak avay .. diverse spreathes and hearshipes, to ther heavie hourt and skeath; 1641 *Acts* V 710/1. To the gryte greif and skaith of the poor laboreris; Spalding I 193.

(*b*) The said Angnes hes sauld, disponit and vaistit his guddis and geir to his greit schait and harschip; 1543-4 *Elgin Rec.* I 79.

(2) Thai wirk it [*sc.* leather] or it be conrait [*pr.* courait] in greit hindering and skaith of the kingis liegis; *Acts* I 336/2. In preiudice or scatht of his nychtburis; *Ib.* 338/2. Gif ony man fyndis a thef in his scath with thift .. he aw to raise cry apon him; *Reg. Maj.* c. 160. In grete skaith dampnage preiudice ande hendring of the said kirk; 1464 *Reg. Episc. Brechin.* II 107. Anent ourlordis that in defraude & skaith of thair .. tenentis deferris till enter to thair landis and superioriteis; 1474 *Acts* II 107/1. The youngar [mouse] wonnyt apon land .. Quhill in the corne an [*Bann.* in] uthir mennis scaith [*Bann.* schecht] As outlawis dois; Henr. *Fab.* 167 (A). Thomas sall nocht .. mak na innovacioun .. concernyng the premisses in his skaith nor contrare; 1496-7 *Acta Conc.* II 64. Thai .. retourit in fraud and skaith of the kingis hienes the landis of Ferny; 1497 *Ib.* 92. For the distruccione of a browme in hir skaith; 1498 *Ib.* 226. In gret skayth and tynsale off the said lord of Ʒestyr; 1525 *Yester Wr.* in *E. Loth. Antiq. Soc.* V 30. The king basterd .. mowit weir aganis King Malcolme in skath of Ingland; Abell 78a.

(*b*) That nan suld be resavyt in hospitalite in hevy schatht of nychtburis; 1460 *Newburgh B. Ct.* 3b.

(*c*) Schecht; Henr. *Fab.* 167 (Bann.) (see (2) above).

(3) That he sall never be at scathe of his secretis .. of the quhilkis his lord haldis him seure; Hay I 122/19. He sall nocht be at his scathe of his moble, gudis na heritagis; *Ib.* 122/27.

d. *To do* .. (*na, the, gret*, etc.) *scathe* (*to, till, upon* a person, place, etc., also const. indirect object, and without const.), to cause (no, the, great, etc.) harm, hurt or damage.

In *passive* construction, also with p.p.

(1) The scaith The dispyt and felny bath That Jhone off Lorne had till him doyn; Barb. x 5. Nane of the lordes beforsayd ne nane of thair boundes sal do skathe to tha of the boundes of the tothir partie; 1385 *Rot. Sc.* II 73/2. Than thai forbare nowcht to do skayth Till abbays; Wynt. vi 1567. The quhilk walde nocht thole thame to pas throu his boundis, thai doand him na scathe; Hay I 7/19. That euill .. He suld haue had that hes done me the skaith; Henr. *Fab.* 2178. And do na scaith to the myrour; *Seven S.* 1785. The said Adam sall mend his saddill .. sa far as he did it scatht; 1516 *Selkirk B. Ct.* (ed.) 38. He ma do ws now nouther skayth no lak; Stewart 16751. And did sober skaith to the court of Rome; ?1539 in *Concilia Sc.* (1866) I cxxx. Without I do my self gret inconvenyent and skayth; 1545 *Corr. M. Lorraine* 149. The aucht command .. techis thé to do na skaith to thi nychtbour in his fame and name; Hamilton *Cat.* 106. For it may fortoun raith, To do hir body skaith; Scott x 18. Gif ony man's beist dois harm and skaith to ony persoun [etc.]; Balfour *Pract.* 490. Enemies that wold doe anie harme or skaith to the countrey; *c*1630 *Macfarlane's Geog. Coll.* II 176.

(*b*) I .. hes never gottin ane ansour fra ʒour L: the quhilk .. dois me verry greite displesour and skeithe; 1581 *Nugae Scoticae* 35.

(*c*) That the said Duncane hed down hym grit skyth in malt makyin; 1549 *St. A. Baxter Bks.* 5.

(*d*) For ʒe to me has done gret schatht; *Leg. S.* i 594. Saiand at he did nathyr schatht, schame na reprufe to the said .. governyoure; 1419 *Wemyss Chart.* 45.

(*e*) That we .. sall in na tym to come do any prejudice hurt or skait to the said lord; 1507-8 *Misc. Spald. C.* II 214.

pl. King Robert for scaithis ser that he to thaim .. Had done off wer .. suld pay [etc.]; Barb. xx 50.

(2) He .. saw a bare .. Defoull the corne with feit .. & do gret skath; *Leg. S.* xxvii 865. Sa in hys tyme gret raynis fell .. That .. The wallys off Rome a part ouryhud .. And mony drownyd .. In Rome this rayne than dyd the skayth [*W.* mekle scaith, *C.* gret skaythe]; Wynt. v 5378. He dyd gret skayth In slawchtyr and in herschype bath; *Ib.* viii 2425. Gyf thai lay ankyr wyth owtyn buy & it do skath thai sal amend it; *Ship Laws* c. 16 (B). Changeing of men that wate the kingis secretis .. may do grete scathe; Hay I 151/29. Teltellaris oft tymes dois grit skaith; Henr. *Hasty Credence* 25. The Inglis ordinaunce schot fast and did greit skaiethe; Leslie 94. Twisday .. was .. ane day of grytt woynd .. quhilkis dyd grytt skayth on the kyrk and howissis; Cullen *Chron. Aberd.* 47. The Erle Douglas .. for al the truble, skaith and cummir that in the realme he done had .. cryes mercie; Dalr. II 64/24. Heiche wyndis .. did gryte skaith be sea; Spalding II 100.

pl. The Phylisteis .. dyde in tyll [*C.* in] it [*sc.* the land of Juda] schathis grete; Wynt. iii 258.

(3) *passive* The wther halff to hym that the skath is doin to; 1460 *Montrose Baillie Ct.* I fol. 13a (26 Jan.). Quhat skath was done .. thair complissis was be the kingis commande; 1499 *Kilravock Chart.* in *Orig. Par.* II ii 528. For the skaith vas fundin donne to the hous; 1523 *Carnwath Baron Ct.* (SHS) 4. For the skaitht don to Mairyone Jonsone sled ii s. to be pait; 1530 *Selkirk B. Ct.* (ed.) 108. That the kingis liegis ma leif in .. pece for ony scaith to be done be the said clan Gregour; 1530 *Acta Conc. Publ. Aff.* 344. Ther wes sckaytht done, bot nocht sa greitt as thay mak the brwt; 1549 *Corr. M. Lorraine* 315. For recompancement of the skaith done to him be umquhile George Erle of Huntleis army; 1562 *Treas. Acc.* XI 215. That thai .. pay .. fourtie schillingis in racumpans of the skayth done; 1580 *Edinb. B. Rec.* IV 190. The availl .. of the guidis stollen or reft or vther skaith done; 1594 *Acts* IV 72/2. Skene *Reg. Maj.* ii 26 (see 1 (1) (*c*) above). The pickman sall be frie ony skathe alledgit downe to the stuf; 1615 *Urie Baron Ct.* 15. Quhat skayth may be done be taking up ane depositioun maid in ane chalmer; 1626 *Justiciary Cases* I 50. It [*sc.* the plate] came flying at her without any skaith done; Sinclair *Satan's Inv. World* 93.

pl. The restauracioune of skathes To hem thruth Troyiens done; *Troy-bk.* ii 97 (D). For skaythtis doyne to Thome Ker to pay vi s.; 1530 *Selkirk B. Ct.* (ed.) 108. The gret thiftis and scathis done be the saidis Egiptianis upoun our soverane lordis liegis; 1541 *Acta Conc. Publ. Aff.* 503.

e. *To tak, have, get, kep*, also *sustene, incur, thole, ressave*, etc. (*the*, etc.) *scathe*, to suffer hurt or damage.

See also Kep *v.* 4 for some further instances.

(1) Sould nane me helpe thocht I tak skaith; *Alex.* ii 6478. Gif .. ony of the said partis sustene tynsale or scathe [etc.]; *c*1400 *Wemyss Chart.* 37. The personis thare in duellande has syndry tymys sufferit disese and skathe; 1412 *Laing Chart.* 24. Sampsone that sa gret scaith had; Wynt. iii 257 (W). The wynys takys scath; *Ship Laws* c.

7 (B). Men that has tholit skath oft; 1445 *Aberd. B. Rec.* 18 Feb. (Jam. Suppl. s.v. *Falk* v.). The gude folk has scathe bathe in body and gudis; HAY I 162/26. The saule .. be the fallouschip that it has with the .. body .. incurris gret skaith as payne deid syn and dampnacioun; IRLAND *Mir.* III 27/17. And to .. suffer blithlie, skaith, persecucione and aduersite; GAU 75/24. The Inglis men well never trast to geit skath; 1548-9 *Corr. M. Lorraine* 292. I hawe skaithe anewche ellis; 1549 *Douglas Corr.* 175. Lat not the man get skatht for any thyng he has of myn; c1550 *Balcarres P.* 314. It is vnderstand that be the geuing of curatouris to minouris .. thair hes bene gret skaith sustenit be the saidis minouris; 1555 *Acts* II 493/2. [As it becomes no] saikles man may incur skayth [for the injuries and demerits of others]; 1559 *Cal. Sc. P.* I 260. That our souerane lord wald gett skaith be ane taid; 1591 *Crim. Trials* I II 245. Giff ony mariner ressaue skaith in the schipis laubour; 1602 *Conv. Burghs* II 137. If they were barefooted, .. they must kep skaith by walking there; HENDERSON *Serm.* 522.

(*b*) Landit men .. sustenit greit skyth in thair guidis; LESLIE 63.

(*c*) And sic lyk in mony owthyr placys schath and hurt we haif; 1456-70 *Liber Aberbr.* II 108.

(2) Schyr Amery that had the skaith Off the bargane I tauld off er [etc.]; BARB. IX 546. The king .. persavand then That he mycht nocht sustene that scaith .. tuke a gret spere [etc.]; *Troy-bk.* II 2121. Thare mayne thai maid Menand the scaith that thai tane had; WYNT. II 1436 (W). The Phylisteis .. [That] tholyt that scaytht and that tynsalle; *Ib.* III 246 (R). That nane haid the scayth onlie bot we; HAY *Alex.* 1797. Assurand ȝow in sick delay oure liegis and subjectis tholis the scaith and dammage; 1529 *St. P. Henry VIII* IV 572. Ve ordand Nychoill to sweir quhat skaytht he had throu tynsaill of hir service; 1534 *Selkirk B. Ct.* (ed.) 145. Thocht ȝe haif gottin the skayth, The schame and lak redundis till ws bayth; STEWART 1237. For the greit skayth and hender and expenses sustenit; 1580 *Inverness Rec.* I 281. 1599 *Acts* IV 188/2 (see OVERTHRAW *n.* (1)).

(*b*) Lat the witnesses descend in speciale quhat dampnage and skeyth the complenaris has sustenit; 1553 *Blackfriars Perth* 234. All the trobell and sketh y haue sostenit; 1604 *Dundee Shipping P.* 80.

(3) That the schyp tak na skath; *Ship Laws* c. 14 (B). Gif sik trespassouris takis ony skaythe in the aresting of thame it salbe imput to thame self; 1424 *Acts* II 4/1. That Theman, goldsmycht, sal sustene na scathe for the brekine of the saide ferthing; 1463 *Aberd. B. Rec.* I 26. That the said Johne, his factouris [etc.] .. sal nocht .. incur ony skayth or danger thairthrow in thair persouns, schippis, merchandice, landis or gudis in onywys; 1515 *Reg. Privy S.* I 400/1. The fyir slaucht vil consume the vyne vitht in ane pipe .. & the pipe vil resaue na skaytht; *Compl.* 60/12. [If] the saidis keparis .. ding the saidis stirkis quhairthrow thai incure ony skaith; 1557 *Peebles B. Rec.* I 238. Gif Thomas obteins ony skaith through downtaking of the stanes [etc.]; 1560 *Old Dundee* II 63. And caus him requeist .. that my lord gaitt na hurtt nor skatht in na matteris; 1561 M. Napier *Mem. J. Napier* 75. The leigis can na greter sckayth sustein; MAITLAND *Maitl. Q.* 166/37. To varrand and keip the said erll skeythles .. anent ony skayth or inconvenient that the said erll may incur; 1578 *Montgomery Mem.* II 220. Beware thou kep na skaith; MONTG. *Ch. & Slae* 386 (W). And the woman sall suffer na skaytht bot be declared innocent; BISSET I 43/12. So that Aboyne his freindis and followeris gat no skaith; SPALDING I 211. For heindering the roofe of the said church to receive any skaith and damage by stormes; 1676 Craven *Ch. in Orkney* 61.

(*b*) The said Jhone and Sanders ar .. bund for his lauchty that the said Rechert sall kep na schacht; 1519 *Selkirk B. Ct.* (ed.) 51.

f. *To cum to (ony,* etc.) *scathe*, to sustain or suffer harm.

And thou may mak of thy frend thy fa .. and cum thé to grete scathe and hyndering; HAY II 153/7. Giff the schip cumis thairthrow [*sc.* overloading] to ony skaith or mistak than is the merchand haldin .. in the half skaithe; 1602 *Conv. Burghs* II 141. Gif ane schip cums to skaith vpoun ane vther schipis anker [etc.]; *Ib.* 144.

g. Freq. in collocation with *schame* or (alliteratively) with *scorne* (also *scorning*). Also *pl.*, = instances of hurt or harm.

See also S(C)HAME *n.* 3 (2) and SCORN *n.* 1 d.

(1) He for ws had skath & scorne; *Leg. S.* xvi 629. But schath & schorne; *Ib.* xxxii 767. Yone knicht to scar with skaitht ye chaip nocht but scorne; *Gol. & Gaw.* 279. And lat no small beist suffir skaith na skornis Of greit beistis; DUNB. (OUP) 144/107. Thus the scorne and the scaith scapit he nothir; Id. *Tua Mar. W.* 358. Than haif ȝe skayth and skornyng ȝow to steir; MERSAR *Bann. MS* 278b/34. Thow hes baithe scayth and scorne; BALNAVIS *Maitl. F.* 356/26. As scorne comes commonlie with skaith; MONTG. *Ch. & Slae* 197 (W). To my skorne, his skaith and youre small honoure; c1595 JAMES VI in *Facs. Nat. MSS* III lxxiv.

(2) He hade ressavyt the malys of the Wemys .. in grete schath and schame; 1419 *Wemyss Chart.* 45. Is nothir scaith, scham na dreid Agan hir wyll may than tak sted; *Ratis R.* 949. Schaith; *Ib.* 1374. Quhar the hyding off suthfastnes ma genner scathe, schame or prejudice till ane innocent man; 1438-9 *Rec. Earld. Orkney* 71. All a hale company may tak bathe scathe and schame; HAY I 150/33. That for his syn gat scath & schame; *Seven S.* 2764. Na man can do thame skaith nor schame; *G. Ball.* 56. He that spendis his gude on ane hure hes bayth skayth and schame; *Maitl. F.* 159/9.

pl. Sin I haue schauin my obediens uith los and schaites and scham; 1593 *Misc. Spald. C.* I 5.

h. Common in various formal expressions of loyalty in bonds of manrent, indentures, etc.

The ath of feautee .. contenis sex poyntis .. the first is that he sall never be quhare scathe is done to his lordis persone; HAY I 122/17. Nane of thaim to heir, se, nor wit harme, scathe, dede, nor disherising to vderis; 1468 *Misc. Abbotsf. C.* 6. I sall neythir heir his skayth nor se it but I sall dyscouir it to my saide lorde and let it at my powar; 1472 *Misc. Spald. C.* II 252. To kep tham in thar personys, and fra sckath in thar gudis; 1474 *Reg. Cupar A.* I 194. I sal never heir se no wyt [*pr.* obyt] to hym no his ayris scayth deid disworschip nor disherysyng; 1474 *Reg. Morton* II 221. Scayt; 1476 *Oliphants* 20. And we sal nother heir nor se schaith nor apperand perell to him bot we sal warne him thairof; 1477 *Misc. Spald. C.* II 253. I sal nouder wit lak dishonour nor scaith to my said lord; 1514 *Bk. Carlaverock* II 456. 1545 *Wemyss Chart.* 171. I sal be lele and trew to ȝou my liege lord and soverane .. and sall not sie ȝour skaith nor heir it; BALFOUR *Pract.* 23. Skeithe; 1611 *Red Bk. Grandtully* I 111. The aith of the burgesses of Aberdene .. I sall warne thame of thair skaith, latt it at my powar; 1616 *Aberd. B. Rec.* II 346. Never to .. see the skaith of any brother or sister of the same without I warn them thereof goodly; 16.. Rankin *St. A. Ch.* 65. He shall not hear nor conceall his said masters hurt skeith nor prejudice but shall tymously reveall it; 1712 *Indenture* in *Sc. Hist. Rev.* XVIII 232.

i. *specif.* Harm brought about by witchcraft or the agency of the Devil.

The said Margaret hed by wischcraft done him great skaith in his gudeis; 1623 *Perth Kirk S.* MS 16 May. Without ony .. apparent caus or midis of reasone to impoirt that skayth except be the devillish sorcerie .. of the pannell; 1629 *Justiciary Cases* I 100. James Rendall .. deponed that he knew that she spak the words and be her words he kept the skaith of the death of his beastis; 1629 Black *Orkn. & Shetl. Folklore* 80. Ane offering to the devill to safe her husband, her offspring and meanes fra skaith; 1650 *Kingarth Par. Rec.* 25. No more skaith was the devil or the witches able to do him; SINCLAIR *Satan's Inv. World* 151.

j. *proverb.*

It apperis that gredy hert followis gret scaith; *Porteous Noblenes* 177/16 (A). Emptie purse dois greit skaith; ROLLAND *Ct. Venus* Prol. 231. Better learn by your neighbors skaith nor by your own; FERG. *Prov.* No. 152. He that is far from his geir is neir his skaith; *Ib.* No. 299. Diverse times a man gets not onlie the skaith bot also the scorne; CARMICHAEL *Prov.* No. 457. Wyne and women gets ay skaith behind baks; *Ib.* No. 1677. Better two skaiths nor ane sorrow; PAPPITY STAMPOY 12.

2. *specif.* Financial loss; loss of profits, fruits or sources of profit.

Cf. 1 above, with which there is some overlapping.

(*a*) Quhatsaever he be that lattis his borch inrin sic scath and walde nocht cum to freyth him; *Acts* I 51/2. The vrangus vseyng and makeyne of ane gait vpone ane part of the said Davidis land .. and the said Davidis is skatht heyrof; 1537-8 *Dunferm. Reg. Ct.* 151. That the deacon .. cause the person complainer therof be satisfied of ther scath through the insufficiencie of ther wark; 1656 *Dumbarton B. Rec.* 74.

(*b*) To preif how mekle scaith thai have sustenit in the eting and distructioun of thair corne; 1483 *Acta Conc.* II cxxviii. To preif the availe of the saidis landis and the skaith that he has sustenit throw the wanting of the sammyn; 1496 *Ib.* 4. John Cowper is .. ordainit to .. upset the skaith of the bred; 1520-3 *Old Dundee* II 26. The said Katrine wes wranguisly .. eieckit furtht of hir saidis landis .. wytht the skaytht of the pastoring of hir cattell; 1521 *Fife Sheriff Ct.* 241. Thai ar namyt executouris .. and mycht report profit or scaith in the said mater; 1522 *Acta Conc. Publ. Aff.* 154. Ten schillingys to the said altar for the costys and skayth it sustenys thairupon; 1531 *Edinb. B. Rec.* II 54. Gif ony skayth cummis to the merchandis or thair guidis in thair default in the nocht fulfilling of the thingis aboue written [etc.]; 1533-4 (*c*1580) *Ib.* 66. It appeiris that his contenwance hes procedit mair of his skaytht nor ony zeill of the comand weil; 1558 *Aberd. B. Rec.* I 312. The intollerabill skaith sustenit .. be the merchandis of hir hienes realme throw granting of letters of marque to [etc.]; 1563 *Acts* II 544/1. The greit skayth and detriment of this commoun weill throw the homebringing of fals corrupt and countirfuttit cunye; 1567 *Reg. Privy C.* I 510. Balfour *Pract.* 494. [If the corn is not well ground] the saidis myllaris sall pay to the awneris of the corneis that hes skaith thair skaithe and inlaik; 1586 *Melrose Reg. Rec.* I 57. Gif they dicht or prepair the flesh not well, they sall restore the skaith to the awner of the beast; Skene *Reg. Maj.* I 127. As thay lost thair crop for the haill yeir, the persewir ought to satisfy them for skaith sustenit; 1630 *Dumbarton B. Rec.* 33. The myllars should pay to the ouners of the cornes that hes the skaithe ther inlaik and skaith; 1666 *Melrose Reg. Rec.* II 162.

(*c*) Quhat sckait that ony nebour takin in hus yaird .. in sa far as he ma prefe the skait; 1523-4 *Stirling B. Rec.* I 18.

(*d*) The vassalls .. sall .. recover the damnage and skeath quhilk they haue susteined be occasion of that poynding; Skene *Reg. Maj.* II 28b.

(*e*) I bind .. me .. to content and pay to the said Duncane .. alsmekle as sal hapin him .. to sustene of schatht be the warding of the saidis landis; 1485 *Breadalbane Doc.* No. 22. That he bring his previs to preif the quantite of the schayth that he has sustenit tharethrow; 1490 *Acta Conc.* I 142/1.

(*f*) Sow[m]ane his chad to ii bollis of salt; 1514 *Wigtown B. Ct.* fol. 31b.

pl. Gif ony man of set purpois .. slayis ane uther man's dog, he sall .. pay to him all his skaithis sustenit be him .. for want of his .. dog; Balfour *Pract.* 492.

3. *specif.* Damage to property, etc., caused by trespassing animals belonging to a neighbour.

In (the) scathe, in or *upon (on)* (another person's) *scathe*, in the act of causing such damage. Also, (*to kepe* animals) *out of (af, from)* .. *skathe*.

(1) Geiff .. Roben Ker .. fand Adam Welkesones swyn doand him scath [etc.]; 1512 *Selkirk B. Ct.* (ed.) 16. For the scatht of suyn; 1518 *Ib.* 46. I sal keyp me my kyne and al that I may let onedowne ony skaith to the said Sir Johne; 1520 *Thanes of Cawdor* 139. Tuecheyng the skatht of certan cornis be his baistis; 1532-3 *Dunferm. Reg. Ct.* 89. That he retenit ane, twa [etc.] .. of the said cattell .. quhill he wer satisfiit of the said skayth; 1574 *Reg. Privy C.* II 417. They tuik in ane kow of myne not doand them skaithe; *c*1590 Fraser *Wigtown* 392. The rest of the beistes .. hald besyd thé .. ay quhil thair maistir repair the skaith; Dalr. I 123/17. Quha .. keipis ony swyne within dykis that the awneris .. tether them that thair nychbouris girse and cornis may be frie of thair skayth; 1603 *Shetland Sheriff Ct.* (ed.) 94. Swyne, hens, geese, gaites doing skaith to ony man may be eschieted and eaten by him to quhom the skaith is done; Skene *Reg. Maj.* II 155. Satisfactioun of the skaith to the partie skaithet; 1652 *Peebles B. Rec.* II 5. And gif .. they [*supra* nolt and scheip] .. brak the fald and do skaith [etc.]; 1661 *Misc. Spald. C.* V 232. For refussall of satisfaction for his corne and skaith; 1665 *Forbes Baron Ct.* 258. The skaith of beasts in the loane; 1670 *Kirkcudbr. Sheriff Ct. Processes* No. 77 (8 Nov.). All the tennants .. to keip ther yairds under sure fence against all beasts for preventing of skaith either to themselves or ther neighbours; 1694 *Stitchill Baron Ct.* 113. The great skaith done by swine by houcking and working up the common grass; 1705 *E. Loth. Antiq. Soc.* II 48.

(*b*) That na swyne be haldin fra fassin ewin to do ony schatth & with a ring in his nose; 1442-3 *Ayr B. Ct.* 28 Jan. That thar katel lonand do na schat to thar nichtburis; 1473 *Reg. Cupar A.* I 180.

(2) Quhar thai [*sc.* the swine] be comprehendit wythin ony scatht [etc.]; 1518 *Selkirk B. Ct.* (ed.) 47. Quhar thar be ony fundin in scatht to follow the best to thaim [that] awn it and to have for thai swyn as oft as thai be in scatht ii s.; 1520 *Ib.* 58, 59. The suyne fundyne in the skeyth .. may weyll be slane and mayd escheyt and etyne; 1554 *Crail B. Ct.* MS 3 Dec. That na tennent .. strik nor ding wthers beastis, how soewer being found in skaithe; 1650 *Misc. Spald. C.* V 230. Haveing found .. his cattelles in the skaith; 1664 *Forbes Baron Ct.* 251. And the persewar depounit he kent no uther guids in the skayth; 1666 *Rothesay B. Rec.* 110. Every calfe that beis fund in skaith to be killed; 1679 *Ib.* 368.

(3) Gif ony .. bestiall beis fund all nycht furtht of houssis in ony persones scaytht on thair stuff [etc.]; 1574 *Glasgow B. Rec.* I 17. Whosoeuer salbe tryit to pick lothe thair nichtbours hors albeit thay be fund in thair skayth [etc.]; 1624 *Misc. Spald. C.* V 223. That nae maner of persone .. suffer any of thair swyne to .. be seene .. in any vther menis skaithe or yairds; 1655 *Glasgow B. Rec.* II 305. Confest his stirks to be in Agnes Kelsows skayth and eating hir corne; 1656 *Rothesay B. Rec.* 10. A naig in my skeath being ane furlett corne; 1671 *Kirkcudbr. Sheriff Ct. Processes* No. 107. The persone in whose skaith the said beast is found hes hereby libertie to pin the same; 1687 *Corshill Baron Ct.* 175. [Any person finding swine] in their skaith [has licence to kill them]; 1702 *Banff Ann.* I 174.

(4) Quhar that ony swyne cane be fundyne upone ane mannes skaytht, it sall be leffull to slay hir; 1527 *Selkirk B. Ct.* (ed.) 85. *Ib.* 91. That nane keip foulls on thair nighbours skayth; 1664 *Rothesay B. Rec.* 82. A proclamation .. discharging all ther inhabitants to suffer any of ther suine to be found upon ther neighboures skaith; 1682 *Musselburgh* 127. That .. non .. sufer horse, nolt nor calves to goe upon their neiburis skeath and prejudice; 1684 *Kirkintilloch B. Ct.* 127. Whether or not George Aitchisons horse was upon the uthers skaith; 1697 *Stitchill Baron Ct.* 125.

(5) Quhatsumever man hawand suyn wythin Selkyrk to kep thaim out of his nychburis scatht; 1518 *Selkirk B. Ct.* (ed.) 46. All scheip .. to be keipit be thair awneris from thair nighbouris skayth; 1669 *Rothesay B. Rec.* 167. All persounis .. to caus keip thair gudis af thair nighbouris skaithe till commoune hirds be gottin; *Ib.* All beasts to be teddirit and keipit out of all skaith both night and day; 1681 *Kirkintilloch B. Ct.* 118.

(*b*) For the wrangus slaying of ane ȝoung swyn out of his shayth pertenying to Johne Crokatt; 1546 *Elgin Rec.* I 88. Giff it hapin ony man to fund ony swyn in ony shayth; 1608 *Inverurie B. Ct.* 18 March.

4. (*It, that,* etc.) *is* (*wes,* etc.) .. *scathe,* (it, that, etc.) is (was, etc.) a cause of harm; a matter for regret. *Quhat skath!* how pitiful! how dreadful!

(*a*) For off his dede wes mekill skath; Wynt. VIII 6036. And that was grete pitee and grete scath; Hay I 32/14. Off the quhilk it was grete scathe to the citee of Rome; *Ib.* 64/17. Than Lowrence leuch .. 'To tak ȝone bud,' quod he, 'it wer na skaith'; Henr. *Fab.* 2249. Sir Eger sayes, it is no skaith That she have quantance with us baith; *Sir Eger* 923. Gret dampnage skaith & imperfeccioun had bene that Jhesus had remanit subiekkit to deid; Irland *Mir.* II 39/32. To gar me se it war na scaith; *Seven S.* 657. It var les skaytht to mak ane iust veyr nor [etc.]; *Compl.* 186/3. Skayth wesd to haif skard him;

Christis Kirk 65 (B). Ȝit than it war na skaith Thocht I ȝow tald the tale and sentence baith; ROLLAND *Seven S.* 2188. — O God, quhat skath! DOUG. II v 41.

(*b*) Angnes Stone vrangit in the saying to Necoll Moressone that shayth [? *for* na shayth] war it to gadder ane dussane of quyins and gar lesch him; 1542 *Elgin Rec.* I 72.

5. Injury, damage or loss for which legal compensation is claimed.

Const. *for (of, in)* the damage sustained; also *to restore, amend,* etc. the loss.

(1) For the qwhilk doyng I soume the schame and the skath of the fornemmit J. and his men apon the R. iiii^c li. of sterlingis; *a*1400 *Reg. Morton* I xl. He sowmis the skaith vj merkis; 1524 *Carnwath Baron Ct.* (SHS) 17. And the party to haue actioun aganis thame for his skayth and damnage; 1552 *Edinb. B. Rec.* II 174. Quhat skaythis his kyll sustenis he may hef tyme and place to recouer the same; 1560 *Inverness Rec.* I 45. And puttis in his clame a sowme of money for harmis and skaithis sustenit be him; BALFOUR *Pract.* 504. The person offender sall pey .. of unlaw xl s. for ilk fault provyding the skaytht be first prysitt; 1604 Murray *Early B. Organ.* II 249. Protestit for his skayth and damnage for the said John for .. hindering his boit; 1655 *Rothesay B. Rec.* 5. Tenentis .. sall .. persew [*ed.* presew] for the skaith be for the court or any other jugditorie; 1668 *Forbes Baron Ct.* 274. Arthure Bryce, one of the compriseris thereof .. declares the said skaith .. wes three peckes; 1673 *Corshill Baron Ct.* 109. The said lose damnage or skaith to be pryset by four newtrall skillfull sworne men; 1674 *Edinb. B. Rec.* X 421. The unlaw to be ten pound and mends to the party, conform to the skaith; MACKENZIE *Laws & C.* I xix 15 (1678) 206. The judge ordaines the birlaymen .. to cast one the skaith proportionallie; 1684 *Corshill Baron Ct.* 168.

(*b*) Extending to the saidis abbot and convent dampnage and schait to the soume of sex markis and als for the wrangwis .. settyng of nettis in thar wattyr .. extendyng be gud estimatione to thar dampnage and schait to tayn markis; 1495 *Paisley B. Rec.* 51.

(2) & the windar is haldin to restor the first scath; *Ship Laws* c. 10 (A). The pond sal be ij d. onforgefyn and mend the scath neuir the les; 1458 *Peebles B. Rec.* I 126. Giff it wer swa, quha suld amend the skaith; HENR. *Fab.* 2864. Thai persons sall caus the scaith and wrang to be amendit; 1475 *Edinb. B. Rec.* I 31. The distroaris of the nychtburis gudis with thare swyne sall amend the scath; 1477 *Prestwick B. Rec.* 27. For and he [*sc.* the notary] do, he aw .. The scaith thairof be ressoun to restor; *Bk. Chess* 1432. Puir folk .. off thair skayth be lawe can get na mending; MAITLAND *Maitl. F.* 429/14. Incaice they be challenged heirefter be way taking ather peitis or kell, ther masteris sall rebate ther skaithe; 1614 *Misc. Spald. C.* V 218.

(3) For the scaitht of his hors ryding & spilling x s.; 1478-9 *Acta Aud.* 82/1. John .. sall restor to .. Robert xiiij boll of mele for the skath of the lying of his land wast; 1479 *Acta Conc.* I 28/2. The chalance of .. the dampnage & skathe clamyt .. for falt biggin of certane hous & dede of his catall [etc.]; 1479-80 *Ib.* 48/2. And for the hurt dampnage & scath sustenit be the said Johne & Jonet in the dovnfalling of the said bakland; 1490 *Ib.* 149/2. 1495 *Paisley B. Rec.* 51 (see 1 (*b*) above). For damnage and skayth sustenit be .. Williame throw the wranguus .. destructioun .. of the hous; 1517 *Misc. Spald. C.* II 80. As to the allegit skatht sustenit be .. David of the said gait the assis absoluit .. thair fray becaus the valer thairof ves nocht .. sufficient prevyn; 1537-8 *Dunferm. Reg. Ct.* 152. Decernis Robert Scheirlaw to pay to William Jamesoun iiij s. for the skaith of his plants; 1578 *Kirkcudbr. B. Rec.* I 73. Fines given to John Briscat .. for the recompans of the skayth sustenit be him of the 27 merks; 1595-6 *Glasgow Burgesses* 22. And ordaines the skayth of the raine weit in the housses to be allowed to Johne Young; 1670 *Kirkintilloch B. Ct.* 27.

b. (The amount ordered to be paid as) compensation for loss or injury; a sum of money paid to restore damage done.

Some quots. may belong in prec. sense.

Giff ony man chalange ane other that he .. wrang has done him & menys his scathis til i c merkis [*B.* put til his scath c s.]; *Burgh Laws* c. 77 (A). He wald nocht strek for ony scaith to get Bot erar wald the haile trespas remyte; *Bk. Chess* 1279. And the skaith of the gudis .. sould be partit amangis the merchandis of baith; BALFOUR *Pract.* 625. In that cais baith the schipis sall pay the skaith; 1602 *Conv. Burghs* II 144. Ane tenant within brugh burneing his house rekleslie should pey the skaith; HOPE *Major Pract.* I 37. Item bee .. burneing of dailes .. and plenisheing demolishing of glasse windowes and iron stanchells, of skaith in that sort thrie hundreth merks; 1649 *Antiq. Aberd. & B.* IV 781.

6. Damages; costs or expenses incurred by the pursuer in a legal action. Chiefly *pl.*

pl. Thre hundyr mark of sterlyngys .. for thaire costis and thaire skathys gyf euyr thai make ony questione or mouyng be the lawch into the contrary of this endenture, in ony thyng, or euyr thai be hard before ony iuge; 1389 *Wemyss Chart.* 24. Qwhyl thai be fullily asythyt of scathes, costages or trawales; 1392 *Lennox Mun.* 47. With his costes & scathes gif ony be in the mene tyme; 1407 Slater *Early Sc. Texts* No. 65. Gyf the maystyr tynys the owre it war skath and than sal he pay the scathis; *Ship Laws* c. 13 (B). Gif the fleschewar graythis ivil flesche he sal restor hym the scathis that aw the bestys; *Acts* I 33/2. The maner of connand & how it wes brokyn & quhat scathis he had be the connand nocht rycht kepyt; *Ib.* 110/2. Al costis scathys trawall .. & expens the qwylkis he or thai sustenys .. in the defawt of the said ten markis ȝerly nocht beand payt; 1447 *Newbattle Coll.* (Somerville) (Reg. H.). He mon deliver all that accioun till his lord gif he will have his scathis hale maid be the said lord; HAY I 136/23. Quhill the merchand of Paris war content bathe of the principale and of the scathis and costis; *Ib.* 208/30. *Ib.* 164/26. Bayth interes and all the scathis haill To be assythit richt as the principall; *Regim. Princ.* 230 (Maitl.). His costis scaithis & dampnagis sustenit be him for falt of his entre to the sammyn; 1476 *Acta Aud.* 43/2. To prufe the avale .. of his costis and scaithis that he has sustenit be the said persons; 1482-3 *Ib.* in *Acta Conc.* II cxxiv. In the naym of costis, skathtis [*ed.* skathlis] .. and expensis; 1486 *Stirlings of Keir* 256. For thar costis skatis & dampnis sustenit tharthrow; 1487 *Fraser P.* 108. For his costis and skathis that the said Thomas as maid for his and hir persoit; 1491 *Dunferm. B. Rec.* I 27. Thomas Leslie forspekar for the said personis protestit for thar costs skaithts and expensis maid be thame; 1504 *Aberd. Sheriff Ct.* I 48. The costis & scathis sustenit be the said Johnne in the defalt of the said dame Katrine; 1519 *Fife Sheriff Ct.* 163. Skathis; 1529 *Reg. Episc. Aberd.* I 397. That day James Brydin vas condamnit efor borrogait of Andreu Schortreid of all and syndry skaythitis, dampningis, roumentis that Thome Prechour incurrit in to be fraud of the forsaid Andreu; 1530-1 *Selkirk B. Ct.* (ed.) 111. Till radres and pay .. all and sindry skayththtis costis .. the said Jhone Campbell .. sustenis; 1549 *Breadalbane Doc.* No. 68. Skaytis; 1585 *Perth B. Ct.* 11 June. Mak equall half of the costis skaythis & expensis; 1589-90 *Prot. Bk. J. Inglis* 10 Feb. The defender .. to pay the debt, with the skaiths taxat and liquidat in the persewers clame; SKENE *Reg. Maj.* I 77. The defender nocht compeirand in propir person .. sall pay coistis and skaithis of the pairtie compeirand; BISSET I 229/20. The said defender .. to mak payment of the coistes and skaithes maid and susteaned be the said persewer; 1662 *Melrose Reg. Rec.* II 2.

(*b*) That the said Sir John was cummyn til his principaile some & his some schathis; 1440-1 *Ayr B. Ct.* 19 Feb. To pay all costis, schathis, expens, trawale and interes at the will of the said Robert; 1469 *Charter* (Reg. H.) No. 419. To gidder with costis schathts that the said Duncan .. sal hapin to sustene; 1485 *Breadalbane Doc.* No. 22. Schaythis; 1497 *Dunferm. B. Rec.* I 80. Thomas Grenschill come in merchiament .. for nocht entris at the instance of James Fairbarn with his schaithis; 1512 *Selkirk B. Ct.* MS fol. 17a. Arrestit at instanch of James Reid to ansuer to him of his costis and schathis; 1519 *Ib.* fol. 76.

sing. The vrangus brakyn off hys sadyll sowman hys schaid to iii croun; 1512 *Wigtown B. Ct.* fol. 25a. For the vrangus scadyng of ane kow summand the scaid to a pek of bear; 1521-2 *Ib.* fol. 121b.

The said Patriyk .. sovmys his schayd to a thovsand povndis; 1530 *Ib.* fol. 227a. Quhilk day comperit Blais Colt and protestit .. for cost skait & expensis; 1563-4 *Perth B. Ct.* 31 (22 Feb.). The other cautioners being obliged to relieve him pro rata of all coast, skeith and damnage; 1665 *Decis. Lords* G. 91. [The procurator protested] for coast, skeath and damnadge; 1705 *Laing Chart.* 699.

Scath(e, Scaith, Schath, *v.* Also: **skath, skaith, scayth, skayth(t, skaycht; skad; (scaithis).** P.p. also **skait, scaith, (skaytit);** Sca. [ME and e.m.E. *skaþenn* (Orm), *scaþien* (*c*1380), *sckathe* (*a*1400), *skathe* (*c*1460), *scathe* (Malory), *scath* (Shakesp.), ON *skaða* impers. to take hurt, corresp. to OE *sc(e)aðian* to injure, rob, OFris. *skathia* to injure.]

1. *tr.* **a.** To damage physically; to harm by physical action; to cause (a person, thing or place) damage which is, at least in part, physical in nature.

There is some overlap with 2 below.

And we sall neir enbuschit be .. Than scath [*E.* na scaith] thame on sum vis we sall; Barb. iv 363 (C). Thai sall nocht scath gretly The Scottis men with schot; *Ib.* xiii 110. They shall .. skaith ȝow all that euire they may; *Troy-bk.* ii 1494 (C). Thay think .. to skaith vs sa That nane sall quick eschaip thame fra; *Alex.* i 709. That it [*sc.* a wall] mycht wythstand Thare fays that thame swa skayth[it] hade; Wynt. v 3249. Scaithis; *Ib.* ix 650 (W). Casting off petis faile na dwuate na nane othir thing that mai rede the erde or skaith the pasture; 1456 *Liber Aberbr.* II 89. With draucht of water quhar euer it ma scath tham [*sc.* lands]; 1473 *Reg. Cupar A.* I 173. [A claim] agains the merchants having guids in his ship for the skaith of the mast and tackle being skaithit in this raise; 1521 *Old Dundee* II 318. Gif it sall happin the .. rebellis to birne ony cornis pertenand to the said lord Hume .. our souerane lord sall .. cause thame sa beand scaithit and brint to be recompensit; 1528 *12th Rep. Hist. MSS* App. viii 181. [He] hes skaythit the kiche of the inland of the forsaid land in the distroying, byrning & away taking of the caberis, tymer, treis & thaik of the said kiche; 1538 *Aberd. B. Rec.* MS XVI 134, 135. Twa webs of his .. mynest be Katrine Henrison or skaithit; 1550-1 *Old Dundee* II 327. I sal .. nevir skaith thair wairis; 1616 *Aberd. B. Rec.* II 345. Thair kirk quhilk wes greatumlie harmit and skaythit throw the great tempest; 1633 *Stirling B. Rec.* I 169.

b. To harm (a person, also his or her character) morally; to damage in reputation.

In the first *Ratis R.* quot., *scaith* may be a further instance of Scath(e *n.* rather than *p.p.*

(1) Robert hes sclanderit the forsaid abbot .. and through his wrang informatioune hes gert skaith the said abbot; 1389-90 Slater *Early Sc. Texts* No. 18. He saw .. Sum rychtwys men and god-lyk baith With wykyt men & wnewyne-lyk scaith; *Ratis R.* 60. Till fle foull cumpanye That thow may ocht be scaithit by; *Ib.* 474. He .. suld nother say nor do thing .. that micht empaire or scaith ane nothire man; *Porteous Noblenes* 181/12 (A).

(2) Scars spending skathis gentris; *Colk. Sow* i 486.

c. *reflex.* To put oneself in physical, or moral, danger.

He saw him .. skaith him throw his great bounte; *Alex.* i 1839. Thane is it misterfull to thé Till have thai placis veil in thocht And se gif thow may scaith thé ocht; *Ratis R.* 130. Do never the thing that ever may scayth thé ocht; *K. Hart* 651.

2. To harm (a person, (group or community of) persons, lands or interests) with specific reference to their rights and, chiefly, their financial and other interests, possessions, etc.

Also const. *of, in* or *to* the (amount of) money, etc. lost or things damaged.

The party scathit, the person injured in the eyes of the law; the complainer; the pursuer.

(1) Gif the forsaide landis .. beis in the mene tyme of his take throuche commone were distroyit that the forsaide Schir Williame [his heirs, etc.] .. sal als lange joyse thaim eftir the oysche of his terme as he or thai ar skathit within the forsaide ten ȝere; 1399 *Melville Chart.* 15. Till assay & gouern all werkis that beis maide .. sua that the kingis liegis be nocht defraudyt & scathyt; 1424 *Acts* II 8/1. Gife A be scathit mar than amendit or beterit; 1449 *Melville Chart.* 33. Sen the fader dois to the sone it that he aw to do .. quha suld be he that suld reprove him na ȝit blame him or scathe him tharfore na do him ony injure? Hay I 225/17. The haill realme .. sulde be gretly hurt and skathyt in the changing and heing of the cours of the mone .. the borowar and the lennare the lorde and the tenande [etc.]; 1467 *Acts* II 89/1. The distroaris of the nychtburis gudis with thare swyne sall amend the scath or the next court to thaim at ar scathit; 1477 *Prestwick B. Rec.* 28. Becaus it is sufficiently previt that the saidis personis has gretly scathit the said Margret and intromettit with the cornez & vitales of the said place .. that .. thai sall .. gif to the said Margrete twentj bollis of malt [etc.]; 1493 *Acta Conc.* I 300/1. One [= un-] hurt harmyt, skathit or trublit; 1523 *Prot. Bk. Sir J. Cristisone* 6. Ane lettre of bailyerie .. be the quhilkis I may be gretlie hurt dampnagit and scathit. The quhilkis .. writingis [etc.] .. hurtand .. me my landis and heretage [etc.]; 1532 *Antiq. Aberd. & B.* IV 231.

(*b*) Sa that innocentis be nocht throu the hiding of verite hurt nor scaithit in our defaltis; 1474 *Edinb. B. Rec.* I 28. The said craft is .. gretly skaithit by the daily markat maid in cremys .. throw the hie street; 1496 *Blue Blanket* 11 (Jam., s.v. *Hedisman*). Nochttheles I was skaythit and wanttyt in hir defalt and sleutht .. item, I wanttyt out of my hous in hir defult and sleutht ane plaid [etc.]; 1541 *Aberd. B. Rec.* I 175. I .. sall neuer wit .. his nor his ayeres skaithit [etc.] .. be neycht nor daye bot I or thai sall warn him [etc.]; 1545 *Montgomery Mem.* II 138. Ye sall nocht noy, hurt or skayth .. ony freindles man or woman; Hamilton *Cat.* 103. Throw the quhilk [false coin] .. this commoun weill hes bene greitlie hurt, and oure soveranis .. defraudit and skaythit; 1566 *Reg. Privy C.* I 468. That none of the freemen mak wark or labour of any ither craft that nane be skaithit thereby; 1570 *Dunferm. Hammermen* 27. The kingis majesties caus and service is greatlie hinderit and skaithit throw the frequent resort of people betuix .. Leith and Edinburgh; 1571 *Reg. Privy C.* II 102. [They] ar grittumlie skaithet and defrawdet be insufficient wark of ignorant persouns; 1586-7 *Edinb. B. Rec.* IV 481. The heven .. did impairt to me .. wisdome, will, And t'aide my freinds and skaith my foes a skill; Garden *Worthies* 108. Paena conventionalis may not be asked bot in swa farr as the partie is interessed or skaithed; Hope *Major Pract.* I 99. Be resone the said communis war preiudgit, hinderit, skaithit and intrest mor nor the said sowme ȝeirlie throwch [etc.]; 1633 *Inverness Rec.* II 173.

(*c*) The nebour nocht keipand him nebourheid sall recompance him at the syicht of nebouris in sa far as he is skait; 1523-4 *Stirling B. Rec.* I 18.

(*d*) At thai resaf thar fee nocht doand the thingis that pertenis to thar .. office that is to say to merk clatht witht the selis of thar office be the quhilkis the byaris ar schathit quhen it mai nocht be knawyn to thaim at thai haf that law; *Acts* I 334/2.

(2) He aw to be payit of the baroune of all his soume of lenth that he war scathit of; Hay I 167/23. Anne Todryk to pay hym als fer as he is skaychtit of his corne as he can preiff upone hir gudis; 1535-6 *Selkirk B. Ct.* (ed.) 163. The said dame Jonet is dampnagit and skaythit of the saidis cornis; 1558 *Inverness Sheriff Ct.* 69b. Gif .. any ane of us does in the contrairie that the .. persons .. hurt or skaithed of thair kyndlie roume and possession, sall complean to mee; 1589 Craig-Brown *Selkirkshire* I 167.

(3) The persons forsaid .. rasit fyre .. and skaithit the sade Wilȝeam in househald geire tharethrou; 1496 *Acta Conc.* II 38. [They] wer damnageit and skaithit .. in the prices and valouris of the saidis oxin; 1562 *Peebles B. Rec.* I 283. That he sall satisfie, content and pay all personis skaythit or hurt in thair gudis be him .. at the jugement, sicht, discretioun and modificatioun of Johnne Erskin of Dun; 1569 *Reg. Privy C.* I 664. The saidis persones .. war havelie .. skaythit in eittin of thair cornes be aither of thair guidis; 1577 *Prot. Bk. J. Scott* 37b. The said Thomas hors skaithit hir in hir corne this last cropt; 1579 *Kirkcudbr. B. Rec.* I 101. He wes suirlie informit befoir of sum guidis that wes appeirand to do skaith him in

the cornes; 1596 *Elgin Rec.* II 43. [An army quartered on estates of Lord Cardross] and his lands & tenents were thereby skaithed in at least 500 lib. sterling; 1690 *Acts* IX 205/2. — I .. sal neuermar .. move plede, raise stryf, distrubill, in quiete, vex, harm, scath or hinder the forsaide Henry .. in the .. joysing of al the forsaid landis; 1453 *Stirlings of Keir* 226. Men of wear maryneris .. committing any enormitie quhairby they think to schorte the weyaige and skaytht the capitanis maisteris and awneris in there purpoise; BISSET II 237/1.

(4) Thai hed scaithit the said Williame .. tharthrow to the soume of xij merkis; 1519 *Fife Sheriff Ct.* 156. Quhilk scaythit the yle .. to the valor of ane hundreth dollors and mair; 1576 *Orkney Oppress.* 47.

(*b*) For the vrangus blesing of his corne skadan him to the sowme of ane firlot of atis; 1516 *Wigtown B. Ct.* fol. 57b.

(5) The wrangar sall assyth bath the party scathit & the aduocatis costis & trauale; 1424 *Acts* II 8/2. The party scathit sall folowe ande the party trespassande sall defende; 1425 *Ib.* 9/2. Gif the pairties skaithed intendis and persewis not within yeir and day [etc.]; 1572-3 Bann. *Memor.* 308. Except .. that the persoun .. eiectit .. had fund souirtie .. for releif of his landislord and to redres partiis skaithit; 1594 *Acts* IV 72/2. Besyde the satisfactioun and redres of the pairtyis skaithit; 1611 *Misc. Maitl. C.* III 28. The contraveiners to pay to the partie .. skaithed as followes; 1615 *Kirkwall Sheriff Ct. Act Bk.* 3. 1635 *Antiq. Aberd. & B.* III 310. But likewayes be debt bound to satisfie the partie skaithed; 1698 *Misc. 3 Spald. C.* II 96.

b. To cause a person (financial loss).

(*a*) He sall .. pay to the said Wil .. sa mekle as he is skathit throu the sawing of the said landis; 1474 *Acta Aud.* 31/1. That the said maister Richard haid scathit the kingis hienes j^c li.; 1494 *Acta Conc.* I 354/2. Quhatsumever the said persewar wes skaithit thairthrow, the said Robert promisit to pay; 1603 *Paisley B. Rec.* 257. John Futhie .. sall .. pay to the said personis so mekill as thay can preif that thay ar skaythit be his myln; 1608 *Douglas Bequest* 30 June.

(*b*) Ilk ȝere at thir ach ȝere skading him vii d.; 1514 *Wigtown B. Ct.* fol. 38a. For the vrangus occuping of ane tenement .. without ony tak or assedation skadand him ȝeirlie xii crounis; 1530 *Ib.* fol. 233b, etc.

c. Of stray or trespassing animals, also *transf.* of their owner: To damage or destroy (a neighbour's corn, etc.).

Suorn .. at he skaithis to the said Jhone v firlottis aittis and decernit to pay the samyn; 1532-3 *Dunferm. Reg. Ct.* 89. Grantit he wes caution for ane Hielandman .. for certain aits skaithit be the Hielandman's guids; 1553-7 *Old Dundee* II 308. They and ather of tham had skaythit Robert Kirlie certane of his cornes be eitting tharof with foullis; 1595 *Paisley B. Rec.* 167. Decerneit to pay thrie pecks quhyte corne skaythit be him to Agnes Kelsow be his stirks; 1656 *Rothesay B. Rec.* 10. Donald M'Kaichen .. decernet to pay to John Kelburn elder of eattin corne skaytit be thair guids in seaven pects corne; 1661 *Ib.* 62.

3. To calculate the fine due for damages caused.

Appar. a *transf.* use of 2 b above.

For the vrangus braking of his arm skadan his sowme to xl cronys; 1517 *Wigtown B. Ct.* fol. 75a.

4. *absol.* or *intr.* To cause damage; to do harm.

It ma nocht scaith, suppos it do na waill; *Wall.* VIII 1132. But scoup or skift, his craft is all to scayth; *K. Hart* 856.

(Scatheabill,) Skathabil, Skaithabill, *adj.* [SCATH(E *v.*] Harmful. = SCATH(E)FULL *adj.* — Na thing is mair skaithabill to the kinrik na inwert discord; ABELL 58b. Batel mast skathabil in the feil of Floudayn; *Ib.* 3*a.

(Scatheand,) Scaithand, Skathand, *ppl. adj.* [SCATH(E *v.*] Harmful. — That raid nocht gretly scaithand [*Au., E.*[2] skathand, *C.* skaythful, *R.* skathfull] was To the cuntre that thai throu raid; WYNT. VIII 6122 (W). All obligaciouns .. that may do gude or proffit to him, or that may be skathand to him; 1480 *Acta Conc.* I 79/2.

Scath(e)full, Scaithful(l, *adj.* Also: **skath-, skaith-, skayth-.** [ME *unnskaþefull* (Orm), e.m.E. *scadefull* (1527), *skathfull* (1586), OE *sceaðfull.*] Harmful, hurtful; that causes damage or injury.

Const. *to (till)* the thing damaged, or without const.

a. Chiefly, of actions or non-material things. Chiefly *predic.*

predic. (1) Giff thé hapnys ony thing That anoyis or scaithfull [*C.* scathfull] be; BARB. V 249. The quhilk tuk [counsel] wpone hand That skaithfull was; *Troy-bk.* I 32. That thou kepe .. all thyne alliaunces, all be thai never sa scathefull na sa hinderand na grevable; HAY II 105/12. Sic wanton latis and mysreul was mare skaithfull becaus [etc.]; *Boece* 578. Alwayis this pest was thus les skaithfull than seditioun was appering to haue bene; BELL. *Livy* II 123/27.

(2) That rad noucht gretly skathfull [*C.* skaythful] was Till the cuntre; WYNT. VIII 6324. Privilegis of law, lettres [etc.] .. the quhilk mycht profit til ony of the said partis or be scaithful or hindering to the tother; 1447 (1449-50) *Reg. Great S.* 68/2. Archibald .. maid his eme our thesaurer .. quhilk wes rycht skaithfull till ws; 1529 *St. P. Henry VIII* IV 548. Scaithfull; BELL. *Boece* II 15. This bargane to baith the pepill sa skaithful; *Boece* 360. The victorie was to him sare & skaithfull; *Ib.* 399. Siclik prayers ar mair skaithful na profetabil to the saluatione of the saul; GAU 85/11. Wordis that ar noysum and skaythfull to thair nychtbouris; HAMILTON *Cat.* 166. And attempt no thing that may [be] skaythfull or dampnable thairto; MAITLAND *Ho. Seytoun* x. The tolbuyth .. is .. hable haistalie to .. fall doun quhilk wil be verray .. skaithfull to the people duelland thairabout; 1561-2 *Reg. Privy C.* I 198. Ane pley .. the persute quhairof wes na les skaithfull and sumptuous to him nor [etc.]; 1570 *Reg. Privy S.* VI 168/2. Quhilk did thame litle good, albeit skaithfull to the owner; SPALDING I 283.

attrib. Sa mony skaithfull iupertyis war made on nychtis and dayis; *Boece* 194b. Amangis mony oder skaithful bukis and fals doctrine; GAU 3/4. Eftir thair langsum and skaythfull troublis; 1572 *Reg. Privy C.* II 173. Quhairon skaithfull & schamefull inconvenientis dalie fallis out; 1581 *Acts* III 222/2.

b. Of material things.

And sum wondring the scaithful gyft beheld; DOUG. II i 34. Winis .. ar richt skaithfull to the nature of Albianis; BELL. *Boece* I xxiv.

c. Of an animal.

To persew .. that beist [*sc.* the wolf] becaus to thare bestiall nane vthir was mare skaithfull; *Boece* 64.

(Scatheles,) Scathles, Scaithles, Schath(t)les, -lace, *adj.* Also: **skath-, scatht-, skatht-, scat-, skat-; skaith(t)-, skayth-, skayt-; skeath-, skeyth-; skith-; skyath-; skad-; schad(e)-, schach-, schaich-, schaith-; sayth-** and **-lese, -less(e, -leis, -lis, -las(e.** [ME *skaþelæs* (Orm), *schaþles* (*c*1350), *scathles* (Chaucer), ON *skaðlauss*; SCATH(E *n.*]

1. Unharmed, undamaged (by (*of, anente*) another person, also rarely by (*of*) a thing); free from injury or violence; unscathed.

Also const. *in* the respect in which the person is harmed.

(1) Of thairis till thame gaf thai Till let thame scathles pas thar vay; BARB. XVIII 200 (C). The fox scathles gat his way; *Ib.* XIX 683. Quylle he fynde souer and sekyr borowys that the kyngis legis and the kynryk be skathlase and kepyt wnhwrt in tyme to come; 1429-30 *Ayr B. Ct.* in *Sc. Hist. Rev.* XXIX 11. Pleg .. that the parte be scathles as the law will; 1440-1 *Ib.* MS 19 Feb. Law burgh that William Vmfray salbe vnscathit .. and .. the said Johne the Rosse salbe scathles in likewise; 1484 *Aberd. B. Rec.* I 40.

(*b*) That skaithles they myght pas the se To thar owne lond in sawyte; *Troy-bk.* II 629 (C). Thay salbe skaythless and without daynger in thair cuming; *Hist. Jas. VI* (1825) 38. Whill thay set sufficient cautioun .. that thay suld be harmles and skaithles, wnder gryte panes; SPALDING II 288.

(*c*) [They] protestit .. that quhat dangeir and damnage cumis thair throw .. that thai be saythles and mair [*sic in pr.*] no preiudice; 1559 *Aberd. B. Rec.* I 316.

(*d*) That Magnus Sinclair .. sall be harmeles and skithles [*blank*] and on nawayis to be trublit; 1612 *Orkney & Shetl. Ct. Bk.* (ed.) 36.

(2) That the sayd Thom suld be skathlas bot as the curs of common law askis of the sayd Doby and his purches; 1458 *Peebles B. Rec.* I 129. That his moder [etc.] .. salbe skathles of him & al that he may let bot as law wil; 1466 *Acta Aud.* 6/1. That .. the tenandis of Elwinston suld be harmles & skathles of of him bot as the cours of common law; 1471 *Ib.* 19/1, etc. Scathtles; 1493 *Antiq. Aberd. & B.* II 141. 1512 *Selkirk B. Ct.* (ed.) 18. That the said Thomas suld be harmeles & scatles of hyme; 1517 *Fife Sheriff Ct.* 83. That .. euir ilkane of thame salbe harmeles and scathles of ws our kin [etc.] .. of bodelie harme or vther ways according to the law; 1520 *Edinb. Chart.* 205. At sayd Androw sall be skathles of all bodely herm of hir .. vnder payne of law; 1536 *Dunferm. Reg. Ct.* 140.

(*b*) Souerte and lawborgh .. that William Dunmure salbe harmeles and scaithles of him .. vnder the pane of xl li.; *c*1456 *Edinb. B. Rec.* MS p. 3. That the Postulat of Ross and his tennandis .. sal be harmeles and skaithles of the sade Huchone his servitoris and anerdance; 1497-8 *Acta Conc.* II 94. That the .. communitie of Abirden .. sal be skaithles aleway of William Urry; 1503 *Aberd. Sheriff Ct.* I 68, etc. 1536 *Carnwath Baron Ct.* (SHS) 184. That Dauid .. suld be harmles and skaithles of thame and all that thai ma lat forther thame; 1538 *Aberd. B. Rec.* I 157. That umquhill Alexander Matho of Kylburnys, his freyndis and servandis, suld be skaithles of the said umquhill Johnne of bodely harme; 1542 *Reg. Privy S.* II 758/1. Skaytles; 1549-50 *Elgin Rec.* I 100. That the saidis Frenchemen suld be harmeles and skaithles of the said Johnne; 1560 *Admir. Ct. Bk.* 141. 1578 *Kirkcudbr. B. Rec.* I 69. That he salbe harmeles & skaithles of him of quhom he dreadis the bodielie harme; Skene *Verb. S.* s.v. *Schireff.* That James Stewin .. salbe harmles and skaythles of the said John .. and nawayes to be molestit nor trublit be the said John, nor na vtheris of his causing, command, assistance, nor ratihabitioun; 1599 *Misc. Spald. C.* II 144. Spalding II 91. — Tailyour salbe scaithles of thaim and harmles of al bodely harme and scaith; 1499-1500 *Exch. R.* XI 448. — The closat .. sal be half ane fut vndir .. the sol of the yeit cled sufficiently witht flagis and at quhilk may keip the said Villiem skaithtles of flovir gif sa beis that the closat be vsit; 1520 *Dunferm. B. Rec.* I 174. Ve ordand it to be disponet uthoucht the toun quhar infektioun is knawn, that nychtboris be skaythles of it; 1534 *Selkirk B. Ct.* (ed.) 146. And that Thomas Menȝeis [etc.] .. sal be harmles and skaithles of all bodelie harme to Sonday callit *Dominica in Albis* inclusive; 1552 *Antiq. Aberd. & B.* III 278. To mak my ease [*pr.* eafe] drops fra quhartrocht my saidis wowttis may be skaythles of weit [etc.]; 1572 *Inverness Rec.* I 215. William Handisyde .. sall be .. skaithlesse .. of his causing sendind hounding out .. or ratihabitione; 1679 *Kelso Baillie Ct.* 78.

(*c*) That the said laird .. and the rest of the leedges may be harmeles and skeathles of the saidis rebelles; 1641 *Acts* V 710/1.

(*d*) The alderman & balȝe suld be skadles of hire; 1512 *Wigtown B. Ct.* fol. 7a.

(*e*) Rob Dykyson fand law borcht for hym Jhon Dykyson that Thomas Haw swld be schathtles of hym; 1467 *Peebles B. Rec.* I 156.

(*f*) That helk ane of tham salbe harwmles & schadeles adir of wdir; 1513 *Wigtown B. Ct.* fol. 22a. Fore Douncan Stevart to keyp the man harmles & schadles of hym; 1529 *Ib.* fol. 223a. — Gylbert Belman was oblist to kepe John Litstar schathlace anente the swethin man for a barrall of salmoundis; 1458-9 *Aberd. B. Rec.* MS V 1 p. 346 (13 Feb.).

(3) That their bairnis tenentis and servandis salbe harmles and skaythles of the said Johne Leslie in thair bodies and landis takkis possessionis guids and geir; 1587 *Aberd. Council Lett.* I 10. That he sauld be harmles and skaithles of him in his body; 1595 *Paisley B. Rec.* 160. That ather of them salbe hermeles and skaythles of uther alswele in thair awin personis as thair servandis, kippages, dependaris [etc.]; 1603 *Shetland Sheriff Ct.* (ed.) 107. That the said Mr. David .. sall be harmeles and skaithles of the said Johnne Littill in thair persones landis guidis and geir; 1631 *Justiciary Cases* I 174. — To repair the said estmest hedge .. that the said Andro and Georg may be skaithless in thair yeard thairthrow; 1605 *Glasgow Merchants House* 94. That the said Alexander Stewart .. and .. his brother .. suld be hermeless and skeythles .. in thair bodeis guidis and geir be the said Neill Stewart; 1608 *Stewart Mem.* 120. Arthour Sinclair .. sall be harmeles and skaithles in his landis .. guidis and geir and on nawayis to be .. molestit thairin; 1612 *Orkney & Shetl. Ct. Bk.* (ed.) 18.

(4) The inqueist ordanis the ley to mak the corn haill of all gud that is put on the ley to haif ony profet of the ley that the corn sal be skaithles of it; 1524 *Carnwath Baron Ct.* (SHS) 17. Cutting and distroying of wodis. .. The wod fund to be skaithles sen the last court; 1592 *Breadalbane Ct. Bk.* 26.

b. Free from damage caused to crops by stray animals (cf. Scath(e *n.* 3).

That na .. bestiale be fund within the yardis .. vnder the pane of warding of the awnariis of the gudis in the tolbuth .. quhill thai fynd causioun that all nychtbouris salbe skaithles in all tymes cuming; 1572-3 *Peebles B. Rec.* I 352. To cais his swyne nocht cum in ony mannis skaith that thei sall be skathles thairof; 1615 *Lanark B. Rec.* 125. [Cattle to be kept within bounds] sua that the inhabitants .. salbe harmeles and skaithles of utheris in thair gers and corne; 1657 *Rothesay B. Rec.* 15.

c. *To kepe* (a person or place) *scatheles*, to preserve from (*of*, also *fra*) damage or harm of any kind exemplified above, also from falling masonry or the like during construction work. Cf. 2 below.

(1) To make [*gap in MS*] sone to the said sir Jamz .. [*gap in MS*] spedifull .. and al thir lelily and solemly to be kept skaithles fra the [*gap in MS*] sir Willame; 1393 *Reg. Morton* II 193. To keip the priour and conwent .. and thair place of Sanct Katrine skaithles of me myne airis .. and of my place; 1476 *Ayr Friars Pr. Chart.* 55. To .. kepe .. the saide James .. harmles & skathles of all maner of intromittyngis mone steddingis [etc.]; 1482 *Reg. Morton* II 245. All wardanis .. sall .. keip all .. trew lieges skaithles of thame [*sc.* thieves, etc.], of all harmis, reifis [etc.]; 1527 (1575) Balfour *Pract.* 598. The uther wardane .. sall relieve and keip him skaithles anent the saidis gudis and anent the freithing .. of his men .. furth of Inglishmen's handis; *Ib.* 599. Prouiding alwaus that the said James mak the said Archebald Douglas .. sickir & skatles of all handis havand interes tharto; 1544 *Fraser P.* 154. We the cautioneris .. sall releve, warrand, freith and keip skaithles our said soverane lord .. anent all poyntis and passis of the said obligatioun at the handis of the said King of Denmark; 1573 *Reg. Privy C.* II 311. From the day and deat heirof .. we sall warrant and skeathles keipe the said lord of Grant .. off all danger or inconuenient he .. might haue incurrit for not obseruing his maiesteis lauis; 1613 *Grant Chart.* 213. Chargeing them to keep the bishop and ministers skaithless of all persons whatsomever; Row 484.

(2) Henre Beg decernit to keip his nychtboris skaythles of thair corns; 1550-1 *Elgin Rec.* I 108. 1669 *Rothesay B. Rec.* 167.

(3) And the saidis David [etc.] .. to tak and to have all thair necessar esimentis .. in the said gavil .. thai kepand the said gavil skathles throw the said bigging; 1507 *Prot. Bk. J. Young* 391. That Johne Burne sall stand content of all ws and vont of five inch of aise drop and inlikvise the landis of Alexander Liddal his nychtbour sall stand ineffek and forme is [= as] biggit and kepit Johne Burne skaithles wythtin the bondis .. of the five inch of ony passaige or occupacione; 1521-2 *Dunferm. B. Rec.* I 7. Skaithles; 1635 *Antiq. Aberd. & B.* III 310. John Stewinsoun .. oblessis him .. to warrand frie and skaithles keip the said Gilbert Hervie, his aires .. and occupeares of his two mercheant boothes [etc.] .. of all danger, hurt, detriment, skaith and inconvenient; 1641 *Aberd. B. Rec.* III 270. The superior tenement was not bound to keep the tenent of the inferior tenement skaithless; 1681 *Decis. Lords* F. 5. Patrik Dickisone to have two windowes to the fleshmercat provyding he .. wyre them close fra head to foot and keep the mercat skaithles; 1686 *Peebles B. Rec.* II 118.

2. Immune from financial loss; not subject to penalty or pecuniary liability.

Chiefly *to kepe (hald)* (a person, also once lands) *scatheles* (*anent*, also *for*, the matter giving rise to expense, etc., or *of* the amount concerned, also without const.).

Common in legal formulae, freq. linked with one or more of DEFEND *v.*, FRETHE *v.*, RELEV(E *v.*[1] (see these words for further instances) and *warrand*.

(1) Anent the warandis of the said landis & to diffend & kepe the said Johne scathles anent schir Gylbert Hering [etc.]; 1476 *Acta Aud.* 55/1. Tak ane instrument thairof for that sal kep thé skaithles anens his respyt; *c*1476 *Lennox Mun.* 108. Johne Wallace .. sal releif and kep skaithles the said persons at the handis of the said Margret .. anent the payment of the said soume; 1482-3 *Acta Conc.* II civ. Bondyne .. to keip scathles the said Robert Scot anentis the band abone wrytting; 1510 *Selkirk B. Ct.* (ed.) 10. Chargit to releif & keip harmles & skaytles the said Johne Makcubbene at the hands of the said Jonet anents the payment of the sovme [etc.]; 1563 *Dumfries B. Ct.* fol. 197. To warand freith releif and keip skaithles the saidis Robert [etc.] .. anent the payment of the foirsaid sowm; 1573 *Crosraguel Chart.* II 31. We .. obleissis us .. to releif and keip skaythles our saidis souirteis respective; 1576 *Reg. Privy C.* II 545. — To hald him skathles insafer as the ladijs thrid extendis to; 1480 *Acta Conc.* I 63/2.

(2) Anent the keping of the said Margret scaithles & harmeles of the malis & fermes of the landis of Dalquhillray of x ȝeris bygane; 1479 *Acta Conc.* I 32/2. Sufficiand souuerte .. to kepe him scaithles at the handis of the laif of executoris of samekle as he ressauis be ressoun of the said obligacioun; 1483-4 *Acta Aud.* *135/1. That the said George Master of Angus sall freith releif and kepe skathtles the said Archibald Erle of Ergile of the said soum; 1510 *Justiciary Rec.* II 253. To warrand defend & keip the said Andro .. skaithles of the said sext part of the landis of Straherly; 1520 *Fife Sheriff Ct.* 195. The saidis William and Katerine oblissis thaim .. till keip, freith, warrand, relief the said Henry harmeles and skaithles .. of the said geir at all handis now and for euir; 1520 *Edinb. B. Rec.* I 202. Offerand that gif the saidis gudis war adjuegeit to be deliverit to him to fynd cautioun that the intromettouris tharwith salbe scaithles tharof at the handis of the said King Frederyk; 1526 *Acta Conc. Publ. Aff.* 254. For non releving and keping of M. Alexander Turing .. skaithles at the handis of Johne Bannavis .. of certane caussis and pleyis betuix thame concernyng the assedatioun of the provestry of Methuen and soumes of money ressavit tharof; 1537 *Reg. Privy S.* II 336/2. Skaythlis; 1564 *Grant Chart.* 130. On na wayis will .. keip the said Johnne .. skyathles of his part of the said sowme; 1577 *Reg. Privy C.* II 651. [To] keip skaythles the said George .. of all damnage, enteres or skayth that may be socht; 1580 *Wemyss of Bogie MSS.* To deffend and keepe the saidis tounes and lands .. free and skeathles of all wards, releiffes [etc.]; 1593 *Grant Chart.* 294. Actet .. to .. keip James Forman, merchant, harmeles and skaythles at the hands of Margaret Reid .. of four elnes of serge of Florence; 1600 *Edinb. B. Rec.* V 268. To warrand, releife, and skaytles keip the said William Kirkland of his becomeing catioun; 1621 *Glasgow Weavers* 67. Skeathlese; 1668 *Galloway P.* 20 March. The counsell .. obleiss them .. to warrand freith releive and skathles keep the said .. dean of gild .. of the said band or oblidgement; 1675 *Edinb. B. Rec.* X 230.

(3) That the said erle [etc.] .. sall content and ples thar awn tenandis .. and to kep the parteis skathlas befor the iustis and the law of satisfaccioun; 1491 *Lennox Mun.* 143. The .. counsall oblist thaim to warrand and keip skaithles the saidis Robert and Thomas and all otheris; 1547-8 *Stirling B. Rec.* I 51. Quhill the earle Bothwell come and was cautioun wpoun his faitht and honour that he sould keip him skaithles; PITSC. II 52/29. The saidis cautioneris obleist thame .. to freyth, releif and keip skaythles vtheris pro rato for the fourt pairt of the said penaltie; 1597 *Crim. Trials* II 20. Quha .. compeired nocht to releaue his borgh and keip him skeathles; SKENE *Reg. Maj.* I 15. Skaithlesse; *Ib.* (Table) 61.

(*b*) He is oblist to pay ilka ȝher to the prioris of Are & to kepe the toun schachles therfor; 1436 *Ayr B. Ct.* 26 Nov. Schaichles; 1452 *Ib.* 16 Oct. Schaithles; 1459-60 *Ib.* 14 March.

(Scathely,) Skaithly, *adj.* [SCATH(E *n.* Cf. ME *scaþli* adv. (*c*1350) with damage or injury.] Harmful. = SCATH(E)FULL *adj.* Also, in later Sc. (1805 in SND). — [The dog] turned skaithly in eating his meall barrell and mad, .. and thairfor he felled him; 1686 *Peebles B. Rec.* II 117.

Scathfull, var. of SCATH(E)FULL *adj.*

Scathing, *vbl. n.* Also: **skath-, scayth-, skaith-, skayth-, scad-; schath-; schad-** and **-inge, -yng, -yne.** [SCATH(E *v.*]

1. The action of harming in various senses of SCATH(E *v.*

(1) Thai mak thaim [*sc.* skinners] maisteris or thai can the craft in grete skaithing of .. the peple; *Acts* I 336/2. That na man sulde byrne ane vther mannys hous bot gif it may be done but .. skathing of his nychtbouris; 1456 *Ib.* II 46/1. That na man tak apon hand to hous sic lyk gudis in pregytis and skathyng of the burgh; 1471 *Peebles B. Rec.* I 166. In hindering or skaithing of the craftis foirsaidis; 1505 (*c*1580) *Edinb. B. Rec.* I 103. Schallans for the schadyng hym vith his sov a pec of bere; 1527 *Wigtown B. Ct.* fol. 200a. For schadyne of tham of nechbovreret to the avayll of vii ȝokenys; 1533 *Ib.* fol. 287b. Schallans fore the wrangus schadyne of hyme in his corn with thare guddis; 1534 *Ib.* fol. 295a. Thou thy selfe full airt had and parte in harming and skaithing of me; SKENE *Reg. Maj.* I 117b.

(2) Ane hewyn spowt of stayne .. to kep the said Mowngous drop of the said Patrikis tenement and skathyne of it in tymis cumyng; 1515 *Glasgow Chart.* II 489. The wronging, hurting or skaithing of his majestys tenement and propperty .. to be punished .. in the severest way; 1665 *Reg. Privy C.* 3 Ser. II 23.

(3) Accusit George Symsone that he haid a kow .. lows & for the principall scaything of vj p[e]kis of ry; 1513 *Prestwick B. Rec.* 45. That quhilk .. dois wrang to utheris in nichtbourheid or skaything ane another in thair gers or corne [etc.]; 1604 Murray *Early B. Organ.* II 249.

(*b*) Scadyng; 1521-2 *Wigtown B. Ct.* fol. 121b (see SCATH(E *n.* 6 *sing.*).

(4) Swa at ilkan of the said partyis may sekirly haf thar awyne but hendering, werring or scathing; 1456 *Wemyss Chart.* 76.

(*b*) Is profytable to led .. warly thi spekinge It may thé nocht to gret schathinge; *Ratis R.* 206.

2. Costs, expenses. = SCATH(E *n.* 6.

For coistis and skathingis and expensis sustenit be hym thair thro; 1518 *Chart. Coupar A.* II 119.

Scatht, var. of SCATH(E *n.* **Scath(t)les,** varr. of SCATHELES *adj.* **Scatland,** var. of SCAT(T)LAND *n.* **Scatles,** var. of SCATHELES *adj.* and SCATTLES *adj.* **Scatring,** var. of SCATTERING *vbl. n.*

Scatse, *n.* [17th c. Eng. *scatse* (1688), MDu. *schaetse* a crutch, stilt, skate, Du. *schaats* a skate, ONF *escache* (13th c. in Godefroy) a stilt.] An ice-skate. — Two pair scatses to go upon the yce; 1652 *Sc. Hist. Rev.* XXXI 9.

Scat(t, Scot(t, *n.* Also: **skat(t, skait, skaytt; skath; schatt; skot(t; scoit, skoit; schoit.** [North. ME and e.m.E. *scat-* in *scathaver* (Durham, 1344-5), *skattegyld* (15th c.), *skaitgyld* (York, 1483), ON *skattr* tribute, a tax, OE *sceat*, MLG, MDu. *schat* treasure. Also ME and e.m.E. *scot* (1297), *scotte* (*c*1430), *skott* (*c*1566), and esp. the phr. *loth and scoth* (1297), *lote & scott* (1320), *scotte and lotte* (1494), OE *sc(e)ot*, ON *skot*, OF *escot* (13th c. in Wartburg) a payment, contribution. Cf. S(C)HOT *n.*[1] 14.

The two words seem to have largely fallen together in Sc., senses 2 and 3 appar. having the closest phonological and semantic links with 'skot', although other instances, e.g. 1598 *Aberd. B. Rec.* II 172 in sense 1 seems to retain some of the notion of a share or contribution rather than a tax.]

1. A tax or levy, passing into **b.** A forced contribution; an exaction.

To liegis within the boundis of Orknay and Schetland .. that nane .. stop .. the rasing of al unlawis, eschetis, proffittis, skattis and dewiteis according to the said regalite; 1501-2 *Reg. Privy S.* I 116/1. Forsamekill as oure soverane lord has send ws his lieutenent .. in thir partis of Ergile [etc.] .. and all pairtis of the Ilis .. heirefor we .. chargis you .. that ye nor nane utheris .. intromett ony maner of way with ony teindis, landis or rentis pertenyng to ony person of halikirk .. bot that the fredome and privelege of halikirk be observit .. eftir the forme of the first foundation without ony scatt, stent, taxation or extortion to be maid in tyme cuming; 1506 *Exch. R.* XII 703. Wrangis to reddres suld wer be vndertane, For na conquest, reif, skat, nor pensioun; Doug. xi Prol. 24. Howbeit the abbot of Halirudhous had gottin ane pretendit infeftment to tak a penny of ilk full cart [passand] throw the Canongait to Edinburgh .. that it suld nocht turne the said toune of Edinburgh .. to na prejudice anent thar rychtis and that thar suld be na scatt nor exactioun put of new to thame; 1535 *Acta Conc. Publ. Aff.* 447. William Logan present ferriour of the ferry boitt of Torrye .. to mak .. guid and redye service to the inhabitantis of Futtie .. and theirfoir ilk howsholder in Futtie paying to the ferriour viii s. at twa termes in the yeir .. but payment of any wther dewtie to be payit be thame induring thair awin will .. and siclyk to pay the skait vsit and wont; 1598 *Aberd. B. Rec.* II 172.

b. For .. compelling of the tennentis of the landis of Strathardill to pay to thame ane skatt of foure hundreth merkis .. committand thairby manifest oppressioune; 1607 *Crim. Trials* II 528.

2. A municipal tax levied on the burgesses for defraying communal expenses; one of the duties required of a burgess. **b.** Freq. in the phrase *scat*, or more commonly *scot and lot*. See also Lot *n.*[1] 2 b.

This sense and sense 3 seem to derive from ON *skot* one's 'shot' or portion but to have fallen together to some extent with the more general sense of 'tax', 'tribute' (ON *skattr*).

[Liberi sint ab omni scotto et gildo et omnibus auxilijs regum; 1235 *Reg. Episc. Aberd.* II 269.] To .. hald stob and staik therein, and walk, ward, pay extentis and skattis within the burch, conforme to thair substance; 1552 *Conv. Burghs* I 4. And sindrie of thame .. ver depauperit and becum vnvorthie to keip skat and stent; 1580 *Kirkcudbr. B. Rec.* I 111. Cautionar for him for skat stent vache vaird and vther portabill chairgis of burgh; 1588 *Ib.* 234.

b. [Ogiluy, Dauid, of Tolmade, services dispensed with at] scott lott waik and warde; 1486-7 *Misc. New Spald. C.* I 32. 1490-1 *Ib.* 35. [The] scheit makers to tak skott and lott with the hammermen; 1521-2 *Stirling B. Rec.* in *Dunferm. Hammermen* 37. [John Leich] oblist him .. that on getting the scot and lot of the merchants [he will] answer for three pounds gret to Sanct James' altar; 1522 *Old Dundee* II 32. The said James hais fund Adam Ahannay pleg to keip scot & loit to the toun & to tak his skair of schippis & wthir necessaris; 1530 *Wigtown B. Ct.* fol. 233b. Skoit; *Ib.* Admytit burges sa that he pay scot and lot, varde and vaik within our fredom; 1530-1 *Selkirk B. Ct.* (ed.) 110. Swa lang as the said Johne kepis scott & lott vach vard siklik as wtheris fre burgessis in the burgh; 1534 *Dumfries B. Ct.* 146b. Scoit; 1535 *Wigtown B. Ct.* fol. 304a. The commond pastuyr beand manurit .. be certan personis quhilkis aucht .. be woyd and giffin to the sustentatioun of the puyr anes gudis and bestiall quhilkis payis scottis and lottis wyth the towneschip; 1570 *Inverness Rec.* I 191. Skot; 1576-7 *Lanark B. Rec.* 72. Plege and souerty for keiping of scot and lot and paying of his burges siluer; 1579 *Kirkcudbr. B. Rec.* I 94. To concur wyth the .. communitie forsaid in scot lott stent and taxt conforme to the tennour of his electioun; 1579-80 *Inverness Rec.* I 274. Such persones who lives nixt adjacent to the toune and who does not pey scot nor lott with the incorporatione and that does make their benefitt within the same; 1666 *Old Ross-shire* II 83. The remanent burgesses inhabitants within the said burgh and territorie therof beiring scott and lott within the samyn; 1671 Ure *Rutherglen* 58. That the haill inhabitants within the colledge bounds .. sall .. be subject to the jurisdictione of Old Abd. .. and shall contribute scott and lott proportionallie .. with the inhabitants of the said citie; 1677 *Rec. Old Aberd.* I 128. Their should be none choicen to represent borrows but merchands within that brugh, .. bearing scott and lot with the rest of the citizens; 1678 Lauder *Observes* 278.

(*b*) That all persones admittit .. thairto sall pay scat and lot taxt and stent to be gadderit and taxit be the deacon; 1483 *Seal of Cause Haddington Hammermen* in *E. Loth. Antiq. Soc.* II 104. [He] hes fund Alexander McClellane in Galtuay cautioun for keiping scat stent and loit conform to vse of burgh; 1582 *Kirkcudbr. B. Rec.* I 173. That personis admittit .. fremen of the saidis craftis sall pay skat and lot taxt and stent; 1647 *Seal of Cause Haddington Masons (etc.)* in *Douglas Bequest* 25 Sept.

(*c*) [He] hais fundin Michell McConnill souerte to keyp schoit & loith; 1515 *Wigtown B. Ct.* fol. 43a.

3. *specif.* A charge made on merchant ships to cover any shortfall in their cargo; esp. a charge made on merchants to ensure that losses due to storm, piracy, etc. are shared equally among those affected.

Grantit .. ane barrell yrne & that he wes awand the fracht & skat thare of to Willem Kinloche; 1551 *Dundee B. Ct.* II fol. 103b (29 Oct.). That the eight tuns desyrit be the queen's grace be deliverit furth of all the ships now within Tay .. and that forty shillings mair be payit for the tun nor the common price .. and that all clerks of the said ships .. gadder the price of ane tun and ane half with the fraucht and the skaths to be taken of the haill wines according to their quality; 1563 *Old Dundee* I 194. That William Rysie, sempleir, be answerit of four tuns of wine of thir present four ships .. and that the clerks of the ships gadder the scatt thereof with the fraucht; 1564 *Ib.* The fraucht vncostis lekkage and skat of certane wynis tane furth of the same schip; 1568 *Edinb. Test.* I 173b. Incais ony schippis be pilleitt the gudes saiff sall contribute scatt and loitt for the relief of the personis dampnefeit, bayth schip and gudes according to thair wairing and gif neid beis ordainis the magistratis of the poirtis within this realme quhair the saidis schippis sall happin te aryve to sett ane scatt and extentt for releif of the personis dampnefeit .. and to arreist poynd or comprise the schippis and gudes of thame that refusis for obeying of the said scatt, euery man according to his pairt as he sall be sett and the samyn ordour to be kepit anent the gudes casten for saiftie of lyfe and gudes vpoun commoun consent; 1575 *Conv. Burghs* I 44, 45. [Complaint that] the commissioners of burrowis .. without ony pouer .. hes .. maid ane act .. that all guidis .. pilleit in tymes bygane [or] .. in tymes cuming be pyrattis fra the subjectis of this realme .. salbe recompansit be ane universall scatt and taxt to be tane alsweill of the schip and fraucht as of the remanent guidis left unpilleit; 1580 *Reg. Privy C.* III 308. Jhonn Robertsoun, bailyie, .. to sett the scatt of the schip callit the Sie Catt .. quhilk wes laitly pilleit in hir vayage frome Londoun and to tak the aythis of the merchants vpoun the quantities of thair .. guids schippet; 1586-7 *Edinb. B. Rec.* IV 486. Giff it happin the schip .. to perisch and the lichter and gear put thairin onely to be saif na contributioun sall be set upon the gear in the boat .. because it is a constant rule in thir caises of casting and lichtning that skatt suld onely be maid quhen schip and guds .. come saife to the port; Welwod 71. He quha man be recompansit be a skatt, man do against the maister to reteine all the guds on his schippe boddome; *Ib.* 75. lxiiij stane wecht of gad irne .. lij lib. xvj s. quhairof to be deducit for ane scatt and fraucht thairof iiij li. vj s.; 1591 *Edinb. Test.* XXIII 50b. The soum .. quhilk was takin .. for payment of ane scatt of the said George his monye .. for contributing with certaine goods castin furth of ane schip of Bruntiland; 1612 *Conv. Burghs* II 339. The said monye .. to be refoundit be .. the persons to quhome they delyverit the samyne in name of scatt .. becaus it was fund be the commissionars .. in thair generall conventione haldin at Craill in the month of Iulie last that no monye sould scatt or contribute with onye goods castin or pilleyit; *Ib.* 340. Gif ane schip .. can nocht eschaip bot gif the guidis be cassin out .. for recovering of the domages the maryneris aw to have ane tun frie, and the uthir aucht to be pairted with the skatt eftir as he [*sc.* the master] hes of quantitie; BISSET II 243/18.

4. *specif.* In Orkney, Shetland and Caithness: A land tax levied on udal property (see the note to OUTHALL *n.*), the unit of valuation being the URISLAND (see PENNY-LAND *n.* 1).

The right to collect this tax, under the Norse occupation enjoyed by the Norwegian crown or the jarls as its representatives, passed to the Scottish crown after 1472.

Orig. paid in butter and malt, supplemented where necessary by a further payment or payments in kind (as barley, meat, oil, etc.) or money (see *scat silver* at 5 (8) below, also FORCOP *n.*). In *pl.*, sometimes referring collectively to these various modes of paying scat.

For further information, and some discussion of the origin of this tax, see H. Marwick *Orkney Farm-Names* (Kirkwall, 1952) pp. 191 ff.

For numerous further instances see *Orkney Rentals* I-IV *passim.*

sing. (1) The Holmes j d. terre quoyland but scat; 1502 *Orkney Rentals* I 13. ix d. terre but scat quia boirland an' therof in land male iij lastis iij s.; *Ib.* 18. Becaus I found in the auld parchment rentall that notwithstanding the land is all the kingis to the male yit ilk iij d. terre is laid by to pay scat to the bischop; *Ib.* 89. Johne of Cragie .. payit scat to the king; *Ib.* 91. That .. this said bordland makis the haill d. terre & payis na scat; 1508 *Ib.* 80. The scatt, teind, and wtheres dewties dew to be payed out of the said lands; *c*1536-7 *Rec. Earld. Orkney* 222. 1545 *Reg. Great S.* 727/1. The skat .. in use and wont to be paid to the king and his chamberlains from the said lands of Feauldquoy; 1560 *Orkney & Shetl. Rec.* 134. *Ib.* 130. Dimidio barrelle butyri pro lie scat de Towat; 1565-6 *Reg. Privy S.* V II 24/1. For the uthel skat and duty thereof usit and wont; 1567 Peterkin *Orkney & Zetl.* App. 8. Without ony .. dewteis to be payit to us, our airis or chalmurlanes quhatsumevir, except scat and teind allanerlie; 1584 *Rec. Earld. Orkney* 305. The haill skatt of the uthail land within the said yle baith butter and wadmell; 1592 (1610) *Reg. Great S.* 118/1. Ane quoy land or outbrek is ane peece of land newly win without the dykis, and payis no scatt; 1595 *Orkney Rentals* II 2. Crowall, na scatt, quia boirland [*ed.* bowland]; 1614 *Ib.* 127. Ewerie vrsland conteining auchtein pennie land possest in smallis by mein men, and labourit be tham, payand for ewerie pennie land, scatt and teind to the king; 1627 *Ib.* III 85. It is creavit be the vddallers of Orknay and Zetland quha hes thir many ages conforme to the dens law .. that they remaine his majesties .. vassels for payment of scatt; 1633 *Acts* V 53/2. Tenstane eighteine d. land, inde the kirk tack the scatt of iiij d. ob. terrae; *c*1642 *Orkney Rentals* V 18. [The country is] called Schetland, because of a kind of custom or tribute called scat, which they payed to their Norvegian masters; BRAND *Orkney & Shetl.* 65. Scat is a kind of rent or due, which is yearly payed to the king or his taxmen; *Ib.* 122.

(2) Ane lysspone and ane halff lysspone buytter in skaytt, as the said erllis rentell and charttur .. beris; 1509 *Rec. Earld. Orkney* 83. Ane just can quhairwith thai mesourit thair vlie payit in scat to the king; 1576 *Orkney Oppress.* 18. [½] setting beir in skat [3 s. 4 d.]; 1612 *Orkney Test. and Inv.* 43. In lie scat 6 merkis butiri; 1631 *Retours* I Inq. Spec. Orkney et Shetland (16).

pl. And the king and erle ever had the scattis of all the bischoppis land in this parrochin quhill of lait that bischop William stoppit the samen; 1497 *Rental* in Hossack *Kirkwall* 63. All the land in Paplay .. conqueist be umquhill Erle William except .. j½ d. terre of the quoyland quhilk is onlie the kingis with the haill scattis of the laif; 1502 *Orkney Rentals* I 11. In ley scattis xlij s. vj d.; *Ib.* 18. Of all thir scattis bigane I Henrie Lord Sinclair hes gottin na payment theirfor. And quhy it is unpayit I knaw not bot becaus the husbandis alledgit that thai aucht nocht to pay the saids scattis; 1503 *Ib.* 42. Schir Robert Sutherland hes haldin the kingis scattis .. on knawin to my Lord Sinclair; *Ib.* 59. The bischop withaldis the hale scattis of ij d. terre; *Ib.* 61. William Tulloh, Bishop of Orknay, had and bruikit .. be gift of umquhile our maist nobil grandsir, the King of Denmark and Norway .. all the skattis, renband duties that pertained to our said grandsir of the said kirklands; 1503 Craven *Ch. in Orkney* III 122. xxv d. terre quhilk payit evir all the scattis to the king for all wes uthale mennis land; 1508 *Orkney Rentals* I 85. Ane lettir in Northin, of the King of Norwayis is [*sic*] of the skattis of Orknay umquhile givin to Bischop Williame Tulloch for his liftyme; *c*1524 *Rec. Earld. Orkney* 423. Richtis, service and skattis usit and wont; 1575-6 *Reg. Privy S.* VII 74/2. Withe the skattis of ane pennyland ane mark mair wthall land in the said towne; 1592 (1610) *Reg. Great S.* 90/1. For payment of ther scattis, callit in our language the fewduties; 1592 *Orkney Oppress.* 101. The said Helein payand ʒeirlie the few melis scattis & vtheris dewteis of the saidis haill landis; 1593 *Prot. Bk. T. Auchinlek* 26b. And that he payit scatis and dewiteis to my lord; 1603 *Shetland Sheriff Ct.* (ed.) 70.

b. Freq. in collocation with *wattell* and *land male* (LAND-MAL(E *n.* 1, q.v. for numerous further examples), payments being in money only.

(1) Ane entire and free tenandry .. with wrak, waith, half-wrak, wattill, cheitrie and scatts; 1572 *Orkney & Shetl. Rec.* 188. Pro lie scatt et wattill juxta ratam; 1595 *Exch. R.* XXIII 441. Bothwell Nicolsoun is decernit to pay his scat and wattill to my lord with the rest of his dewiteis of underfold; 1603 *Shetland Sheriff Ct.* (ed.) 74. As his awin proper land .. haldin frielie of the king for payment of scat and wattell; 1610 *Rec. Earld. Orkney* 185. The cornteynd thairof is payit in butter and oylie diversly according to the divers portions of the same with the skatt and wattell to the king; *Rep. Parishes* 232.

(2) The thrid part blawin [land] quhilk payis nother scat nor land male; 1508 *Orkney Rentals* I 80. Thair clayth, callit wadmell, .. is ane dewitie thai pay to the kingis maiestie for thair scat and landmales ʒeirlie; 1576 *Orkney Oppress.* 18. Thair haill buttir, bayth of scat and landmales; *Ib.* For the thrie pennyland of kinges land called Cursetter in scat, land meallis and teind ane barrell butter; 1587 *Orkney & Shetl. Rec.* 212. Pwle .. payis in but. scat i lesp. xii mks. in scat silver xxi d. in scat land male and teynd; 1595 *Orkney Rentals* II 7. Scattis land maillis [etc.] .. and vtheris rychtis and dewties quhatsomeuir pertening to .. Orknay and .. Zetland; 1612 *Acts* IV 481/2. The whole scatt and land mealis is payit yeirlie to the Bishop of Orknay as ane pairt of the patrimonie thairof; 1627 *Orkney Rentals* III 41.

(3) Cum .. lie wattill, foircoip, landsettirtun, ansterscope, scattis, landmaillis, wrak, wair et waif et aliis juribus; 1589 *Reg. Great S.* 565/2.

c. Defined by a preceding noun in *attrib.* use, see the note on modes of payment above.

And the laif of the butter scatt of unblawn land to be payit in pennyworthis sic as growis upon the ground; 1492 H. Marwick *Orkney Farm-names* (1952) 197. Grymesetter iij farding terre uthall land an' in butter scat v d. j farding .. et in malt scat j m. xviij merk; 1502 *Orkney Rentals* I 3. Becaus it payis nather malt scat nor butter scat we ken nocht quhy; *Ib.* 6. The half of the malt scat wes gevin quyt be umquhile Erle William; *Ib.* 7. Et in cost scat an' vj m. jam tantum & na forcop; *Ib.* 47. Et in corne scat of the uris terre an' xxj m. beer; *Ib.* 87. Beer scat; *Ib.* 88. 1503 *Ib.* 26 (see d below). *Ib.* 36 (see d below). *Ib.* 40 (see d below). 1595 *Ib.* II 7 (see b (2) above). Houbuster sex penny land vdall payis in butter scatt ij leis. in scat silver foircoipe viij s. in scat malt iiij m.; *c*1642 *Ib.* V 5.

d. *Richt scat*, scat levied at what was seen as the normal rate as opposed to *wrang scat*; also *doubill scat*, when the tax was increased because of an improvement in the value of land.

Richt scat; 1492 *Rental* in H. Marwick *Orkney Farm-Names* (1952) 197. & na forcop quia double malt scat; 1502 *Orkney Rentals* I 8. Kirkbustar ix d. terre an' in butter scat iij span viz. richt scat; 1503 *Ib.* 26. In butter scat xv span quia the tane of thir uris terre payis wrang scat viz. x d. j½ d. on ilk pennie terre; *Ib.* 36. Inde stent vj leisp. quhair it payis doubill buttir scat; *Ib.* 40.

5. *attrib.* and *comb.* In sense 4 above: Paid by way of scat.

Scat butter, see also LOWS *adj.* 7 b for further instances.

Scat silver, money payments of scat, as FORCOP *n.*

attrib. (1) Pro Worsatter in lie scatt-bear; 1632 *Reg. Great S.* 644/1.

(2) For the skatt buttir of the wthall land of Clestrane; 1592 (1610) *Reg. Great S.* 90/2. Ane merk scat butter; 1597 *Edinb. Test.* XXX 54. Summa exonerationis of scatt and louse butter is 34 leispd.; 1613 *Orkney Rentals* App. 97.

(3) 1 meill 3 settingis scatt-flesche; 1632 *Reg. Great S.* 644/1.

(4) 27 bollas skat malt; 1539 (1542-3) *Reg. Great S.* 675/2. 48 sol. lie skait-silver et 13 bollis 2 firlotis lie skait-malt; 1583 (1610) *Ib.* 150/2. Twa settinges malt for the scat malt of ane half pennie land .. in Aichstoun; 1613 *Orkney Rentals* App. 95. *c*1642 *Ib.* V 5 (see 4 c above). 1660-75 *Ch. in Orkney* 187. A firlot scat malt; 1697 *Old-lore Misc.* IX 46.

(5) All that the bischop takis is .. ilk yeir ane scat mairt for the pryce; 1502 *Orkney Rentals* I 6. vij scat mairtis; 1503 *Ib.* 63.

(6) Scat oiel; 1595 *Orkney Rentals* II 42. *Ib.* 103 (see (8) below). The peopill wer not abil to beir it, payand scat silver, scat wylle and garsum, as said is; 1627 *Ib.* III 55. 1662 *Retours* I Inq. Spec. Orkney et Shetland (96).

(7) Et in scat order [*erron. for* ordei (L. *(h)ordeum* = barley)] of the vj d. terre; 1502 *Orkney Rentals* I 89. In butter ix span. Et in scat ordei an' xxxj m. ½; *Ib.* 91.

(8) Skat silvir extenden. annuatim ad 4 libras 16 sol.; 1539 (1542-3) *Reg. Great S.* 675/2. 1577 *Reg. Privy S.* VII 144/2. 1583 (1610) *Reg. Great S.* 150/2 (see (4) above). Lie scattis-silver et land-maillis; 1587 *Ib.* 468/1. Reservand to my lord the flesche debt, scat silver, scat olie [etc.]; 1595 *Orkney Rentals* II 103. Thrie pund fyve penneis of scat siluer; 1597 *Edinb. Test.* XXX 53b. Togither with 9 s. monie for the scatt siluer of the 18 d. land of the Skanabie; 1613 *Orkney Rentals* App. 93. Ness ob. terre, pro rege, quoyland, butt. scatt, payis in land-maill .. 6 d. scatt silver, outcast teind and ane hak dog yeirlie; 1614 *Ib.* II 131. Ane part [of the rowms] belongs to his majestie .. and payis .. in victuall, butter, flesche, schatt silver .. according to the rentell; 1627 *Ib.* III 94. *c*1642 *Ib.* V 5 (see 4 c above). E. 19 d. scat silver &c.; 1662 *Retours* I Inq. Spec. Orkney et Shetland (92).

(9) To the king, in vittaill called skat vittaill; *c*1627 *Orkney Rentals* III 71. 2 firlots scat victtual; 1697 *Old-lore Misc.* VIII 207.

comb. And now sett to John Burvick for the skatt dewtye; 1613 *Orkney Rentals* App. 94. The skat dutie is alwayes reckoned in the stocke secluding the teynde; *c*1627 *Ib.* III 74. The scatt dewtie contenit in the rentall ȝeirlie at the termes of payment wsitt; 1639 *Sc. Hist. Rev.* XXI 131. He .. payis a skatdutie conteyned in the rentall; *c*1644 *Orkney Rentals* III 17.

b. Pertaining to 'scatland' SCAT(T)LAND *n.*

The skatt dyk of Toop that passis fra the overmylne dams of Sabay; 1519 *Rec. Earld. Orkney* 93. 1583-4 *Ib.* 159.

6. As a place-name element.

Scatnes in the parichin of Sanct Gregorii; 1545-6 *Rec. Earld. Orkney* 229. Scatstay; 1603 *Shetland Sheriff Ct.* (ed.) 71.

Scat(t, Scot(t, *v.* Also: **skat(t, scait, skait.** [e.m.E. *scatte* (Caxton), ON *skatta*, MDu. *schatten*; SCAT(T *n.*]

1. *tr.* To exact (a bribe, or a payment due).

And swa ane bud at boith the parteis skat; HENR. *Fab.* 1278. Sayand that he suld iniuriuslie haue skattit tollit and wranguslie tane certane hunderis of heryng fra fywe batis; 1540-1 *Linlithgow B. Ct.* 17 Jan.

b. To oppress by exactions.

The toune is hauely murmurit be the landmen that the vittell byaris of the merkat scattis thame grytlie in taking of sampillis [etc.]; 1543 *Aberd. B. Rec.* I 191. [To satisfy some of his friends who were] skattit [the last year] at the haldin of Murrayis airis and for susteyning of the souldioris; 1570 *Cal. Sc. P.* III 219.

2. Only in Orkney: To assess (land) for scat or land-tax. Cf. SCAT(T *n.* 4.

& na forcop quia double malt scattit; 1502 *Orkney Rentals* I 9. Cara, of old not contained in the rental, and lately scatit .. payis [etc.]; 1595 *Ib.* II 94.

3. *intr.* **a.** To contribute to the municipal charge payable by the freemen of a burgh, esp. in the phrase *to scot and lot*, etc. (SCAT(T *n.* 2, see also LOT *v.* 1). Also (once) *tr.*, to pay or contribute to (a craft guild). **b.** To contribute to a charge made on shipping to defray losses (SCAT(T *n.* 3).

a. (1) Compeirit Mr. Johnne Scharp [etc.] .. and alledgeit that [etc.] .. becaus thay ar burgessis of Edinburghe and stent, watcht and waird with the burgessis thairof .. (May 31) Interlocutor anent Burgessis [etc.] .. that thay wer burgessis of Edinburghe and that thay stentit and skattit with the toune thairof; 1587 *Crim. Trials* I II 156.

(2) Made burges .. and wes svorne to scot and lot witht the nychtbouris; 1503 *Dunferm. B. Rec.* I 136. All frie and unfre to loit and scot and pay thair part tharto [*sc.* the cost of the procession] as thai are extentit to be the deknys of thair craftis; 1512-13 *Aberd. B. Rec.* I 442. That thair sal saile bot ix personis now in thir thre schips .. quhilkis sal leif nyne men for thame weil .. bodin as efferis, to lot, scot, waik and ward for thame; 1513 *Ib.* 87. Mony sundry persouns .. quhilk .. nother scotts lotts extents walkis nor wairds .. siclyke as thai awcht to do; 1517 (*c*1580) *Edinb. B. Rec.* I 172. The dekin and craftismen of the hammirmen to prefe gife scheit makaris scottis and lottis with thar craft in uder borouis; 1521-2 *Stirling B. Rec.* I 14. Because the saidis vnfremen nowthir scottis, lottis, walkis nor wardis within our said fredome; 1531 *Edinb. B. Rec.* II 53. Qualifyit .. to scot, lot, walk and ward with the laif of the nychbouris; 1545 (1546) *Reg. Great S.* 757/2. He hes nocht .. scottit nor lottit .. at nay tyme; 1559-60 *Inverness Rec.* I 42. Gif ony wedow buy and sell within burgh with the nichtbouris, sche sall scot and lot with thame in taxatiounis and utheris helpis; and siclike gif sche dois scot and lot, sche sould have libertie to buy and sell; BALFOUR *Pract.* 49. Crestene to cum to the .. myll & to ressawe hir part .. thairof .. & scot & lot & mak equall half of the costis skaythis & expensis that it aucht to be furnesit to the said myll as scho aucht [etc.]; 1589-90 *Prot. Bk. J. Inglis* 10 Feb. Sic as hes nocht hors sall scot and lott with the toun for thair p[ar]t [*sc.* of the small customs]; 1616 *Rec. Old Aberd.* I 56. To repair heir to the samen, and heir keip ther forum et larem and scott, lott, wache, wark & warde with the place; 1664 *Inverness Rec.* II 219. They ar by ther burgess oath suorne to scot lot watch wake and ward with the inhabitants of the brughe; 1677 *Aberd. Council Lett.* VI 133.

(*b*) Archbald Oliphant .. hes .. oblist hym to scat and lot and pay with the merchandis of Pertht in all taxationis quhen ony occurris; 1550 *Perth Guildry* p. 268 (30 May). Scattis; 1560 *Edinb. B. Rec.* III 87. The merchandis are not haldin .. to scat and lot thairfoir; BALFOUR *Pract.* 623. That thai .. sall scatt and lott, walk and waird, and pay stentis as burgessis suld do; 1581 *Glasgow B. Rec.* I 84. The comburgessis and merchands of the said burge quha skattis and lattis; 1613 *Inverness Rec.* II 108. The counsell declaires James Stevensone .. is onlie ane calsay burges untill he find cawtioun to scatt, loit, watche and waird; 1653 *Peebles B. Rec.* II 13. He sall watch, waird, scatt and lott with the counsall; 1658 *Kirkintilloch B. Ct.* 2.

(*c*) The said Dauid .. sall skait and lot with the gudtowne; 1540 *Linlithgow B. Ct.* 4 Oct. Nevir in tyme bygane scho .. skaittit and lotit with the said craft as use is; 1556 *Haddington B. Rec.* (Robb) 22 Sept. That all .. that .. exersis the libertie and privielege of merchandis .. stent scait lott and beir chargis .. with the saidis merchandis; 1558 *Edinb. B. Rec.* III 19.

tr. Thom Watsoun allegit that he aucht nocht apon rasoun to scot nor lot the craftismen of the hammirmen becaus he occupies na warklomis that pertenis to thair said craft; 1521-2 *Stirling B. Rec.* I 14.

b. (1) That the schip be nocht haldin to scatt or contribute, but onely hir fraucht; 1581 *Conv. Burghs* I 117. The schip and the rest of gudds sal skatt to the upset thairof; WELWOD 71. 1612 *Conv. Burghs* II 340 (see SCAT(T *n.* 3).

(2) Na gudis vnder the ourlop to scot nor lot with tha gudis in case thai be castin; 1466-7 *Acts* II 87/1. Ale the haile salt being within the Inglis schip sal scott [*pr.* stott] lott & upricht the ix bollis;

1492-3 Kennedy *Aberd. Ann.* II 475. [That the whole goods on board Nicolas Rynart's ship .. should scott and lot with the ship or freight thereof] in recompensation of the skaith [of the rigging of the ship, which was cutt in the storm]; 1560 *Ib.* 486.

(*b*) Nottheles diverse merchandis havand guddis in the said schip by the commoun ordour in sic caissis refussit to skat and lott wit the gudis lost; 1570 *Edinb. B. Rec.* in Smit *Bronnen* II 1012. That no cloithis .. within sey kistis cassin or pilleit sall scatt and lot with the vther guidis .. cassin or pilleit; 1579 *Conv. Burghs* I 100. Gudis imput in the schip at any vther port, vther nor at hir ladinning port, salbe haldin to scat and lot; *Ib.* Andrew Craig merchant .. oblist to procure and schaw befor thame ane sufficient certificate .. testefeying the guids quhilk he schippett .. at Burdeulx wes pilleit furth thairof and quhilk hes scattit and littet with the rest with sufficient customes in the said ports; 1586 *Edinb. B. Rec.* in *Bk. Old Edinb. C.* XV 18. The merchandice under the overloft sall nocht skatt lott nor contribute with the said cassin guidis; BISSET II 205/23.

Scat(t)ald, Skattall, -ell, *n.* Also: **scattill, scathald.** [Only Shetland; SCAT(T *n.* 4. The second syllable, OED *et alia* conjecture, may be f. *hald*, here = HALDING *vbl. n.* 2, but this 'presents difficulties, both as regards form and semantic development' (SND). Also in the mod. dial. of Shetland, only in reference to common grazing.] **a.** A group of inhabited townships and the lands pertaining to them. **b.** The lands, in some instances *specif.* the pasture-land or outfield, of such a township. **c.** The members of such a community, collectively.

In *Orkney Oppress.*, twice equated with *scatland*, see SCAT(T)LAND *n.* c, and in sense c appar. equated with PAROCHIN *n.* 3 (= the body of parishioners).

See Brian Smith 'What is a Scattald?' in Barbara E. Crawford (ed.) *Essays in Shetland History* (1984), W. P. L. Thomson 'Ouncelands and Pennylands in Orkney and Shetland' in L. J. Macgregor and Barbara E. Crawford (eds.) *Ouncelands and Pennylands* (Univ. of St. Andrews St. John's House Papers No. 3) and A. Fenton *The Northern Isles* (1978), Index s.v. *Scattald.*

a., b. For ellevin scathaldis that is within that yle [*sc.* Yell]; 1576 *Orkney Oppress.* 47. For in uther parochinis and yles he tuik bot for everie scathald thre dolors, bot in this parochin [*sc.* Dunrossness] he gart everie man particularlie pay thre dolors; *Ib.* 48. Amangis nichtbouris that dwellis within ane scathald; *Ib.* 49. In name and behalff of the haill scattilmen of Gulberweik and Quarff that thay salbe all hermeles and skaythles of the said Magnus and all utheris within the said skattell in thair girse and cornis; 1602 *Shetland Sheriff Ct.* (ed.) 24. Magnus Daile, officiar, is fund to halfe riddin ane mere .. without leive over foure scatellis of land; *Ib.* 34. The stowth of ane lamb .. tane away afe the scattell of Colvadaile; 1603 *Ib.* 70. Magnus Clerkis guid sone hes riddin ane horse .. ovir the twa skattallis of Scatnes and Lie; *Ib.* 88. The ryding of .. [a] horse throche the skattell; 1604 *Ib.* 137. It sall not be lesum to ony persone .. to go throuch his nychtbouris scattell or comontie with ane sheip dog except [etc.]; 1615 *Orkney & Shetl. Ct. Bks.* 67. Thay offirit thame to prove that the scattellis and comounties of Hildiswick and Urafirth ar severall scattellis and devydit be meithis and merche stanes; *Ib.* 96. Quhatsumever persone .. sal happin to slay ane earne and delyver the samen to the bailyie of the parochin within the quhilk scho sal be slaine, the said bailyie sall caus the persones indwelleris within the scattell quharin sche beis slaine to give the slayer thairof ane scheip or pryce thairof; 1616 *Shetland Sheriff Ct.* (ed.) 165. The dewties receavit at Laxfirth .. for the crope 1652 payable 1653 .. scattill of Stenswale, receavit from the tenants per money £11 9s. Rests Chene's aires 5s., which makis upe the scattill; 1652 *Shetland Archives* MS GD. 150/2015/B/1. The description of the marches of the several [*i.e.* 11] scattalds in the island of Yell; 1667 *Old-lore Misc.* IX 140. Houlland Scattald. Being the first northmost scattald, is bounded to the west and north with the sea, to the east and south with Brugh Scattald and the sea. The first march mark dividing these scattalds stands upon the .. hillock be-east the little piece of dyke; *Ib.* 141, etc. That every scatald have a sufficient pund, under the pain of ten pounds Scots; *a*1733 *Shetl. Country Acts* in *Soc. Ant.* XXVI 201. That none contemptuously pasture upon, .. cut floss, or cast peats in their neighbour's scatald .. nor .. cut floss before Lammas-day in their own scatald without due advertising of their neighbours belonging to the said scatald; *Ib.*

c. Quhair the use was that the haill parochin, callit ane scathald, payit for the unlaw of swyne worting .. xl babeis; 1576 *Orkney Oppress.* 46. The lairdis awin man .. decernit ilk scathald within the said yle [*sc.* Unst] to pay to the laird thre dolors for the swyne ruting; *Ib.* 47. Quhairas be the auld custome of our said cuntrie that for breiking of nichtbourheid, the haill parochin, callit the scattald, payit fyve gudlingis; *Ib.* 49.

(Scattaldman,) Scattilman, *n.* [SCAT(T)ALD *n.*] A member of a community as in SCAT(T)ALD *n.* — 1602 *Shetland Sheriff Ct.* (ed.) 24 (see SCAT(T)ALD *n.* a).

Scattell, var. of SCAT(T)ALD *n.*

Scatter, -ir, *v.* Also: **skat(t)-, scat-** and **-yr; skaitter.** P.t. and p.p. also **skattirte, skatyrt;** CHATTER. [ME and e.m.E. *scater(en* (1154), *skater(en* (*a*1300), *schatre(n* (*c*1330), *scatir(en* (Wyclif), *sketer* (16th c.), of obscure origin.]

1. *tr.* To separate and drive off in different directions; to disperse (persons or things) far and wide. Also with some *fig.* implication.

Also with locative compl.

Gif that the wynd thame [*sc.* leaves] skattir [*Sm.* scattir] and blaw away; DOUG. VI i 147. And furth our al the place .. The fyre blesys, thame semyt, skattirris [*Sm.* skaitteris] sche; *Ib.* VII i 116. Than the impriour .. skatterit the Turkis sua at thai come neuir to giddir; ABELL 118b. Thir knichtis .. the mailȝeis scatterit in the feild; *Clar.* II 42. He [*sc.* God] scatteris all men of proude intent; *G. Ball.* 143. Thair sall ane hunder ryse .. and sall skatter the proude pak of ȝow hiepocreitis; PITSC. II 136/10. The quene .. lyk a certane sone, new risen to skail and skattir the cloudis of al tumulte; DALR. II 453/7.

b. *passive.* To be dispersed or have suffered dispersal by some agency, a superior power, the elements, or some other circumstance stated or implied. Also *fig.*

(*a*) In that desert dispers in sonder skatterit Wer bewis bair; DOUG. *Pal. Hon.* 346. The wyndis weltris the sey continualy .. Skatterit widequhar our the fame ful oft War our schippis; Id. *Æn.* III iii 92. I sal smyte the schephird and the schepe of the flok salbe scattirit; NISBET *Matth.* xxvi 31. Scaterit; Id. *Acts* xi 19. All the gret hors .. war sa effrayit with the noyis .. that na renyeis micht hald thame bot ran skatterit throw the cuntre; BELL. *Boece* II 454. Sic hestie tempestis .. that hir schippis was scatterit; PITSC. I 156/10. That the Romanis now skattired had na tyme to gather agane; DALR. I 165/21. Quhill now the men of weir in cruell battel ar all skattirte throuch and throuch; *Ib.* 298/15. Defend thai flok .. Dispersit to the warldis end & bring thame hame Into thay fald that now is wyd scatteritt abrod; *Montg. Suppl.* xxxi 25. Ane fire in the midde flure .. being skattered, the miln is brunt; SKENE *Reg. Maj.* II 41b.

(*b*) So the ost was dispersit and skaitterit abrode and the night was growand mark wpoun thame; PITSC. I 405/33.

fig. The hail warld than scatterit be a cruel storme of a suddane hæresie; WINȜET II 23/30.

c. *passive.* To be in a dispersed condition; to be spread out; to be separated from others of the same kind; to be broken up into small groups or quantities.

Mony ilis .. Skatterit in the sey yclepit Cyclades; DOUG. III ii 116. On far the cite wallis se thai With towris and the hows hedis on raw Skatterit dispers [L. *rara*] and bot a few to knaw; *Ib.* VIII iii 6. Quhar as the army of the Troiane syde Was thynnast scattyrrit; *Ib.* IX viii 118. The said Equis .. culd led batall bettir quhen thai war skatterit in mony handis than quhen thai war all iunit togidder vnder ane body; Bell. *Livy* II 290 (BM). The haill wynis being within the schip sall remane undisponit or scatterit for the said space of four dayis; 1565-6 *Reg. Privy C.* I 427. Our countrie clerkis .. Wald draw thame hame from all countries .. Quhair thay ar scatterit in all airtis; 1573 Davidson *Sat. P.* xlii 342. The vther tuo horss regimentis .. wes scatterit heir and thair; Spalding II 449.

d. *intr.* Of a group: To disperse, to break up.

It is ordanit .. that .. scho and all hir barnis .. [be] bannist this toune .. and at thai remane still on the muire quhill thai be clengit and nocht to scatter in the cuntre; 1530 *Edinb. B. Rec.* II 30. The nolt begowth till skatter, The ky ran startling to the corne; *Sym & Bruder* 132.

e. Of stuff: ? To come apart, so that holes appear between the threads.

Taffata sarsenet, such as will not run & scatter in the threads for thinnes; 1642 *Caldwell P.* 94.

2. *tr.* **a.** To spread by scattering, to strew.

The purpour flowris I sal skattir and pull That I may strow with sik rewardis at lest My nevoys sawle; Doug. vi xv 88.

b. *absol.* To lose or dissipate resources, or the like, by allowing their dispersal.

He that gaderis nocht togiddir with me scatteris on breed; Nisbet *Matth.* xii 30.

c. To shed (blood), to kill.

Them that settis ther felicite to skattir & to skail blude; *Compl.* 27/16.

d. To break up (a language) into mutually incomprehensible dialects.

For the quhilk God scatterit the tung of Hebrewe quhar with all folkis spak till that day in [72] .. langagis sa that nane wnderstud vther; *Asl. MS* I 311/15.

3. To distribute, to place here and there at (irregular) intervals.

Sum othir per ordour caldronys gan vpset, And, skatyrt endlang the greyn, the colis het Vndir the spetis swakkis; Doug. v ii 116. We .. skatered it [*sc.* animal flesh] wpon and down abow and about the gait .. quhair the lairdis .. wold most haunt; 1662 *Crim. Trials* III 618.

4. To spread abroad (a rumour).

Some persones .. have scatered false reports traduceing there magistrats; 1675 *Edinb. B. Rec.* X 231.

Scattering, *vbl. n.* Also: **skattir-, scater-, scatr-.** [ME and e.m.E. *scatiryng* (Rolle), *-eringe* (Wyclif); Scatter *v.*] **a.** Those who are scattered; the condition of being scattered or dispersed. **b.** The action of scattering, in sense 4 of Scatter *v.* — **a.** He wil ga into the scattering of hethin men; Nisbet *John* vii 35. James .. to the xii kinredis that ar in scatring on breed, hele; Id. *Epist. James* i 1. Scatering; Id. *1 Peter* i 1. Because of the far scattering of the paroch fra the kirk; 1614 Cramond *Ch. Aberdour* 12. — **b.** Nocht lang efter, throuch skattiring of [not] vane clattiris, Jhone Borthuik knycht .. is callit to Edinburgh .. to gyue an counte of his hæresie; Dalr. II 247/16.

(Scattfall,) Scatfall, *n.* Also: **skat-** and **-fa(a.** [ON *skattr* Scat(t *n.* For the second element, cf. ON *(leiðangrs)fall* n. 'a fine for default as to the levy or tax' (Marwick).] Only in *to fall in scatfall*, to be subject to confiscation in consequence of failure to pay scat. — Land .. fallin to me in skatfaa for none payment off the skait thairoff; 1590 *Rec. Earld. Orkney* 317. 1595 *Orkney Rentals* ii 17. 1 d. land pertaining to the Fressiris falling in scat fall; *Ib.* 18. $\frac{1}{2}$ mk. terræ fallen in scat fa; *Ib.* 41. A penny land lyand in Hestwall, fallin in skatfall to the earle; 1613 *Ib.* App. 93.

(Scattfalling,) Scatfalling, *vbl. n.* [Scattfall *n.*] Confiscation because of failure to pay scat. — And als frae all escheittis, scatfalling, .. setting of tak .. befoir the date of thir presents; 1564-5 *Rec. Earld. Orkney* 279.

(Scatt-fre,) Scat-free, *adj.* [Scat(t *n.* 4.] Not subject to scat. — By reason the masters of these houses did suffer a loss this way, they were declared to be scat-free; Brand *Orkney & Shetl.* 122.

Scattill, var. of Scat(t)ald *n.* **Scattilman,** var. of Scattaldman *n.*

Scatting, Scottin(g, *vbl. n.* Also: **skatting, scoting, scottyn(g.** [Scat(t *v.*] The action of contributing, or of being required to contribute, in senses 1 b and 3 of Scat(t *v.*

(1) He conselled thame [to] exerceis skarting [*I.* scatting] and oppressioun wpoun the realme as he had himself wpoun the poore tennentis of Glasgow; Pitsc. I 67/4.

(2) Tom Dekyson is borch for scottyng lottyn .. and borou mallis payn; 1456-7 *Peebles B. Rec.* I 116. Scottyn; 1457-8 *Ib.* 124. That na schippar .. fure nor stow na merchandise abon the ourloftis [.. otherwise they shall be responsible for loss] without ony scottin or lottin with the gudis that beis uncastin; 1507-8 *Acta Conc. Publ. Aff.* lxiii. In amerciament for the non scottin and lottin with the maltmen .. in the geving of his penny to the repairing .. of the altar of St. Mathew; 1523 *Stirling Ant.* IV 211. Thair scotting & lotting, with the furing of his gudis furth of Aberdeen to Leyth; 1545 *Aberd. B. Rec.* MS (Jam.). Being ane unfrieman bearing no chairge within burghe nor scotting and lotting with the inhabitants thairof; 1642 *Dunferm. B. Rec.* II 193. — That hereafter thaj [*sc.* the officers] sal not truble the town be scoting any thing fra the neighbours but once in the ȝeir; 1610 *S. Leith Rec.* 9/1.

(*b*) Cautioun .. in respect he interit burges and gild for walking and warding scatting lotting .. as nychtbouris dois; 1580-1 *Perth Guildry* MS p. 401 (19 Jan.). Anent the scatting and lotting for pilleit and cassin guids; 1581 *Conv. Burghs* I 117. That thai [*sc.* testimonials] be gevin only to sic as actually remanis within the burch, walking wairding scatting and lotting; *Ib.* 123. Oure said soverane lord .. hes exemit the saidis fameleis .. of all taxationis and impositionis quhatsumevir, watcheing, wairding, skatting, loitting, or ony uther maner of burding; 1600 *Reg. Privy C.* VI 124. Statute, skatting of money .. the complent maid .. agains Sir Robert Denninstoun .. quha compellit his monye .. to contribute with the guids that was cassin; 1610 *Conv. Burghs* II 297 marg.

(Scattingabil,) Scotinyabil, *adj.* [Scatting *vbl. n.*] Of property: Susceptible to 'scatting' (Scatting *vbl. n.*), able to supply or produce revenue. — Plegis John Red for scottyn lottyn [etc.] .. and drauis his land in borch qwil he get a land scotinyabil of his an; 1457 *Peebles B. Rec.* I 125.

Scattir, var. of Scatter *v.*

Scat(t)land, *n.* Also: **scait-.** [Scat(t *n.* 4 and Land *n.*[1] Written as one word or two.] **a.** Only in Orkney (cf. b below): A quarter of an urisland, a $4\frac{1}{2}$ d. land. **b.** Only in the rentals of Paplay and Greenwall: Land subject to scat. **c.** In certain references to Shetland, equated with Scat(t)ald *n.*

a. Lyngrow, 1 scattland, viz. 4 d. terre et a half .. Camstaith, a scattland; 1492 *Rental* in H. Marwick *Orkney Farm Names* (1952) 207. Brusgarth ix d. terre inde the haill scattis of iiij d. terre & a half ar fre in the handis of William of Clat .. for als meikill land male of ane uther scat land quhilk wes coffit for the said scattis; 1502 *Orkney Rentals* i 99. And of the uther iij scat land of Tenstayth .. the uthale men payit yeirlie the butter scat [etc.]; 1503 *Ib.* 54. [Malkcum Movat of Swynnenig to infeft the Earl in $5\frac{1}{2}$ scatlands of the lands of Scaill; 1548 *Prot. Bk. Gilbert Grote* 67.] Grobuster 2 scat land, in butter scat [etc.]; 1595 *Orkney Rentals* ii 86. 9 denariatis vel duabus lie scatland vocatis Grobuster; 1630 *Retours* I Inq. Spec. Orkney et Shetland (14).

b. All the land in Paplay and Grenewall is bocht & conqueist be umquhill Erle William except allanerlie j d. terre scat land & j$\frac{1}{2}$ d. terre of the quoyland quhilk is onlie the kingis with the haill scattis of the laif; 1502 *Orkney Rentals* i 11. The Bull of Skaile v d. terre scat land; *Ib.* 13. iiij d. terre callit the scatland of Greinvall; *Ib.* 14. Braguyis 2 d. terræ scat land and 1 d. terræ quoyland payis for the scat land 16 merks butter [etc.]; 1595 *Ib.* ii 11. Ane penny scatland; 1642 *Ib.* iv 9. [A full exoneration of all his feu-duties preceding 1655, for his scait and udal lands; 1684 M. P. Brown *Suppl. Decis.* III 509.]

c. In Unst there is xxi scatlandis vtherwayis callit scathaldis; 1576 *Orkney Oppress.* 46. The inhabitaris of Fetlar .. grant .. that thai have payit for ten scatlandis contenit in their yle .. to the laird for the swyne ruting .. xxx dolors; *Ib.* 47.

(Scattles,) Scatles, *adj.* [SCAT(T *n.*] Not subject to scat. Cf. SCATT-FRE *adj.* — j farding terre scatles; 1508 *Orkney Rentals* I 85.

Scattyr, var. of SCATTER *v.*

(Scaubert,) Scalb(o)urd(e, Scabard, *n.* Also: **scawbert, scau-, scawbart, skawbert; scal-, skalbert, -bart; skealbert; scabert, -berd, skabert, -bart, -bort, scabbert, skabbert(t, -bartt, scabbord, scabboird, -boord; scobert; scha(w)bert.** [ME and e.m.E. *scauberc* (1297), *skawbard* (Cursor M.), *scaubert* (c1325), *skaberte* (c1450), *scaberd* (1546), *skabard* (a1548), *scabbert, -erd* (1600-1), AF **escauberc* (recorded only in pl. *escaubers, -erz*), *escauberge* (all 13th c. in OED) vagina.] A scabbard.

See also SAVE-SKALBERT *n.*

(*a*) For .. binding of Wallas sword with cordis of silk, new hilt and plomet, new scaubart, new belt to the said sword, xxvj s.; 1505 *Treas. Acc.* III 175. Scawbertis; 1506-7 *Ib.* 367. A swerd .. Of quham the scawbart [*Sm.* schawbert] with brown iasp wes pight; DOUG. IV v 160. Abowt hys gorget .. Was hung hys swerd with evor scawbart fyne; *Ib.* XI i 27. Ane swerde having plwmet and hiltis of gold with ane purpure skawbert; *Boece* 118. For graithing and dichting of the kingis suerdis, Jedburgh stavis, and making of skawbertis to thame; 1539 *Treas. Acc.* VII 179. *Ib.* 177. Ane Frence repar with ane Scottis skawbert thairone; 1543 *Aberd. B. Rec.* MS XVIII (Jam., s.v. *Skawbert*). Skawberttis; 1549 *Treas. Acc.* IX 301.

(*b*) Jhesu Crist .. bad Sanct Petir .. that he suld put agayne the suerd in the scalburde; HAY I 110/18. And syne suld he .. kys the croce of the suerd .. and sa put it in the scalbourd; *Ib.* II 43/13. Skalbertis; 1511-12 *Treas. Acc.* IV 328. To be scalbartis .. to the king ane elne .. veluett; 1531 *Ib.* VI 21. The scalbart wes of purpure poleist new; STEWART 44442. For making of scalbertis to the kingis suerdis and cramppettis; 1539 *Treas. Acc.* VII 191. To .. suerd slipper for his fe in making of the kingis grace skalbertis; 1541 *Ib.* 480. For skaillis of scalbertis; 1573-4 *Elgin Rec.* I 144. [Ordered not to] make skalberts or verneis gairdis nor graith swords with unfreemen; 1586 *Edinb. Hammermen* in Whitelaw *Sc. Arms Makers* 126. Tuentie barkit huddrone skins appointit to be skalbert of suordis; 1611 *Brechin Test.* III 220. Ane braid mountit suord with ane skalbart graythit; 1618 *Edinb. Hammermen* in Whitelaw *Sc. Arms Makers* 74.

(*c*) Scabertis; 1537 *Treas. Acc.* VI 300. Skabort; 1545-6 *Ib.* VIII 432. Skabart; 1561-2 *Ib.* XI 107. Ane sword, having the hiltis and skabert of gold; LESLIE 75. And striking him with suerd being withtin the skabars [? *erron. for* skabart] on the hed; 1590-1 *Aberd. B. Rec.* II 69. JAMES VI *Poems* II 155/57. My lord .. rysing agane .. drew his tua suordis furth of ane skabert; 1600 *Crim. Trials* II 156. Two fyne gilt swordis withe fyne wellvet skabertis and guid clips; c1614 *Laing MSS* 138. Suord skabertis; 1616 *M. Works Acc.* (ed.) II 5. The rapper .. was so rousted in the skabert that skarse two men could pull it out perforce; CALDERWOOD VI 73. Scabard; 1681 *Foulis Acc. Bk.* 92.

(*d*) Scabbertis to his grace raperis; 1540 *Treas. Acc.* VII 407. They saw .. James Watt strik .. Willeam Ingram with ane sword, skabbertt and all .. athort the scholderis; 1605 *Aberd. B. Rec.* II 277. Steilling out his suerd and putting ȝokis of egis in the skabbert; *Chron. Kings* 159. Skabbartt; *Ib.* 162.

(*e*) [A] dule scabbord [of cloth]; 1600-1 *Cal. Sc. P.* XIII 765. For a new scabboord [*MS* scabboird] to my lord sword £1 4 s.; 1664 *Household Bks. Archb. Sharp* in *Misc. Maitl. C.* II 510.

(*f*) [Essay for an armourer:] Ane old shable .. dight and ane new scobert made therto; 16.. *Glasgow Hammermen* 10.

(*g*) Admitted freeman as a sword dresser making skealberts and scheathis; 1660 Whitelaw *Sc. Arms Makers* 273.

attrib. Blew purpure velvet to be skawbert beltis to the kingis grace rapparis; 1538 *Treas. Acc.* VI 411.

comb. Androv Cove, alias Skabertmaker; 1547 *Elgin Rec.* I 92. Schabertmakar; *Ib.* Skalbert makar; 1573-4 *Ib.* 144. Scabbertmacker; 1580 *Ib.* 155.

proverb. A sword holds ane uther in the scaberd; Ferg. *Prov.* MS No. 173. Ye hav lyen in your skabert as many good sword dois; *Ib.* No. 1629.

(Scauburne,) Skawburne. Var. of SCAUBERT *n.* with altered ending. — A pirne of gold for a skawburne to the sammyn swerd; 1474 *Treas. Acc.* I 25.

Scauld, var. of SCALD *n.*[1] and *adj.* **Scaulding,** var. of SCALDING *vbl. n.* **Scaunce,** var. of SCANSE *v.* **Scaunt,** var. of SCANT *n.* **Scaupe,** var. of SCALP *n.* **Scaur,** var. of SCAR *v.*[1] **Scaut,** var. of SCOUT *n.*[1]

(Scavanger,) Scawanger, Scafanger, Scaffinger, *n.* Also: **skaffinger, skavienger.** [e.m.E. *scavenger* (1530), *scauinger* (1563-83), AF *scawageour* (15th c. in OED).] A street-sweeper or rubbish collector. Also *possess.* with *tax.*

Al the scawangeris of the tone ar in amerciment for faut of aperans; 1466 *Newburgh B. Ct.* 8b. That .. skaviengeris may be appointit .. for carying furth .. all sort of filth; 1608 *Conv. Burghs* II 253. Skaffingeris; 1616 *Irvine Mun.* II 48. To meett .. with Robert Leslie whom they have a mynd to elect as generall scaffinger [*cf.* Robert Leslie generall undertaker for cleangeing the streetts of this cittie; *Ib.* 84]; 1670 *Edinb. B. Rec.* X 82. To the chamberman tuenty merks and to the scaffinger tuenty merks; 1690-3 *Glasg. Univ. Mun.* I 435.

possess. For a years scafangers tax; 1716 *Household Bk. Gr. Baillie* 47. Scaffingers tax; *Ib.*

(Scavangerie,) Skaffingerie, *n.* [e.m.E. *scavengery* (1656); SCAVANGER *n.*] Municipal street cleaning and refuse collection. — To meitt with Mr. Barrenger .. anent his undertaking the skaffingerie of the toun; 1655 *Edinb. B. Rec.* VIII 388.

Scaw, var. of SCAL(L *n.* **Scawanger,** var. of SCAVANGER *n.* **Scawbart, -ert,** varr. of SCAUBERT *n.* **Scawde,** var. of SCALD *v.*[1] **Scawding,** var. of SCALDING *n.* **Scawe,** see S(C)HAW *n.*[1] **Scaweling,** var. of S(C)HAVEL(L)ING *n.* **Scawpe,** var. of SCALP *n.*

Scayland, var. of SCALAND *ppl. adj.* **Scayle, Scayll-,** varr. of SCAIL *v.* **Scayre,** var. of SCAR(E *n.*[1] **Scayt(h, Scayth(e, Scaytht,** varr. of SCATH(E *n.* and *v.* **Scaything,** var. of SCATHING *vbl. n.*

Sccletter, var. of SCLATAR *n.*

Scearche, var. of SERCH(E *v.* **Scearching,** var. of SERCH(E)ING *vbl. n.* **Scebles,** var. of SIBLES *adj.* **Sced,** var. of SED(E *n.* **Scedicious,** var. of SEDITI(O)US *adj.* **Scedule, -ull,** varr. of SEDULE *n.* **Scedusueus,** var. of SEDITI(O)US *adj.* **Sceeiring,** var. of S(C)HERING *vbl. n.* **Sceether,** var. of SITER *n.* **Sceil(l,** varr. of SELE *n.*[2] and SKILL *n.* **Sceirlein,** var. of S(C)HORLING *n.* **Sceirtein,** see S(C)HORLING *n.* **Sceis,** var. of SES(E *v.*[2] **Sceit,** var. of SETE *n.* **Sceith,** var. of SETH(E *v.* **Scele,** var. of SCHELE *n.*

Scelerat, -ait, *adj.* [e.m.E. *scelerat* (a1513), L. *scelerātus*, p.p. of *scelerāre*, f. *scelus* wickedness.] Excessively wicked; atrociously bad; vile. — So terribill thay ar be apperance, .. So scelerait, and ingrait for to chois; ROLLAND *Ct. Venus* II 296. That he micht be exempill scelerat Till all vtheris; *Ib.* 367.

Scelet, var. of SKEL(L)ET *n.* **Scelf,** var. of SKELF *n.*[1] **Scell,** var. of SCHELE *n.* **Scellair,** var. of SELLAR *n.*[1] **Scellar,** var. of SELLAR *n.*[2] **Scellat,** var. of SKELLAT *n.*

Scellatour, Scellitour, Skellatar, *n.* [? Var. of SKELLIT *n.*] A pot or pan. — Tua scellitouris or aquavytie pottis; 1614 *Edinb. Test.* XLVIII 133. Ane pewdar skellatar pryce £12; ?1666 *Kirkcudbr. Sheriff Ct. Processes* No. 8. A timber backet. Ane scellatour; 1685 *Soc. Ant.* LVIII 362.

Sceme, var. of SCHEM(E *n.*

(Scene,) Schene, Shaene, *n.* [e.m.E. *scene* (1592).] *fig.* (The location of) an event or (part of) an event or incident. — The last counsell-day, Thursday .. was act[ed] the last part of our schene. The twelve petitioners came. .. My Lord Loudon delivered this speech [etc.]; 1637 BAILLIE I 45. Angus .. extorts the grate seall from the archbischope .. and playes Rex in all the schenes off that comedey; BALFOUR *Ann.* I 258. O Edinburrow .. the sun which hath those blind works seen .. May see the shaene changed, which discover shall The mystery wherein infolded all; 1661 *Fugitive Poetry* II xxxii 12/5. A gallant jurie we, condemne we must .. Into the schene, what ere they please wee'll act .. we can ne're be blamed; 1673 *Ib.* xxxiv 3/41.

Scenick, *adj.* and *n.* Also: **schen-.** [17th c. Eng. *scenicke (poet)* (1623) dramatic (poet), MF *scénique* (1375 in Larousse), L. *scēnicus, scænicus.*] **a.** *adj.* Only in derogatory sense: Marked by theatricality. **b.** *noun.* One concerned with theatrical production; an ? actor or ? dramatist. — **a.** [Bishop Lesley] is a mad scenick railer; 1640 BAILLIE I 243. Any who had perused your former schenick writs, that comedie of your seven sages; 1640 Id. *The Caunterburians self-conviction* Postcr. 3. — **b.** Canonists .. declared all scenicks and stage players infamous; 1681 LAUDER *Observes* 51.

Scenographie, *n.* (17th c. Eng. *scenographie* (1645) the representation of a building or other object in perspective, F. *scénographie* (1545 in Larousse), L. *scēnographia,* f. the Gk.) — It is one of the best contryved litle houses .. whereof I have drawen the scenographie so neere as I could; 1646 J. HOPE *Diary* (1958) 186.

Scenophegia, *n.* (ME *senofegia* the Jewish Feast of Tabernacles (Wyclif), L. *scēnopēgia,* f. the Gk. Cf. F. *scénopégie.*) Only in Nisbet, after Purvey. — A feest day of the Jewis, Scenophegia; NISBET *John* vii 2.

Scentries, var. of SENTREIS *n.* **Scep,** var. of S(C)HEP(E *n.* and SKEP(E *n.* **Scepit,** p.p. of S(C)HAPE *v.*

S(c)epter, -re, *n.* Also: **-yr(e, -or(e, -ur(e, -our(e, -ar; -eour; seipter; s(c)hepter, -o(u)r; schapter; (secepter);** CEPTRE. [ME and e.m.E. *sceptire* (Rolle), *septre* (Wyclif), *-ur* (*c*1400), *sceptre* (Lydgate), OF *sceptre* (1080 in Larousse), L. *scēptrum, scæptrum,* f. the Gk.] A sceptre, the ornamental rod held or carried by a monarch to symbolise his or her authority.

To geve (bere, have) the sceptour (*of* a land, etc.), to give (hold) authority (as symbolised by the sceptre).

In mention of the honours, or regalia, of Scotland, frequently coupled with *crown* and *sword* (*of honour*).

The ceremony of touching an act of parliament with the sceptre, orig. by the king himself, symbolised royal approval or ratification of the act (cf. quot. 1689 from *London Gazette* in OED, s.v. *Sceptre*).

(1) Ȝet brocht he furth his dochtir .. With crowne one hed & septre in hande; *Leg. S.* xxxiii 167. Quhare-wyth kyngis and empryowris Are oyhntyd, takand thare honowris, Thaire ryng, thaire sceptyre and thare crownys, With haly prayere; WYNT. VI 115. Sceptyr; *Ib.* 1530. No langer that I ȝeme My crowne, my septure, nor my dyademe; *Lanc.* 665. Ane septour with ane grete bereall and ane perle in the heid of it; 1539 *Soc. Ant.* XXIV 49. Ane cais to the quenis septur; 1540 *Treas. Acc.* VII 304. My said lord governour deliuerit to thaim [*sc.* the Douglases] the sceptour & batoun in parliament in signe & takin of thar restorance; 1542 *Acts* II 419/1. Death .. Hes .. him bereft his septoure and his croun; *Duncan Laideus Test.* 159. To the sweard and sceptour [*sc.* the fact that they were not carried into parliament] .. we ansuere nothing; for oure adversaries know weall yneuche that [it] .. is rather a pompe and gloriouse vane ceremonye, than a .. point of necessitie; KNOX II 127. In administration of justice be the wand of equitie (quhilk is callit the scepture) .. he [*sc.* the king] may euir appeir to all men ane just .. judge; *c*1580 *Sc. Ant.* XIII 3. Triumphant kingis and empreouris .. Whair ar thair gems and pretious stones and sceptaris of estait? FOWLER I 81/104. And Bothuell was promovit to beir the sceptor from the queynis palice to the parliament-hous; *Hist. Jas. VI* (1825) 8. Septeour; MOYSIE 35.

(*b*) Scheptour; 1612 *Aberd. Council Lett.* I 117.

(*c*) We came nixt and saw the honors, wheir we saw the sword and seipter of honor [*sc.* in the Tower of London]; 1665-7 LAUDER *Jrnl.* 1.

(*d*) The last tyme when I reid the parliment, I cearied the secepter; 1681 *Red Bk. Menteith* II 196.

transf. Muse, yow fair dame .. : Your browe, your hair .. Tua starns, a mouthe with perle and rubyeis dekt Whyte hands which suld weire sceptars of estate; FOWLER I 191/11.

(2) That Pirrus oure all the laif, Of Thesalie suld sceptour haif; *Troy-bk.* II 2636. For thou the sceptoure geuis me in hand Of all this realme; DOUG. I ii 44 (Ruddim.). Helenus .. Rang king our mony cities in Greke land, Berand thairof the septour and the wand; *Ib.* III v 25 (Sm.). Abone all kingis that euer buir sceptour; ROLLAND *Ct. Venus* III 374.

(3) Bathe scepter, swerd, crowne and ryng Fra this Jhon [Ballioll] .. he tuk thare, And made hym off the kynryk bare; WYNT. VIII 1959. In that throne thair sat ane wild lyoun In rob royall with sceptour, swerd and croun; HENR. *Fab.* 879. Ferichare .. ressauit princelie ornamentis quhilkis war ane nakit swerde, .. ane evin wand (now callit ane scepture), with ane crowne of gold; *Boece* 60. To Malcom Gourlay quha past to Edinburght with the sweird of honour, croun and septour to the parliament; 1545 *Treas. Acc.* VIII 409. Ane king at euin, with sceptur, sword & crown, At morne bot ane deformit lumpe of clay; 1567 *Sat. P.* iii 6. [The parliament holden by the regent] without crowne, septer and sworde [wherein he was confirmed regent]; 1570 *Cal. Sc. P.* III 713. In time of parliament the .. regent sall wear .. ane crowne of ane duke; the sceptour, in signe of regiment and governament of the realme; and the sword of honour, in signe of justice; BALFOUR *Pract.* 265. The imperiall crowne of Scotland togidder with the septure sword and rob royall; PITSC. II 127/20. [That] the croune, septar and suord sould pase to France that the Dollphin sould be crounitt thairwitht and proclammitt King of Scottland; *Ib.* 129/28. Manifald atestatiounis in parliament secreitlie taine and solemnilie sworne and establissit with sword septer and diadame; 1600 *Aberd. Council Lett.* I 88. That your petitioner .. did .. secure your majesties crowne, sword and scepter, being the honours of that auncient kingdom; 1661 *Laing MSS* 285.

(*b*) A parlement halden in Edinburgh .. be the Duke of Shatillo, Huntly and utheris, with croun, sheptour and swourd; 1571 *Misc. Bann. C.* III 131. Scheptour; *Diurn. Occurr.* 118.

(4) The king in proper persone come to the tolbuith and be the twiching of the sceptore did ratifie all thingis done befoir; BANN. *Memor.* 186. Seing the necessar solemnities requisite to doomes of forfaultures before the parliament, such as voteing approveing toucheing with the royall sceptre, intimateing by sound of trumpet and teareing of armes were not adhibite; 1690 *Acts* IX 222/1.

(*b*) The act being drawen, read and touched with the shepter; 1689 *Reg. Privy C.* 3 Ser. XIII 412.

b. A representation or replica of a sceptre, or something carried to represent a sceptre.

Behald how thi king furth is led, .. Septour in hand, with croun of thorn on heid; KENNEDY *Pass. Christ* 619. For ane new septour, .. for the mending of the dewillis cot and play claythis; 1522 *Perth Hammermen* 18. For the collouring .. of the effesay croun and septur; 1542 *Treas. Acc.* VIII 142. The Scottis croune havand on the ane syde the croune sceptour & sword makand in the middis the forme of Sanctandrois croce; 1579 *Acts* III 150/1. Ilk pece .. havand .. on the vther syde ane lyoun crownit sitting in ane chyre halding ane scepture; 1591 *Ib.* 526/1. Couuetusness .. doth sceptres als in painterie carie ay; JAMES VI *Poems* I 183/652. [Statue of James VI at the Netherbow Port:] To Henrye Mylne goldsmyth for making the sceptre [etc.]; 1616 *Treas. Acc.* in Mylne *Master Masons* 106. For gilting .. the thrissell, crowne and scheptour [*Misc. Maitl. C. III 370 n.* -or] abone the kingis armes; 1630 *Glasgow B. Rec.* I 374.

c. In *fig.* or allusive use: An instrument of power or authority; the power or authority of the monarch symbolised by the sceptre.

They haiff .. thrawin the sword of the word of God and schepter of ecclesiasticall government .. from the hands of Chryst Jesus; 1584 MELVILL 176. Haiff we the authoritie and powar of his [*sc.* Christ's] schapter? MELVILL 284. [Your majesty] commanding at your first entrie to your roial scepter to reform the grammar; HUME *Orthog.* 3. Albeit .. Scotland and England be now most happelie united undir the royall sceptour of oure most gratious soverane; 1610 *Reg. Privy C.* IX 8. He sall with scepter hard You smyt; 1639 *Fugitive Poetry* II xv 2/4. That .. line which hath sweyed the scepter of this kingdom for many hundreds of yeers past; 1649 *Rec. Kirk Scotl.* 546.

d. *attrib.* With *wand.*

That the septer wand suld nocht be away tane fra the princis of Jowry quhill the saynde of God; HAY I 189/29. To conquest Italie .. Tharin to bruke the croun and sceptre wand; DOUG. IX V 103 (Ruddim.).

Scepter. Var. of CHEPTOUR *n.*[2] — That the battlement round the scepter house be mended; 1721 *Glasgow B. Rec.* IV 105.

Scepterlie. Var. of CHEPTOURLY *adv.* — The quhilk day the magistrates and counsell above named beeng scepterlie convened for treating of materes tending to the advantage of this brugh [etc.]; 1680 *Inverness Rec.* II 285.

Sceptor(e, -our(e, -re, -ur(e, -yr(e, varr. of S(C)EPTER *n.*

Scer, var. of SCAR *adj.* **Scercement,** var. of SCARSMENT *n.* **Scerching,** var. of SERCH(E)ING *vbl. n.* **Scerlane,** var. of SKIRLING *vbl. n.* **Scerse,** var. of SERCH(E *v.* **Scersly,** var. of SCARSLY *adv.* **Scet,** var. of S(C)HETE *n.*[1] **Sceuitt,** p.p. of SEW *v.* **Scevin,** var. of SEVIN *num.* and *n.* **Scew,** var. of S(C)HAW *v.* **Scewing,** var. of SEWIN(G *vbl. n.*[1] **Sceyp,** var. of S(C)HEP(E *n.*

Schaam, var. of S(C)HAME *v.* **Schaan,** p.t. of S(C)HINE *v.*

Schab(b)it, varr. of SCABBIT *ppl. adj.* **Schabe,** var. of SCAB *n.* **Schabert,** var. of SCAUBERT *n.*

S(c)hable, Schabill, *n.* Also: **sheabill, chable, shabel, shabble, sabble.** [Obs. Germ. *schabel, sabel,* local varr. of Germ. *säbel,* Polish *szabla,* Magyar *szablya.* Cf. Ital. *sciab(o)la.* Cf. also 17th c. Eng. *sabel* (1617), of which the form *sabble* (see (*b*) below) may be a single instance, Germ. or Du. *sabel,* of the same ulterior origin, whence also the F. altered form *sabre.*] A sabre.

Also *attrib.* with *blade.*

(*a*) The onely best shables, or short crooked swords, that be in the world are made here [*sc.* Damascus]; LITHGOW *Trav.* V 208. [The defender had bought] ane gune and sheabill [for £6]; 1637 *Aberd. Sheriff Ct.* II 436. The Hollanderis canon and schables; SPALDING I 238. For .. blooding a man buy shabellis; 1654 *Dumfries Kirk S.* 16 March. Schabill; 1667 *Edinb. Test.* LXXIII 86. And .. smote him on the cheek with his shabel; KIRKTON *Hist.* 418. The general .. struck him with the pomel of his shable on his face; 1681 Fountainhall *Decis.* I 159. A sea-captain offered to strike off my head with a shable; COLVIL *Whig's Suppl.* (1681) Sig. A 5b. The Persians, Turks, Russians, Polonians, and Hungarians, for most part wear scimiters and shables; 1683 TURNER *Pallas Armata* 171. Chable; 1685 *Soc. Ant.* LVIII 356. [Supplied] 250 mounted shables; 1699 Whitelaw *Sc. Arms Makers* 221. He was .. repute an Egyptian, and did wear a pistol and shable; 1715 Macritchie *Gypsies* 118.

(*b*) The commander .. with a shabble struck him on the face; 1679 Wodrow *Hist.* III (1829) 44. That he [*sc.* the jailer] .. wear constantly a sword or sabble; 1683 *Haddington B. Rec.* (Robb) 20 Jan.

attrib. Shable blades; 1699 Whitelaw *Sc. Arms Makers* 221.

Schabyose, var. of SCABIOSE *n.* **Schacar,** var. of SCHAKAR *n.* **Schachles,** var. of SCATHELES *adj.* **Schacht,** var. of SCATH(E *n.* and S(C)HAFT *n.*[2] **Schack,** var. of S(C)HAKE *v.* **Schackert, -irt,** varr. of SCHAKERT *adj.*

S(c)had, *n.* Also: **s(c)hade, shed.** [ME and e.m.E. *shade* (Chaucer), *shad* (Lydgate), OE *sceadu* str. fem. beside *scead* neut.; see also S(C)HAD(D)OW *n.*]

1. Shelter from the rays of the sun; shade. = S(C)HAD(D)OW *n.* 2.

Thou who .. lay alone .. Amid those greene and grouie shads; CRAIG II 52. Quhill Beutie .. reposes With fairest schads of trees o'rschadoued wnder; MURE *Early Misc. P.* ii 2. The pleasant shed Of oake and plaine oft serves them for a bed; DRUMMOND II 119/53.

2. a. A shadow. = S(C)HAD(D)OW *n.* 4. **b.** Outward seeming; appearance. = S(C)HAD(D)OW *n.* 6c. **c.** A ghost, a spirit. = S(C)HAD(D)OW *n.* 7.

a. The schad of euerie towre and tree Extended is in length; HUME 31/179.

b. Anonder pittyeis schade schee does desplay The fulnes of her rigour; FOWLER I 180/5. Of types, of shads, the body true; MURE *Spirituall Hymne* 67.

c. Glorefied amidst the schads dewyne; MURE *Early Misc. P.* xvii 26.

3. With reference to the custom of starting to divide land from either the 'dark' or the 'sunny' side. See S(C)HAD(D)OW *n.* 1.

The sheriff or bailie must also .. ken the relict to her terce, which is ordinarily done by the sun or the shade; that is, whether the division shall begin at the east or the west; STAIR *Inst.* II vi § 14.

Schad, var. of S(C)HED *n.*

Schaddell, *n.* (ME and e.m.E. *chetel* (Ancr. R.), *chatel* (*a*1240) non-count noun = property, *chattell* (1549) count noun = a movable possession, used of slaves or bondmen, OF *chatel, chetel* (ONF *catel*) (all in Godefroy), late L. *captāle,* L. *capitāle.* Cf. CATALE *n.*) — [He is compelled to serve other men's turns, who use him for a] schaddell; 1571 *Cal. Sc. P.* IV 143.

Schadderar, *n.* [See next.] A person who breaks up ore. — Schadderaris and budleris of the lang heap; 1609 *Hilderstoun Silver Mines* I 132.

(Schaddering,) Shaddering, *vbl. n.* [17th c. Eng. *shadder* (1622) to break up the larger pieces of ore, of unknown origin. Perh. connected with e.m.E. *shalder* (1577) to crumble, or ME and e.m.E. *schateren* (*c*1330), *shatter* (*a*1530) to break in pieces by a sudden blow, etc.] The action of breaking up ore. — For sax dayis wark in shaddering of mettell at vj s. viij d. in the day; 1609 *Hilderstoun Silver Mines* I 112b.

Schadding, var. of S(C)HEDDIN(G *vbl. n.*

S(c)had(d)ow, *n.* and *adj.* Also: **sadow; s(c)hed(d)-, schaud-, schawd-** and **-o(u, -owe, -aw, -ew.** [ME and e.m.E. *sceadewe* (*c*1175), *scadewe* (*c*1200), *sadue* (*a*1220), *scheadewe* (Ancr. R.), *schadewe* (*a*1272), *scaudu* (Cursor M.), *shadow* (Rolle), *schadue* (*c*1350), OE *scead(u)we,* oblique case of *sceadu* S(C)HAD *n.*]

A. *noun.* **I. 1.** The 'dark' or west side, or end, of a piece of land, as opposed to the 'sunny' or east end or side.

Cavilling be sone and schadow, drawing lots to decide whether, when dividing land, to begin the division at the east ('sone') or west ('schadow') end of the land.

For a very full discussion of the 'sun-division' system of allotting parcels of land, see R. A. Dodgshon 'Scandinavian 'Solskifte' and the Sunwise Division of Land in Eastern Scotland' in *Scottish Studies* XIX (1975) 1-14.

Strekand .. allssua in tyl thar toftis the forsayd Gylbert at the sadow and the forsayd Alysonder at the sone as alde departyson wes; 1388 *Bamff Chart.* 22. I kest thaim [*supra* twa kabillis], the tane to the son, the tothir to the schadow & thus it [*sc.* the land] was departit; 1428 *Liber Melros* 521. The actione .. tuecheing .. the cawelleyng be sone and schadow of thairis haill landis; 1537-8 *Dunferm. Reg. Ct.* 151. Johne Robb .. schiref deput .. kaist cavillis for sone & schaddow of the saidis landis; 1549 *Lanark Sheriff Ct.* MS 30. All byrun proffeitis .. of the saidis half landis lyand at the schado; 1556 *Dundee B. Ct.* III fol. 65b (16 Aug.). And cawillis being cassin that scho fell hir terce to the schadow; 1557 *Inverness Sheriff Ct.* 59b. Anent the kenning [etc.] .. of the landis of Waterstoun be half and quarteris, sone and schaddow; 1557 *Reg. Privy S.* V I 37/2.

2. Shade, esp. viewed as protection from the rays of the sun; shelter.

Const. *of* the thing providing such shade, also *of* the sun or other luminary causing the shadow.

Me to conserue than fra the sonis heit, Vnder the schaddow off ane hawthorne grene, I lenit doun; HENR. *Fab.* 1343. A wod abufe .. castis a plesand schaddow our the clewis; DOUG. I iv 18. Ane semly schaw, With hys maist plesand sobir schaddowis; *Ib.* vii 43. Sum [monsters] that liand furth with the schaddo of thare braid fut thai ar kepit fra the rayne or the heit of the sone; ABELL 3b. Quhen the sune castis oure grite heyt, ve pas vndir the vmbre or the schaddou; *Compl.* 34/19. The plesant plane-trie .. With fairest schaddow to save the sone in symmer; 1584 SEMPILL *Sat. P.* xlv Pref. 46.

3. *sing.* and *pl.* Darkness caused by absence of light or obstruction of a luminary.

Schadow of death, after English translations of the Bible, rendering L. *umbra mortis*. Also *fig.*

(1) Me smyte ful deip onto the schaddoys dern, Amang pail gastis of hellis holl cavern; DOUG. IV i 51. Amang the schaddowys and the skuggis mark The hell hundys hard thai .. bark; *Ib.* VI iv 43. Bot a blak sop of myst, als dyrk as nycht, With drery schaddo bilappis his hed; *Ib.* XV 51. Quhilk goddes .. Was wont to .. heild hir self alsso in schaddois vayn; *Ib.* XII i 132. As schadow in the sonnis beme .. So all my dayis is bot ane dreme; *Maitl. F.* 206/17.

(2) Quhil sonnys schaddow circlys hillis about, And the firmament starris doith conteyn; DOUG. I ix 52. [They] Past vnder the shadow of the mone .. and stude awaittand vpon bodelie harme to haif done to the said James; 1563-4 *Inverness Rec.* I 110.

(3) And quhile men sat in the cuntre of schadou of deid licht raise to tham; NISBET *Matth.* iv 16. Whiles thy foes in schadows of death wer swounding; 1639 *Fugitive Poetry* II xv 5/18.

(4) *fig.* Silit he [*sc.* Adam] wes vnder schaddew of syn; KENNEDY *Pass. Christ* 92. The schaddois of pertrublans Was dryve away; DOUG. XII xi 119.

b. *specif.* The darkness of night.

The dyrk nycht hyd ws with cloys schaddowis meit; DOUG. II vi 116. The dyrk nycht With hir donk schaddow hydis of the erth the sycht; *Ib.* IV vi 142. Suteing nights shaddowes did tempestuous showe; 1634 *Fugitive Poetry* II ix 3.

4. The dark outline of a person or thing cast upon a surface by his or its interception of the rays of the sun or other luminary. Also *proverb.*

And alse the mune is maid obscure quhen it clips be rason that the vmbre and schaddou of the eird empeschis hyr to resaue lycht fra the soune; *Compl.* 56/15. Quhen the sonne is at the hycht .. The schaddow of that hydduous strenth [*sc.* the tower of Babel] Sax myle and more it is of lenth; LYND. *Mon.* 1751. Ane dog .. ane collap staw .. the schadow saw, That it was mair, larger, the flesche leit faw; ROLLAND *Seven S.* 1694. Ane covertour of the tries .. quhilk makis ane glansing shaddow on the watter; BUCH. *Comm. on Virgil* Æn. I 164. Chasaris dog That for the sheddow sene he loist ane ben; SEMPILL *Warning* v. And mak your awin particulars to follow efter as the servant and sheddow, and nocht go befor as the maister and body; 1584 MELVILL 180. Upoun the wyndo .. he affixt blak cloths, that his shaddow sould not be sene .. when he went to or fro; *Hist. Jas. VI* (1825) 46.

proverb. He is fleid at his awin schadow; CARMICHAEL *Prov.* No. 694.

b. As a type of what is transient or fleeting.

This present pilgremage, quhilk passis as a schadow daily; *Wisd. Sol.* (STS) 312.

5. A reflected image.

I sawe ane howlat .. Lukand the laike throwe, And saw his awne schadowe; *Howlat* 50 (A). The schadow off the mone schone in the well; HENR. *Fab.* 2392. Scho .. tuik Ane poleist glas and hir schaddow culd luik; Id. *Test. Cress.* 348. DOUG. *Pal. Hon.* 1490. For as thow come sa sall thow pas, Lyk as ane schaddow in ane glas; DUNB. (OUP) 176/5. For ane noble of blud that hes no vertousnes .. Is evin bot as a schaddow in a glas; *Bann. MS* 92b/72. The chirries .. Quhais schaddowes in the river schew; MONTG. *Ch. & Slae* 312 (W). *Ib.* 51 (L). Lett youre awin lyfe be .. a mirroure to youre people, that .. thairin thay maye see be youre shaddou quhat lyfe thay soulde leade; JAMES VI *Basil. Doron* 104/9.

6. In various *transf.* or *fig.* extensions of the above senses. **a.** Covering, protection.

He said the wertu of God in hire suld tak ane schadow defendand hire fra all troubile and hyde hire in the gret licht of hevin that was the son of God .. And be the manere of a schadou the blist sone of God and hire haly lif fyrst suld be hid fra the waurld; IRLAND *Mir.* I 135/32, 34. WINȜET II 64/19 (see S(C)HAD(D)OW *v.* 2 (2)). The childring of men trust under the schadow of Thy weingis; 1614 in Melvill lxii.

b. (Outward) appearance; image.

The IRLAND quots. here and above translate and explicate the Latin *obumbrabit.*

The angell promittit .. that .. [God's] derrast sone suld in hir blissit wame tak a schadow that is a westament and clethinge of hire nature and humanite; IRLAND *Mir.* I 135/30. Remember .. be youre outuairde using of youre office to testifie the inuarde uprichtnes of youre hairte & be youre behauioure in all indifferent things to sett furth the treu shaddou [*v.r.* image] of youre uertuouse disposition; JAMES VI *Basil. Doron* 201/2.

c. Outward show, as opposed to an inner and invisible reality; a delusive semblance; an unreal appearance; a pretence.

Thai may haue the schadow of honour and for a litle tyme waurdly and fleschly plesaunce; IRLAND *Mir.* II 72/15. He has nocht bot ane schawdow of ane knycht; *Bk. Chess* 1103. He is for ladyis in luf a right lusty schadow; DUNB. *Tua Mar. W.* 191. Dreames .. ar deceatful, and quha sa euer regardis sic, takis hald of a schedow and followis efter the wynd; NISBET II 22 marg. Siclik our life without activite Giffis na frut, howbeit ane schado blume; BELL. *Boece* I xii. Thay ar bot ane schaddow of pepill, and may thairfore be slane like miserabill beistis; *Ib.* II 351. Those that have a vane schaddow .. content thame selfis with the same, .. thinking that thai have a minister, quhen in verray deid thai have none; *Bk. Disc.* 194. He tuke nocht our natuir on Him finȝetlie and vnder a schaddow, bot verelie and manifestlie; WINȜET II 42/20. To schaw ane schaddow quhair my hairt hes bene; *Bann. MS* 223b/12. DALR. I 258/5. To abuse thame selfis with the delytis of the vayne schaddowes of this lyfe; 1611 *Crim. Trials* III 160. Yea, some-where there's lord Baillie, and men must Vpon his sheep-drawne shaddow lordship thrust; 1633 LITHGOW *Poet. Remains* 112. Neither is father nor mother [etc.] .. in this over-lustred world .. anything else .. but feathers, shadows, night-dreams, and straws; 1637 RUTHERFORD *Lett.* (1894) 176. Thus ended this circuit, being but a ludibrie and shadow of justice; 1679 LAUDER *Notices Affairs* I 243.

d. A foreshadowing or prefiguration.

The synagoge .. wes bot ane schaddou and vmbre of the trew kirk; 1573 *Cath. Tr.* 26/28. The neu testament .. beand the veritie, & the auld testament the vmbre and schaddou onlie; BURNE *Disput.* 25b.

e. An adumbration; an obscure indication.

The evangell was ay the mair obscurely preached under dark types and shadow; R. BRUCE *Serm.* 130. Men .. may perceiue the rest of our cuntrey .. be this schadow that heir we haue schawin and similitude sett afore thair eyne; DALR. I 128/1.

f. A faint appearance; a trace.

It weschis away all schaddow it allane; *Bann. MS* 77b/81. [I cannot see] quhat schaddow of any possebelitie thair is heirin; ?1591 *Warrender P.* (SHS) II 160.

g. A misrepresentation.

As your majestie will hawe me free of that schaddowe be the quhilk my ondeserwit enemye crawis my lyiff and land; 1609 *Crim. Trials* III 65.

7. An apparition, wraith, phantom; an incorporeal being.

Than a schaudow rycht nerehand As of a man saw by hym stand; *Leg. S.* ii 1159. Schedow; *Ib.* 1165. He schaw a schedaw one his rycht hand In forme of mane semand to be; *Ib.* xviii 212. Gousty schaddois of eild and grisly ded; Doug. vii Prol. 46. Bot this ilk schaddo .. gaif the bak and fled; *Ib.* x xi 105. Thydder went this wrath or schaddo of Ene .. And hyd hyr dern .. tharin; *Ib.* 127. That in this my decripit aige I salbe compellit to fight against schaddowes and howlatis that dare not abyde the lycht; 1571 Bann. *Memor.* 103. These kindes of spirites, when they appeare in the shaddow of a person newlie dead, or to die, to his friends, .. are called wraithes in our language; James VI *Dæmonol.* (STS) 42/13. For the Deuill durst neuer haue borrowed their shaddow or similitude .. if their consent had not bene at it; *Ib.* 55/16. Thow affermis that the elphis .. ar bot schaddowis, bot ar starker nor men; 1597-8 *Misc. Spald. C.* I 121.

II. 8. An ornamental eyeshade for affording a woman's face some protection from the rays of the sun.

Ane perill schadow estimat to [£66 13 s. 4 d.] .. ane litle neck chayne; 1605 Paterson *Ayr & Wigton* III 197. Twa .. perlit chenȝeis ane pair of bracelletis of gold and thrie perlit schaddowis; 1610 *Glasgow Test.* VII 17 Dec. Ane schaddow of orien pairill, estimat to x lib.; 1612 Pont *Cunningham* App. 197.

III. 9. *attrib.* **a.** In sense 1, chiefly with terms used in describing divisions or apportionments made in a piece of land.

? Only in Fife: *Schadow cavill*, the lot which when drawn indicated that division of land should begin at the 'shadow' end. See the note to 1 above.

[Litera regressus .. super umbrali dimedietate terrarum de Litill Folay et Blakwattir; 1531 *Reg. Privy S.* II 112/2. Incipiendo vel a parte solari [*i.e.* oriente] vel ab umbrali [*i.e.* occidente] prout sors evenerit; Hope *Major Pract.* I 202.]

[The west third part of the west or shadow half of the lands and manor-place of Crottie]; 1590 *Bamff Chart.* 138. [Two oxgait of the two sunny ploughs of Percock. One oxgait of the sunny plough of the shadow half thereof]; 1619 *Aberd. Sheriff Ct.* II 77. [The shadow half of the east third part toun and lands of Lytill Banff, with the shadow half of the houses and pertinents [etc.]]; 1631 *Bamff Chart.* 226.

(1) 6 mercatas terrarum .. de Smythtoun de Noth, viz., le schaddow-half earundem; 1504-5 *Reg. Great S.* 600/2. The schedowe half landis of Ruthiristoun; 1508-9 *Aberd. Sheriff Ct.* I 79. The schaddow half of the toun and landis of Gylstoun; 1539-40 *Reg. Privy S.* II 484/1. Scheddow; 1563 *Prot. Bk. D. Gray* 57b. [Cowstoun] baith sonny half and schaddow half; 1568 *Reg. Privy S.* VI 59/2. The .. schaddow half of Dikmountlaw. .. the sonnie half of Dikmountlaw; 1592 *Acts* III 568/2. [The] myddill pleuch of the schaddou half lands; 1603 *Aberd. Sheriff Ct.* II 31. The shaddow half of the uther third pairt of the lands of Ester Gellets; 1653 *Retours* I Inq. Spec. Fife (822). Alse weill sunny as shadow halfes; *Ib.* Forfar (322). The lands of the west or shaddow-halfe of Mylnetoune of Connone; 1654 *Ib.* (338). My spous be our matrimoniall contract is provyded to my shadowe halff lands of Glenboy; 1656 *Brechin Test.* VI 12. Tua pairt of that shadow halfe of oxingait land of Pitkeithlie; 1658 *Retours* II Inq. Spec. Perth (659). The town and lands of Barnhill of Garvock, as well sun as shadow half thereof; 1695 Fraser *Hist. Laurencekirk* 52.

(2) Binas partes umbrales (the schaddow tua part) viz. 2 aratrata seu tua plewis ville de Jakkistoun .. vic. Abirdene; 1533 (1534) *Reg. Great S.* 306/2. The schaddow-pleuch of Sonny-syde; 1544 (1587) *Ib.* 402/1. The scheddow part of the samyn [lands]; 1565 *Prot. Bk. J. Drummond* 58. Enterit .. Eufame Wauchop .. to the schaddo thrid rig, toft and croft of the landis of Edmanstoun; 1566 *Prot. Bk. James Foulis* 116b. Hir schadow rig and landis .. pertenyng to hir be ressone of terce; 1568 *Prot. Bk. J. Drummond* 73. The schaddow or myd thrid part and how schaddow thrid part; 1574 (1585) *Reg. Great S.* 263/1. The schaddow ten oxingait of land of Blairmad; 1577-8 *Reg. Privy S.* VII 212/2. The schaddow quarter of the sonie half landis; 1584 *Acts* III 372/1. The half schaddow lands of Auchtoreles; 1586 *Digest Justiciary Proc.* L 21. The teynd schavis of the west schado quarter landis of Drumphingallie; 1603 *Inchaffray Reg.* 134. The shadow 4 oxingait of land, of the shadow pleugh; 1655 *Retours* I Inq. Spec. Aberdeen (326). The pleuch, shadow third pairt, and Ord of Monykebbock; 1657 *Ib.* (339). The west shadow quarter of the toune and lands of Drumphingallie; 1657 *Ib.* II Inq. Spec. Perth (654).

(3) To hir part thairoff being vpon the schaddow syd; 1566 *Prot. Bk. W. Douglas* 81b. That ane of the bailȝeis .. cawill and diwyd the sone syd of tham fra the schaddow syd thairof that .. William Ȝoung may .. intromet with the ane auchten part thairof of quhilk lyis on the sune syd thairof; 1572 *Elgin Rec.* I 130. Ane pleuch of land on the sone syid of the auld toune of Pettindrech, ane vthir pleuch of land on the schaddow syid of the .. toune; 1590 *Prot. Bk. J. Inglis* 20 July.

(4) It was fundin .. that Margrete Forrate .. of Caldhame .. is befallyne the schadow cavill of the samyn; 1516 *Fife Sheriff Ct.* 27. Cristiane Spens .. wes .. seruit of hir ressonable terce of .. the landis of Ramkelour nethir .. to the quhilk scho befell the schaddo cawill; 1521 *Ib.* 209. *Ib.* 237.

b. In sense 2, rendering L. *umbrosa* in a place-name.

Fra the name of the monttanes that is called wallis umbra or umbrosa the schaddo valey; Bisset II 414/23. *Ib.* 122/13.

B. *adj.* Faint, as if obscured by cloud.

The glymerand monys schaddo lycht; Doug. viii i 12. The sunne .. was seene to shine with a fant beame, yeelding a dime and shaddow light ewen in a cleare heaven; P. Gordon *Brit. Dist.* 62.

S(c)had(d)ow, *v.* Also: **schado(u.** [ME and e.m.E. *schadow* (*a*1300), also *sseduy* (1340), OE *sceadwian*, f. *sceadu*; S(c)had(d)ow *n.*]

1. a. *tr.* To reflect. **b.** *intr.* To cast a shadow. **c.** *tr.* To provide (a painting) with 'shadows'.

a. All the lake as lamp did leme of licht, Quhilk schadovit all about wyth twynkling glemis; Dunb. *G. Targe* 31 (Ch. & M.).

b. Bricht Cytherea, Quhilk only schaddowist amang starris lyte; Doug. iv Prol. 2.

c. Like a painter shadowing .. a portrait .. for giuing it greter viuenes, so I eike or paire to the circumstaunces of the actions; James VI *Poems* I 200/40.

2. *transf.* or *fig.* To enfold protectively; to protect. Also *to schaddow oure.*

(1) Quhar Tendir mariolyne and sweit flowris thar out With thar dulce smell hym schaddowit rownd about; Doug. i x 70. Boithuile .. schadoand his murther and vtheris wickit enormiteis fra pvnisment with the cloik of hir auctoritie; 1567 *Facs. Nat. MSS* III lv. Scheltred and schaddowit by a pretendit auchtoritie of Parliament; Spalding II 197.

(2) Na thing of thair awin almaist euir produce thai, quhilk thai intend nocht to schaddow oure also with the wordis of Scriptuir .. Bot samekle mair ar thai to be .. ferit, the mair quietlie thai lurk wnder the schadowis of Godis law; Winȝet II 64/11.

b. *To schaddow up*, to conceal, hide away.

All .. worthy to be imparted to the aige present and posteritie, and not so to be schaddowed up in perrisying scrollis; 1610 *Reg. Privy C.* IX 572.

(Schaddowing,) Shaddowing, *ppl. adj.* [e.m.E. *schadowynge* (1552).] That affords shelter. — That all the shaddowing mirtell woods wer filled with thair tranes; Fowler I 34/210.

(Schaddowles,) Shaddowles, *adj.* [S(c)had(d)ow *n.* 6.] ? Without a trace. — When that a moment of ane houer sall shaddowles leawe voyd That vpon which so many yeares yow hawe your panis employid; Fowler I 129/85.

(Schaddy,) Shaddy, -ie, *adj.* [S(c)had *n.*] **a.** Affording shade. **b.** = S(c)had(d)ow *n.* 9 a. — **a.** We shall on the mountains go, In shaddie vmbers too and fro: In vallies low, and on the bray; Craig ii 151. A shaddy groave amidst this citty grew; Mure *Dido & Æneas* i 523. — **b.** [The lands of Tillienaid as well the sunny as the shady half]; 1677 *Antiq. Aberd. & B.* III 37.

Schade, var. of S(c)had *n.* and S(c)hed *n.* and *v.* **Schad(e)les,** varr. of Scatheles *adj.* **Schado(u, -ow(e,** varr. of S(c)had(d)ow *n.*, *adj.* and *v.* **Schadyne, -yng,** varr. of Scathing *vbl. n.* **Schaeff,** var. of

SCHEF(E *n.*[2] **Schaeling,** var. of SCHELING *n.* **Schaem,** var. of S(C)HAME *n.* **Schaf,** var. of SCHAVE *v.* and SCHEF(E *n.*[2]

Schafe, Schaife, Scheaf, *n.* Also: **scheav-.** [Only in Easter Ross, in the early 17th c. Perh. the same word as SCHEF(E *n.*[2], but see *Soc. Ant.* (1943-4) LXXVIII 74 for an argument that it is altered from *sceth* SCHETH *n.*] A local land measure, appar. equal to about $\frac{1}{7}$ of an oxgang. — Payand yeirlie .. for the saids vi schafis lands of Culkenzie; 1603 *Old Ross-shire* I 171. 3 bovatis et 1 garbata terrarum vulgariter appelatis the thrie oxgang and schaifeland villæ et terrarum de Culkenzie in baronia de Delnie; 1615 *Retours* II Inq. Spec. Ross et Cromarty (39). 7 garbatis terrarum vulgo appellatis 7 scheavis of land .. extendentibus ad 1 bovatam terræ vulgariter nuncupatam an oxgait of land unamque garbatam lie scheaf dictarum terrarum et villæ de Cowlichmoir infra baroniam de Line; 1623 *Ib.* (66). $10\frac{1}{2}$ garbatis villæ et davatæ terrarum .. vulgo nuncupatis ten scheaffis and ane half scheaf of land infra baroniam de Delny noviter erectam; 1635 *Ib.* (83). 7 garbatis lie sevin scheaffis land .. extendentibus ad $1\frac{1}{6}$ bovatam terrarum de Culkenzie; 1635 *Ib.*

Schafe, Schaff, varr. of SCHEF(E *n.*[2] **Schaferrn,** var. of SCHAFFRO(U)N *n.* **Schafeté,** var. of SAFETY *n.* **Schaffaldin(g, -altin, -eltin,** varr. of SCAFFALTING *vbl. n.* **Schaffel(d,** varr. of SCAFFALD *n.*

Schaffit, *p.p.* [*Schafe* SCHEF(E *n.*[2]] Provided with a 'sheaf' (bundle of arrows), in sense 2 of SCHEF(E *n.*[2] — All vthir ȝemen .. salbe sufficiandly bowit & schaffit, with suerde buklare & knyfe; 1425 *Acts* II 10/2.

Schaffold, var. of SCAFFALD *n.*

Schaffro(u)n, -rone, *n.* Also: **schaferrn, saferon; schaif(f)ron, chaiffroune, cheffroun, (schraifron); (s)chaiffer;** CHAFFERON, CHAUFFRON, CHAMFERON. [ME and e.m.E. *shawfron* (1465) in sense 2 below var. of *chaffron* and *champrain.* Of unknown origin.]

1. Some ornamental part of, or addition to, a woman's hood (perh., a front border).

(*a*) To be schaiffronis to the sam hudis, half ane elne crammesy; 1511 *Treas. Acc.* IV 210. To be hir ane pair of sclewez and ane schaffrone v quartaris crammesy satyne; *Ib.* 219. For the schaffronis and necessaris to the said hud; *Ib.* 221. Thir jowellis and baggis heir writtin .. war deliverit. . . ane saferon with ane chenȝe of gold of blak velvos .. ; ane saferon with punctis of gold with lxi perle of crammasy welvot .. ; ane saferon with xxi rubies [etc.]; 1516 *Acta Conc. Publ. Aff.* 70. Ane schaffroun with ane burd of gold .. ; ane schaffroun with ane chenȝe of gold quhite anamelet velvot .. ; ane cheffroun with ane punct with settis of perle; *Ib.* 71. For goldin wark .. to geve in New ȝere giftis .. in chenȝeis, tabullattis, ringis, stanis, carkannis, schaferrnis and uther goldin wark; 1539 *Treas. Acc.* VII 265. Ane grete chaiffroune of gold, ane litill chaiffroune; 1540 *Ib.* 400.

(*b*) Crammesy satin to be ane schaiffer to hir hud. . . for making of the hud and chaiffer; 1511 *Treas. Acc.* IV 230.

2. The frontlet of a barded horse.

For the dichting .. of l steill sadillis, schraifronis, roundellis and .. justing harnes; 1541 *Treas. Acc.* VII 469. For dichting of the kingis grace justing harnes [etc.] .. schaifronis for hors; 1542 *Ib.* VIII 92.

S(c)haft, *n.*[1] Also: **s(c)hafft, schafete, shaf(f, schaift, (schffat), cheft;** CHAFT. [ME and e.m.E. *scaft* (Layamon), *saft* (*c*1250), *ssafte* (1297), *schaft* (Wyclif), *shaft* (*c*1400), OE *sceaft.* Cf. ON *skapt.*]

1. The staff or pole or rod forming the body of certain, usu. long-handled, weapons, tools or implements. Also in *fig.* context.

Freq. as the second element of a compound or quasi-compound, the weapon or tool concerned being specified by the first element, see AX-SCHAFT *n.*, ARROW *n.*, CHARGEOUR *n.*[2], MAT(T)OK *n.*, MEL(L *n.*, MOPPAT *n.*, PIK *n.*[2], etc., for further examples.

(1) BARB. XII 57 (see HAND-AX *n.*). Quhill that the hand-ax schaft .. brak; *Alex.* II 4497. For xij staf schaftis ilk pece ij s.; 1506 *Treas. Acc.* III 358. For .. iii pikkis and mattokkis schaftis; 1530 *M. Works Acc.* (ed.) I 38. Sextene hundreth arrow schaftis price of the hundreth vj s. viij d. . . xl speir schaftis price of the pece xl d. . . ane hundreth stalf schaftis price of the scoir xx s.; 1578 *Edinb. Test.* VI 322b. For foirstalling of certene speir schaftis; 1579 *Perth Guildry* MS p. 391 (3 Sept.). Delywerit .. of fork schaftis iiij; 1588 *Kinnaird Farm Bk.* fol. 108b. Veluet to be handis of the rakkettis and club schaftis; 1600 *Treas. Acc.* MS 50. Four hundreth arrow schaftis .. and thrie scoir club schaftis; 1612 *Edinb. Test.* XLVII 123b. Ane fork shaft 1 l.; 1666 *Melrose Reg. Rec.* II 148. Tuo pick shafts 6 s. 8 d.; 1678 *Fawside Coal Compt* 33.

(*b*) James Johnnesoun bowar in Edinburght for thre scoir staf shaffis & speiris; 1564 *Perth B. Ct.* 167 (26 Sept.).

(2) Thai .. The Dowglas hit and brak his spere; And a sclys off the schafft that brak, In till his hand a wounde can mak; WYNT. VIII 5155. The grounden sper throuch his body schar The shafft to schonkit off the fruschand tre; *Wall.* III 147. Off schafftis part Wallace in sondyr schayr; *Ib.* X 411. Throw hys gardy sone The grundyn hed and bludy schaft ar done; DOUG. X vi 114. v dosane shaftis to Jedburcht stavis; 1533-4 *Treas. Acc.* VI 188. For shaftis to thame [*supra* irne muschis]; 1540 *Ib.* VII 355. For spaddis scholis skippis with all thair heidis and schaftis; 1542 *Ayr Common Good Acc.* Girth stingis to be schaftis to the saidis chargeouris; 1545-6 *Treas. Acc.* VIII 359. iiij mells with thair schaftis; 1557-8 *Edinb. Old Acc.* I 248. xxix pece of eschin tymmar to be schaftis to pikis and matokis; 1561 *Treas. Acc.* XI 115. To ane wrycht for his laubouris in making of schaftis to the saidis pickis, mattokis and mellis; 1571 *Ib.* XII 290. Irne graippis with thair schaftis; 1572 *Ib.* 339. MONTG. *Misc. P.* xxviii 54, 55. For four schafts of poleaxes to be maid mell schaftis; 1651 *Stirling B. Rec.* II 305.

(*b*) Twa schaiftis of gedwart stavis; 1566 *Prot. Bk. Thomas Johnsoun* 101.

fig. I sall ger fasone weile a flane .. The schaft sel be of sorowefule men The hede of panis strang; *a*1510 *Aberd. Univ. Rev.* XXXVI 1 44.

b. The stem, or handle, of a domestic knife or spoon.

Ane dosone of silwer sponis with braid schaftis; 1619 *Edinb. Test.* L 217b. 2 silver spoons one of them with a short cheft the uther with a falded-in one; 1685 *Soc. Ant.* LVIII 354. A caise with a dozen of knyfs with silver chefts; *Ib.*

c. ? The upright part or stock of an anchor.

For sawing ane anker soke iii grts. . . For ane anker soke xxiiii grts. . . For ane botis sake schafete ii grts.; 1597 *Skipper's Acc.* (*Morton*) 22a.

2. Only in verse: **a.** A spear, lance or javelin. **b.** An arrow.

a. The heuy dintis .. Scheildis to frushe and shaftis to shaik; *Alex.* I 2444. That steid and knycht baith schaft and scheild He bare doun; *Ib.* II 1637. Ane man .. That for the maistry dar mell With schaft and with scheild; *Gol. & Gaw.* 544. Thair men micht heir schukin of schaftis; *Peblis to Play* 253. The schaft flaw furth belyve; DOUG. X viii 106. The schaft he schuke, and branglys lustely; *Ib.* XII ii 111. The schaft .. throu the hawbrik skyrtis persyt has; *Ib.* xiv 91.

b. Bot evir the dedly schaft [L. *sagitta*] stykkis in hir syde; DOUG. IV ii 46. Phebus .. Vnto the eist schutand his schaftis schare; *Bann. MS* 249b/4. The armes, the bow, the shafts of lowe; FOWLER I 77/12. He tuik a shaft, and suddently me shot; MONTG. *Misc. P.* xxiv 25. The bow the schafts the quaver and the brace; MURE *Early Misc. P.* ii 40.

3. A pillar. [Cf. late ME *schafte of a pylar* (1483).]

Blasphemaris .. his [= is] ordand to be put on a schffat [*sic*] 7 Sondays abvn the kyrk dvir [etc.]; ABELL 64a.

4. a. A heddle-bar for weaving.

Ane pece of narro clayth quyt of four schaftis, of lent 8 ellis 3 quarteris .. to be littit skarlet in London; 1611 *Wedderb. Compt Bk.* 56.

b. ? A shaft of a cart.

14 s. for 8 pair shaffts; 1666 *Melrose Reg. Rec.* II 148.

5. In sense 1 or 2 a, used in measuring land.

In SKENE quot. perh. of indeterminate length, or ? put for SCHAFTMONTHE *n.*

With the quots. relating to Orkney, cf. ON *skipta jörðu með skapti*, to measure land with a spear.

Also *attrib.* (see SKENE quot. below).

Cf. *CHAFT *n.*[3]

See also SCHOFTLONG *n.*

(1) Of the vulgar people there is bot ane forme of metting vsed .. be rod and raip .. be ane rod .. of sex elnes lang: or be ane .. coard .. stented betuixt twa staues. The coarde being ane schaft length abone the pykes, or nether endes of the staues; SKENE *Verb. S.* s.v. *Particata.*

(2) Hes part of the penie land .. mett with schaft and lyne; 1605 *Rec. Earld. Orkney* 181. [In Hoxa they began by dividing] the hill balkis of the foresaid 2 d. land in halferis, laying fyve scoir nine shaftis to ilk pennyland, the lenth of ilk shaft being seven futtis of ane futt in measure and four inches mair; 1645 *Sc. Hist. Rev.* XVII 30, 31.

6. ? A shaft or beam of sunlight as seen as the sun becomes lower in the sky.

[Cf. ME *schaftez* (14th c.) in this sense and *schaft* to set in *til the sunne schafted* (Gawain & Green Knight.]

Be that the sone wes settand schaftis And neir done wes the day Thair men micht heir schukin of schaftis Quhen that thai went thair way; *Peblis to Play* 251.

S(c)haft, *n.*[2] Also: **chaft; schacht; (sheft), shaffe.** [Late ME and e.m.E *shaft* (1433-4, north.), MLG *schacht*, prob. specif. application of prec.] A pit-shaft, the well-like excavation made to allow working of a mine and to give access to it. Also *attrib.*

To draw levallis, watergangis, sinkis, conductis, shaftis and uthiris necessaris for the same [*sc.* mines]; 1607 *Reg. Privy C.* VII 358. Brome to mak covers for the schaftis to cover the windes men fra the iniurie of the wedder; 1608 *Hilderstoun Silver Mines* I 12b. Hadder for cradilling of the new schaftis; *Ib.* 30b. Of these pickmen thair is onlie sevin in the day and sevin in the nicht employed in working in that schaft whairin the mettell is found; 1609 *Mining Rec.* 146. [The principal shafts at Hilderstoun are mentioned under the following names:-] Blissing shaft... Germans West shaft... A new shaft set down by David the German callit—Serve them all. Harlies shaft [etc.]; ?*c*1610 *Ib.* 156. There [*sc.* alum mines] sc[h]achtes or fosses .. are .. timbred with wands and small roddes. .. Wee went doune into one of them... in returneing towards the schacht I [etc.]; 1646 J. HOPE *Diary* (1958) 178. For fliting the windises from one chaft to ane other; 1683 *Sheriffhall Coal Accompt* 23 June. 20 laids of hedder and kinling for ane eye shaffe; 1683 *Ellon Presb.* 222. [We can raise in two minutes' time a ton of water, provided the pit or sheft will admit of two such casks to pass one another]; 1695 *Reg. Privy C.* MS in R. Chambers *Domestic Annals of Scotland* III (1861) 102.

attrib. The earth about it .. lay all in litle heapes as if it had beene new coupped out of hurlebarrowes just as my pryon does about the schacht mouths; 1653 J. HOPE *Diary* (1919) 155.

Schaft. Var. of CHAFT *n.*[1] (jaw). *To semble with* (one's) *schaft*, to make faces. Also *attrib.* in *schaft bleid*, jaw-bone. — Sum ledys langis on the land, for love or for lak, To sembyll with thar schaftis and set apon sys; DOUG. VIII Prol. 60. In the harn pan the schaft he hes affixt, Quhil blude and brane al togidder mixt; *Ib.* IX vii 113. And he cum neir me I sall lay on his schaftis; 1608 *Glasgow B. Rec.* I 291. Dinging of hir .. and gewing of hir sindrie cuiffis in the schaftis; 1637 *Elgin Rec.* I 256. Neuer man shall get the doore dung in his schafts that would be in; 1638 DICKSON in *Sel. Biog.* II 25. — *attrib.* As his schaft-bleid can witnes beir; *Montg. Suppl.* xv 24.

Schaft, *v.* [17th c. Eng. *shaft* (1611) to fit (a tool) with a shaft, 19th c. Eng. *shaft out* (1862) to shoot as an arrow or shaft, cf. also mod. Eng. *shaft* to copulate with (Partridge *Dict. of Slang*); S(C)HAFT *n.*[1]] **a.** To spear with a shaft (S(C)HAFT *n.*[1] 2 a). **b.** To fit (a tool) with a shaft, to prepare a shaft for (a tool). — **a.** Quha strak doun quham, tha[r] was na man that wist For [quhen] men cummys in-to sa grete melle Thare may na tent till vther takin be Bot everie man to haue watt on his spere And schaip to schaft him that him schapis to dere; HAY *Alex.* 11012. — **b.** For ane axe to schaft the picks; 1681 *Fawside Coal Compt* 88.

Schafted, var. of SCHAFTIT *ppl. adj.*

S(c)haftin(g, *vbl. n.* [SCHAFT *v.*] The action of furnishing (a tool or implement) with a shaft. — For twa mattok schaftis, and schaftin of thame, xvj d.; 1559-60 *Edinb. Old Acc.* I 325. For schafting of the saidis pickis and mattokis; 1572 *Treas. Acc.* XII 331. For ane harrill and schafting of it; 1626 (?1625) *Glasgow Trades House* 113. To the wright for picke shafting; 1685 *Fawside Coal Compt* 153.

Schaftit, Shafted, *ppl. adj.* [e.m.E. *shafted* (1586); SCHAFT *v.*] Furnished with a shaft or handle. — Tuentie schaftit pikis; 1585 *Elphinstone Mun.* 193/1. He should hav ane long shafted spoone that sups caill with the devill; Ferg. *Prov.* MS No. 521.

Schaftmonthe, *n.* [ME and e.m.E. *schaftmonde* (?*a*1400), *shaftmount* (*a*1483), *shaftemente* (1545), *shaftman* (1598), OE *sceaftmund*, f. *sceaft* (see S(C)HAFT *n.*[1]) and *mund* hand, handbreadth ('probably *sceaft* was intended to denote the extended thumb' (OED)).] The distance from the end of the extended thumb to the opposite side of the hand, used as a measure (= about 6 inches). — The tong salbe the lynth off a schaftmonthe and j ynche; *Acts* I 387/2.

S(c)hag(e, *n.* Also: **schaig; s(c)hagg.** [e.m.E. *shage* (1592) in sense 1 below, *shag* (1598), also (1601) = rough matted hair, wool, after OE *sceacga.*]

1. A woollen stuff with a long rough nap on one side; a kind or variety of this.

Chiefly, as a non-count noun.

(1) For 5 yairds half of vailsch shag to lyne cloik at 7 s. the yaird; 1614 *Laing MSS* 135. Quhyt shag to lyne the doublitt; *Ib.* Mixt schag at xxvj s. aucht pennyis the elne; 1618 *Edinb. Test.* L 164. My .. cloke linit in the lapis with blak schage; 1623 *Ib.* LI 337b. xl ellis of schage at l s. the elne; 1624 *Ib.* LII 236b. xlij ellis of cullouris of schage at x li. the elne; 1627 *Ib.* LIV 49b. Ane elne reid schaig xx s.; 1631 *Brechin Test.* V 51b. Scarlet reid schage to be lyning to the caisses of the kingis pictoures; 1633 *M. Works Acc.* (ed.) II 333. To my lady to be ane gowne 7 ells of double skarlet schag 4 lib. 6 s. 8 d. ell; 1635 *Galloway P.* 11 Aug. Everie sute consisting of .. breikis and clokis dowble with schag or velvott; 1640 *Bk. Carlaverock* II 504. Ane large white Spanish blankett of shagg; 1653 *Soc. Ant.* XXIII 302. A quarter shagg; 1659 Craven *Ch. in Orkney* II 198. For .. shage, 6 qrtr. broad to be ane long coatt to Sir James; 1661 NICOLSON *Diurnals* 11 Jan.

(2) Certane remanes .. of schagis and blak burratis; 1620 *Edinb. Test.* L 352. Fyve haill peisis of schagis at xxiiij lib. the peis; *Ib.* 352b. Five hundreth ten [*sc.* 510] elns of shages Londoun at tuentie shillinges the eln; 1632 *Ib.* LV 254.

2. *attrib.* Consisting of, or made of, shag; hence, having a long, rough nap.

(1) Schag velvot; 1620 *Edinb. Test.* L 353. Schag freis at xl s. the elne; 1625 *Ib.* LIII 32. Schage veluot at x li. the elne; 1627 *Ib.* LIV 49. Aucht elnes and ane quarter shag bais and freis; 1647 *Ib.* LXIII 34. Stof, red shage £3 12 s.; *c*1680 *Bk. Dunvegan* I 195.

(2) Thrie schage Irland coveringes; 1634 *Edinb. Test.* LVI 367b. — Ane reid schag coitt estimat to twa pund xiiij s.; 1642 *Edinb. Test.* LX 144. Ane schagg vallicoatt; 1660 *Rothesay B. Rec.* 53. Ane shag coat of barrogone; *c*1670 *Old Ross-shire* I 143. Ane red shag peticoat; 1675 *Argyll Justic. Rec.* 53. A reid shag coat; 1691 *Brechin Test.* VIII 23b.

3. ? A fleece.

As to my schiep I can not tell quhat it can be vurthe to me as yit, for if she thrayf she vill multiple and incries, iff uther vays she be but bairane I heff lost mor nor tuente schage can be [vurthe]; 1604 *Misc. Hist. Soc.* II 246.

(Schage,) Shag, *v.* (e.m.E. *shag* (1596) *intr.* to be shaggy, (1612) *tr.* to render rough or shaggy, (1671) to make a long or rough nap or pile on (a cloth).) *tr.* — To mount .. souldiours sufficientlie with new reid coates .. for that effect .. to buy plaideing caus make lilt and shag the samen and caus mack the saids coats; 1685 *Rec. Old Aberd.* I 140.

Schagg, var. of S(C)HAG(E *n.*

(Schagrin,) Shagrin, Shagareen, *n.* Also: **shaggerine.** [17th c. Eng. *shagrin-leather* (1677), F. *sagrin* (1606), *chagrain* (1648, both in Larousse).] **a.** Shagreen, a type of leather with a granular surface, freq. dyed green. Also *attrib.* with *case.* **b.** A silk fabric. — **a.** To buy .. 6 .. bibles .. and to cause bind them in shaggerine; 1646 J. HOPE *Diary* (1958) 165. Two steel seals with shagareen cases ij s.; 1725 *Edinb. Test.* XC 12 July. — **b.** 9⅛ ells of grein silk shagrin for lyning; 1701 *Fam. Rose* 389. For 5½ ells black shagrin .. £18 9 s.; 1702 *Household Bk. Gr. Baillie* 197.

Schaheme, var. of S(C)HAME *n.* **Schaichles,** var. of SCATHELES *adj.* **Schaid,** var. of SCATH(E *n.* and p.t. of S(C)HED *v.* **Schaif(e, Schaiff,** varr. of SCHAFE *n.*, SCHAVE *v.*, S(C)HAW *v.* and SCHEFE *n.*[1] and *n.*[2]

Schaiffar. Var. of CHAFER *n.* (a chafing-dish). — Twa medling panis .., ane auld schaiffar; 1562 *6th Rep. Hist. MSS* App. 648/2.

Schaiffer, Schaiffron, varr. of SCHAFFRO(U)N *n.* **Schaiffyne,** p.p. of SCHAVE *v.* **Schaifling,** var. of S(C)HAVEL(L)ING *n.*

Schaifpeet, *n.* (?) — Quarta parte molendini de Skryne cum terris molendinariis, multuris lie schaifpeet et sequelis eiusdem; 1565-6 *Reg. Privy S.* V II 33/2.

Schaifron, var. of SCHAFFRO(U)N *n.* **Schaift,** var. of S(C)HAFT *n.*[1] **Schaig,** var. of S(C)HAG(E *n.* **Schaik-, Schaik(e,** varr. and p.p. of SCHAK(E-, S(C)HAKE *v.* and SCHALK *n.* **Schaiken, -in(g,** p.p. of S(C)HAKE *v.* **Schaik(h)ill, -le,** varr. of SCHAKILL *n.* **Schaikill-bane, Schaikle-bone,** varr. of SCHAKIL(L)-BANE *n.* **Schailbane,** var. of SCHAKIL(L)-BANE *n.* **Schaild,** var. of SCHALD(E *adj.* **Schaill,** var. of SCAIL *v.*, SCALE *n.*[1] and SCHELE *v.*[1]

Schaim, Schaim-, varr. of S(C)HAME *v.* and S(C)HAME-. **Schaink,** var. of S(C)HANK *n.* **Schaip(e,** varr. and p.p. of SCAP(E *v.*, S(C)HAP(E *n.*[1], SCHAPE *n.*[2] and S(C)HAPE *v.* **Schaipell, -ill,** varr. of SCHAPEL(L *n.* **Schaipin,** var. of SCHAPEN *ppl. adj.* **Schaipin(e,** p.p. of S(C)HAPE *v.* **Schaiping,** var. of SCHAPIN(G *vbl. n.* **Schaiplayn,** var. of SCHAPLANE. **Schaiplie,** var. of SCHAPLY *adj.* **Schaippell, -ill,** varr. of SCHAPEL(L *n.* **Schaippin,** var. of SCHAPEN *ppl. adj.*

Schair, S(c)heir, S(c)hire, *n.* Also: **s(c)hyre, schyire.** Varr. of CHAIR *n.*[1], CHEAR(E *n.* and CHYRE *n.*[1] (= a chair). Also *comb.*

(*a*) Ane arme schair; 1585 *Elgin Rec.* I 177.

(*b*) Ane auld scheir; 1550 *Rec. Earld. Orkney* 242. Ane scheir; 1605 *Edinb. Test.* XL 154. Ane four fittit scheir; 1621 *Barcaldine Inv.* Four plush velvet scheirs; 1703 *E. Loth. Antiq. Soc.* IV 27.

(*c*) Twa schyris; 1565 *Prot. Bk. Gilbert Grote* 68. The schyre of stait; *Bann. MS* 86a/49. Ane mekle schyre; 1572 *Prot. Bk. J. Robeson* 166. I leve to .. my gud sister .. the schyre & cuscheonis & vther small insicht of the chalmer; 1574 *Edinb. Test.* III 48. I desyre my brother the king of Scottland .. that he wald sit in his awin schyre and lat me and the Frinchemen pairt amang ws; PITSC. I 254/9. Out of hir schyire [*Misc. P. li 22* schyne] scho [*sc.* Diana] slippis; MONTG. *Maitl. Q.* 197/22. For mending of the glob and for making of ane new schyre of irne vnto; 1589 *Edinb. Treas. Acc.* in Mill *Mediæv. Plays* 196. The midd scaffold, where sall be placed two shires for the king and queene; 1589 Calderwood V 61. Ane faldand schyre price thairof xiij s. iiij d.; 1606 *Edinb. Test.* XLI 37b. The schyre quhairin I satt coupit, and I gatt ane heavy fall; 1642 T. HOPE *Diary* 157. Ane vide resting rushie lether shyre; 1662 *Edinb. Test.* LXX 317. Four shued schires; 1668 *Ib.* LXXIII 168b. 1674 *Melrose Reg. Rec.* III 84.

comb. John Broun, sheirmaker; 1602 *Edinb. Test. Index* II 50.

Schair, var. of SCARE *v.*, SCHARE *v.* and S(C)HIRE *n.* and var. and p.t. of SCHER(E *v.* **Schairce,** var. of SCARS(E *adj.* and *adv.*

S(c)haird, She(a)rd, *n.* Also: **shard.** [ME and e.m.E. *sherd* (Wyclif), also *shord* (Id.), *shard* (1488), OE *sceard*, MDu. *scharde.*] *lit.* and in *fig.* context: A shard, a fragment of broken earthenware.

To break, or *ding, in shairds*, to smash; to break into pieces or fragments.

(1) That rod without regaird, Shall brecke & breiffe them, as an pottars shaird; NORVELL *Meroure* 20b. Quhairvith thou sall Evin bruis thame all In peces, lyk a potters shaird; MONTG. *Devot. P.* ii 33. Quhair the pig broke let the schairds lie; CARMICHAEL *Prov.* No. 1285. Ingentemque manu sherdam levavit; *Polemo-Mid.* 132. Sheards; LAW *Memor.* 232.

(2) Had the wind been as it was the day before, .. our veshell had been dung in shards; 1641 BAILLIE I 355. To breake the greatest in schairds, as earthen vessels; CALDERWOOD VI 337.

Schaire, var. of SCHERE *n.*[2] and p.t. of SCHER(E *v.* **Schairg,** var. of SCHARGE. **Schairge,** var. of SARG(E *n.* (*adj.*). **Schairing,** var. of S(C)HERAND *ppl. adj.* and S(C)HERING *vbl. n.* **Schairkill-bane,** var. of SCHAKIL(L)-BANE *n.* **Schairne,** var. of S(C)HARN *n.* **Schairnie,** var. of SCHARNY *adj.* **Schairp-, Schairp(e,** varr. of S(C)HARP- and S(C)HARP *adj.*, *adv.* and *n.*[2] and *v.* **Schairs,** var. of SCARS(E *adj.* and *adv.* **Schais,** pl. of S(C)HAW *n.*[1] **Schait(e, Schaith,** varr. of SCATH(E *n.* **Schaithles,** var. of SCATHELES *adj.* **Schaitmaker,** var. of SCHETHEMAKER *n.* **Schaive,** var. of SCHEF(E *n.*[2] **Schaiver,** var. of SCHAVER *n.*

(Schak,) Shak. ? Var. of CHAK *v.*[1] to check. — This youngling boy [*sc.* Cupid] his puissance maid him prowe When in Thessalia he him [*sc.* Apollo] shakt and made him for to lowe; FOWLER I 34/218.

Schak, var. of SCHAKE *n.* and *v.* and SCHALK *n.* **Schakall,** var. of SCHAKILL *n.*

(Schakand,) Schaikand, *ppl. adj.* Also: **(shaithand).** [S(C)HAKE *v.*] Shaking, quivering. — Thocht deip dissait and dowble taill .. Resembles weill thy shaithand kneis, Thy wavill feit, thy reland eis; *Rob Stene* 8. God set .. a fyrie schaikand sword to keip the way of the trie of lyf; HAMILTON *Facile Tr.* 209.

Schakar, *n.* Also: **schaker, schacar, shaker, s(c)haiker.** [Late ME and e.m.E. *schakare* (Prompt. Parv.); S(C)HAKE *v.*]

1. One who shakes (something). Only in HAL(L)AND-SCHAIKER, q.v.

2. *pl.* An ornamentation or trimming, composed of thin plates of metal, attached to or suspended from a garment or head-dress, and intended to vibrate with the movements of the wearer.

Also, once, attached to a bed or bed-covering.

For xv goldin skinnis to stomois for thaim, and schakaris and bordouris to the tailes; 1506 *Treas. Acc.* III 313. The pryce .. a bul, and al hys hed Of goldyn schakaris [*Sm.* schacaris] and roys garlandis red Buskyt full weil; DOUG. V vii 8. *Ib.* IX x 88. He [*sc.* Clariodus] schynit .. With goldin schaikeris abone his plumes greine; *Clar.* I 27. Ane coit of blew reid and quheit tauffateis hingand full of schakaris .. ane howd of blew reid and quheit tauffateis with schakaris; 1569 *Inv. Q. Mary* 186. Shakers; MONTG. *Ch. & Slae* 49 (Wr.). [Ministers to] refraine from tockens of lightness in their apparell, such as shaikers at the knees; 1650 Cramond *Presb. Fordyce* 22.

b. *transf.* Vibrating drops of dew.

And syluer schakaris gan fra levys hyng, With crystal sprayngis on the verdour ȝyng; DOUG. XII Prol. 131. With schakeris of the schene dew schour; *Tayis Bank* 21.

c. In DOUG., rendering L. *bractea.*

Siklyke was of this gold the figur brycht, That burgionyt fayr on the rank akis hycht. Evir as the branch for pypand wynd reboundit, The goldin schakeris ratlis and resoundit; DOUG. VI iii 102.

Schakar, var. of SCHEKER.

(Schakarstane,) Schakerstane, *n.* (Mod. Sc. dial. *stane-cha(c)ker, sten-shakker*, etc., applied to several birds viz. the stone-chat, the wheatear or the whin-chat.) — The stainȝell and the schakerstane, Behind the laue wer left alane; BUREL *Pilgr.* I 365.

Schakart, var. of SCHAKERT *adj.*

(Schake,) Schak, *n.* [ME and e.m.E. *schak* (c1380), *schakke* (?c1475), *shake* (1581); f. next.] The action of shaking, in sense 5 of

S(C)HAKE *v*. — Scottis kingis .. sufferit thameselffis be sa oft deponit without straik or schak of swerd; LAMB *Resonyng* 87/9.

S(c)hake, S(c)haik, S(c)ha(c)k, *v.* Also: **schaike, schayke, chaik, sheake, shecke.** P.t. **s(c)huk(e; s(c)huik(e; s(c)heuk; schouk, schowk; schuck; s(c)hoke, shook(e, schoak; schake; schakit, shaiked.** P.p. **s(c)hakin, -ine, -yn, -en; s(c)haikin(g, -en, -ne; schakkin, shauken, shoken, shooken; schake, schaik.** [ME and e.m.E. *scæken* (Layamon), *saken* (*a*1220), *scak* (Cursor M.), *schake(n* (*c*1320). P.t. *scok* (Cursor M.), *schoke* (Manning), etc., *shakide* (Wyclif). P.p. *i-shake* (*a*1300), *shake* (Chaucer), *shaken* (*a*1534), etc. OE *scacan*, p.t. *scók*, p.p. *scacen*.]

I. 1. *intr.* A poetical word for: To go, pass, move.

Quhare the assembleis togidder schuke [F. *La ou il s'entr'aprochent*] The play vox wery for mony man But lauching losit thare lyues than; *Alex.* II 9754.

II. To vibrate, tremble.

2. Of things normally stable or still: To vibrate irregularly; to tremble, as a result of impact. Hence, to totter, lose stability, become weakened.

pres. The erd sould trymbill, the firmament sould schaik; DUNB. *Flyt.* 9. LYND. *Mon.* 1414. The thundring doun of cannounis .. maid hevin and erthe to schaik; *Maitl. Q.* 201/14. MELVILL 525.

p.t. The erde steryt sa felloun[l]y, That al the cyte in til hy Schuke & to-giddire strake; *Leg. S.* xlii 261. The erd dyndlyt, and all the cite schuke; DOUG. IX xii 61. STEWART 7194.

(*b*) The erd trimblit and all the housis schoke [: quuke]; HAY *Alex.* 153.

(*c*) Hillis trimlit, erd schowk, rochis claif; *Makc. MS* iv 58. Schouk; J. STEWART 51/108.

(*d*) Schuike [*ed.* schoke]; *Clar.* V 2017. The parpan wall of the hous schuik and trymblit; 1597 *Misc. Spald. C.* I 148.

b. With a complement expressing a resultant situation or condition.

His bost & brag .. Maid all the heuin most lyk to schaik in schonder; LYND. *Dreme* 446. Ay till his sword did shaik in sunder; Id. *Meldrum* 156. Thair bains ar lyke to schaik sindrie; BUCH. *Comm. on Virgil* Eclogues III 102.

3. Of persons or beasts, or their limbs: To quake or tremble with infirmity; to quiver with emotion, shiver with cold, quake with fear.

Thane begane his kneis to quake, & al his body for to schake; *Leg. S.* xvi 846. Belive Eneas membris schuk for cald; DOUG. I iii 1. *Ib.* V viii 21. Tha trymlit and tha shuke; STEWART 22764. *Ib.* 14916. ROLLAND *Seven S.* 7401. 1572 *Sat. P.* xxxiii 78. The lyoun, with ane awfull brow, Maid every beist to chaik and bow; *Rob Stene* 3. Thou sall schaik and trimbill at his countenance; ROLLOCK *Wks.* I 335. My hands did shaik, that I him held withall; E. MELVILLE *Godlie Dreame* 316.

b. To stagger or reel in consequence of a blow or violent impact. With and without complement.

And in the stour sa hardyly He ruschyt that all the semble schuk; BARB. II 383. Wallas with it fast on the cheik him tuk Wyth so gud will, quhill of his feit he schuk; *Wall.* I 404.

4. Of things having some freedom of movement: To move irregularly and quickly to and fro, etc.; to quiver, quake, vibrate, waver; also, to be brandished or shaken.

In athiris hand yfeir The braid steil heid schuke on the huntyng speir; DOUG. I vi 16. Down on the grond scho fell, Hyr targe trymlyng, and schakyng fast hir speir; *Ib.* II iii 59. Quharthrou the sey and al the fludis schuke; *Ib.* III x 33. *Ib.* II x 120, X ix 39. The ledder schuke, he lape, and gat one fall; LYND. *Test. Pap.* 552. The schawis trymlit all and schuke; STEWART 57051. Ʒeill fart fast in Baquhidder or the corne schaik; MONTG. *Misc. P.* liv 4. And evill wedder tak theme in sic maner that the taikling schaik; BISSET II 244/30.

b. To tremble or fall in a shivering manner. Also with *doune*.

The perly droppis schake in [*M.* schuk in to] silvir schouris Quhill all in balme did branch and levis flete; DUNB. *G. Targe* 14. Siluer schouris doune schuke as the schene cristall; Id. *Tua Mar. W.* 515 (Ch. & M.).

III. To cause to vibrate, agitate.

5. *tr.* To brandish or flourish threateningly (a weapon). Also, to flourish, wave (something) in ostentation or triumph.

The dart he tuke And felly at Vlixes schuke; *Troy-bk.* II 2958. Scorne .. oft on me his babill schuke; DUNB. (OUP) 26/36. He schuke and branglit fast his speir that tyde; DOUG. XII vii 117. ROLLAND *Seven S.* 6194. When I so did he threatned and abused me with his tungu and shaiked his staffe at me; 1673 *Kirkcudbr. Sheriff Ct. Processes* No. 188.

6. To cause to move irregularly to and fro; to agitate; to toss about. Also with adv. complement.

pres. [The wind] drownis schippis, schakis cornis .. and brekis treis; HAY II 101/2.

(*b*) Thocht storm of temptatioun my schip oft schaik, Fra swelth of Sylla [etc.]; DOUG. III Prol. 43. To schaik the brydle to encourage the horse; BUCH. *Comm. on Virgil* Æn. VI 100. Thai be bot leavis that wind dois schaik; *Maitl. F.* 288/34. That the person shakeing or treaming other mens netts shall satisfie the pairtie his wholl skaith; 1688 *E. Fife Admir. Depute Ct. Bk.* 28 July.

(*c*) Be reason whereof she did gnaw her finger and shack the kist and the crook; 1650 *Inverurie* 306.

p.t. Thay gaif ane cry, and schuik coppis gude speid; HENR. *Test. Cress.* 492. Throuch the tempest of a terrible storme, that sa vehementlie schuke thame; DALR. I 203/26. Ʒe said to Isobel Wilson .. that the said Henry shuik the sheet well enough yesterday; 1662 *Soc. Ant.* XXII 221. It [*sc.* an evil spirit] .. shoke men back and forward; 1696 T. Davidson *Rowan Tree* (1949) 266.

p.p. His schippis thare That with the tempest schaken war; *Troy-bk.* II 2394.

(*b*) Quhou that Ene was with the tempest schaik And quhou Neptune his navy salvyt fra wraik; DOUG. I iii heading.

b. *fig.* ?To throw (troops) into disorder or confusion.

Cut them in peices .. so great was the execution which they made efter the horse had shauken [*pr.* shanken] and quyt astonished them by perseing rudly throw them; P. GORDON *Brit. Dist.* 126.

c. *To schake* (one's) *lap*, to display scorn or the like.

And Hamiltowne was but a knave, When he his lap did shake; *c*1665 *Anal. Scot.* II 91. At the Lord's Supper he used to shak his lap against them who would break the Covenant; *c*1679 *Ib.* 91 n.

7. To move to and fro irregularly or tremulously; to agitate (some part of the body).

This senȝeour .. Schakand his berd, he said, 'My bairne [etc.]'; HENR. *Age & Yowth* 26. [The fox] Syne schuk his taill with quhinge and ȝelp; DUNB. (OUP) 113/10. The snaill Schakand hir coppit schell; DOUG. XIII iv 80.

b. *To schake* (one's) *head* (*on* or *apoun* a person), to express disapproval, dissent or doubt.

(1) The duke Betys the king hes sene And shuke his heid for proper tene; *Alex.* I 2272. Schaikand his heid, he said, 'My sone, lat be'; HENR. *Fab.* 1388. Bot quhat wes ȝone the carll cryit on hie And schuke his heid quhen that he saw thow fell? *Ib.* 2112 (Ch.). DOUG. XII xiv 21. And becaus sche refusit, he schuke his heid, and said that he suld caus hir forthink it; 1576 *Crim. Trials* I II 57. MONTG. *Ch. & Slae* 927 (W). Rychard .. declairit thair .. that scho [and two others] .. sould be thre of the doaris of itt; quhilk quhen scho hard, scho schuik hir heid; 1591 *Crim. Trials* I II 245.

(2) Thir folk for scorne apoun him schuk thair heid; KENNEDY *Pass. Christ* 857. Than all the court on me thayr hedis schuke [: ruke, bruke, quuke; *E.* schuik]; DOUG. *Pal. Hon.* 648 (L).